Elementary Differential Equations

This book is written to innovate the standard course and to modernize it. There has been a dramatic change in the past century in how solutions to differential equations are calculated. However, the way the topic has been taught in introductory courses has barely changed to reflect these advances, which leaves students at a disadvantage.

Elementary Differential Equations, Third Edition builds on the previous editions, continuing to innovate with the addition of a new topic: combinatorics. The goal is to assist instructors as they facilitate new teaching methods and utilize the latest technology and to provide flexibility in presenting a new topic to encourage students to explore mathematics and not only its applications.

The text is designed to help instructors who want to use computers in their classrooms. It accomplishes this by emphasizing and integrating computers. Many examples and exercises included in the text require the use of computer software to solve problems. It should be noted that since instructors use their own preferred software, this book has been written to be independent of any specific software package.

Features:

- Focuses on numerical methods and computing to generate solutions
- Features extensive coverage of nonlinear differential equations and nonlinear systems
- Contains a wider variety of non-mathematical models than any competing textbook
- In the new Chapter 11, readers will learn notions of set partitions, graphs, and rooted trees, the concept of exponential generating functions, and see how differential equations can be surprisingly useful in solving complex counting problems.

The third edition continues to serve as an excellent introductory textbook for undergraduate students majoring in applied mathematics, computer science, various engineering disciplines, and other sciences. They also will find that the textbook will aid them greatly in their professional careers because of its instructions on how to use computers to solve equations.

Textbooks in Mathematics

Series editors:
Al Boggess, Kenneth H. Rosen

Elements of Algebraic Topology, Second Edition
James R. Munkres, Steven G. Krantz, and Harold R. Parks

One Complex Variable from the Several Variable Point of View
Peter V. Dovbush and Steven G. Krantz

Math Anxiety How to Beat It
Brian Cafarella

Lectures on Differential Geometry with Maple
Mayer Humi

Numerical Analysis for Engineers
Methods and Applications
Bilal M. Ayyub and Richard H. McCuen

An Invitation to Real Analysis
Andrew D. Hwang

Fourier Series and Boundary Value Problems with Engineering Applications
Youssef N. Raffoul

A Course in Real Analysis, Second Edition
Hugo D. Junghenn

Real and Functional Analysis
Kenneth Kuttler

Mathematics for Artificial Intelligence
Jane Hawkins

Elements of Topology, Second Edition
Theory and Practice
Steven G. Krantz

Convex Analysis, Second Edition
Steven G. Krantz

A Second Course in Complex Analysis
Peter V. Dovbush and Steven G. Krantz

Real and Complex Analysis, Second Edition
Volume 1
Christopher Apelian and Steve Surace

Elementary Differential Equations, Third Edition
Charles E. Roberts, Jr. and Miklós Bóna

https://www.routledge.com/Textbooks-in-Mathematics/book-series/CANDHTEX-BOOMTH

Elementary Differential Equations
Third Edition

Charles E. Roberts Jr. and Miklós Bóna

CRC Press
Taylor & Francis Group
Boca Raton London New York

CRC Press is an imprint of the
Taylor & Francis Group, an **informa** business
A CHAPMAN & HALL BOOK

Designed cover image: Shutterstock

Third edition published 2026
by CRC Press
2385 NW Executive Center Drive, Suite 320, Boca Raton FL 33431

and by CRC Press
4 Park Square, Milton Park, Abingdon, Oxon, OX14 4RN

CRC Press is an imprint of Taylor & Francis Group, LLC

First edition published by CRC Press 2011
Second edition published by CRC Press 2019

Library of Congress Cataloging-in-Publication Data
Names: Roberts, Charles E., 1942- author | Bóna, Miklós author
Title: Elementary differential equations / Charles E. Roberts Jr. and
Miklós Bóna.
Description: Third edition. | Boca Raton, FL : CRC Press, 2026. | Series:
Textbooks in mathematics | First edition published by CRC Press 2011.
Second edition published by CRC Press 2019. | Includes bibliographical
references and index.
Identifiers: LCCN 2026000688 (print) | LCCN 2026000689 (ebook) | ISBN
9781041151463 hardback | ISBN 9781041151470 paperback | ISBN
9781003678014 ebook
Subjects: LCSH: Differential equations--Textbooks
Classification: LCC QA372 .R7495 2026 (print) | LCC QA372 (ebook)
LC record available at https://lccn.loc.gov/2026000688
LC ebook record available at https://lccn.loc.gov/2026000689

ISBN: 978-1-041-15146-3 (hbk)
ISBN: 978-1-041-15147-0 (pbk)
ISBN: 978-1-003-67801-4 (ebk)

DOI: 10.1201/9781003678014

Typeset in Latin Modern font
by KnowledgeWorks Global Ltd.

Publisher's note: This book has been prepared from camera-ready copy provided by the authors.

To the memory of my mother and father,
Evelyn and Charles

And to those who "light up my life,"

my wife, Imogene

my children, Eric and Natalie

and my grandsons, Tristan and Luke

Contents

About the Authors

Charles E. Roberts, Jr. (deceased) was Professor Emeritus in the Department of Mathematics and Computer Science at Indiana State University. He is remembered as a master teacher and prolific researcher. His book, *A Mathematical Introduction to Proofs*, is also published by CRC Press.

Miklós Bóna received his Ph.D. in mathematics from the Massachusetts Institute of Technology in 1997. Since 1999, he has taught at the University of Florida, where, in 2010, he was inducted into the Academy of Distinguished Teaching Scholars. Professor Bóna has mentored numerous graduate and undergraduate students. He is the author of five books and more than 100 research articles, mostly focusing on enumerative and analytic combinatorics. His book, *Combinatorics of Permutations*, won a 2006 Outstanding Title Award from Choice, the journal of the American Library Association. He is also an Editor-in-Chief for the *Electronic Journal of Combinatorics*, and for two book series at CRC Press.

Preface to the Second Edition

Elementary Differential Equations is an introductory level textbook for undergraduate students majoring in mathematics, applied mathematics, computer science, one of the various fields of engineering, or one of the physical or social sciences. During the past century, the manner in which solutions to differential equations have been calculated has changed dramatically. We have advanced from paper and pencil solution, to calculator and programmable calculator solution, to high speed computer calculation. Yet, in the past fifty years, there has been very little change in the topics taught in an introductory differential equations course, in the order in which the topics are taught, or in the methods by which the topics are taught. The "age of computing" is upon us, and we need to develop new courses and new methods for teaching differential equations. This text is an attempt to facilitate some changes. It is designed for instructors who wish to bring the computer into the classroom and emphasize and integrate the use of computers in the teaching of differential equations. In the traditional curriculum, students study few nonlinear differential equations and almost no nonlinear systems due to the difficulty or impossibility of computing explicit solutions manually. The theory associated with nonlinear systems may be considered advanced, but generating a numerical solution with a computer and interpreting that solution is fairly elementary. The computer has put the study of nonlinear systems well within our grasp.

In this text, several examples and exercises require the use of some computer software to solve a problem. Consequently, the reader needs to have either computer software or artificial intelligence tools available which can perform, at least, the following functions:

1. Graph a given function on a specified rectangle.

2. Graph the direction field of the first-order differential equation $y' = f(x, y)$ in a specified rectangle.

3. Solve the first-order initial value problem $y' = f(x, y)$; $y(c) = d$ on an interval $[a, b]$ which contains c.

4. Find all roots of a polynomial with complex coefficients.

5. Calculate the eigenvalues and eigenvectors of a real $n \times n$ matrix A where $2 \leq n \leq 6$.

6. Solve on an interval $[a, b]$ a vector initial value problem consisting of a system of n first-order differential equations and n initial conditions where $2 \leq n \leq 6$.

Many computer software packages are readily available which include these features and usually many additional features. Three of the best known and most widely used packages are MAPLE, *Mathematica*, and MATLAB®. In general, each instructor already has his or her own favorite differential equations software package or combination of packages. For this reason, the text was written to be independent of any particular software package.

It is assumed the reader has completed calculus at least up to and including the concept of partial derivatives and knows how to add, subtract, multiply, and divide complex numbers. Concepts with which the reader may not already be familiar are introduced and explained to the degree necessary for use within the text at the location where the concept is first used.

Students who enroll in ordinary differential equations courses normally do so for only one or two semesters as an undergraduate. In addition, few of these students ever enroll in a numerical analysis course. However, most students who complete a differential equations course find employment in business, industry, or government and will use a computer and numerical methods to solve mathematical problems almost exclusively. Consequently, one objective of this text is to solve ordinary differential equations in the same way they are solved in many professions—by computer. Thus, the single most useful and distinguishing feature of this text is the use of computer software throughout the entire text to numerically solve various types of ordinary differential equations. Prior to generating a numerical solution, applicable theory must be considered; therefore, we state (but usually do not prove) existence, uniqueness, and continuation theorems for initial value problems at various points in the text. Numerical case studies illustrate the possible pitfalls of computing a numerical solution without first considering the appropriate theory.

Differential equations are an important tool in constructing mathematical models for physical phenomena. Throughout the text, we show how to numerically solve many interesting mathematical models—such as population growth models, epidemic models, mixture problems, curves of pursuit, the Richardson's arms race model, Lanchester's combat models, Lotka–Volterra prey-predator models, pendulum problems, and the restricted three-body problem. When feasible we develop models entirely within separate sections. This gives the instructor more flexibility in selecting the material to be covered in the course. We hope to enrich and enliven the study of differential equations by including several biographical sketches and historical comments. In this text, we also attempt to provide an even balance between theory, computer solution, and application.

We recommend that the core of a one-quarter or one-semester course consist of material from Chapters 1, 2, 4, 5, 7, 8, and Sections 10.1, 10.2, and 10.3. The remainder of material covered in the course should come from the applications and models in Chapters 3, 6, 9, and Sections 10.4 through 10.11. The selection of applications and models to be included in the course will depend on the time available, on the intent of the course, on the student audience, and, of course, on the preferences of the instructor. The following is a summary of the material to be found in each chapter.

In **Chapter 1** we present a very brief history of the development of calculus and differential equations. We introduce essential definitions and terminology. And we define and discuss initial value problems and boundary value problems.

In **Chapter 2** we discuss in detail the first-order initial value problem $y' = f(x, y);\ y(c) = d$. First, we define the direction field for the differential equation $y' = f(x, y)$, we discuss the significance of the direction field, and we show how to use a computer program to produce a graph of the direction field. Next, we state a fundamental existence theorem, a fundamental existence and uniqueness theorem, and a continuation theorem for the initial value problem. We show how to apply these theorems to a variety of initial value problems, and we illustrate and emphasize the importance of these theorems. Then we discuss how to find explicit solutions of simple first-order differential equations such as separable equations and linear equations. Next, we present simple applications of linear first-order differential equations. Finally, we present some of the simpler single-step methods for computing a numerical approximation to the solution of an initial value problem. We explain how to use a computer program to generate approximate, numerical solutions to initial value problems. (Appendix A contains additional single-step, multistep, and predictor-corrector methods for computing numerical approximations to initial value problems.) We illustrate and interpret the various kinds of results which the computer produces. Furthermore, we reiterate the importance of performing a thorough mathematical analysis, which includes applying the fundamental theorems to the problem, prior to generating a numerical solution.

In **Chapter 3** we consider a variety of applications of the initial value problem $y' = f(x, y);\ y(c) = d$. First, in the section Calculus Revisited, we show that the solution to the special initial value problem $y' = f(x);\ y(a) = 0$ on the interval $[a, b]$ is equivalent to the definite integral $\int_a^b f(x)\,dx$. Then we show how to use computer software to calculate an approximation to the definite integral $\int_a^b f(x)\,dx$. This will allow one to numerically solve many problems from calculus. In the sections Learning Theory Models, Population Models, Simple Epidemic Models, Falling Bodies, Mixture Problems, Curves of Pursuit, and Chemical Reactions, we examine physical problems from a number of diverse disciplines which can be written as initial value problems and then solved using numerical integration software. At the end of the chapter, we present additional applications.

In **Chapter 4** we discuss the basic theory for n-th order linear differential equations. We present a history of the attempts of mathematicians to find roots of polynomials. Then we illustrate how to use computer software to approximate the roots of polynomials numerically. Next, we show how to find the general solution of an n-th order homogeneous linear differential equation with constant coefficients by finding the roots of an n-th degree polynomial. Finally, we indicate how to find the general solution of a nonhomogeneous linear differential equation with constant coefficients using the method of undetermined coefficients.

In **Chapter 5** we define the Laplace transform and examine its properties. Next, we show how to solve homogeneous and nonhomogeneous linear differential equations with constant coefficients and their corresponding initial value problems using the Laplace transform method. Then we define the convolution of two functions and prove the convolution theorem. Finally, we show how to solve nonhomogeneous linear differential equations with constant coefficients in which the nonhomogeneity is a discontinuous function, a time-delay function, or an impulse function.

In **Chapter 6** we examine several linear differential equations with constant coefficients which arise in the study of various physical and electrical systems.

In **Chapter 7** we define a system of first-order differential equations. We state a fundamental existence and uniqueness theorem and a continuation theorem for the system initial value problem. Then, we show how to apply these theorems to several initial value problems. Next, we show how to rewrite an n-th order differential equation as an equivalent system of n first-order equations.

In **Chapter 8** we discuss linear systems of first-order differential equations. First, we introduce matrix notation and terminology, we review fundamental facts from matrix theory and linear algebra, and we discuss some computational techniques. Next, we define the concepts of eigenvalues and eigenvectors of a constant matrix, we show how to manually compute eigenvalues and eigenvectors, and we illustrate how to use computer software to calculate eigenvalues and eigenvectors. We show how to write a system of linear first-order differential equations with constant coefficients using matrix-vector notation, we state existence and representation theorems regarding the general solution of both homogeneous and nonhomogeneous linear systems, and we show how to write the general solution in terms of eigenvalues and eigenvectors.

In **Chapter 9** we examine a few linear systems with constant coefficients which arise in various physical systems such as coupled spring-mass systems, pendulum systems, the path of an electron, and mixture problems.

In **Chapter 10** we present techniques for determining the behavior of solutions to systems of first-order differential equations without first finding the solutions. To this end, we define and discuss equilibrium points (critical points), various types of stability and instability, and phase-plane graphs. Next, we show how to use computer software to solve systems of first-order differential equations numerically, how to graph the solution components, and how to produce phase-plane graphs. We also state stability theorems for systems of first-order differential equations. Throughout this chapter we develop and discuss a wide variety of models and applications which can be written as vector initial value problems and then solved numerically.

Acknowledgments

This text evolved from lecture notes for a course which I have taught at Indiana State University for a number of years. I would like to thank my students and my colleagues for their support, encouragement, and constructive criticisms. In particular, I would like to thank Professor Robert Sternfeld who helped with computing matters. Also, I would like to thank Mr. Robert Ross of Taylor & Francis/CRC Press for his assistance in bringing this version of the text to fruition.

Charles E. Roberts, Jr.

Preface to the Third Edition

I feel honored to have been asked to expand the comprehensive book of the late Professor Charles E. Roberts. Introduction to differential equations has always been my favorite second-year class to teach. When teaching this course, I often wanted to show my students how differential equations were useful in my research. This has been a challenge, since I am an enumerative combinatorialist, so I work in a field to which most students in an introductory differential equations have not encountered. So I had to define the objects, the problems, and the methods, so that the discussion is rigorous enough to be interesting, but quick enough to serve its purpose of illustrating large family of applications. As I faced this problem often, I did not have to think long about the direction in which I would extend the previous edition of the present textbook. In Chapter 11, readers will learn notions of set partitions, graphs, and rooted trees, the concept of exponential generating functions, and see how differential equations can be surprisingly useful in solving complex counting problems. They will see famous combinatorial sequences such as the Bell numbers or the Euler numbers and understand how a problem that is entirely about nonnegative integers can nevertheless have its answer encoded by a function like $f(x) = e^{e^x-1}$ or $F(x) = \sec x + \tan x$, and how differential equations can lead us to these surprising solutions. This will hopefully both increase readers' confidence in the power of their newly acquired knowledge of differential equations and encourage them to study the exciting field of combinatorics further. The new chapter will also give instructors teaching a one-semester course from the book additional choices of topics, and it will make teaching a two-semester course from the book easier.

Miklós Bóna
Gainesville, Florida
December 2025

1

Introduction

1.1 Historical Prologue

The singular concept which characterizes calculus and simultaneously sets it apart from arithmetic, algebra, geometry, and trigonometry is the notion of a limit. The idea of a limit originated with the ancient Greek philosophers and mathematicians. However, they failed to fully develop and exploit this concept. It was not until the latter half of the 17th century, when the English mathematician Isaac Newton (1642-1727) and the German mathematician Gottfried Wilhelm Leibniz (1646-1716) independently and almost simultaneously invented differential and integral calculus, that the concept of a limit was revived and developed more fully.

Calculus, as presently taught, begins with differential calculus, continues with the consideration of integral calculus, and then analyzes the relationship between the two. Historically, however, integral calculus was developed much earlier than differential calculus. The idea of integration arose first in conjunction with attempts by the ancient Greeks to compute areas of plane figures, volumes of solids, and arc lengths of plane curves.

Archimedes was born about 287 B.C. in the Greek city-state of Syracuse on the island of Sicily. He was the first person to determine the area and the circumference of a circle. Archimedes determined the volume of a sphere and the surface areas of a sphere, cylinder, and cone. In addition, he calculated areas of ellipses, parabolic segments, and sectors of spirals. However, somewhat surprisingly, no Greek mathematician continued the work of Archimedes, and the ideas which he had advanced regarding integration lay dormant until about the beginning of the 17th century. Using the present-day theory of limits, Archimedes' ingenious method of equilibrium can be shown to be equivalent to our definition of integration.

Early in the 17th century a significant development, which would affect calculus dramatically, was taking place in mathematics—the invention of analytic geometry. Credit for this innovation is given to both René Descartes (1596-1650) and Pierre de Fermat (c. 1601-1665). In 1637, Descartes published the philosophical treatise on universal science, *A Discourse on the Method of Rightly Conducting the Reason and Seeking for Truth in the Sciences.* The third and last appendix to his *Discourse* is titled *La géométrie.* In *La géométrie,* Descartes discusses finding normals to algebraic curves—which is

DOI: 10.1201/9781003678014-1

equivalent to finding tangents to the curves; he also introduces into mathematics the custom of using letters which appear first in the alphabet for constants and letters which appear last for variables; and he formulates our present system of exponents in which x^2 denotes $x \cdot x$, in which x^3 denotes $x \cdot x \cdot x$, etc. Pierre de Fermat's claim to priority in the invention of analytic geometry is based on a letter he wrote to Gilles Roberval in September 1636, in which he claims that the ideas he is advancing are seven years old. Fermat's method for finding a tangent to a curve was devised in conjunction with his procedure for determining maxima and minima. Thus, Fermat was the first mathematician to develop the central idea of differential calculus—the notion of a derivative. Fermat also had great success in the theory of integration. By 1636 or earlier, Fermat had discovered and proved by geometrical means the power formula for positive integer exponents—that is, Fermat had proved for positive integers n

$$\int_0^a x^n\,dx = \frac{a^{n+1}}{n+1}.$$

Later, Fermat generalized this result to rational exponents $n \neq -1$. In many respects, the work of Descartes and Fermat was antipodal. Generally speaking, Descartes started with a locus and then derived its equation, whereas Fermat began with an equation and then found the locus. Their combined efforts illustrate the two fundamental and inverse properties of analytic geometry.

Until approximately the middle of the 17th century, integral and differential calculus appeared to be two distinct branches of mathematics. Integration, in the case of calculating the area under a curve, consisted of finding the limit approached by the sum of a very large number of extremely thin rectangles as the number of rectangles increased indefinitely and the width of each rectangle approached zero. Differentiation, on the other hand, consisted of finding the limit of a difference quotient. About 1646, Evangelista Torricelli (1608-1647), a student of Galileo and inventor of the barometer (1643), showed that integration and differentiation were inverse operations for equations of the form $y = x^n$ where n is a positive integer. That is, Torricelli showed for n a positive integer

$$\frac{d}{dx}\int_0^x t^n\,dt = \frac{d}{dx}\left(\frac{x^{n+1}}{n+1}\right) = x^n.$$

Isaac Barrow (1630-1677) is usually given credit for being the first mathematician to recognize in its fullest generality that differentiation and integration are inverse operations. In his *Lectiones*, Barrow essentially stated and proved geometrically the **fundamental theorem of calculus**—that is, if $f(x)$ is a continuous function on the interval $[a, b]$ and if x is in the interval $[a, b]$, then

$$\frac{d}{dx}\int_a^x f(t)\,dt = f(x).$$

By 1670 the idea of a limit had been conceived; integration had been defined; many integrals had been calculated to find the areas under curves, the volumes of solids, and the arc lengths of curves; differentiation had been defined; tangents to many curves had been effected; many minima and maxima problems had been solved; and the relationship between integration and differentiation had been discovered and proved. What remained to be done? And why should Isaac Newton and Gottfried Wilhelm Leibniz be given credit for inventing the calculus? The answers to these questions are: A general symbolism for integration and differentiation needed to be invented, and strictly formal rules, independent of geometric meaning, for analytic operations needed to be discovered. Working independently of each other, Newton and Leibniz both developed the required symbolism and rules for operation. Newton's "fluxional calculus" was invented as early as 1665, but he did not publish his work until 1687. Leibniz, on the other hand, formulated his "differential calculus" about 1676, ten years later than Newton, but published his results in 1684, thus provoking a bitter priority dispute. It is noteworthy that Leibniz's notation is superior to Newton's, and it is Leibniz's notation which we use today.

In 1661, at the age of 18, Isaac Newton was admitted to Trinity College in Cambridge. The Great Plague of 1664-1665 (a bubonic plague) closed the university in 1665, and Newton returned home. During the next two years, 1665-1667, Newton discovered the binomial theorem, invented differential calculus (his fluxional calculus), proved that white light is composed of all the spectral colors, and began work on what would later evolve into the universal law of gravitation. In 1670-1671, Isaac Newton wrote his *Methodus fluxionum et serierum infinitorum*, but it was not published until 1736—nine years after his death. Newton's approach to differential calculus was a physical one. He considered a curve to be generated by the continuous motion of a point. He called a quantity which changes with respect to time a *fluent* (a flowing quantity). And the rate of change of a fluent with respect to time he called a *fluxion* of the fluent. If a fluent is represented by y, then the fluxion of the fluent y is represented by $\dot{y}$. The fluxion of $\dot{y}$ is denoted by $\ddot{y}$ and so on. The fluent of y was denoted by $\boxed{y}$ or $\acute{y}$. Thus, to Newton $\dot{y}$ was the derivative of y and $\boxed{y}$ or $\acute{y}$ was the integral. Newton considered two different types of problems. The first problem, which is equivalent to differentiation, is to find a relation connecting fluents and their fluxions given some relation connecting the fluents. The inverse problem, which is equivalent to solving a differential equation, is to find a relation between fluents alone given a relation between the fluents and their fluxions. Using his method of fluxions, Newton determined tangents to curves, maxima and minima, points of inflection, curvature, and convexity and concavity of curves; he calculated numerous quadratures; and he computed the arc length of many curves. Newton was the first to systematically use results of differentiation to obtain antiderivatives—that is, Newton used his results from differentiation problems to solve integration problems.

Sometime between 1673 and 1676, Leibniz invented his calculus. In 1675 he introduced the modern integral sign, an elongated letter S to denote the first letter of the Latin word *summa* (sum). After some trial and error in selecting notation, Leibniz settled on dx and dy for small differences in x and y.

He first used these two notations in conjunction late in 1675 when he wrote $\int y\,dy = y^2/2$. In 1676, Leibniz used the term "differential equation" to denote a relationship between two differentials dx and dy. Thus, the branch of mathematics which deals with equations involving differentials or derivatives was christened. To find tangents to curves, Leibniz used the *calculus differentialis* from which we derive the phrase "differential calculus," and to find quadratures, he used the *calculus summatorius* or the *calculus integralis* from which we derive the phrase "integral calculus."

Initially, it was believed that the elementary functions[1] would be sufficient for representing the solutions of differential equations arising from problems in geometry and mechanics. So early attempts at solving differential equations were directed toward finding explicit solutions or reducing the solution to a finite number of quadratures. By the end of the 17th century, most of the calculus which appears in current undergraduate textbooks had been discovered along with some more advanced topics such as the calculus of variation.

Until the beginning of the 19th century, the central theme of differential equations was to find the general solution of a specified equation or class of equations. However, it was becoming increasingly clear that solving a differential equation by quadrature was possible in only a few exceptional cases. Thus, the emphasis shifted to obtaining approximate solutions—in particular, to finding series solutions. About 1820, the French mathematician Augustin-Louis Cauchy (1789-1857) made the solution of the initial value problem $y' = f(x, y)$; $y(x_0) = y_0$ the cornerstone in his theory of differential equations. Prior to the lectures developed and presented by Cauchy at the Paris École Polytechnique in the 1820s, no adequate discussion of differential equations as a unified topic existed. Cauchy presented the first existence and uniqueness theorems for first-order differential equations in these lectures. Later, he extended his theory to include a system of n first-order differential equations in n dependent variables.

There are two fundamental subdivisions in the study of differential equations: quantitative theory and qualitative theory. The object of quantitative theory is (i) to find an explicit solution of a given differential equation or system of equations, (ii) to express the solution as a finite number of quadratures, or (iii) to compute an approximate solution. Early in the development of the subject of differential equations, it was thought that elementary functions were sufficient for representing the solutions of differential equations. However, in 1725, Daniel Bernoulli published results which showed that even a first-order, ordinary differential equation does not necessarily have a solution which is finitely expressible in terms of elementary functions. And in the 1880s, Picard proved that the general linear differential equation of order n is not integrable

[1]Let a be a constant, and let $f(x)$ and $g(x)$ be functions. The following operations are called *elementary operations*: $f(x) \pm g(x)$, $f(x) \cdot g(x)$, $f(x)/g(x)$, $(f(x))^a$, $a^{f(x)}$, $\log_a f(x)$, and $T(f(x))$, where T is any trigonometric or inverse trigonometric function. *Elementary functions* are those functions that can be generated using constants, the independent variable, and a finite number of elementary operations.

by quadratures. In a series of papers published between 1880 and 1886, Henri Poincaré (1854-1912) initiated the qualitative theory of differential equations. The object of this theory is to obtain information about an entire set of solutions without actually solving the differential equation or system of equations. For example, one tries to determine the behavior of a solution with respect to that of one of its neighbors. That is, one wants to know whether or not a solution $v(t)$ which is "near" another solution $w(t)$ at time $t = t_0$ remains "near" $w(t)$ for all $t \geq t_0$.

1.2 Definitions and Terminology

One purpose of this section is to discuss the meaning of the statement:

"Solve the differential equation $y'' + y = 0$."

Several questions must be asked, discussed, and answered before we can fully understand the meaning of the statement above. Some of those questions are

"What is a differential equation?"

"What is a solution of a differential equation?"

"Given a particular differential equation and some appropriate constraints, how do we know if there is a solution?"

"Given a particular differential equation and some appropriate constraints, how do we know how many solutions there are?"

"How do we find a solution to a differential equation?"

We will answer the first two questions in this section and devote much of the remainder of the text to answering the last three questions.

At this point in your study of mathematics, you probably completely understand the meaning of the statement:

"Solve the equation $2x^4 - 3x^3 - 13x^2 + 37x - 15 = 0$."

You recognize this is an algebraic equation. More specifically, you recognize this is a polynomial of degree four. Furthermore, you know there are exactly four roots to this equation, and they can all be found in the set of complex numbers. This last fact is due to the **fundamental theorem of algebra** which states:

"Every polynomial of degree $n \geq 1$ with complex coefficients has n roots (not necessarily distinct) among the complex numbers."

The fundamental theorem of algebra is an existence theorem, since it states that there exist n roots to a polynomial equation of degree n. Of course, the set of roots of a polynomial equation is unique. So the solution of the equation $2x^4 - 3x^3 - 13x^2 + 37x - 15 = 0$ is a set of four complex numbers and that set is unique. Can you solve this equation?

Throughout this text, we will state existence and uniqueness theorems for various types of problems involving differential equations and systems of differential equations. Before generating a numerical solution to any such problem, it is necessary to verify that the hypotheses of appropriate existence and uniqueness theorems are satisfied. Otherwise, the computer may generate a "solution" where none exists, or it may generate a single solution in a region where the solution is not unique—that is, the computer may generate a single solution where there are multiple solutions. Sometimes we present examples which illustrate the erroneous results one may obtain if a numerical solution is produced without regard for the appropriate theory.

An equation that contains one or more derivatives of an unknown function or functions or that contains differentials is called a **differential equation** (DE). When a differential equation contains one or more derivatives with respect to a particular variable, that variable is called an **independent variable.** A variable is said to be a **dependent variable** if some derivative of the variable appears in the differential equation.

In order to systematically study differential equations, it is convenient and advantageous to classify the equations into different categories. Two broad, general categories used to classify differential equations are ordinary differential equations and partial differential equations. This classification is based on the type of unknown function appearing in the differential equation. If the unknown function depends on only one independent variable and the differential equation contains only ordinary derivatives, then the differential equation is called an **ordinary differential equation** (ODE). If the unknown function depends on two or more independent variables and the differential equation contains partial derivatives, then the differential equation is called a **partial differential equation** (PDE). The **order** of a differential equation, ordinary or partial, is the order of the highest derivative occurring in the equation.

For example,

$$\frac{dy}{dx} = \cos y \tag{1}$$

is a first-order, ordinary differential equation. The dependent variable is y and the independent variable is x—that is, the unknown function is $y(x)$.

The equation

$$\frac{d^3y}{dx^3} - x\left(\frac{dy}{dx}\right)^2 + x^2y = \tan x \tag{2}$$

is a third-order, ordinary differential equation.

The equation

$$\left(\frac{d^4y}{dt^4}\right)^3 + ty\frac{d^2y}{dt^2} - y^5 = e^t \tag{3}$$

is a fourth-order, ordinary differential equation. The dependent variable is y and the independent variable is t—thus, the unknown function is $y(t)$.

The equation

$$\frac{\partial z}{\partial x} + \frac{\partial z}{\partial y} = z \tag{4}$$

is a first-order, partial differential equation. The dependent variable is z and the independent variables are x and y. Hence, the unknown function is $z(x, y)$.

The equation

$$\alpha\left(\frac{\partial^2 u}{\partial x^2} + \frac{\partial^2 u}{\partial y^2} + \frac{\partial^2 u}{\partial z^2}\right) = \frac{\partial u}{\partial t} \tag{5}$$

is a second-order, partial differential equation. The unknown function is $u(x, y, z, t)$.

In calculus, you used the notations $\dfrac{d^2y}{dx^2}$ and y'' to represent the second derivative of y. In the first notation, it is clear that x is the independent variable. In the second notation, the prime notation, it is not obvious what the independent variable is. The first notation is somewhat cumbersome, while the second is convenient only for writing lower-order derivatives. Reading and writing the tenth derivative using prime notation would be tedious, so we introduce another notation, $y^{(2)}$, for the second derivative. The tenth derivative in this notation is $y^{(10)}$, which is both easy to read and write. The kth derivative of y is written $y^{(k)}$. Throughout the text, we will use all three notations for the derivative of a function. (Caution: When first using this new notation, one sometimes mistakenly writes y^2 for the second derivative instead of $y^{(2)}$. Of course, y^2 is "y squared" and not the second derivative of y.)

In this text, we will deal mainly with ordinary differential equations. However, ordinary and partial differential equations are both subdivided into two large classes, linear equations and nonlinear equations, depending on whether the differential equation is linear or nonlinear in the unknown function and its derivatives.

The **general n-th order ordinary differential equation** can be written symbolically as

$$F(x, y, y^{(1)}, \ldots, y^{(n)}) = 0. \tag{6}$$

An n-th order ordinary differential equation is **linear**, if it can be written in the form

$$a_0(x)y^{(n)} + a_1(x)y^{(n-1)} + \cdots + a_n(x)y = g(x). \tag{7}$$

The functions $a_k(x)$ are called the **coefficient functions**. A **nonlinear ordinary differential equation** is an ordinary differential equation which is not linear.

It follows from the definition that for an ordinary differential equation to be linear it is necessary that:

1. Each coefficient function $a_k(x)$ depends only on the independent variable x and not on the dependent variable y.

2. The dependent variable y and all of its derivatives $y^{(k)}$ occur algebraically to the first degree only. That is, the power of each term involving y and its derivatives is 1.

3. There are no terms which involve the product of either the dependent variable y and any of its derivatives or two or more of its derivatives.

4. Functions of y or any of its derivatives such as e^y or $\cos y'$ cannot appear in the equation.

Equations (1), (2), and (3) are all nonlinear ordinary differential equations. Equation (1) is nonlinear because of the term $\cos y$. Equation (2) is nonlinear because of the term $(dy/dx)^2$. And equation (3) is nonlinear because of the terms $(d^4y/dt^4)^3$ and $ty(d^2y/dt^2)$. The equations

$$y^{(2)} - 3y^{(1)} + 2y = x \tag{8}$$

$$x^3y^{(4)} - (\sin x)y^{(3)} + 2y^{(1)} - e^x y = x + 1 \tag{9}$$

are both linear ordinary differential equations.

A solution of the n-th order ordinary differential equation

$$F(x, y^{(1)}, \ldots, y^{(n)}) = 0$$

on an interval $\mathbf{I = (a, b)}$ is a function $y = f(x)$ which is defined on I, which is at least n times differentiable on I, and which satisfies the equation $F(x, f^{(1)}, \ldots, f^{(n)}) = 0$ for all x in I. Since a solution $y = f(x)$ is at least n times differentiable on the interval I, the functions $f(x), f^{(1)}(x), \ldots, f^{(n-1)}(x)$ are all continuous on I. Usually the interval I is not specified explicitly, but it is understood to be the largest possible interval on which $y = f(x)$ is a solution.

Let $p(x)$ be a polynomial. By definition, a solution of the equation $p(x) = 0$ is a root of the polynomial. To determine if any particular complex number s is a root of $p(x)$ or not, we compute $p(s)$ and see if the value is 0 or not. That is, to determine if a number s is a root of a polynomial, we do not need to be able to "solve the polynomial" by factoring, by using some formula such as the quadratic formula, or by using a numerical technique such as Newton's method to find the roots of the polynomial. All we need to do is substitute the number s into the polynomial $p(x)$ and see if $p(s) = 0$ or not. Similarly, if we write a differential equation so that 0 is the right-hand side of the equation, then to determine if some function $y(x)$ is a solution of the differential equation or not, all we need to do is (1) determine the order n of the given differential equation, (2) differentiate the function $y(x)$ n times, (3) substitute $y(x)$ and its n derivatives into the given differential equation, and (4) see if the result is

0 or not. If the result is not 0, then the function is not a solution of the given differential equation. If the result is 0, then the function $y(x)$ is a solution on any interval on which it and its n derivatives are simultaneously defined.

Example 1 Verifying that a Function is a Solution of a DE Verify that $y = xe^{2x}$ is a solution of the linear second-order differential equation $y'' - 4y' + 4y = 0$ on the interval $(-\infty, \infty)$.

Solution

Differentiating $y = xe^{2x}$ twice, we find

$$\begin{aligned} y' &= 2xe^{2x} + e^{2x} \\ y'' &= 2(2xe^{2x} + e^{2x}) + 2e^{2x} = 4xe^{2x} + 4e^{2x}. \end{aligned}$$

The three functions y, y', and y'' are defined on $(-\infty, \infty)$ and

$$\begin{aligned} y'' - 4y' + 4y &= (4xe^{2x} + 4e^{2x}) - 4(2xe^{2x} + e^{2x}) + 4xe^{2x} \\ &= (4 - 8 + 4)xe^{2x} + (4 - 4)e^{2x} = 0 \end{aligned}$$

for all x in $(-\infty, \infty)$. So $y = xe^{2x}$ is a solution of $y'' - 4y' + 4y = 0$ on $(-\infty, \infty)$. ■

Example 2 Being a Solution Depends on the Interval

a. Show that $y = x + \frac{1}{x}$ is not a solution of $x^2y'' + xy' - y = 0$ on the interval $(-\infty, \infty)$.

b. Verify that $y = x + \frac{1}{x}$ is a solution of $x^2y'' + xy' - y = 0$ on the intervals $(-\infty, 0)$ and $(0, \infty)$.

Solution

a. The function $y = x + \frac{1}{x}$ is not defined at $x = 0$, so it cannot be the solution of any differential equation on any interval which contains the point $x = 0$.

b. Differentiating $y = x + \dfrac{1}{x} = x + x^{-1}$ twice, we find

$$y' = 1 - x^{-2} = 1 - \frac{1}{x^2} \qquad \text{and} \qquad y'' = 2x^{-3} = \frac{2}{x^3}.$$

The three functions y, y', and y'' are defined on the intervals $(-\infty, 0)$ and $(0, \infty)$. And for $x \neq 0$

$$\begin{aligned} x^2y'' + xy' - y &= x^2\left(\frac{2}{x^3}\right) + x\left(1 - \frac{1}{x^2}\right) - \left(x + \frac{1}{x}\right) \\ &= \frac{2}{x} + x - \frac{1}{x} - x - \frac{1}{x} = 0. \end{aligned}$$

So the function $y = x + \dfrac{1}{x}$ is a solution of $x^2y'' + xy' - y = 0$ on $(-\infty, 0)$, on $(0, \infty)$, and on any subinterval of these intervals, but $y = x + \frac{1}{x}$ is not a solution of the differential equation on any interval which includes the point $x = 0$. ■

In calculus, you learned the definition of the derivative of a function, and you calculated the derivative of a few functions from the definition. Also, you may have seen the definition of the left-hand derivative and right-hand derivative of a function in calculus. Since we want to determine where piecewise functions are differentiable, we provide those definitions again.

Let $f(x)$ be a real-valued function of a real variable defined on some interval I and let c be in the interval I.

The **derivative of** f **at** c is

$$f'(c) = \lim_{h \to 0} \frac{f(c+h) - f(c)}{h}, \quad \text{provided the limit exists.}$$

The **left-hand derivative of** f **at** c is

$$f'_-(c) = \lim_{h \to 0^-} \frac{f(c+h) - f(c)}{h}, \quad \text{provided the limit exists.}$$

The **right-hand derivative of** f **at** c is

$$f'_+(c) = \lim_{h \to 0^+} \frac{f(c+h) - f(c)}{h}, \quad \text{provided the limit exists.}$$

Remarks: The derivative of the function f at c exists if and only if both the left-hand derivative at c and the right-hand derivative at c exist and are equal. Recall from calculus, if f is differentiable at c, then f is continuous at c. In order to show that a function f is not differentiable at c, we can show that (1) the function f is not continuous at c, or (2) either the left-hand derivative at c or the right-hand derivative at c does not exist, or (3) the left-hand derivative at c and the right-hand derivative at c both exist, but they are not equal.

For example, the piecewise defined function

$$y(x) = \begin{cases} x^3 + 2, & x < 0 \\ x^2 - 1, & 0 \le x \end{cases}$$

is defined on the interval $(-\infty, \infty)$. It is continuous on the intervals $(-\infty, 0)$ and $(0, \infty)$, but it is not continuous at $x = 0$, since

$$\lim_{x \to 0^-} y(x) = 2 \ne -1 = \lim_{x \to 0^+} y(x) = y(0).$$

Because $y(x)$ is not continuous at $x = 0$, the function $y(x)$ is not differentiable at $x = 0$; however, $y(x)$ is differentiable on $(-\infty, 0)$ and $(0, \infty)$. In fact,

$$y'(x) = \begin{cases} 3x^2, & x < 0 \\ 2x, & 0 < x. \end{cases}$$

The absolute value function

$$y(x) = |x| = \begin{cases} -x, & x < 0 \\ x, & 0 \le x \end{cases}$$

is a piecewise defined function which is continuous on $(-\infty, \infty)$. Computing the left-hand derivative of $y(x) = |x|$ at $x = 0$, we find

$$y'_-(0) = \lim_{h \to 0^-} \frac{|0+h| - |0|}{h} = \lim_{h \to 0^-} \frac{|h|}{h} = \lim_{h \to 0^-} \frac{-h}{h} = -1.$$

And computing the right-hand derivative of $y(x) = |x|$ at $x = 0$, we find

$$y'_+(0) = \lim_{h \to 0^+} \frac{|0+h| - |0|}{h} = \lim_{h \to 0^+} \frac{|h|}{h} = \lim_{h \to 0^-} \frac{h}{h} = 1.$$

Since $y'_-(0) = -1 \neq 1 = y'_+(0)$, the absolute value function, $y(x) = |x|$, is not differentiable at $x = 0$. However, the absolute value function is differentiable on $(-\infty, 0)$ and $(0, \infty)$, and its derivative is

$$y'(x) = \frac{d|x|}{dx} = \left\{ \begin{array}{rl} -1, & x < 0 \\ 1, & 0 < x \end{array} \right\} = \frac{|x|}{x} = \operatorname{sgn}(x)$$

where $\operatorname{sgn}(x)$ is an abbreviation for the *signum* function.

The previous two examples illustrate that points where piecewise defined functions may not be continuous or may not be differentiable are points at which the definition of the function changes from one mathematical expression to another.

Example 3 A Solution that is Defined Piecewise Verify that the piecewise defined function

$$y(x) = \begin{cases} -x^2, & x < 0 \\ x^2, & 0 \le x \end{cases}$$

is differentiable on the interval $(-\infty, \infty)$ and is a solution of the differential equation $xy' - 2y = 0$ on $(-\infty, \infty)$.

Solution

On the interval $(-\infty, 0)$, $y(x) = -x^2$; therefore, on $(-\infty, 0)$ its derivative is $y'(x) = -2x$. On the interval $(0, \infty)$, $y(x) = x^2$; hence, on $(0, \infty)$ its derivative is $y'(x) = 2x$. To determine if $y(x)$ is differentiable at $x = 0$, we compute the left-hand derivative at 0 and the right-hand derivative at 0. Doing so, we find

$$y'_-(0) = \lim_{h \to 0^-} \frac{-(0+h)^2 + 0^2}{h} = \lim_{h \to 0^-} \frac{-h^2}{h} = \lim_{h \to 0^-} -h = 0$$

and

$$y'_+(0) = \lim_{h \to 0^+} \frac{(0+h)^2 - 0^2}{h} = \lim_{h \to 0^+} \frac{h^2}{h} = \lim_{h \to 0^+} h = 0.$$

Since $y'_-(0) = 0 = y'_+(0)$, the function $y(x)$ is differentiable at $x = 0$. Since $y(x)$ is differentiable on $(-\infty, 0)$, on $(0, \infty)$, and at $x = 0$, the function $y(x)$ is differentiable on the interval $(-\infty, \infty)$, and its derivative is

$$y'(x) = \begin{cases} -2x, & x < 0 \\ 2x, & 0 \le x. \end{cases}$$

A graph of $y(x)$ and $y'(x)$ is shown in Figure 1.1. For $x < 0$, $xy' - 2y = x(-2x) - 2(-x^2) = -2x^2 + 2x^2 = 0$, so $y(x)$ satisfies the given differential equation on the interval $(-\infty, 0)$. For $x > 0$, $xy' - 2y = x(2x) - 2(x^2) = 2x^2 - 2x^2 = 0$, so $y(x)$ satisfies the differential equation on the interval $(0, \infty)$. At $x = 0$, $xy' - 2y = 0(0) - 2(0) = 0$, so $y(x)$ satisfies the differential equation at $x = 0$. Therefore, the piecewise defined, differentiable function $y(x)$ satisfies the differential equation $xy' - 2y = 0$ on the interval $(-\infty, \infty)$.

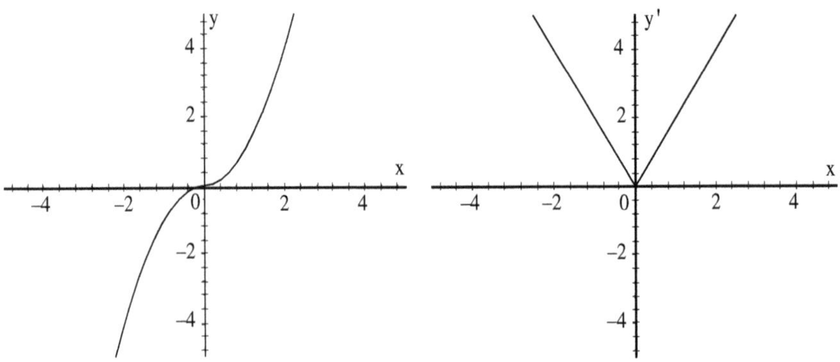

FIGURE 1.1
Graph of y(x) and y'(x). ■

1.2.1 Exercises

In Exercises 1–14 determine the order of the given ordinary differential equation and state whether the equation is linear or nonlinear.

1. $y' + x^2 y = 3\cos x$
2. $y' + a(x)y = b(x)$
3. $y' - 2e^x y = y^2$
4. $y' + a(x)y^n = b(x)$ (where $n \neq 0$ and $n \neq 1$)
5. $(y')^3 - xy^2 = \sin x$
6. $3x^2 dx - 4y dy = 0$ (Hint: Divide by dx.)
7. $ydx - xdy = 0$
8. $y(1 + (y')^2) = 5$
9. $y'' - 3y'y = 4$
10. $y'' + x^2 y' - x^3 y = \tan x$
11. $y'' + y\sin x = 3$
12. $y'' + x\sin y = 3$
13. $(y^{(3)})^2 - 4(y^{(2)})^4 + x^2 y = 0$

14. $y^{(4)} + x^2 y^{(3)} - (\sin x) y^{(2)} = y$

15. For what values of x is $y = \sqrt{x^2 - 1}$ a solution of the differential equation $x - yy' = 0$?

16. Show that the differential equation $|y'| + y^2 + 5 = 0$ has no solution.

17. What is the unique solution of the differential equation $|y'| + 5y^2 = 0$ on the interval $(-\infty, \infty)$?

18. Show that y $= e^{x^2} + e^{x^2} \int_0^x e^{t^2}\, dt$ is a solution of the differential equation $y' = 2xy + 1$.

19. Is $y(x) = 1/x$ a solution of the differential equation $x^2 y'' + xy' - y = 0$
    a. on the interval $(-1, 1)$? Why?
    b. on the interval $(0, \infty)$? Why?
    c. on the interval $(-\infty, 0)$? Why?

20. Is $y = |x|$ a solution of the differential equation $xy' - y = 0$
    a. on the interval $(-1, 1)$? Why?
    b. on the interval $(0, \infty)$? Why?
    c. on the interval $(-\infty, 0)$? Why?

21. Is $y = \sqrt{x}$ a solution of the differential equation $2x^2 y'' + 3xy' - y = 0$
    a. on the interval $(-1, 1)$? Why?
    b. on the interval $(0, \infty)$? Why?
    c. on the interval $(-\infty, 0)$? Why?

In Exercises 22–30 verify that the given function or functions is a solution of the given differential equation and specify the interval or intervals on which the solution exists.

22. $y'' - 3y' + 2y = 0; \qquad y_1(x) = 3e^x, \qquad y_2(x) = e^{2x}$

23. $x^2 y'' - 2y = 0; \qquad y_1(x) = x^2 - 1/x$

24. $y' + 1/(2y) = 0; \qquad y_1(x) = \sqrt{3 - x}$

25. $y' - y/x = 1; \qquad y_1(x) = x \ln x$

26. $y' - 2\sqrt{|y|} = 0; \qquad y_1(x) = x|x|$

27. $x^2 dy + 2xy dx = 0; \qquad y_1(x) = -1/x^2$

28. $y' - y^2 = 1; \qquad y_1(x) = \tan x$

29. $2x^2y'' + xy' - y = 0;\qquad y_1(x) = x,\qquad y_2(x) = 1/\sqrt{x}$

30. $xy' - \sin x = 0;\qquad y_1(x) = \int_0^x \frac{\sin t\, dt}{t},\qquad y_2(x) = -\int_x^\pi \frac{\sin t\, dt}{t}$

In Exercises 31–34 find the value or values of r for which the function $y(x) = e^{rx}$ is a solution of the given differential equation.

31. $y' + 3y = 0$

32. $y'' - 3y' - 10y = 0$

33. $y'' + 2y' + y = 0$

34. $y''' - 7y'' + 12y' = 0$

In Exercises 35–38 find the value or values of r for which the function $y(x)=x^r$ is a solution of the given differential equation.

35. $2xy' - y = 0$

36. $x^2y'' - xy' = 0$

37. $x^2y'' + 6xy' + 4y = 0$

38. $x^2y'' - 5xy' + 9y = 0$

39. Find the values of r for which the function $y = rx^3$ is a solution of the following differential equations.

    a. $x^2y'' + 6xy' + 5y = 0$

    b. $x^2y'' + 6xy' + 5y = 2x^3$

    c. $x^2y'' + 6xy' + 5y = x^3$

40. Verify that the piecewise defined function

$$y(x) = \begin{cases} 0, & x < 0 \\ x^2, & 0 \le x \end{cases}$$

is differentiable on the interval $(-\infty, \infty)$ and is a solution of the differential equation $(y')^2 - 4y = 0$ on $(-\infty, \infty)$.

41. Verify that the piecewise defined function

$$y(x) = \begin{cases} x^3, & x < 0 \\ 0, & 0 \le x \end{cases}$$

is differentiable on $(-\infty, \infty)$ and is a solution of the differential equation $(y')^2 - 9xy = 0$ on $(-\infty, \infty)$.

42. The differential equation

$$\text{(L)}\qquad y' = x^3$$

is linear, but the differential equation

$$\text{(N)}\qquad (y')^2 = x^6$$

is nonlinear.

a. Verify that the derivative of the piecewise defined function

$$y(x) = \begin{cases} -x^4/4, & x < 0 \\ x^4/4, & 0 \leq x \end{cases}$$

is

$$y'(x) = \begin{cases} -x^3, & x < 0 \\ x^3, & 0 \leq x. \end{cases}$$

b. Show that $y(x)$ is a solution on $(-\infty, \infty)$ of (N) $(y')^2 = x^6$, but $y(x)$ is not a solution on $(-\infty, \infty)$ of either (L) $y' = x^3$ or $y' = -x^3$.

43. a. Show that the one-parameter family of lines $y(x) = Cx + f(C)$, where C is a constant, is a solution of **Clairaut's equation** $y = xy' + f(y')$.

b. For $f(y') = -\frac{1}{4}(y')^2$ Clairaut equation becomes (*) $y = xy' - \frac{1}{4}(y')^2$; and the one-parameter family of lines $y(x) = Cx - \frac{1}{4}C^2$ is a solution of (*) for arbitrary C. Show for any constant C that the line $y(x) = Cx - \frac{1}{4}C^2$ is tangent to the parabola $y = x^2$ at the point $(C/2, C^2/4)$ and show that the parabola $y = x^2$ is a solution of equation (*).

44. Let f and g be any two twice differentiable functions of x and t. Show that $u(x,t) = f(x+at) + g(x-at)$ is a solution of the partial differential equation

$$a^2 \frac{\partial^2 u}{\partial x^2} = \frac{\partial^2 u}{\partial t^2}.$$

This equation is called the **wave equation**.

1.3 Solutions and Problems

In the previous section, we learned what a solution to a differential equation is and how to check if a particular function is a solution to a specified differential equation or not. At this point, you might ask: "How many solutions does a differential equation have?" The short answer is "It depends on the differential equation." For instance, the differential equation $|y'| + 1 = 0$ has no solution, while the differential equation $(y')^2 + y^2 = 0$ has exactly one solution—the function $y(x) = 0$. By integration, we find that the solution of the differential equation $y' = 2x$ is

$$y(x) = x^2 + C \tag{1}$$

where C is an arbitrary constant. Hence, the differential equation $y' = 2x$ has an infinite number of solutions—one solution for each choice of the value of the constant C. The set of solutions $y(x) = x^2 + C$ is called a **one-parameter family of solutions** of the differential equation $y' = 2x$. The graph of this family of solutions is called the **integral curves** or **solution curves** of the differential equation. A solution of a differential equation which contains no arbitrary constants is called a **particular solution.** Choosing $C = 1$ in equation (1), we obtain the particular solution $y(x) = x^2 + 1$ of the differential equation $y' = 2x$. A graph of the particular solutions obtained from (1) by choosing $C = -2$, $C = -1$, $C = 0$, $C = 1$, and $C = 2$ are shown in Figure 1.2. The graphs of these solution curves are parabolas which open upward, have vertices at $(0, C)$, and have the y-axis as their axes of symmetry. Thus, Figure 1.2 represents the solution curves of the differential equation $y' = 2x$.

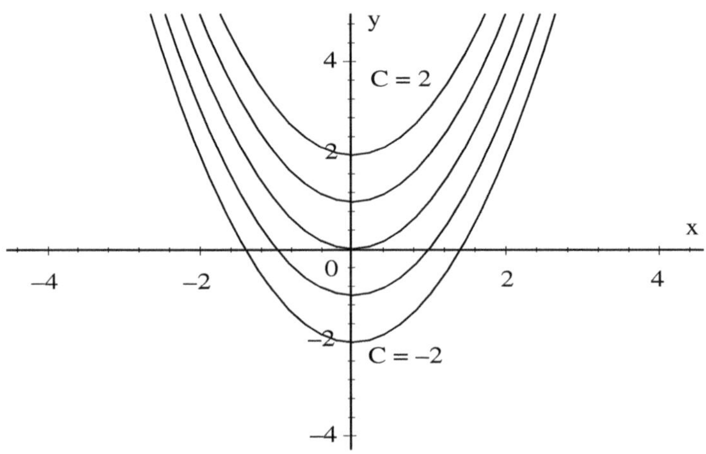

FIGURE 1.2
Solution Curves of the Differential Equation $y' = 2x$.

Integrating the second-order differential equation $y'' = 2x + 1$ twice, we find $y' = x^2 + x + c_1$ and

(2) $$y(x) = \frac{1}{3}x^3 + \frac{1}{2}x^2 + c_1 x + c_2.$$

Hence, (2) is a two-parameter family of solutions of the DE $y'' = 2x + 1$.

When specifying the number of parameters in a family of functions, we must be careful to avoid calling every constant a parameter. For example, the family of functions $y = c_1 e^{x+c_2}$ contains two constants c_1 and c_2, but this family is not a two-parameter family, since

$$y = c_1 e^{x+c_2} = c_1 e^x e^{c_2} = ke^x.$$

That is, $y = c_1 e^{x+c_2}$ is actually the one-parameter family of functions $y = ke^x$ disguised as a two-parameter family. When a set of constants $\{c_1, c_2, \ldots, c_n\}$

in a family of functions cannot be reduced to a smaller number by algebraic manipulation, then the constants are called **essential parameters.** Henceforth, when we say a family of functions is an n-parameter family, we will assume that it has n essential parameters.

Example 1 Verifying a Two-Parameter Family is a Solution Verify that

$$y(x) = c_1 e^x + c_2 e^{-x} \tag{3}$$

is a two-parameter family of solutions of the second-order, linear differential equation

$$y'' - y = 0. \tag{4}$$

Solution

Differentiating (3) twice, we get $y' = c_1 e^x - c_2 e^{-x}$ and $y'' = c_1 e^x + c_2 e^{-x}$. Substituting these expressions for y'' and y into (4), we find

$$y'' - y = (c_1 e^x + c_2 e^{-x}) - (c_1 e^x + c_2 e^{-x}) = 0$$

for all x and all choices of c_1 and c_2. Hence, (3) is a solution of the DE (4) on the interval $(-\infty, \infty)$ for all choices of c_1 and c_2. ■

If **every** solution on an interval I of the n-th order differential equation

$$F(x, y, y^{(1)}, y^{(2)}, \ldots, y^{(n)}) = 0 \tag{5}$$

is obtainable from an n-parameter family of functions

$$G(x, y, c_1, c_2, \ldots, c_n) = 0 \tag{6}$$

by an appropriate choice of the parameters $c_1, c_2, \ldots, c_n$, then the family of functions (6) is called the **general solution** of the differential equation (5). A solution of an n-th order differential equation which cannot be obtained from an n-parameter family of solutions is called a **singular solution.**

As we shall see later in the text, when the coefficient functions of a linear differential equation satisfy fairly simple conditions on an interval, then solutions exist on the interval, and all solutions are obtainable from a family of functions. With the exception of a few first-order equations, nonlinear differential equations are difficult or impossible to solve explicitly. So for all practical purposes, the term general solution is used only in conjunction with linear differential equations. The general solution of a differential equation may not have a unique representation as an n-parameter family of functions. That is, there may be more than one function of the form (6) which is the general solution of (5). For example, both of the two-parameter families

$$y_1 = c_1 e^{-x} + c_2 e^{x} \qquad \text{and} \qquad y_2 = k_1 \sinh x + k_2 \cosh x$$

are general solutions of the differential equation $y'' - y = 0$. They are simply two different ways to represent the same solutions, since

$$\sinh x = \frac{1}{2}e^{x} - \frac{1}{2}e^{-x} \quad \text{and} \quad \cosh x = \frac{1}{2}e^{x} + \frac{1}{2}e^{-x}.$$

Example 2 Singular Solution of a Differential Equation

a. Verify that

$$\text{(7)} \qquad y(x) = cx + c^2$$

is a one-parameter family of solutions of the first-order nonlinear differential equation

$$\text{(8)} \qquad (y')^2 + xy' - y = 0.$$

b. Show that

$$y(x) = -x^2/4$$

is a singular solution of the DE (8).

Solution

a. Differentiating $y(x) = cx + c^2$, we get $y' = c$. Substituting for y and y' in (8), yields

$$(y')^2 + xy' - y = c^2 + xc - (cx + c^2) = 0.$$

Thus, (7) is a one-parameter family of solutions of the DE (8).

b. The function $y(x) = -x^2/4$ cannot be obtained from the one-parameter family $y(x) = cx + c^2$ by any choice of the constant c. Differentiating $y(x) = -x^2/4$, we find $y' = -x/2$. Substituting for y and y' in (8), we find

$$(y')^2 + xy' - y = (-\frac{x}{2})^2 + x(-\frac{x}{2}) - (-\frac{x^2}{4}) = (\frac{1}{4} - \frac{1}{2} + \frac{1}{4})x^2 = 0.$$

Thus, $y(x) = -x^2/4$ is a solution of the DE (8) and it is not a member of the one-parameter family (7). So by definition $y(x) = -x^2/4$ is a singular solution of (8). A graph of the singular solution $y(x) = -x^2/4$ and members of the one-parameter family of solutions $y(x) = cx + c^2$ obtained by choosing $c = -2$, $c = -1$, $c = 0$, $c = 1$, and $c = 2$ are displayed in Figure 1.3. Observe that the one-parameter family of lines $y(x) = cx + c^2$ are tangent to the singular solution (the parabola) $y(x) = -x^2/4$.

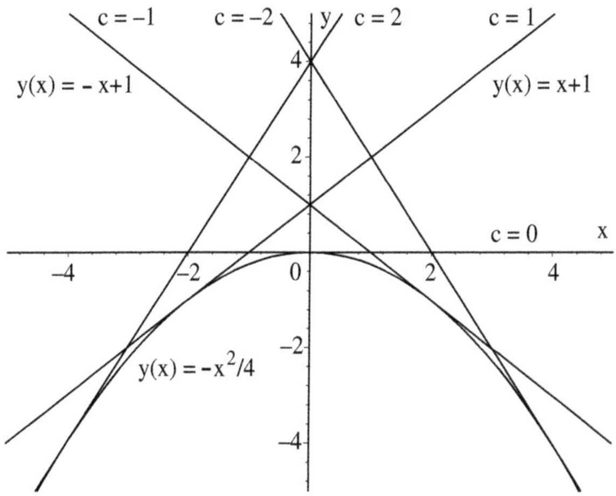

FIGURE 1.3
A One-Parameter Family of Solutions and the Singular Solution of $(y')^2 + xy' - y = 0$. ■

Given a function $y(x)$, we are usually able to determine where the function is differentiable and to calculate its derivative. For example, the function $y(x) = x^3 - 4x + 1$ has derivative $y'(x) = 3x^2 - 4$. Given the differential equation $y'(x) = 3x^2 - 4$, how do we recover the original function $y(x) = x^3 - 4x + 1$? Integrating the differential equation $y'(x) = 3x^2 - 4$ we get, not a single function, but the one-parameter family of functions, $y(x) = x^3 - 4x + C$ where C is an arbitrary constant. In order to select the specific function $y(x) = x^3 - 4x + 1$ from the family of functions $y(x) = x^3 - 4x + C$, we must specify some condition to be satisfied in addition to the differential equation. The original curve $y(x) = x^3 - 4x + 1$ passes through the point $(2, 1)$. Requiring $y(x) = x^3 - 4x + C$ to satisfy the condition $y(2) = 1$, we see that C must satisfy the equation $y(2) = 1 = 2^3 - 4(2) + C$. Hence, $C = 1$ and we recover the original function.

Proceeding one step further, we differentiate $y(x) = x^3 - 4x + 1$ twice and find $y''(x) = 6x$. Integrating this differential equation twice, we obtain the two-parameter family of solutions $y(x) = x^3 + Ax + B$ where A and B are arbitrary constants. To recover the original function, in this case, we must specify two additional conditions which will require us to choose $A = -4$ and $B = 1$. For instance, we could require $y(2) = 1$ and $y'(2) = 8$, since the original curve passes through the point $(2, 1)$ and has slope 8 at $(2, 1)$—that is,

$$\left.\frac{dy}{dx}\right|_{x=2} = y'(2) = 8.$$

Or, we could require $y(1) = -2$ and $y(2) = 1$, since the original curve passes through both $(1, -2)$ and $(2, 1)$.

The problem of solving the differential equation $y'' = 6x$ subject to the two conditions

$$(9) \qquad y(2) = 1 \quad \text{and} \quad y'(2) = 8$$

is called an **initial value problem** and the conditions (9) are called **initial conditions.**

The problem of solving the differential equation $y'' = 6x$ subject to the two conditions

$$(10) \qquad y(1) = -2 \quad \text{and} \quad y(2) = 1$$

is called a **boundary value problem** and the conditions (10) are called **boundary conditions.**

So two types of problems to be solved in the study of differential equations are initial value problems and boundary value problems. A precise statement of these two types of problems for n-th order ordinary differential equations follows.

The problem of finding a solution $y = f(x)$ of the n-th order differential equation

$$\text{(DE)} \qquad y^{(n)} = F(x, y, y^{(1)}, \ldots, y^{(n-1)})$$

subject to a set of n conditions, called **initial conditions,**

$$\text{(IC)} \qquad y(x_0) = c_0,\ y^{(1)}(x_0) = c_1, \ldots,\ y^{(n-1)}(x_0) = c_{n-1}$$

where x_0, c_0, $c_1, \ldots, c_{n-1}$ are real constants is called an **initial value problem (IVP)**. Notice that in an initial value problem, all n conditions to be satisfied are specified at a single value of the independent variable—namely, at x_0.

The problem of finding a solution $y = f(x)$ to the differential equation

$$\text{(DE)} \qquad y^{(n)} = F(x, y, y^{(1)}, \ldots, y^{(n-1)})$$

subject to a set of n conditions, called **boundary conditions (BC)**, which specify values of the function y or some of its derivatives at two or more distinct values of the independent variable is called a **boundary value problem (BVP)**.

The theory associated with initial value problems is well established and relatively simple. As a consequence, throughout this text, we will state various theorems which guarantee the existence of solutions and theorems which guarantee the uniqueness of solutions to different types of initial value problems—such as initial value problems in which the differential equation is first-order, initial value problems in which the differential equation is linear with constant coefficients, and initial value problems in which the differential equation is linear with variable coefficients. We will also define an analogous initial value problem for systems of first-order differential equations and state an existence and uniqueness theorem for this problem.

At this point, we need to examine a few initial value problems and boundary value problems to determine whether all initial value problems and all boundary value problems have a solution or not. That is, we need to examine the question of existence of solutions to initial value problems and the question of existence of solutions to boundary value problems. We also need to discover if any initial value problem or any boundary value problem has more than one solution. That is, we need to consider the question of uniqueness of solutions to initial value problems and the question of uniqueness of solutions to boundary value problems.

Let us examine the differential equation

$$xy' - y = 0. \tag{11}$$

The function $y(x) = cx$, where c is an arbitrary constant, and its derivative $y'(x) = c$ are both defined on the interval $(-\infty, \infty)$. Substituting y and y' into the DE (11), we find

$$xy' - y = xc - cx = 0$$

for all x and for all choices of c. Thus, $y(x) = cx$ is a one-parameter family of solutions to the DE (11) on $(-\infty, \infty)$. This family of functions is the set of all lines which pass through the origin except for the line $x = 0$ (the y-axis).

Example 3 An Initial Value Problem with No Solution Consider the initial value problem

$$xy' - y = 0; \quad y(0) = 1. \tag{12}$$

For $x = 0$ the differential equation $xy' - y = 0$ reduces to $-y = 0$. Consequently, for $x = 0$ the differential equation $xy' - y = 0$ is undefined unless $y(0) = 0$ also. Hence, the IVP (12) is an example of an initial value problem which does not have a solution. Furthermore, there is no solution to any initial value problem of the form $xy' - y = 0$; $y(0) = d \neq 0$. In geometric terms, there is no solution of the differential equation $xy' - y = 0$ which passes through any point on the positive or negative y-axis—that is, any point of the form $(0, d)$ where $d \neq 0$. ■

Example 4 An Initial Value Problem with an Infinite Number of Solutions

The differential equation of the initial value problem

$$xy' - y = 0; \quad y(0) = 0 \tag{13}$$

is the same as the differential equation in the previous example. As we noted earlier, $y = cx$ is a one-parameter family of solutions on $(-\infty, \infty)$. Substituting the initial condition $y(0) = 0$ into the solution $y(x) = cx$ results in the equation $y(0) = 0 = c(0)$ which is satisfied by any constant c. Thus, $y(x) = cx$ is a solution of the IVP (13) on the interval $(-\infty, \infty)$ for any choice of the constant c. Consequently, the IVP (13) is an example of an initial value problem with an infinite number of solutions. ■

Example 5 An Initial Value Problem with a Unique Solution Now consider the initial value problem

$$xy' - y = 0; \quad y(1) = 4. \tag{14}$$

Again, the function $y(x) = cx$, where c is an arbitrary constant, is a one-parameter family of solutions of the differential equation $xy' - y = 0$ on the interval $(-\infty, \infty)$. Imposing the initial condition $y(1) = 4 = c(1)$, we find $c = 4$. So the IVP (14) has the unique solution $y(x) = 4x$ on $(-\infty, \infty)$. ■

Let us examine initial value problems of the form $xy' - y = 0$; $y(a) = b$ where $a \neq 0$. The solution of the differential equation is $y(x) = cx$. Imposing the initial condition $y(a) = b$, we find c must satisfy the equation $y(a) = b = ca$. Since we have assumed $a \neq 0$, the unique solution of this equation is $c = b/a$; and, therefore, the unique solution of the initial value problem $xy' - y = 0$; $y(a) = b$ where $a \neq 0$ is $y(x) = bx/a$. Interpreted geometrically, this means that for any point (a, b) where $a \neq 0$—that is, for any point which is not on the y-axis—there is a unique solution of the differential equation $xy' - y = 0$ which passes through (a, b).

Example 6 An Initial Value Problem with an Infinite Number of Piecewise Defined Solutions One solution on the interval $(-\infty, \infty)$ of the initial value problem

$$y' = 3y^{2/3}; \quad y(0) = 0 \tag{15}$$

is clearly $y_1(x) = 0$. A second solution is $y_2(x) = x^3$. (Verify this fact.) As a matter of fact, there are an infinite number of solutions to this initial value problem. In Exercise 6 at the end of this section, you are asked to show that

$$y_{ab}(x) = \begin{cases} (x-a)^3, & x < a \leq 0 \\ 0, & a \leq x \leq b \\ (x-b)^3, & 0 \leq b < x \end{cases}$$

is a solution on $(-\infty, \infty)$ for every choice of $a \leq 0$ and $b \geq 0$. This example illustrates that if an initial value problem has a solution, the solution may not be unique. ■

In the following three examples of boundary value problems, we will use the same differential equation and vary only one boundary condition. First, we verify that a two-parameter family of solutions of the differential equation $y'' + y = 0$ is $y = A\sin x + B\cos x$, where A and B are arbitrary constants. Differentiating $y = A\sin x + B\cos x$ twice, we find

$$y' = A\cos x - B\sin x \qquad \text{and} \qquad y'' = -A\sin x - B\cos x.$$

Substitution into the differential equation, yields

$$y'' + y = (-A\sin x - B\cos x) + (A\sin x + B\cos x) = 0$$

for all real x and for all A and B. Since y, y', and y'' are all defined on $(-\infty, \infty)$, the function $y = A\sin x + B\cos x$ is a solution of $y'' + y = 0$ on the interval $(-\infty, \infty)$. Suppose in addition to satisfying $y'' + y = 0$, we require that $y(0) = 0$. Imposing this condition results in the equation

$$y(0) = 0 = A\sin 0 + B\cos 0 = B.$$

Thus, $B = 0$ and A is arbitrary. Consequently, the solution of the differential equation $y'' + y = 0$ which satisfies the condition $y(0) = 0$ is the one-parameter family of functions $y = A\sin x$. A graph of this family on the interval $(-5, 5)$ is shown in Figure 1.4.

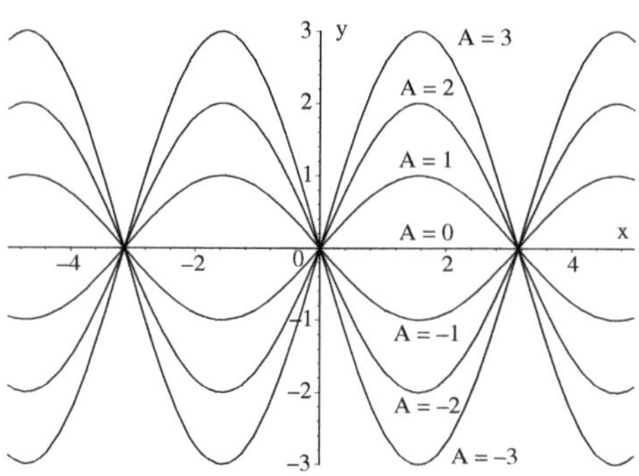

FIGURE 1.4
Solutions Curves $y = A\sin x$ of $y'' + y = 0; y(0) = 0$.

Example 7 A Boundary Value Problem with No Solution Consider the boundary value problem

$$y'' + y = 0; \quad y(0) = 0, \quad y(\pi) = 1. \tag{16}$$

We already know from the previous discussion that if there is a solution to this boundary value problem, it must be of the form $y = A\sin x$. Imposing the second boundary condition, $y(\pi) = 1$, results in the following equation and contradiction

$$y(\pi) = 1 = A\sin\pi = 0 \quad \text{or} \quad 1 = 0.$$

Hence, there is no solution to the boundary value problem (16). ■

Example 8 A Boundary Value Problem with a Unique Solution Now consider the boundary value problem

$$y'' + y = 0; \quad y(0) = 0, \quad y(\pi/2) = 1. \tag{17}$$

Again, if a solution of this boundary value problem exists, it must be of the form $y = A\sin x$. Imposing the second boundary condition, $y(\pi/2) = 1$, yields

$$y(\pi/2) = 1 = A\sin(\pi/2) = A.$$

So $A = 1$ and the unique solution of the boundary value problem (17) is $y = \sin x$. ■

Example 9 A Boundary Value Problem with an Infinite Number of Solutions

As in the previous two examples, if the boundary value problem

$$y'' + y = 0; \quad y(0) = 0, \quad y(\pi) = 0 \tag{18}$$

is to have a solution, it must be of the form $y = A\sin x$. Imposing the second boundary condition, $y(\pi) = 0$, results in the equation $A\sin\pi = A \cdot 0 = 0$, which is satisfied by all choices of the constant A. So the boundary value problem (18) has an infinite number of solutions—the one-parameter family of functions $y = A\sin x$, where A is an arbitrary constant. ■

In the previous three examples, the differential equation and one boundary condition were identical. However, the results which we obtained were radically different because of the second boundary condition. These examples illustrate the complexities inherent in the study of boundary value problems. These complexities can, at least partially, be attributed to the interaction of the boundary conditions with the differential equation. Because of these inherent complexities, we shall not present any general theory for boundary value problems.

1.3.1 Exercises

1. Verify that $y(x) = c_1e^{x^2}$ is a one-parameter family of solutions of the differential equation $y' - 2xy = 0$ on the interval $(-\infty, \infty)$.

2. Show that $y = c_1e^{-x} + x^2 - 1$ is a one-parameter family of solutions of the differential equation $y' + y = x^2 + 2x - 1$ on $(-\infty, \infty)$.

3. Show that $y = c_1e^{-2x} + c_2e^{3x}$ is a two-parameter family of solutions of the differential equation $y'' - y' - 6y = 0$ on $(-\infty, \infty)$.

4. Show that $y = (\frac{1}{4}x^2 + c^2)^2$ is a one-parameter family of solutions of the differential equation $y' = xy^{1/2}$ on $(-\infty, \infty)$ and verify that $y(x) = 0$ is a singular solution.

5. Consider the differential equation (19) $y'' - y = 0$.

   a. Show that $y = c_1 e^{-x} + c_2 e^x$ is a two-parameter family of solutions of the DE (19).

   b. Show that $y = k_1 \sinh x + k_2 \cosh x$ is also a two-parameter family of solutions of the DE (19).

6. Show that for all $a \le 0$ and all $b \ge 0$

$$y_{ab}(x) = \begin{cases} (x-a)^3, & x < a \le 0 \\ 0, & a \le x \le b \\ (x-b)^3, & 0 \le b < x \end{cases}$$

   is a solution of the initial value problem (20) $y' = 3y^{2/3}$; $y(0) = 0$ on the interval $(-\infty, \infty)$. This proves that the initial value problem (20) has an infinite number of solutions.

7. Verify that $y = c_1 x \ln x$ is a one-parameter family of solutions of the differential equation (21) $(x \ln x)y' - (1 + \ln x)y = 0$ on the interval $(0, \infty)$. Solve, if you can, the three initial value problems consisting of the DE (21) and the following three sets of initial conditions

   a. $y(2) = 4$ b. $y(1) = 0$ c. $y(1) = 2$

8. Verify that $y = c_1 e^{-x} + c_2 e^{2x}$ is a two-parameter family of solutions of the differential equation (22) $y'' - y' - 2y = 0$ on $(-\infty, \infty)$.

   a. Solve the initial value problems consisting of the DE (22) and the following two sets of initial conditions.

      (i) $y(0) = 2, \quad y'(0) = -5$ (ii) $y(1) = 3, \quad y'(1) = -1$

   b. Solve the boundary value problems consisting of the DE (22) and the following two sets of boundary conditions

      (i) $y(0) = 1, \quad y(2) = 0$ (ii) $y(0) = 0, \quad y'(2) = 1$

9. Verify that $y = c_1 x + c_2 x^2 + c_3 x^3$ is a three-parameter family of solutions of the differential equation (23) $x^3 y^{(3)} - 3x^2 y^{(2)} + 6xy^{(1)} - 6y = 0$ on $(-\infty, \infty)$. Find the solution of the initial value problem consisting of the DE (23) and the initial conditions $y(1) = 2, \quad y'(1) = 3, \quad y''(1) = 4$.

10. Verify that $y = c_1 x^2 + c_2 x^3$ is a two-parameter family solution of the differential equation (24) $x^2 y'' - 4xy' + 6y = 0$ on $(-\infty, \infty)$. Solve, if you can, the boundary value problems consisting of the DE (24) and the following sets of boundary conditions.

    a. $y(1) = 0, \quad y(2) = -4$ b. $y'(1) = 0, \quad y(2) = 4$

    c. $y(1) = 1, \quad y'(2) = -12$ d. $y'(1) = 3, \quad y'(2) = 0$

    e. $y(0) = 0, \quad y(2) = 4$ f. $y(0) = 2, \quad y'(2) = -1$

1.4 A Nobel Prize Winning Application

Generally a mathematical model is constructed to approximate a physical problem. Often this model includes a differential equation. Unless the model was poorly constructed, the solution of the differential equation, if one exists, will usually approximate the solution of the physical problem.

Radioactive Decay. Physical experimentation has shown that radioactive substances decompose at a rate which is proportional to the quantity of radioactive substance present. If we let $Q(t)$ represent the quantity of radioactive substance present at time t, then the statement above may be expressed mathematically by the differential equation

$$\frac{dQ}{dt} = kQ \tag{1}$$

where k is the constant of proportionality. Multiplying equation (1) by dt and dividing by Q, we obtain

$$\frac{dQ}{Q} = k dt. \tag{2}$$

The variables Q and t are "separated" in this equation in the sense that Q and its differential dQ appear on the left-hand side of the equation, while the differential of t, dt, appears on the right-hand side of the equation. Integrating equation (2), we find

$$\int \frac{dQ}{Q} = \int k\,dt \qquad \text{or} \qquad \ln |Q| = kt + C$$

where C is an arbitrary constant of integration. Exponentiating the right-hand equation, we see

$$|Q| = e^{kt+C} = e^C e^{kt}.$$

Since Q and e^C are positive constants, we may rewrite this last equation as

$$Q(t) = Pe^{kt} \tag{3}$$

where P is a positive constant. To determine the two constants k and P in equation (3), we need to specify two physical conditions to be satisfied. Suppose at time t_0 the amount of radioactive substance present is Q_0 and suppose at some later time t_1, the amount of substance present is Q_1. Then stated mathematically, the two conditions to be satisfied are

$$Q(t_0) = Q_0 \tag{4}$$

and

$$Q(t_1) = Q_1. \tag{5}$$

Evaluating equation (3) at $t = t_0$ and imposing the condition (4), we see the constant P must be chosen to satisfy $Q(t_0) = Q_0 = Pe^{kt_0}$. Hence, $P = Q_0e^{-kt_0}$. Substituting this expression into equation (3), we find the solution of the DE (1) which satisfies condition (4) has the form

$$(6) \qquad Q(t) = Q_0e^{k(t-t_0)}.$$

Evaluating equation (6) at $t = t_1$ and imposing the condition (5), we see the constant k must satisfy

$$(7) \qquad Q(t_1) = Q_1 = Q_0e^{k(t_1-t_0)}.$$

Solving equation (7) for k, we find the constant of proportionality is

$$(8) \qquad k = \frac{\ln Q_1 - \ln Q_0}{t_1 - t_0}.$$

In the radioactive decay process, the amount of substance present decreases with increasing time. Since we have assumed $t_1 > t_0$, it follows that $Q_1 < Q_0$. Hence, $t_1 - t_0 > 0$, and since $\ln Q_1 < \ln Q_0$, we have $\ln Q_1 - \ln Q_0 < 0$. It then follows from equation (8) that $k < 0$.

The rate of decay of a radioactive substance is often expressed in terms of **half-life**—that is, the time required for any given quantity of the substance to be reduced by a factor of one-half. If in equation (8) we let $Q_1 = \frac{1}{2}Q_0$, then the half-life $T = t_1 - t_0$ satisfies the equation

$$(9) \qquad kT = -\ln 2.$$

Consequently, if either k or T is known or can be determined experimentally, then the other variable can be determined from equation (9).

The half-life of uranium 238 is 4.5 billion years, the half-life of potassium 40 is 1.4 billion years, and the half-life of rubidium 87 is 60 billion years. By checking the ratio of elements such as these to the elements into which they decay radioactively, geologists and archaeologists can reliably estimate dates of significant events that occurred millions and even billions of years ago. However, since they decay so slowly, radioactive elements with half-lives of millions or billions of years are not suitable for dating events which took place relatively recently.

In the late 1940s and early 1950s, the American chemist Willard F. Libby (1908-1980) developed the technique of radiocarbon dating, which can be used to estimate the dates of events that occurred up to 50,000 years ago. In 1960, Libby was awarded the Nobel Prize in Chemistry for this achievement. Libby's technique is based on a phenomenon which involves the radioactive isotope carbon 14, ^{14}C, which is called *radiocarbon* and has a half-life of 5568 years. Radioactive carbon is constantly being produced in the earth's upper atmosphere by incoming cosmic rays. These rays produce neutrons, which in turn collide with nitrogen 14 to produce carbon 14. The radioactive carbon is oxidized and forms radioactive carbon dioxide, which circulates in the earth's atmosphere. Plants which "breathe" carbon dioxide also breathe radioactive carbon dioxide and, through their life processes, absorb radiocarbon in their tissue. Likewise, animals which eat these plants absorb radiocarbon in their tissue. The rate of absorption of radiocarbon by living tissue is in equilibrium with the rate of disintegration. However, when a plant or animal dies, it ceases to absorb radiocarbon, and only the process of disintegration continues. The age of a substance of organic origin can be estimated by measuring the radioactivity of carbon 14 of a sample of that substance. For example, a piece of charcoal that has one-half the radioactivity of a living tree died approximately 5568 years ago, and a piece of charcoal that has one-fourth the radioactivity of a living tree died approximately 11,136 years ago.

Solving equation (9), $kT = -\ln 2$, for the decay constant k, and substituting 5568 for the half-life, T, we find the decay constant for radioactive carbon, ^{14}C, is

$$k = -\frac{\ln 2}{5568 \text{ years}} = -.00012449/\text{years}.$$

From equation (6) we see that the amount, $Q(t)$, of ^{14}C present at time $t \geq t_0$ in some organic substance is

$$Q(t) = Q_0 e^{k(t-t_0)}$$

where Q_0 is the amount that was present at the time t_0 when the substance died. Differentiating this equation, we find that the rate of disintegration, $R(t)$, of ^{14}C at any time $t \geq t_0$ is

$$R(t) = kQ_0 e^{k(t-t_0)}.$$

At time $t = t_0$ the rate of disintegration is

$$R(t_0) = kQ_0.$$

So the ratio of the disintegration rate at time t to the disintegration rate at time t_0 is

$$\frac{R(t)}{R(t_0)} = e^{k(t-t_0)}.$$

Solving for the time since the death of the substance, $t - t_0$, we find

$$t - t_0 = \frac{1}{k} \ln\left(\frac{R(t)}{R(t_0)}\right). \tag{10}$$

Assuming that for any particular living substance the rate of disintegration of ^{14}C is a constant (that is, the rate of disintegration is the same now as it was in the past), the time that a particular sample of the same substance died can be calculated from equation (10). For example, in 1950, the rate of radioactive disintegration of ^{14}C from a piece of charcoal found in the Lascaux cave in France was .97 disintegrations per minute per gram. Tissue from living wood has a disintegration rate of 6.68 disintegrations per minute per gram. So the tree from which the charcoal came died

$$t - t_0 = \frac{1}{-.00012449} \ln\left(\frac{.97}{6.68}\right) = 15{,}500 \text{ years before 1950.}$$

We have only discussed radioactive substances with relative long half-lives. However, the reader should be aware that there are radioactive substances with half-lives on the order of a few years, a year, a day, a second, and even a very small fraction of a second.

1.4.1 Exercises

1. If 5% of a radioactive substance decomposes in 50 years, what percentage will be present at the end of 500 years? 1000 years? What is the half-life of the substance?

2. If the half-life of a radioactive substance is 1800 years, what percentage is present at the end of 100 years? In how many years does only 10% of the substance remain?

3. If 100 grams of a radioactive substance are present one year after the substance was produced and 75 grams are present two years after the substance was produced, how much was produced and what is the half-life of the substance?

4. In 1977 the rate of carbon 14 radioactivity of a piece of charcoal found at Stonehenge in southern England was 4.16 disintegrations per minute per gram. Given that the rate of carbon 14 radioactivity of a living tree is 6.68 disintegrations per minute per gram and assuming the tree which was burned to produce the charcoal was cut during the construction of Stonehenge, estimate the date of the construction of Stonehenge.

5. During the 1950 excavation of the Babylonian city of Nippur, charcoal from a roof beam was discovered which had a rate of carbon 14 radioactivity of 4.09 disintegrations per minute per gram. Given that the rate of radioactivity of a living tree is 6.68 disintegrations per minute per gram and assuming the charcoal was created during the reign of Hammurabi, approximately when was the reign of Hammurabi?

2

The Initial Value Problem $y' = f(x, y);\ \ y(c) = d$

In this chapter, we discuss in detail the first-order initial value problem $y' = f(x, y);\ \ y(c) = d$. First, we define the direction field for the differential equation $y' = f(x, y)$, we discuss the significance of the direction field, and we show how to use a computer program to produce a graph of the direction field. Next, we state a fundamental existence theorem, a fundamental existence and uniqueness theorem, and a continuation theorem for the initial value problem. We show how to apply these theorems to a variety of initial value problems and we illustrate and emphasize the importance of these theorems. Then we discuss how to obtain the general solution to first-order differential equations which are separable or linear and, thereby, solve initial value problems in which the differential equation is separable or linear. Next, we present some simple numerical techniques for solving first-order initial value problems. Finally, we explain how to use a computer program to generate approximate, numerical solutions to first-order initial value problems. We illustrate and interpret the various kinds of results which computer software may produce. Furthermore, we reiterate the importance of performing a thorough mathematical analysis, which includes applying the fundamental theorems to the problem, prior to generating a numerical solution.

The **first-order initial value problem** is to solve the differential equation (DE)

$$y' = f(x, y) \tag{1}$$

subject to the constraint, called an **initial condition** (IC),

$$y(c) = d. \tag{2}$$

It is customary to write this initial value problem (IVP) more compactly as

$$y' = f(x, y);\quad y(c) = d. \tag{3}$$

In Chapter 1 we gave examples of initial value problems with no solution, with a unique solution, and with multiple solutions. Later, we will state a theorem which will guarantee the existence of a solution to an initial value problem of the form (3), and we will state a second theorem which will guarantee the existence and uniqueness of a solution.

DOI: 10.1201/9781003678014-2

2.1 Direction Fields

First, let us examine the geometric significance of the differential equation (1) $y' = f(x, y)$. At each point (x, y) in the xy-plane for which the function f is defined, the differential equation defines a real value, $f(x, y)$. This value is the slope of the tangent line to every solution of the differential equation which passes through the point (x, y). Thus, the differential equation specifies the direction that a solution must have at every point (x, y) in the domain of f. Imagine passing a short line segment of slope $f(x, y)$ through each point (x, y) in the domain of f. The set of all such line segments is called the **direction field** of the differential equation $y' = f(x, y)$. Usually, the domain of f contains an infinite number of points; and, therefore, we cannot possibly draw the direction field. Instead, we choose some rectangle

$$R = \{(x, y) |\ \text{Xmin} \le x \le \text{Xmax} \ \text{ and } \ \text{Ymin} \le y \le \text{Ymax}\}$$

which contains points of the domain of f; we select a set of points (x_i, y_i) contained in R; and for those points (x_i, y_i) in the domain of f, we construct a short line segment at (x_i, y_i) with slope $f(x_i, y_i)$. We will call a graph constructed in this manner the direction field of $y' = f(x, y)$ in the rectangle R. The direction field indicates subregions in R in which solutions are increasing and decreasing, it often reveals maxima and minima of solutions in R, it sometimes indicates the asymptotic behavior of solutions, and it illustrates the dependence of solutions on the initial conditions.

Let (x, y) be a fixed point in the rectangle R at which $f(x, y)$ is defined.

- If $y' = f(x, y) > 0$, then the solution which passes through the point (x, y) is increasing.
- Similarly, if $y' = f(x, y) < 0$, then the solution which passes through the point (x, y) is decreasing.
- When $y' = f(x, y) = 0$, we must consider the direction lines near (x, y), and there are five distinct cases to consider.

(1) If the direction lines of points immediately to the left and right of (x, y) are also horizontal ($y' = 0$ to the right and left of (x, y)), then the solution through (x, y) is constant on some interval containing x.

(2) If the direction lines immediately below and to the left of (x, y) have positive slope ($y' > 0$ below and to the left of (x, y)), then the solution through the point (x, y) increases as it approaches (x, y) from the left.

(3) If the direction lines immediately above and to the left of (x, y) have negative slope ($y' < 0$ above and to the left of (x, y)), then the solution through the point (x, y) decreases as it approaches (x, y) from the left.

(4) If below and to the right of (x, y), the direction lines have negative slope, then the solution through (x, y) decreases to the right of (x, y).

(5) If above and to the right of (x, y), the direction lines have positive slope, then the solution through (x, y) increases to the right of (x, y).

- When $f(x, y) = 0$ and (2) and (4) both occur, then there is a relative maximum at, or near, (x, y).
- When $f(x, y) = 0$ and (3) and (5) both occur, then there is a relative minimum at, or near, (x, y).
- When $f(x, y) = 0$ and either (2) and (5) both occur or (3) and (4) both occur, then there is an inflection point at, or near, (x, y).

In order to use a computer program to graph the direction field of $y' = f(x, y)$ in a rectangular region R of the xy-plane bounded by the lines $x =$ Xmin, $x =$ Xmax, $y =$ Ymin, and $y =$ Ymax, you must enter the function f and the values for Xmin, Xmax, Ymin, and Ymax. The way in which you do this depends on the software you are using. The following example illustrates the typical output of such computer programs.

Example 1 Direction Field for $y' = x - y$ Graph the direction field of the differential equation $y' = x - y = f(x, y)$ on the rectangle $R = \{(x, y) | -5 \leq x \leq 5 \text{ and } -5 \leq y \leq 5\}$.

Solution

We input the function $f(x, y)$—namely, $x - y$ and indicated Xmin $= -5$, Xmax $= 5$, Ymin $= -5$, and Ymax $= 5$. The direction field shown in Figure 2.1 is the output from MAPLE.

Notice that on the line $y = x$, $y' = x - y = 0$. Above the line $y = x$, $y > x$ so $y' = x - y$ is negative. Thus, any solution which passes through some point above the line $y = x$ decreases until it reaches a minimum, which occurs when the solution crosses the line $y = x$ (where $y' = 0$). Once the solution crosses the line $y = x$, $y' = x - y$ becomes positive, since $x > y$ and the solution increases.

Look carefully at the direction field in Figure 2.1. You should be able to pick out the line $y = x - 1$. That is, from Figure 2.1, it looks like $y = x - 1$ may be a solution of the differential equation $y' = x - y$. To determine if it is, we differentiate $y = x - 1$ and find $y' = 1$. Since for $y = x - 1$, $x - y = x - (x - 1) = 1$ also, the function $y = x - 1$ is a solution of the differential equation $y' = x - y$.

Notice that a solution which passes through any point below the line $y = x - 1$ is strictly increasing, since below $y = x - 1$, $y < x$ and $y' = x - y > 0$. We also claim that as x increases all of the solutions approach the line $y = x - 1$. That is, all solutions approach the line $y = x - 1$ asymptotically.

Verify that the general solution of the differential equation $y' = x - y$ is $y = Ce^{-x} + x - 1$, where C is any arbitrary constant. Observe that $y = x - 1$ is

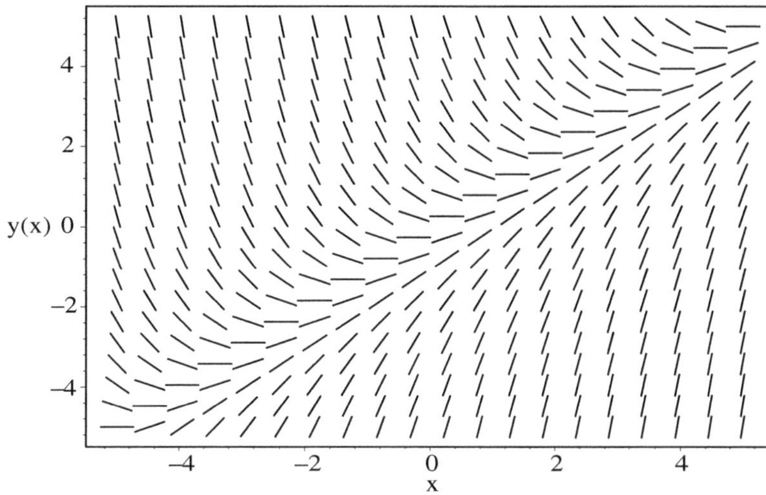

FIGURE 2.1
Direction Field for $y' = x$ - y.

the solution corresponding to $C = 0$ and that for any $C \neq 0$, $y = Ce^{-x} + x - 1$ approaches $x - 1$ as x approaches $+\infty$. Figure 2.2 is a graph of the direction field for $y' = x - y$ and the solution curves $y = Ce^{-x} + x - 1$ for $C = -2$, $C = -1$, $C = 0$, $C = 1$, and $C = 2$.

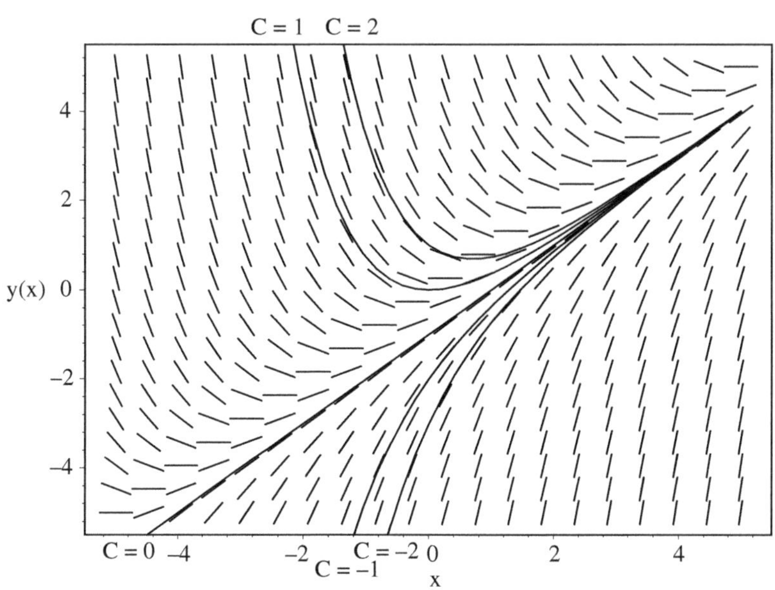

FIGURE 2.2
Direction Field and Solution Curves for $y' = x$ - y. ■

Perhaps the following analogy will help you better understand the relationship of the direction field and a solution to an initial value problem. Think of the rectangle R as a football field and imagine that you are floating above the football field in a hot air balloon and looking down upon the field. Suppose flags have been placed on the field to form a grid work. And suppose a gentle breeze starts to blow across the field. Each flag will point in the direction the breeze is blowing at that point on the field. Since we are above the field, we see only the top of the flag which will be a straight line. If a dandelion seed is released at some point on the field (the initial point), then the seed will float across the field. The path of the dandelion seed corresponds to the solution of the initial value problem (3) consisting of the differential equation (1), which corresponds to the direction field, and the initial condition (2), which corresponds to the point at which the seed was released. At any point where the path of the seed touches a flag, the path will be tangent to the flag.

2.1.1 Comments on Computing

You may use your favorite software package or artificiall intelligence tools to graph the direction field of the differential equation $y' = f(x, y)$ in the rectangular region R of the xy-plane bounded by the lines $x =$ Xmin, $x =$ Xmax, $y =$ Ymin, and $y =$ Ymax. To graph the direction field you will typically need to enter an expression for the function $f(x, y)$ and the values for Xmin, Xmax, Ymin, and Ymax. The output will be similar to the output shown in Figure 2.1, which was produced using the following four MAPLE statements.

```
with(DEtools): with(plots):
de:=diff(y(x), x) = x - y(x):
p:=DEplot(de, y(x), x = -5..5, y = -5..5, arrows=LINE, axes=BOXED):
display(p);
```

The first statement informs MAPLE what software packages are required to run the program. The second statement specifies the differential equation whose direction field is to be plotted—in this instance, $\dfrac{dy}{dx} = x - y$. To specify any other differential equation, the expression $x - y(x)$ which appears in the second statement must be replaced and written in the appropriate MAPLE syntax. The third statement specifies that the x values for the graph are to vary from -5 to 5, the y values for the graph are to vary from -5 to 5, the direction field elements are to be graphed as line segments (arrows = LINE), and the x and y axes are to appear as a box outside of the rectangle (axes = BOXED) instead of passing through the origin perpendicular to one another. You can change the size and location of the rectangle by changing the range of the variables x and y. The fourth statement causes the graph shown in Figure 2.1 to be displayed on the computer monitor.

Example 2 Direction Field for $\mathbf{y' = -x/y}$ Graph the direction field of the differential equation $y' = -x/y = f(x, y)$ on the rectangle $R = \{(x, y) |\ -5 \leq x \leq 5 \ \text{ and } \ -5 \leq y \leq 5\}$.

Solution

Notice that $f(x, y)$ is undefined when $y = 0$—that is, f is undefined on the x-axis. So there is no solution which passes through $(x, 0)$ for any x. To graph the direction field on the rectangle R, we set $f(x, y) = -x/y$, Xmin $= -5$, Xmax $= 5$, Ymin $= -5$, and Ymax $= 5$. The resulting graph is displayed in Figure 2.3. From the graph, we see that the solution curves above the x-axis increase in the second quadrant, have a maximum at the y-axis, and decrease in the first quadrant. We also see from the graph that solution curves below the x-axis decrease in the third quadrant, have a minimum at the y-axis, and increase in the fourth quadrant. The solution of the differential equation $y' = -x/y$ is discussed in detail in Example 4 of the following section.

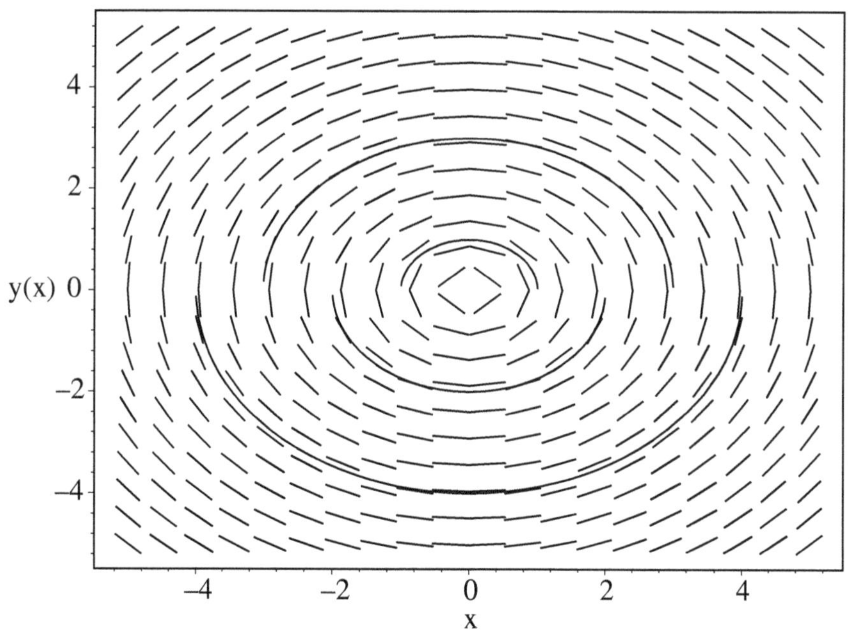

FIGURE 2.3
Direction Field for $y' = -x/y$. ■

Example 3 Direction Field for $\mathbf{y' = \sqrt{xy}}$ Graph the direction field of the differential equation $y' = \sqrt{xy} = f(x, y)$ on the rectangle $R = \{(x, y) |\ -5 \leq x \leq 5 \ \text{ and } \ -5 \leq y \leq 5\}$.

Solution

To graph this direction field, we entered $f(x,y) = \sqrt{xy}$ and set Xmin = -5, Xmax = 5, Ymin = -5, and Ymax = 5. The graph of this direction field and some of its solution curves are shown in Figure 2.4. Observe that the function $f(x,y) = \sqrt{xy}$ is undefined in the second and fourth quadrants where $xy < 0$. Since the differential equation $y' = \sqrt{xy}$ is undefined for $xy < 0$, there can be no solution to an initial value problem consisting of this differential equation and an initial condition which corresponds to a point in the second or fourth quadrant. Notice that the curve $y = 0$ (the x-axis) is a solution of the differential equation. In the first and third quadrants $xy > 0$, so $y' = \sqrt{xy} > 0$; and, therefore, all solutions in the first or third quadrant are increasing functions. A solution which is in the third quadrant increases until it terminates at the y-axis, or it increases until it reaches the x-axis. If a solution in the third quadrant reaches the x-axis, it may continue along the x-axis indefinitely or, once x becomes positive, it may increase into and through the first quadrant. This is an example of a differential equation with an infinite number of solutions passing through some points. Any point in the third quadrant above the curve $y = x^3/9$ and any point in the first quadrant below the curve $y = x^3/9$ has an infinite number of solutions passing through it.

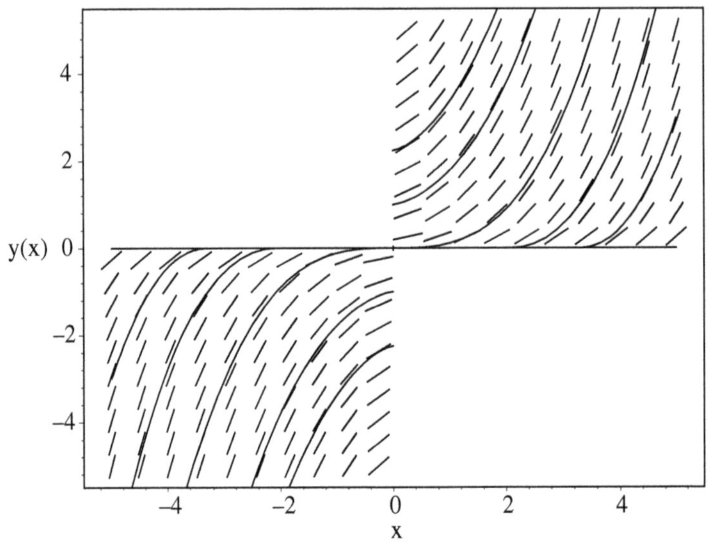

FIGURE 2.4
Direction Field and Solution Curves for $y' = \sqrt{xy}$. ■

2.1.2 Exercises

In Exercises 1–5 match the given differential equations with their direction fields without using any computer software. Then use your computer software to verify that you have matched them correctly. Finally, on the direction field of each differential equation sketch a few solution curves.

1. A. $y' = 1 - x$ B. $y' = x - 1$

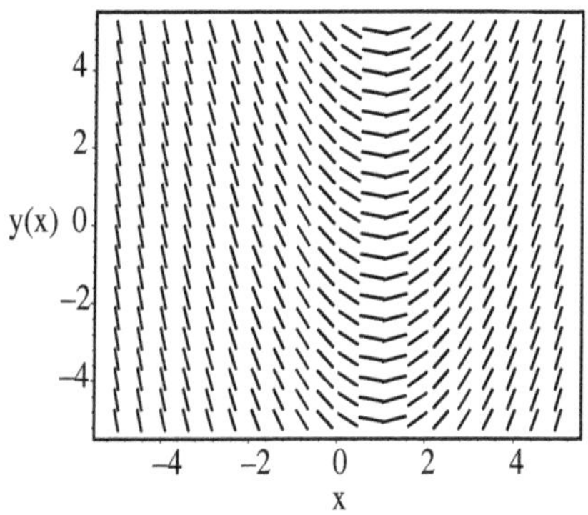

a

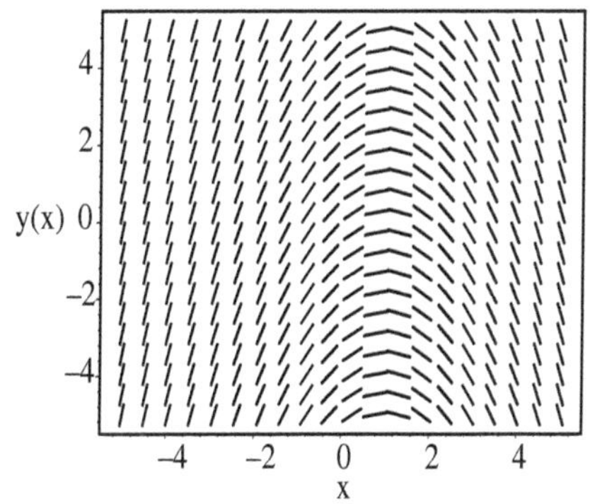

b

2. C. $y' = 1 - y$ D. $y' = 1 + y$

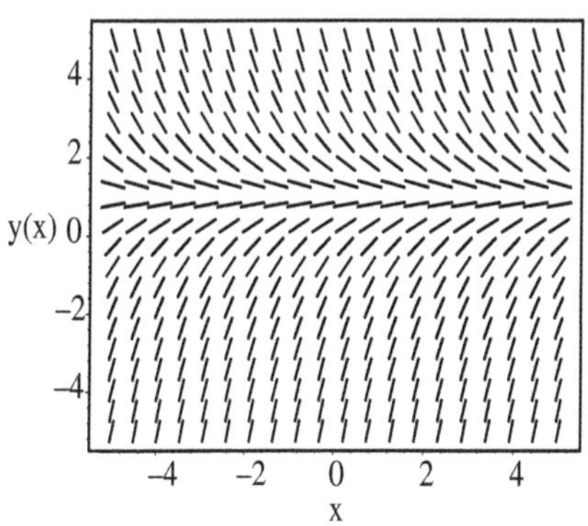

c

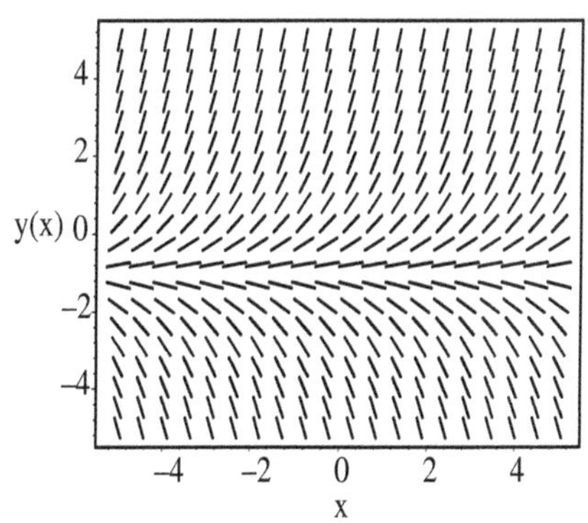

d

3. E. $y' = y^2 - 4$ F. $y' = 4 - y^2$

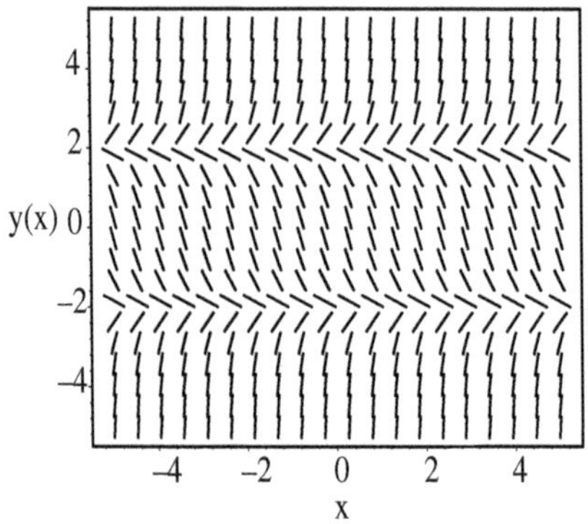

e

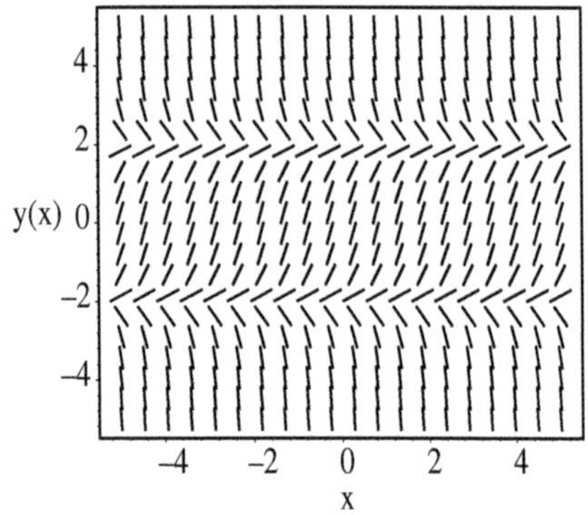

f

4. G. $y' = xy$ H. $y' = -xy$

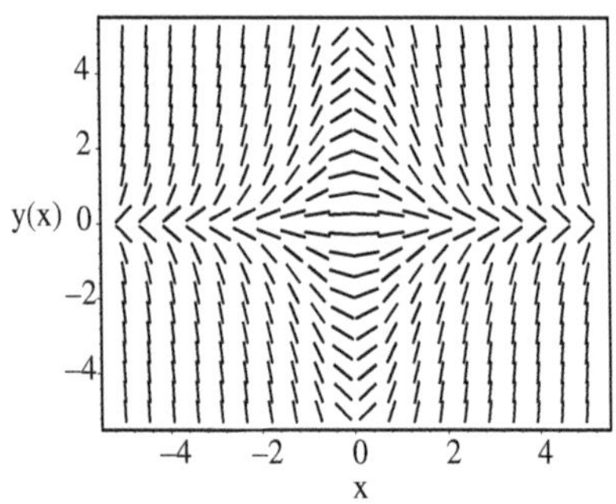

g

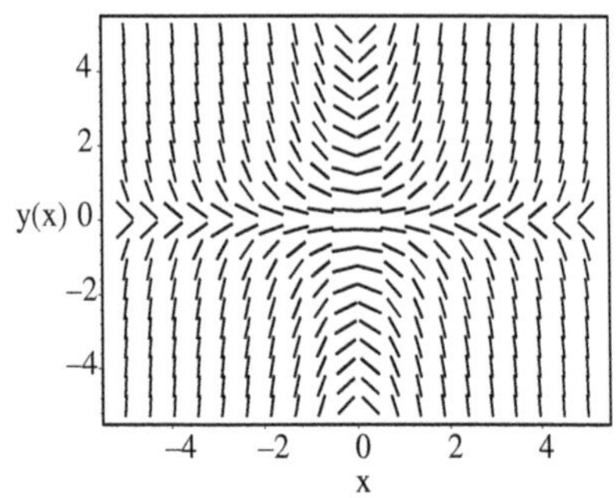

h

5. I. $y' = x^2 - y^2$ J. $y' = y^2 - x^2$

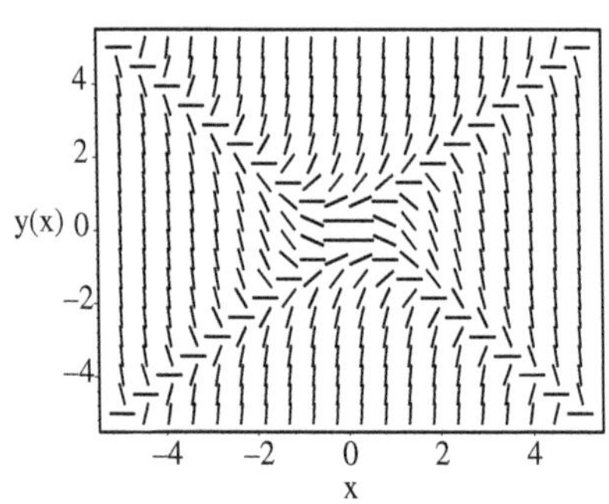

i

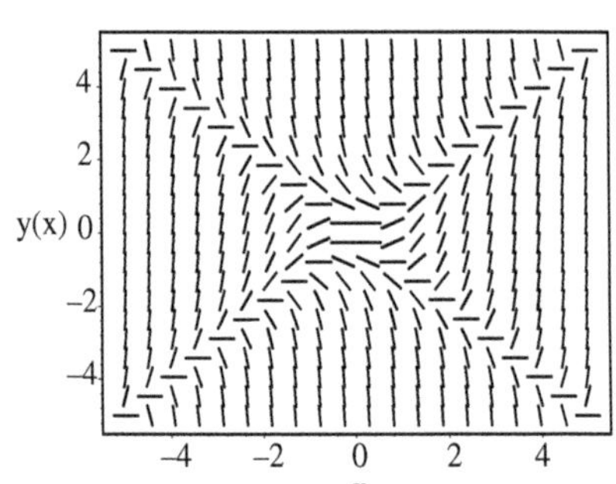

j

In Exercises 6–17 graph the direction field of the given differential equation on the rectangle

$$\mathbf{R}=\{(\mathbf{x},\mathbf{y}) \mid -5 \le \mathbf{x} \le 5 \text{ and } -5 \le \mathbf{y} \le 5\}.$$

When possible, indicate where the direction field is undefined, where solutions are increasing and decreasing, where relative maxima and relative minima occur, and the asymptotic behavior of solutions.

6. $y' = x + y$
7. $y' = xy$
8. $y' = x/y$
9. $y' = y/x$
10. $y' = 1 + y^2$
11. $y' = y^2 - 3y$
12. $y' = x^3 + y^3$
13. $y' = |y|$
14. $y' = e^{x-y}$
15. $y' = \ln(x + y)$
16. $y' = \dfrac{2x - y}{x + 3y}$
17. $y' = \dfrac{1}{\sqrt{15 - x^2 - y^2}}$
18. Graph the direction field for $y' = 3y^{2/3}$ on the rectangle

$$R = \{(x,y) \mid -5 \le x \le 5 \text{ and } -5 \le y \le 5\}.$$

(Hint: Enter $y^{2/3}$ as y^(2/3) and as (y^2)^(1/3). Notice the difference in the graphs. Which graph is the correct direction field for $y' = 3y^{2/3}$?)

2.2 Fundamental Theorems

By stating and discussing three fundamental theorems regarding the initial value problem

$$y' = f(x,y); \quad y(c) = d \tag{1}$$

we hope to answer, at least in part, the following three questions:

"Under what conditions does a solution to the IVP (1) exist?"

"Under what conditions is the solution to the IVP (1) unique?"

"Where—that is, on what interval or what region—does the solution to the IVP (1) exist and where is the solution unique?"

We state the following existence theorem without proof. This theorem is due to the Italian mathematician and logician Giuseppe Peano (1858-1932).

2.2.1 Fundamental Existence Theorem

Let $R = \{(x,y) \mid \alpha < x < \beta \text{ and } \gamma < y < \delta\}$ where α, β, γ, and δ are finite real constants. If $f(x,y)$ is a continuous function of x and y in the finite rectangle R and if $(c, d) \in R$, then there exists a solution to the initial value problem $y' = f(x, y)$; $y(c) = d$ on some interval $I = (c - h, c + h)$ where I is a subinterval of the interval (α, β).

The geometry of the fundamental existence theorem is depicted in Figure 2.5. The theorem itself states a fairly simple condition—namely, continuity of $f(x, y)$—which guarantees the existence of a solution to the initial value problem (1) $y' = f(x, y)$; $y(c) = d$. However, neither the theorem nor its proof provides a method for producing the solution or for satisfactorily calculating the value of h which, in theory, determines the interval on which the solution exists. The theorem simply states that there is an interval $(c - h, c + h)$, which is a subinterval of (α, β), on which the solution exists. But the length of the interval is not specified. As the following example illustrates, there are instances in which the interval of existence depends to a greater extent upon the initial condition $y(c) = d$ than it does upon the function $f(x, y)$.

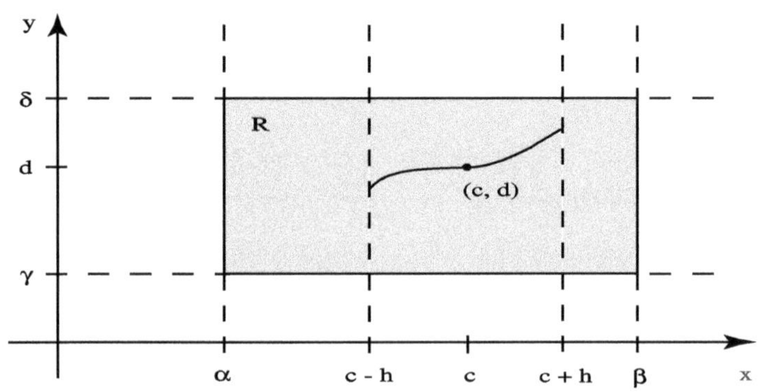

FIGURE 2.5
Geometry of the Fundamental Existence Theorem.

Example 1 An Initial Value Problem for which the Interval of Existence Depends on the Initial Condition Consider the initial value problem

(2) $$y' = y^2; \quad y(0) = d.$$

Let R be any finite rectangle containing the point $(0, d)$, which lies on the y-axis. The function $f(x, y) = y^2$ is continuous on R. So by the fundamental existence theorem, there exists a solution to the IVP (2). In this case, we can easily verify that

(3) $$y(x) = \frac{1}{K - x},$$

where K is an arbitrary constant, is a one-parameter family of solutions of the differential equation $y' = y^2$. Differentiating (3), we find

$$y'(x) = \frac{1}{(K - x)^2} = y^2(x)$$

for any K. So, indeed, $y(x) = 1/(K - x)$ is the solution of $y' = y^2$. Notice that $y(x)$ is defined, continuous, and differentiable for all real $x \neq K$. That is, $y(x) = 1/(K - x)$ is defined, continuous, differentiable, and the solution of $y' = y^2$ on the intervals $(-\infty, K)$ and (K, ∞). To solve the IVP (2), K must be chosen to satisfy the initial condition $y(0) = d = 1/K$. Consequently, $K = 1/d$; and, therefore, the interval of existence depends solely on the initial value d and not on the function $f(x, y) = y^2$.

Notice that when $d = 2$, the solution of the IVP $y' = y^2;\ y(0) = 2$ on the interval $(-\infty, 1/2)$ is

$$y(x) = \frac{1}{\frac{1}{2} - x}.$$

Since the interval of existence specified in the fundamental existence theorem is symmetric with respect to the initial point c, which in this example is the point 0, the value of h specified in the theorem must of necessity be less than or equal to $1/2$ when $d = 2$. Likewise, for arbitrary $d > 0$ the solution of the IVP (2) exists on the interval $(-\infty, 1/d)$; and, therefore, the value of h specified by the existence theorem must be less than or equal to $1/d$. The direction field for $y' = y^2$ and the solution of the IVP $y' = y^2;\ y(0) = 2$ are displayed in Figure 2.6.

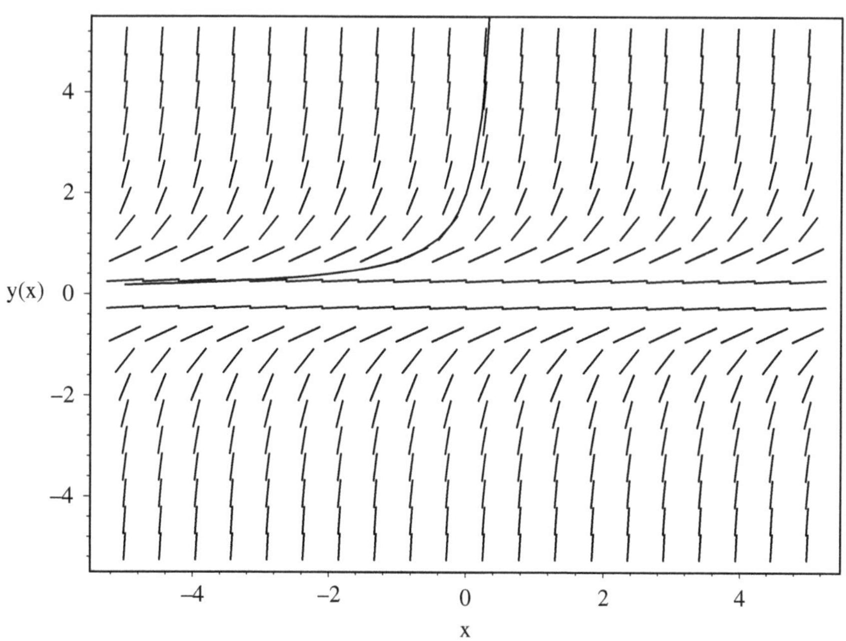

FIGURE 2.6
Direction Field for the DE $y' = y^2$ and the Solution of the IVP $y' = y^2$; y(0) = 2. ■

You might well ask: "What value is the fundamental existence theorem? It does not provide a technique for finding the solution to an initial value problem. Nor does it provide a means of determining the interval on which the solution exists." The answer is quite simple. The fundamental existence theorem assures us that there *is* a solution. It is a waste of time, energy, and often money to search for a solution when there is none.

In Chapter 1, we saw that both $y_1(x) = 0$ and $y_2(x) = x^3$ were solutions of the initial value problem

$$(4) \qquad y' = 3y^{2/3}; \quad y(0) = 0.$$

Notice that $f(x, y) = 3y^{2/3}$ is a continuous function of x and y in any finite rectangle R which contains $(0, 0)$. So the IVP (4) satisfies the hypotheses of the fundamental existence theorem, and as guaranteed by the theorem, there is a solution to the initial value problem. But, in this instance, there is no unique solution. Consequently, some additional condition stronger than continuity of $f(x, y)$ must be required in order to guarantee the uniqueness of a solution to an initial value problem.

We state the following existence and uniqueness theorem, again without proof. This theorem is due to the French mathematician Charles Émile Picard (1856-1941).

2.2.2 Fundamental Existence and Uniqueness Theorem

Let $R = \{(x, y) | \, \alpha < x < \beta \text{ and } \gamma < y < \delta\}$ where α, β, γ, and δ are finite real constants. If $f(x, y)$ is a continuous function of x and y in the finite rectangle R, if $f_y(x, y)$ is a continuous function of x and y in R, and if $(c, d) \in R$, then there exists a unique solution to the initial value problem (1) $y' = f(x, y)$; $y(c) = d$ on some interval $I = (c - h, c + h)$ where I is a subinterval of the interval (α, β).

This existence and uniqueness theorem tells us two things. First, when the hypotheses are satisfied, **there is a solution** of the IVP (1) on some "small" interval about c. And second, **the solution is unique**. That is, there is only one solution of the IVP (1). Geometrically, the theorem states there is one and only one solution of the differential equation in (1) which passes through any point (c, d) in the rectangle R.

Returning to the IVP (4) $y' = 3y^{2/3}$; $y(0) = 0$ and taking the partial derivative of $f(x, y) = 3y^{2/3}$ with respect to y, we find $f_y(x, y) = 2y^{-1/3}$. Observe that $f_y(x, y) = 2y^{-1/3}$ is defined and continuous for all x and for all $y \neq 0$. So f_y is continuous in any finite rectangle R which does not contain any point of the x-axis (where $y = 0$). Thus, the fundamental existence and uniqueness theorem just stated guarantees the existence of a unique solution to the initial value problem

$$y' = 3y^{2/3}; \quad y(c) = d$$

on some interval centered about c provided $d \neq 0$. Notice that the IVP (4) does not satisfy the hypotheses of the fundamental existence and uniqueness theorem because $d = 0$.

The conditions (hypotheses) of the fundamental existence theorem are sufficient to guarantee the existence of a solution to an initial value problem. And the conditions of the fundamental existence and uniqueness theorem are sufficient to guarantee the existence of a unique solution to an initial value problem. However, the conditions are not necessary conditions. Hence, if $f(x,y)$ does not satisfy the hypothesis of the fundamental existence theorem—that is, if $f(x,y)$ is not a continuous function of x and y in R, then we cannot conclude that there is no solution to the initial value problem. If $f(x,y)$ is not continuous, the initial value problem may have a solution, or it may not have a solution. Likewise, if $f(x,y)$ does not satisfy the hypotheses of the fundamental existence and uniqueness theorem—that is, if $f(x,y)$ is not a continuous function of x and y or if $f_y(x,y)$ is not a continuous function of x and y, then the initial value problem may have no solution, it may have a unique solution, or it may have multiple solutions.

Example 2 The IVP $\mathbf{y' = x\sqrt{y};\ \ y(c) = d}$ Analyze initial value problems of the form

(5) $$y' = x\sqrt{y};\quad y(c) = d.$$

Solution

In this example, the function $f(x,y) = x\sqrt{y}$ is defined and a continuous function of x and y on any finite rectangle in the xy-plane where $y \geq 0$. Thus, by the fundamental existence theorem, for every point (c,d) where $d \geq 0$, there is some interval I with center c in which the initial value problem (5) has a solution. Calculating the partial derivative of f with respect to y, we find $f_y(x,y) = x/(2\sqrt{y})$. The function $f_y(x,y)$ is defined and continuous for $y > 0$. Thus, by the fundamental existence and uniqueness theorem, when $d > 0$ the solution of the IVP (5) is guaranteed to be unique. Hence, the solution of the initial value problem

(6) $$y' = x\sqrt{y};\quad y(1) = 2$$

is unique on some interval with center 1. However, the solution to the initial value problem

(7) $$y' = x\sqrt{y};\quad y(1) = 0$$

may or may not be unique, since $d = 0$. In this case, we easily find the solution of the IVP (7) is not unique, since both $y_1(x) = 0$ and $y_2(x) = (x^2 - 1)^2/16$ are solutions of (7). ■

The fundamental existence and uniqueness theorem is called a local theorem, because the solution is guaranteed to exist and be unique only on a "small" interval. The following theorem, which we again state without proof, is called a continuation theorem.

2.2.3 Continuation Theorem

If $f(x, y)$ and $f_y(x, y)$ are both continuous functions of x and y in a finite rectangle R and if $(c, d) \in R$, then the solution of the initial value problem

$$(8) \qquad y' = f(x, y); \quad y(c) = d$$

can be extended uniquely until the boundary of R is reached.

If $f(x, y)$ and $f_y(x, y)$ are both defined and continuous on a finite rectangle, the rectangle can be enlarged until either $f(x, y)$ or $f_y(x, y)$ is not defined or not continuous on bounding sides of the rectangle or until the bounding side of the rectangle approaches infinity. Thus, by the continuation theorem the solution to the IVP $y' = f(x, y); \ y(c) = d$ has a unique solution which extends from one boundary of the enlarged rectangle to another (although perhaps, the same) boundary of the enlarged rectangle. Suppose, for instance, the enlarged rectangle R for a particular differential equation $y' = f(x, y)$ has left boundary $x = L$, right boundary $x = +\infty$, bottom boundary $y = B$, and top boundary $y = T$, where L, B, and T are real numbers and $B < T$. That is, suppose $f(x, y)$ and $f_y(x, y)$ are both defined and continuous inside the "infinite" rectangle

$$R = \{(x, y) \mid L < x \text{ and } B < y < T\}$$

and that either $f(x, y)$ or $f_y(x, y)$ is not defined or not continuous at some point or set of points on all of the lines $x = L, y = B$, and $y = T$. Displayed in Figure 2.7 is the "infinite" rectangle R and the solutions $y_i(x)$ of the initial value problems $y' = f(x, y); \ y(c) = d_i$ for $i = 1, 2, 3, 4$.

Notice that the solution $y_1(x)$ exists and is unique on the interval (a, b) and that both endpoints (a, B) and (b, B) of $y_1(x)$ lie on the bottom boundary of R. If $f(x, y)$ and $f_y(x, y)$ are both defined and continuous at (a, B), then the solution $y_1(x)$ can be extended uniquely to the left. If $f(x, y)$ is continuous at (a, B) but $f_y(x, y)$ is not defined or not continuous at (a, B), then the solution can be extended to the left, but the extension may not be unique. Furthermore, if $f(x, y)$ is not continuous at (a, B), the solution may or may not be extendable to the left. And if extendable, the solution may or may not be unique. Similar comments apply regarding extending $y_1(x)$ to the right from the point (b, B). The solution $y_2(x)$ is unique on the interval $(e, +\infty)$. The endpoint (e, B) of $y_2(x)$ lies on the bottom boundary of R. If $f(x, y)$ and $f_y(x, y)$ are both defined and continuous at (e, B), then $y_2(x)$ can be extended uniquely to the left. If $f(x, y)$ is continuous at (e, B) but $f_y(x, y)$ is not defined or not continuous at (e, B), then the solution can be extended to the left, but the extension may not be unique. If $f(x, y)$ is not continuous at (e, B), then the solution $y_2(x)$ may or may not be extendable to the left.

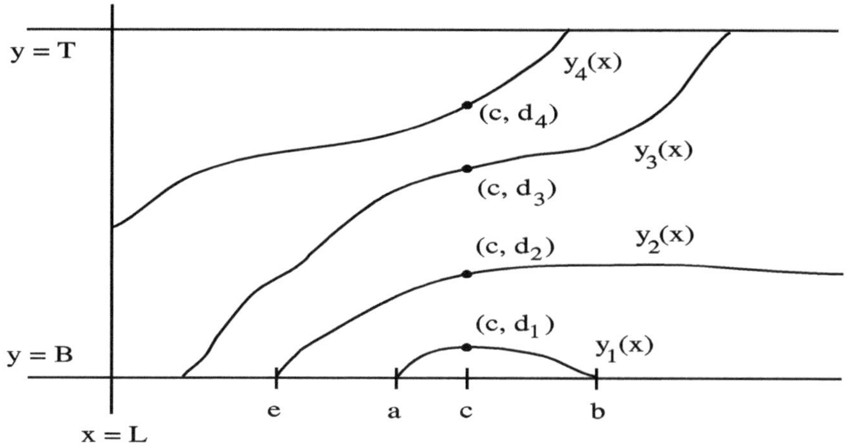

FIGURE 2.7
Solutions to an Initial Value Problem on an Infinite Rectangle.

Moreover, if the solution $y_2(x)$ can be extended to the left, the extension may or may not be unique. Notice that the solution $y_3(x)$ extends from the bottom boundary of R to the top boundary and may or may not be extendable to the left or right in a unique or nonunique fashion. Likewise, the solution $y_4(x)$ extends from the left boundary of R to the top boundary and may or may not be extendable to the left or right in a unique or nonunique manner. The following two examples should help further clarify the results that can be obtained by using the continuation theorem.

Example 3 At Every Point in the xy-plane a Unique Solution Exists Yet the Interval of Existence of the Solution is Finite Analyze the initial value problem

(9) $$y' = 1 + y^2; \quad y(\pi/4) = 1.$$

Solution

Here $f(x, y) = 1 + y^2$ and $f_y(x, y) = 2y$ are both defined and continuous on any finite rectangle in the xy-plane which contains the point $(\pi/4, 1)$. By the fundamental existence and uniqueness theorem, there exists a unique solution to the IVP (9) on some interval with center $\pi/4$. By the continuation theorem the solution can be extended uniquely until the boundary of the rectangle—the xy-plane, in this case—is reached. Thus, the solution can be extended until two of the following four things occur: $x \to -\infty$, $x \to \infty$, $y(x) \to -\infty$, $y(x) \to \infty$. It is tempting to erroneously jump to the conclusion that since the functions $f(x, y)$ and $f_y(x, y)$ are continuous on the entire plane, the solution to the IVP (9) should be valid for all real x. However, this is not what the continuation theorem states. Verify that $y(x) = \tan x$ is the unique solution of the IVP (9). Notice that $y(x) = \tan x$ is defined, continuous, and differentiable on the interval $(-\pi/2, \pi/2)$, but it is not defined at $x = -\pi/2$ or at $x = \pi/2$. So $(-\pi/2, \pi/2)$ is the largest interval on which the IVP (9) has a solution. Observe that as $x \to -\pi^+/2$, $y(x) = \tan x \to -\infty$ and as $x \to \pi^-/2$, $y(x) \to +\infty$. Notice that these results satisfy the conclusion of the continuation theorem. The direction field for the differential equation $y' = 1 + y^2$ and the solution of the IVP (9) are displayed in Figure 2.8.

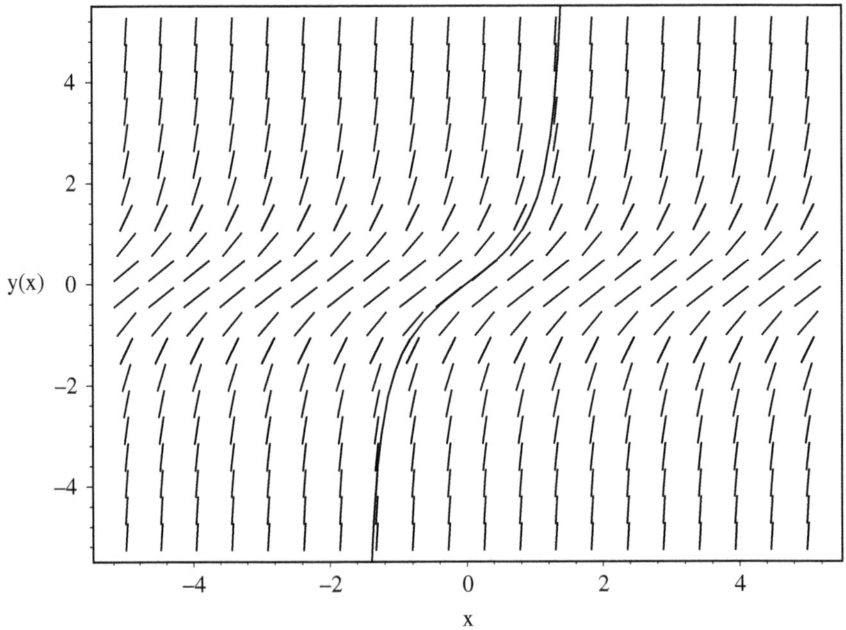

FIGURE 2.8
Direction Field for the DE $y' = 1 + y^2$ and the Solution of the IVP $y' = 1 + y^2; y(\pi/4) = 1$. ■

Example 4 The IVP $\mathbf{y' = -x/y;\ \ y(-1) = 1}$ Analyze the initial value problem

$$y' = -x/y; \quad y(-1) = 1. \tag{10}$$

Solution

The functions $f(x, y) = -x/y$ and $f_y(x, y) = x/y^2$ are defined and continuous except for $y = 0$. Since the point $(-1, 1)$ is in the upper half plane (where $y > 0$), and since f and f_y are defined and continuous for $y > 0$, by the continuation theorem the solution of the IVP (10) can be extended uniquely until two of the following four things occur: $x \to -\infty$, $x \to \infty$, $y(x) \to 0$, or $y(x) \to +\infty$.

Verify that $y(x) = \sqrt{2 - x^2}$ is the unique solution of the IVP (10) and that $y(x)$ is defined, continuous, and differentiable on the interval $(-\sqrt{2}, \sqrt{2})$. Since $y(x)$ is not defined outside the interval $[-\sqrt{2}, \sqrt{2}]$, $(-\sqrt{2}, \sqrt{2})$ is the largest interval on which the IVP (10) has a solution. Observe that as $x \to -\sqrt{2}^{+}$, $y(x) \to 0^+$ and as $x \to \sqrt{2}^{-}$, $y(x) \to 0^+$. Notice that these results satisfy the conclusion of the continuation theorem. In this example, we have $y(x) \to 0^+$ for two different values of x. So in this case, the solution approaches the same boundary of the rectangle twice instead of two different boundaries. The direction field for the differential equation $y' = -x/y$ and the solution of the IVP (10) are displayed in Figure 2.9.

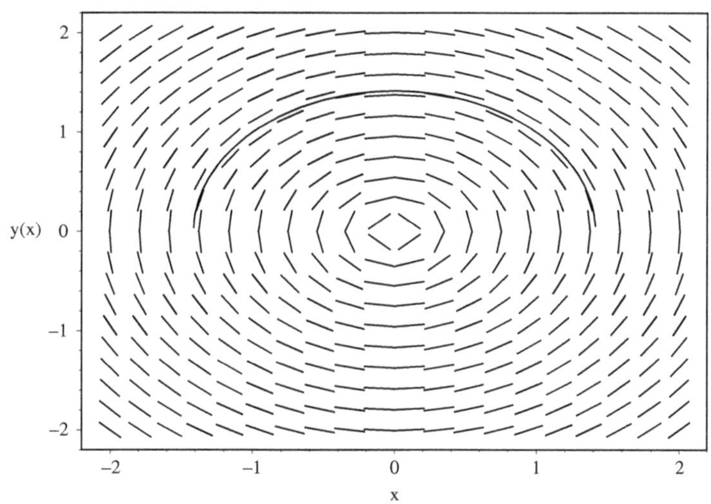

FIGURE 2.9
Direction Field for the DE $y' = -x/y$ and the Solution of the IVP $y' = -x/y$; y(-1) = 1. ■

When the function $f(x,y)$ in the initial value problem (1) $y' = f(x,y)$; $y(c) = d$ has the form $f(x,y) = a(x)y + b(x)$, the initial value problem is called a **linear first-order initial value problem**; otherwise, the initial value problem is said to be **nonlinear**. The initial value problems (2), (5), and (9) which we examined in Examples 1, 2, and 3, respectively, are all nonlinear first-order initial value problems. The intervals on which the solutions to these problems existed and were unique exhibited no discernible pattern, mainly because the initial value problems were nonlinear. However, when we apply the fundamental existence and uniqueness theorem to the linear first-order initial value problem

$$(11) \qquad y' = a(x)y + b(x) = f(x, y); \quad y(c) = d$$

we obtain an existence and uniqueness theorem in which the interval of existence and uniqueness is completely determined by the functions $a(x)$ and $b(x)$. Differentiating $f(x, y) = a(x)y + b(x)$ partially with respect to y, we find $f_y(x, y) = a(x)$. Hence, when the functions $a(x)$ and $b(x)$ are both defined and continuous functions of x on the interval (α, β), the functions $f(x, y)$ and $f_y(x, y)$ are both defined and continuous functions of x and y in any finite rectangle $R = \{(x, y) | \ \alpha < x < \beta \text{ and } \ \gamma < y < \delta\}$. So the linear IVP (11) has a unique solution and this solution can be extended uniquely until the boundary of the largest rectangle on which both $a(x)$ and $b(x)$ are defined and continuous functions of x on (α, β) is reached. For linear initial value problems, it is always the left-hand and right-hand boundary of the largest rectangle that the solution approaches. Hence, for linear initial value problems, we have the following theorem.

2.2.4 An Existence and Uniqueness Theorem for Linear First-Order Initial Value Problems

If the functions $a(x)$ and $b(x)$ are defined and continuous on the interval (α, β) and if $c \in (\alpha, \beta)$, then the linear initial value problem

$$(11) \qquad y' = a(x)y + b(x); \quad y(c) = d$$

has a unique solution on the interval (α, β).

The existence and uniqueness theorem for the linear IVP (11) states that it **always** has a solution on any interval (α, β) containing c on which both $a(x)$ and $b(x)$ are continuous for any choice of the initial value d. Furthermore, the linear IVP (11) has only one solution on the entire interval (α, β). Moreover, because of the uniqueness of solutions throughout the rectangle R, no two distinct solution curves can cross one another. This existence and uniqueness theorem and the previous two examples illustrate one of the important differences between the types of results we can expect for linear initial value problems versus nonlinear initial value problems. For the linear initial value problem, we are able to explicitly determine the interval of existence and uniqueness of the solution from the differential equation and initial condition prior to producing a solution; whereas, for the nonlinear initial value problem we are, in general, unable to do so. For this reason when producing a numerical solution to a nonlinear initial value problem, it is often desirable to be able to monitor the solution as it is being generated.

Example 5 Determination of the Interval of Existence and Uniqueness On what interval does each of the following linear initial value problems have a unique solution?

a. $y' = 3x^2y + e^x; \quad y(-2) = 4$

b. $y' = \dfrac{y}{(x-1)(x+2)} + \dfrac{1}{x}; \quad y(-1) = 2$

c. $y' = (\cot x)y + \tan x; \quad y(2) = 3$ ■

Solution

a. The functions $a(x) = 3x^2$ and $b(x) = e^x$ are both defined and continuous on $(-\infty, \infty)$. So by the previous existence and uniqueness theorem the given initial value problem has a unique solution on the interval $(-\infty, \infty)$.

b. In this instance, $a(x) = 1/((x-1)(x+2))$ is defined and continuous for $x \neq -2$ and $x \neq 1$. So $a(x)$ is defined and continuous on the intervals $(-\infty, -2)$, $(-2, 1)$, and $(1, \infty)$. The function $b(x) = 1/x$ is defined and continuous for $x \neq 0$. Thus, $b(x)$ is defined and continuous on the intervals $(-\infty, 0)$ and $(0, \infty)$. Taking the intersection of these intervals with the intervals on which $a(x)$ is defined and continuous, we find $a(x)$ and $b(x)$ are simultaneously defined and continuous on the intervals $(-\infty, -2)$, $(-2, 0)$, $(0, 1)$, and $(1, \infty)$. Since -1 is in the interval $(-2, 0)$, the initial value problem b. has a unique solution on $(-2, 0)$. If the initial condition $y(-1) = 2$ were changed to $y(1) = 2$, then there would be no solution to the corresponding new initial value problem, since $a(x)$—and, therefore, the differential equation—is not defined at $x = 1$. If the initial condition were changed to $y(2) = 2$, then the unique solution to the corresponding initial value problem would exist on the interval $(1, \infty)$.

c. In this case, $a(x) = \cot x$ is defined and continuous on the inter- vals $(n\pi, (n+1)\pi)$ for $n = 0, \pm 1, \pm 2, \ldots$ and $b(x) = \tan x$ is defined and continuous on the intervals $((2n-1)\pi/2, (2n+1)\pi/2)$ for $n = 0, \pm 1, \pm 2, \ldots$. The intersection of these sets of intervals is the set of intervals $(n\pi/2, (n+1)\pi/2)$ for $n = 0, \pm 1, \pm 2, \ldots$. Since $2 \in (\pi/2, \pi)$, the initial value problem c. has a unique solution on the interval $(\pi/2, \pi)$.

2.2.5 Exercises

In Exercises 1–6 find all points (c, d) where solutions to the initial value problem consisting of the given differential equation and the initial condition $y(c) = d$ may not exist.

1. $y' = \dfrac{3y}{(x-5)(x+3)} + e^{-x}$

2. $y' = \dfrac{xy}{(x^2 + y^2)}$

3. $y' = \dfrac{1}{xy}$

4. $y' = \ln(y - 1)$

5. $y' = \sqrt{(y+2)(y-1)}$

6. $y' = \dfrac{y}{y - x}$

In Exercises 7–14 find all points (c, d) where solutions to the initial value problem consisting of the given differential equation and the initial condition y(c) = d may not exist or may not be unique.

7. $y' = x/y^2$

8. $y' = \sqrt{y}/x$

9. $(4 - y^2)y' = x^2$

10. $(x^2 + y^2)y' = y^2$

11. $y' = xy/(1 - y)$

12. $y' = (xy)^{1/3}$

13. $y' = \sqrt{(y - 4)/x}$

14. $y' = -y/x + y^{1/4}$

In Exercises 15–24 state the interval on which the solution to the linear initial value problem exists and is unique.

15. $y' = 4y - 5; \quad y(1) = 4$

16. $y' + 3y = 1; \quad y(-2) = 1$

17. $y' = ay + b; \quad y(c) = d$ where a, b, c, and d are real constants.

18. $y' = x^2 + e^x - \sin x; \quad y(2) = -1$

19. $y' = xy + \dfrac{1}{1 + x^2}; \quad y(-5) = 0$

20. $y' = \dfrac{y}{x} + \cos x; \quad y(-1) = 0$

21. $y' = \dfrac{y}{x} + \tan x; \quad y(\pi) = 0$

22. $y' = \dfrac{y}{4 - x^2} + \sqrt{x}; \quad y(3) = 4$

23. $y' = \dfrac{y}{4 - x^2} + \sqrt{x}; \quad y(1) = -3$

24. $y' = (\cot x)y + \csc x; \quad y(\pi/2) = 1$

25. Show that $y(x) = -1$ is the unique solution of the initial value problem $y' = x(1 + y); \quad y(0) = -1$.

26. a. Verify that $y = \tan x$ is the solution of the initial value problem $y' = 1 + y^2; \quad y(0) = 1$.

    b. On what interval is $y = \tan x$ the solution?

27. Does the initial value problem $y' = y/x; \; y(0) = 2$ have a solution?

28. Verify that $y_1(x) = 1$ and $y_2(x) = \sin(\frac{x^2+\pi}{2})$ are both solutions on the interval $(-\sqrt{\pi}, \sqrt{\pi})$ of the initial value problem

$$y' = -x\sqrt{1-y^2}; \quad y(0) = 1.$$

Does this violate the fundamental existence and uniqueness theorem? Explain.

29. Verify that $y_1(x) = 9 - 3x$ and $y_2(x) = -x^2/4$ are both solutions of the initial value problem

$$y' = (-x + \sqrt{x^2+4y})/2; \quad y(6) = -9.$$

Does this violate the fundamental existence and uniqueness theorem? Explain.

2.3 Solution of Simple First-Order Differential Equations

Sometimes one can solve the initial value problem $y' = f(x,y);\ y(c) = d$ by finding the solution $\phi(x,y,C) = 0$ of the differential equation $y' = f(x,y)$ and then determining the value of C which satisfies the initial condition $y(c) = d$. Perhaps in calculus, you examined various elementary techniques for finding the solution of the differential equation $y' = f(x,y)$, such as separating variables, performing a change of variable, testing for exactness and solving those equations which are exact, and using integrating factors. These techniques for solving first-order differential equations were all developed and employed prior to 1800. However, relatively few differential equations can actually be solved using these techniques. In Chapter 1, we showed how to verify that a given function is the solution of a specific differential equation; however, we did not discuss how to actually find a solution. In this section, we shall show how to solve the differential equation $y' = f(x,y)$ when $f(x,y)$ has one of the three forms:

(i) $f(x,y) = g(x)$, (ii) $f(x,y) = g(x)/h(y)$, and (iii) $f(x,y) = a(x)y + b(x)$.

2.3.1 Solution of $\mathbf{y' = g(x)}$

In calculus, you solved many linear differential equations of the form

$$y' = g(x) \tag{1}$$

by finding the antiderivative of g. Integrating equation (1) symbolically, we find the general solution to be

$$(2) \qquad y(x) = \int^x g(t)\,dt + C$$

where C is an arbitrary constant. By the existence and uniqueness theorem for first-order linear initial value problems, this solution exists and is unique on any interval on which the function $g(x)$ is defined and continuous.

Example 1 Solving an Initial Value Problem by Finding the General Solution

a. Find the general solution of the differential equation

$$(3) \qquad y' = \frac{1}{(x+2)}.$$

b. Solve the initial value problem

$$y' = \frac{1}{(x+2)}; \quad y(1) = 4$$

and specify the interval on which the solution exists and is unique.

c. Solve the initial value problem

$$y' = \frac{1}{(x+2)}; \quad y(-7) = 3$$

and specify the interval on which the solution exists and is unique.

Solution

1. Integrating equation (3), we find the general solution to be

$$(4) \qquad y(x) = \int^x \frac{1}{(t+2)}\,dt + C = \ln|x+2| + C.$$

Since the function $g(x) = 1/(x+2)$ is defined and continuous on the intervals $(-\infty, -2)$ and $(-2, \infty)$, equation (4) is the general solution of differential equation (3) on the intervals $(-\infty, -2)$ and $(-2, \infty)$.

2. To solve the initial value problem

$$(5) \qquad y' = \frac{1}{(x+2)}; \quad y(1) = 4$$

all we need to do is find the value of C in the general solution (4) which will yield $y(1) = 4$. Setting $x = 1$ in (4), we see that C must satisfy

$$y(1) = \ln|1+2| + C = 4.$$

Solving for C, we get $C = 4 - \ln 3$. Substituting this value into (4) produces the following general solution to the IVP (5)

$$(6) \qquad y(x) = \ln|x+2| + 4 - \ln 3 = \ln\frac{|x+2|}{3} + 4.$$

Since the initial condition, $y(1) = 4$, is specified at $x = 1 \in (-2, \infty)$, equation (6) is the unique solution of the IVP (5) on the interval $(-2, \infty)$.

3. To solve the initial value problem

$$(7) \qquad y' = \frac{1}{(x+2)}; \quad y(-7) = 3$$

we set $x = -7$ in the general solution (4), solve the resulting equation for C, and substitute this value into (4). Doing so, we find the solution to the IVP (7) to be

$$(8) \qquad y(x) = \ln\frac{|x+2|}{5} + 3.$$

Since the initial condition, $y(-7) = 3$, is specified at $x = -7 \in (-\infty, -2)$, equation (8) is the unique solution of the IVP (7) on the interval $(-\infty, -2)$. ■

Many pages in calculus texts and much student effort are devoted to methods for calculating and expressing antiderivatives of continuous functions in terms of elementary functions. Expressions for antiderivatives have been collected and appear in tables of integrals. However, the reader should be aware that there are many relatively simple continuous functions whose antiderivatives cannot be expressed as an elementary function. For example, the following list of functions which are defined and continuous on certain intervals do not have antiderivatives that can be expressed as elementary functions:

$$e^{-x^2}, \quad \frac{e^x}{x}, \quad e^x \ln x, \quad \frac{1}{\ln x}, \quad \sin x^2, \quad \frac{\sin x}{x}, \quad \frac{\sin^2 x}{x}, \quad x \tan x, \quad \frac{1}{\sqrt{1-x^3}}$$

When it is impossible or impractical to express the antiderivative of a function $f(x)$ as an elementary function, then one must use series or numerical integration techniques to calculate the integral of $f(x)$. In some cases, even if an antiderivative for $f(x)$ can be expressed as an elementary function, it may still be simpler to approximate the integral of $f(x)$ than to evaluate the antiderivative.

2.3.2 Exercises

In Exercises 1–16 find the general solution of the differential equation and specify the interval(s) on which the solution exists.

1. $y' = x^{1/2}$
2. $y' = 2^{-x}$
3. $\dfrac{dy}{dx} = \dfrac{2}{x-3}$
4. $\dfrac{ds}{dt} = e^{-\pi t}$
5. $y' = \dfrac{1}{x+1}$
6. $y' = \dfrac{x}{x+1}$
7. $y' = \ln x$
8. $y' = x \ln x$
9. $y' = \dfrac{dx}{1-\cos x}$
10. $y' = \dfrac{1}{\sin 2x}$
11. $y' = \dfrac{1}{x^2+1}$
12. $y' = \dfrac{1}{\sqrt{x^2+1}}$
13. $y' = \dfrac{1}{\sqrt{x^2-1}}$
14. $y' = e^{-x}\sin 2x$
15. $y' = \dfrac{1}{x^2-4x+5}$
16. $y' = \dfrac{x}{x^2-4x+5}$

In Exercises 17–26 solve the given initial value problem by first finding the general solution of the differential equation and then determining the value of the constant of integration which satisfies the initial condition. Also specify the interval(s) on which each solution exists.

17. $y' = 3x+1;\quad y(1) = 2$
18. $y' = x + \dfrac{1}{x};\quad y(1) = 2$
19. $y' = 2\sin x;\quad y(\pi) = 1$
20. $y' = x\sin x;\quad y(\pi/2) = 1$
21. $y' = \dfrac{1}{x-1};\quad y(2) = 1$
22. $y' = \dfrac{1}{x-1};\quad y(0) = 1$
23. $y' = \dfrac{1}{x^2-1};\quad y(2) = 1$
24. $y' = \dfrac{1}{x^2-1};\quad y(0) = 1$
25. $y' = \tan x;\quad y(0) = 0$
26. $y' = \tan x;\quad y(\pi) = 0$

2.3.3 Solution of the Separable Equation $\mathbf{y' = g(x)/h(y)}$

From your study of calculus, you are already familiar with the concept of explicit and implicit functions and you know how to differentiate both explicitly and implicitly. In differential equations, we distinguish between explicit and implicit solutions as well. Thus far in the text, we have encountered only explicit solutions to differential equations; therefore, we have referred to them simply as solutions. The definitions of explicit solution and implicit solution for first-order differential equations follow.

An **explicit solution** of the differential equation $y' = f(x,y)$ on an interval $I = (a,b)$ is a function $y = \phi(x)$ which is differentiable at least once on I and satisfies $\phi' = f(x, \phi(x))$.

A relation $R(x, y) = 0$ is an **implicit solution** of the differential equation $y' = f(x, y)$ on an interval I, if the relation defines at least one function $y_1(x)$ on I such that $y_1(x)$ is an explicit solution of $y' = f(x, y)$ on I.

Usually we refer to both explicit and implicit solutions simply as solutions. However, when we want to be specific about the kind of solution we are talking about, we will include the designation explicit or implicit.

We solved the radioactive differential equation (1) of Section 1.4 by "separating" the variables. Differential equations which can be solved using this technique are called separable equations and are defined as follows.

The differential equation $y' = f(x, y)$ is a **separable equation**, if it can be written in the form

$$y' = \frac{g(x)}{h(y)}. \tag{9}$$

In most cases, a separable differential equation will not be in the form (9) initially. Some algebraic manipulation will usually be required in order to write the given equation in the form (9). Of course, most differential equations are not separable; and, therefore, no amount of valid algebraic manipulation will produce an equation of the form (9). For example, the differential equation $y' = (x - y)/x$ is not separable. When possible, the algebraic process of converting $y' = f(x, y)$ into the form $y' = g(x)/h(y)$ is called "separating variables." Once an equation is in the form (9), the process of separating variables is completed by replacing y' by dy/dx and then multiplying the resulting equation by $h(y)\,dx$ to obtain

$$h(y)\,dy = g(x)\,dx. \tag{10}$$

In this form, you can clearly see that the variables x and y are "separated" and associated with their respective differentials. Symbolically integrating equation (10) yields

$$\int h(y)\,dy = \int g(x)\,dx. \tag{11}$$

The explicit solution of the separable differential equation (9) and the determination of the interval of existence of the solution depends upon being able to represent both integrals appearing in equation (11) as elementary functions and upon being able to solve the resulting equation for y.

Example 2 Solving a Separable Differential Equation Solve the differential equation

$$y' = -x/y. \tag{12}$$

Solution

The direction field for the differential equation $y' = -x/y$ appears in Figure 2.9. The equation $y' = -x/y$ is already in the form of a separable differential equation. Replacing y' by dy/dx and multiplying the resulting equation by $y\,dx$, we get

$$y\,dy = -x\,dx.$$

Integration yields

$$\int y\,dy = \int -x\,dx$$

or

$$\frac{y^2}{2} = \frac{-x^2}{2} + C, \tag{13}$$

where C is an arbitrary constant.

(Note: When integrating a separable equation, there is no need for two constants of integration—one constant, c_1, for the left-hand side of the equation and another constant, c_2, for the right-hand side of the equation, since these constants can be combined by subtraction into the single constant, $C = c_2 - c_1$. In addition, since a multiple of a constant, combinations of constants, and functions of constants are constant, when solving differential equations we will often replace such expressions by a new constant.)

Multiplying the equation (13) by 2 and then adding x^2 to the resulting equation, we obtain the following relation for the solution to the DE (12)

$$y^2 + x^2 = 2C. \tag{14}$$

For $C < 0$, equation (14) yields no real solution to the DE (12). For $C = 0$, equation (14) produces the point $(0,0)$ which is not a solution to the DE (12). Assuming $C > 0$ and replacing $2C$ in equation (14) by the new constant K^2, where we may also assume $K > 0$, equation (14) becomes

$$y^2 + x^2 = K^2. \tag{15}$$

From equation (15) we see that geometrically the general solution to the DE (12) is a family of concentric circles with centers at the origin and radii $K > 0$. To verify that (15) is actually an implicit solution, we must show that (15) defines at least one function $y_1(x)$ which is an explicit solution of (12). Solving equation (15) for y, we get $y(x) = \pm\sqrt{K^2 - x^2}$. The functions

$$y_1(x) = \sqrt{K^2 - x^2} \quad \text{and} \quad y_2(x) = -\sqrt{K^2 - x^2}$$

are both defined and real for $x \in [-K, K]$. Differentiating $y_1(x)$ and $y_2(x)$, we obtain

$$y_1'(x) = \frac{-x}{\sqrt{K^2 - x^2}} \quad \text{and} \quad y_2'(x) = \frac{x}{\sqrt{K^2 - x^2}}.$$

Both $y_1'(x)$ and $y_2'(x)$ are defined and real for $x \in (-K, K)$. Substituting y_1 and y_1' into the DE (12) $y' = -x/y$, we obtain the identity

$$\frac{-x}{\sqrt{K^2 - x^2}} = -x/\sqrt{K^2 - x^2}$$

for $x \in (-K, K)$. Thus, $y_1(x)$ is an explicit solution of (12) on the interval $(-K, K)$. And consequently, by definition, equation (15) is an implicit solution of (12) on $(-K, K)$. Likewise, $y_2(x)$ can be shown to be an explicit solution of (12) on the interval $(-K, K)$. Hence, the implicit solution (15) defines at least two explicit solutions of (12) on the interval $(-K, K)$. The graph of the solution $y_1(x)$ is shown in Figure 2.10 (a) and the graph of the solution $y_2(x)$ is shown in Figure 2.10 (b).

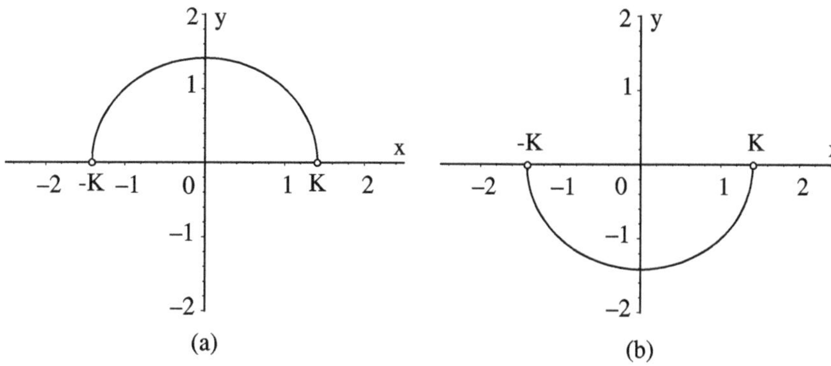

FIGURE 2.10
Graph of Explicit Solutions $y_1(x)$ and $y_2(x)$ of the DE (12). ■

In the previous example, it was fairly easy to determine an explicit solution from the implicit solution and to determine the interval on which the solution exists. However, this will not be the case in general. Normally, we will not be able to solve the given relation in x and y explicitly for either x or y. Therefore, usually we will obtain a relation in x and y by some means, verify that this relation formally satisfies the differential equation under consideration, and say that the relation is an implicit solution. For example, we will say that the relation

$$y^3 + 2xy - x^2 = C, \tag{16}$$

where C is a constant, is an implicit solution of the differential equation

$$y' = \frac{2x - 2y}{3y^2 + 2x}. \tag{17}$$

In order to verify formally that (16) is a solution of (17), we implicitly differentiate (16) with respect to x and obtain

$$3y^2y' + 2(xy' + y) - 2x = 0.$$

Solving this equation for y' yields the differential equation (17).

We will soon discover that, in theory, we will be able to explicitly solve linear differential equations and determine the interval on which the solution exists directly from the differential equation itself. However, the best we will be able to accomplish usually for nonlinear differential equations is to obtain an implicit solution, a series solution, or a numerical solution. This is one of the primary differences between the kinds of results that we can expect to obtain for linear differential equations versus nonlinear differential equations.

Example 3 Solving an IVP Using Separation of Variables Solve the initial value problem

$$y' = xy + 2x; \quad y(0) = 3. \tag{18}$$

Solution

Writing the derivative y' as the ratio of differentials, dy/dx, and factoring the right-hand side of the differential equation in (18), we obtain the equivalent equation

$$\frac{dy}{dx} = x(y+2).$$

Multiplying by dx and dividing by $(y+2)$, we get

$$\frac{dy}{y+2} = x\,dx, \quad \text{provided} \quad y \neq -2.$$

Integration yields

$$\int \frac{dy}{y+2} = \int x\,dx$$

and then the implicit solution

$$\ln|y+2| = \frac{x^2}{2} + C, \quad \text{provided} \quad y \neq -2$$

and where C is an arbitrary constant. Exponentiating, we find

$$|y+2| = e^{\ln|y+2|} = e^{x^2/2+C} = e^C e^{x^2/2} \quad \text{for} \quad y \neq -2. \tag{19}$$

Both the left-hand side and the right-hand side of (19) are positive. Since $|y+2| = \pm(y+2)$ and e^C is an arbitrary positive constant, we may remove the absolute value appearing in equation (19), if we replace e^C by a new arbitrary constant K which may be positive or negative. Doing so, we obtain the following solution of the differential equation $y' = xy + 2x$:

$$y + 2 = Ke^{x^2/2}$$

where $K \neq 0$. Noting that $y = -2$ is a particular solution of the differential equation $y' = xy + 2x$, we can remove the restriction $K \neq 0$ and obtain the explicit solution of $y' = xy + 2x$

$$y + 2 = Ke^{x^2/2} \tag{20}$$

where K is any arbitrary real constant. Since the initial condition, $y(0) = 3$, in the IVP (18) is specified at $x = 0$, we set $x = 0$ in equation (20) and find that K must satisfy

$$y(0) + 2 = Ke^0 \qquad \text{or} \qquad 3 + 2 = 5 = K.$$

So the explicit solution of the IVP (18) is

$$y(x) = 5e^{x^2/2} - 2.$$

Notice that this solution exists on the interval $(-\infty, \infty)$. ■

2.3.4 Exercises

In Exercises 1–12 use the technique of separation of variables to solve the given differential equation.

1. $y' = x^2/y$
2. $\dfrac{dV}{dP} = -\dfrac{V}{P}$
3. $\dfrac{ds}{dt} = ts^3$
4. $4dt = t\sqrt{t^2 - 4}\,ds$
5. $y' = e^{x+y}$
6. $y' = x^2 \sin y$
7. $(x^2y + y)\,dx - dy = 0$
8. $xy\,dx + y^2\,dy = 0$
9. $xy\,dx + (1 + x^2)\,dy = 0$
10. $dr = e(r \sin\theta\, d\theta - \cos\theta\, dr)$
11. $y \ln x \ln y\,dx + dy = 0$
12. $\dfrac{dq}{dt} + \dfrac{t \sin t}{q} = 0$

In Exercises 13–24 solve the given initial value problem by finding the solution of the separable differential equation and then determining the value of the constant of integration which satisfies the initial condition. Specify the interval on which each solution exists.

13. $y' = 3y; \quad y(0) = -1$
14. $y' = -y + 1; \quad y(0) = 1$
15. $y' = -y + 1; \quad y(0) = 2$
16. $y' = xe^{y-x^2}; \quad y(0) = 0$
17. $y' = y/x; \quad y(-1) = 2$
18. $y' = 2x/y; \quad y(0) = 2$
19. $y' = -2y + y^2; \quad y(0) = 1$
20. $y' = xy + x; \quad y(1) = 2$
21. $xe^y\,dx + dy = 0; \quad y(0) = 0$
22. $y\,dx - x^2\,dy = 0; \quad y(1) = 1$
23. $\dfrac{dr}{dt} = -4rt; \quad r(0) = v_0$
24. $v\dfrac{dv}{dx} = g; \quad v(x_0) = v_0$
25. Verify that $y^2 - x = 1$ is an implicit solution of the differential equation $2yy' = 1$ on the interval $(-1, \infty)$.
26. Verify that $xy^2 + x = 1$ is an implicit solution of the differential equation $2xyy' + y^2 = -1$ on the interval $(0, 1)$.

27. Verify that $x = e^{xy}$ is an implicit solution of the differential equation $y' = (1 - xy)/x^2$ on the interval $(0, \infty)$.

28. Verify that the relation $xy^2 + yx^2 = 1$ formally satisfies the differential equation $y' = -y(2x + y)/x(2y + x)$.

29. Verify that the relation $y = e^{xy}$ formally satisfies the differential equation $y' = y^2/(1 - xy)$.

30. **A snowplow problem** One morning it started snowing at a heavy and constant rate. A snowplow started out at 8 a.m. By 9 a.m. the snowplow had gone 2 miles. By 10 a.m. the snowplow had gone 3 miles. Assuming that the snowplow removes a constant volume of snow per hour, determine the time at which it started snowing.

2.3.5 Solution of the Linear Equation $\mathbf{y' = a(x)y + b(x)}$

Throughout this subsection, we will assume that the functions $a(x)$ and $b(x)$ are both defined and continuous on some interval (α, β). By the existence and uniqueness theorem for linear first-order initial value problems stated in Section 2.2, we are assured that for any $c \in (\alpha, \beta)$ the initial value problem

$$y' = a(x)y + b(x); \quad y(c) = d \tag{21}$$

has a unique solution on (α, β). The purpose of this section is to present a technique for finding the solution of the differential equation appearing in (21), namely,

$$y' = a(x)y + b(x) \tag{22}$$

and to express the solution of (22) as an equation which involves two quadratures (integrals).

If $b(x)$ is identically equal to zero on the interval (α, β), then equation (22) reduces to

$$y' = a(x)y. \tag{23}$$

Equation (23) is called a **homogeneous** linear first-order differential equation. (To indicate mathematically that "$b(x)$ is identically equal to zero on the interval (α, β)," we write "$b(x) \equiv 0$ on the interval (α, β).") Notice that equation (23) is separable. We rewrite y' as dy/dx, multiply by dx, and divide by y to obtain the separation of variables

$$\frac{dy}{y} = a(x)\,dx \quad \text{provided} \quad y \neq 0.$$

Integrating, we find

$$\int \frac{dy}{y} = \int a(x)\,dx \quad \text{or} \quad \ln|y| = \int a(x)\,dx + C \quad \text{provided} \quad y \neq 0$$

and where C is an arbitrary constant. Exponentiation yields

$$|y| = e^{\int a(x)\,dx + C} = e^C e^{\int a(x)\,dx}. \tag{24}$$

The left-hand side and the right-hand side of equation (24) are both positive. Since $|y| = \pm y$ and since e^C is an arbitrary positive constant, we may remove the absolute value appearing in (24), if we also replace e^C by a new constant K which may be positive or negative. Observing that $y(x) = 0$ is a solution of equation (23) $y' = a(x)y$, we also may allow K to assume the value zero. Thus, we find the explicit solution of (23) on the interval (α, β) is

$$y(x) = Ke^{\int a(x)\,dx} \tag{25}$$

where K is an arbitrary constant. Thus, we have proved the following theorem.

An Existence Theorem for the Homogeneous Linear First-Order Differential Equation

If $a(x)$ is a continuous function on the interval (α, β), then the solution on (α, β) of the homogeneous linear first-order differential equation

$$y' = a(x)y \tag{23}$$

is

$$y(x) = Ke^{\int a(x)\,dx} \tag{25}$$

where K is an arbitrary constant.

Example 4 Solution of a Homogeneous Linear First-Order IVP

a. Find the solution of $y' = (\tan x)y$.

b. Solve the initial value problem

$$y' = (\tan x)y; \quad y(2) = 3 \tag{26}$$

and specify the interval on which the solution exists and is unique.

Solution

1. In this example, the function $a(x) = \tan x$ is defined and continu- ous on each of the intervals $I_n = ((2n-1)\pi/2, (2n+1)\pi/2)$ for $n = 0, \pm 1, \pm 2, \ldots$. By the previous theorem the solution of $y' = (\tan x)y$ on any interval I_n is

$$y(x) = Ke^{\int \tan x\,dx} = Ke^{-\ln|\cos x|} = \frac{K}{|\cos x|} = C\sec x \tag{27}$$

where C is an arbitrary constant.

2. Setting $x = 2$ in equation (27) and imposing the initial condition, $y(2) = 3$, requires that C satisfy the equation $y(2) = 3 = C \sec 2$. Thus, $C = 3/\sec 2$ and the solution of the IVP (26) is $y = 3\sec x/\sec 2$. Since $x = 2 \in I_1 = (\pi/2, 3\pi/2)$, this solution exists and is unique on the interval $(\pi/2, 3\pi/2)$. ■

Next, we consider the nonhomogeneous linear first-order differential equation

$$(28) \qquad y' = a(x)y + b(x)$$

where $a(x)$ and $b(x)$ are assumed to be defined and continuous on the interval (α, β) and $b(x) \not\equiv 0$ on (α, β). (Here "$b(x) \not\equiv 0$ on (α, β)" is read "$b(x)$ is not identically equal to zero on the interval (α, β)" and equation (28) is called **nonhomogeneous** because $b(x) \not\equiv 0$ on (α, β).) Let $y_K(x)$ denote the solution of the homogeneous linear differential equation $y' = a(x)y$ associated with the nonhomogeneous linear differential equation (28). The exact expression for $y_K(x)$, which contains one arbitrary constant K, is given explicitly by equation (25). On the interval (α, β), the function $y_K(x)$ satisfies $y_K' = a(x)y_K$. Suppose that $y_p(x)$ is any particular solution of the nonhomogeneous differential equation $y' = a(x)y + b(x)$ on the interval (α, β). Since $y_p(x)$ is a particular solution, it contains no arbitrary constants and it satisfies $y_p'(x) = a(x)y_p(x) + b(x)$ on the interval (α, β). Assume that $z(x)$ is any solution of the nonhomogeneous differential equation (28) $y' = a(x)y + b(x)$ which is defined on the interval (α, β). Since $z(x)$ satisfies (28), $z'(x) = a(x)z(x) + b(x)$ on (α, β). Now consider the function $w(x) = z(x) - y_p(x)$. Differentiating, we find for all $x \in (\alpha, \beta)$,

$$\begin{aligned} w'(x) = z'(x) - y_p'(x) &= \{a(x)z(x) + b(x)\} - \{a(x)y_p(x) + b(x)\} \\ &= a(x)(z(x) - y_p(x)) = a(x)w(x). \end{aligned}$$

That is, $w(x)$ is a solution of the associated homogeneous differential equation (23) $y' = a(x)y$ on (α, β). Since equation (25) is the solution of (23), there is a specific value of K such that for all $x \in (\alpha, \beta)$

$$w(x) = z(x) - y_p(x) = Ke^{\int a(x)\,dx}.$$

Hence, any solution $z(x)$ of the nonhomogeneous linear differential equation (28) $y' = a(x)y + b(x)$ on the interval (α, β) has the form

$$(29) \qquad z(x) = y_K(x) + y_p(x) = Ke^{\int a(x)\,dx} + y_p(x)$$

where K is an arbitrary constant and $y_p(x)$ is any particular solution of the associated homogeneous differential equation (28).

It would be nice if we had an expression for the particular solution $y_p(x)$ appearing in equation (29) or a procedure for computing $y_p(x)$ given specific functions $a(x)$ and $b(x)$. Since the function $a(x)$ appears explicitly in the first term of (29), we anticipate that the function $b(x)$ will appear explicitly in the particular solution, the second term of (29). Let us assume that $a(x)$ and $b(x)$ are both continuous on the interval (α, β). Letting $K = 1$ in equation (25), we obtain the function

$$y_1(x) = e^{\int a(x)\,dx}. \tag{30}$$

Remember that $y_1(x)$ is a solution of the homogeneous differential equation $y' = a(x)y$ on the interval (α, β). Observe that $y_1(x)$ is positive on (α, β) and that any constant times $y_1(x)$ is also a solution of $y' = a(x)y$ on (α, β). Let us see if it is possible to find a nonconstant function $v(x)$ defined on (α, β) such that the function

$$y(x) = v(x)y_1(x) \tag{31}$$

is a particular solution on (α, β) of the nonhomogeneous differential equation (28) $y' = a(x)y + b(x)$. Differentiating (31), we find

$$y'(x) = v'(x)y_1(x) + v(x)y_1'(x). \tag{32}$$

Substituting (31) and (32) into (28), we see that $v(x)$ must satisfy

$$v'(x)y_1(x) + v(x)y_1'(x) = a(x)(v(x)y_1(x)) + b(x). \tag{33}$$

Since the function $y_1(x)$ satisfies the homogeneous linear differential equation $y_1'(x) = a(x)y_1(x)$, it follows that $v(x)y_1'(x) = v(x)a(x)y_1(x)$. Consequently, (33) reduces to

$$v'(x)y_1(x) = b(x). \tag{34}$$

Since $y_1(x)$ is positive on the interval (α, β), the function $v'(x) = b(x)/y_1(x)$ is defined for all $x \in (\alpha, \beta)$ and since $b(x)$ and $y_1(x)$ are continuous on (α, β),

$$v(x) = \int^x \frac{b(t)}{y_1(t)}\,dt$$

exists on (α, β). Therefore, a particular solution to the nonhomogeneous differential equation (28) is

$$y_p(x) = v(x)y_1(x) = y_1(x)\int^x \frac{b(t)}{y_1(t)}\,dt. \tag{35}$$

Substituting (30) and (35) into (29), we find the solution of the nonhomogeneous differential equation (28) is

$$z(x) = Ky_1(x) + v(x)y_1(x) = y_1(x)(K + v(x))$$

where K is any arbitrary constant, $y_1(x) = e^{\int^x a(t)\,dt}$, and $v(x) = \int^x \dfrac{b(t)}{y_1(t)}\,dt$. Consequently, we have proved the following theorem.

An Existence Theorem for the Nonhomogeneous Linear First-Order Differential Equation

If $a(x)$ and $b(x)$ are continuous functions on the interval (α, β), then the solution on (α, β) of the nonhomogeneous linear first-order differential equation

$$(28) \qquad y' = a(x)y + b(x)$$

is

$$(36) \qquad y(x) = y_1(x)(K + v(x))$$

where K is an arbitrary constant, $y_1(x) = e^{\int^x a(t)\,dt}$, and $v(x) = \int^x \dfrac{b(t)}{y_1(t)}dt$.

From equation (36), we observe that in order to write the solution of the nonhomogeneous linear first-order differential equation (28) as an elementary function, we must be able to write $y_1(x) = e^{\int^x a(t)\,dt}$ and $v(x) = \int^x \dfrac{b(t)}{y_1(t)}dt$ as elementary functions. Of course, being able to write both of these integrals in terms of elementary functions is often very difficult or even impossible. Also notice by setting $b(x) \equiv 0$ in equation (36) reduces this equation to equation (25), the solution of the homogeneous linear differential equation. Thus, equation (36) is an expression for the solution of both the homogeneous and nonhomogeneous linear first-order differential equation.

Example 5 Solution of a Nonhomogeneous Linear IVP

a. Find the solution of the linear differential equation

$$(37) \qquad y' = (\tan x)y + \sin x.$$

b. Solve the initial value problem

$$(38) \qquad y' = (\tan x)y + \sin x; \quad y(\frac{\pi}{4}) = \sqrt{2}$$

and specify the interval on which the solution exists.

Solution

a. In this instance, the function $a(x) = \tan x$ is defined and continuous on the intervals $I_n = ((2n-1)\pi/2, (2n+1)\pi/2)$ where n is an integer and undefined at the endpoints of the intervals. The function $b(x) = \sin x$ is defined and continuous for all real x. So the solution to the differential equation (37) will exist only on the intervals I_n. Integrating, we find

$$y_1(x) = e^{\int^x a(t)\,dt} = e^{\int^x \tan t\,dt} = e^{-\ln|\cos x|} = \frac{1}{|\cos x|} = |\sec x|.$$

One often obtains $y_1(x)$ as the absolute value of some function, as we did in this instance. We may choose $y_1(x)$ to be $\sec x$ or $-\sec x$. We select $y_1(x) = \sec x$. Then, integrating, we find

$$v(x) = \int^x \frac{b(t)}{y_1(t)}dt = \int^x \frac{\sin t}{\sec t}dt = \int^x \sin t \cos t\, dt = \frac{1}{2}\sin^2 x + C.$$

For computational convenience, we select $C = 0$ and obtain the following solution of (37)

$$\text{(39)} \qquad y(x) = y_1(x)(K + v(x)) = (\sec x)(K + \frac{1}{2}\sin^2 x)$$
$$= K \sec x + \frac{1}{2}(\sin x)(\tan x).$$

b. The solution to the IVP (38) is the member of the one-parameter family (39) which satisfies the initial condition $y(\pi/4) = \sqrt{2}$. Imposing the initial condition, we see that K must satisfy the equation

$$\sqrt{2} = K \sec\left(\frac{\pi}{4}\right) + \frac{1}{2}\sin\left(\frac{\pi}{4}\right)\tan\left(\frac{\pi}{4}\right) = K\sqrt{2} + \left(\frac{1}{2}\right)\left(\frac{1}{\sqrt{2}}\right)(1).$$

Solving for K, we find $K = 3/4$. Hence, the solution of the IVP (38) is

$$y(x) = \frac{3\sec x}{4} + \frac{1}{2}(\sin x)(\tan x).$$

Since the initial condition is specified at $\pi/4 \in I_0 = (-\pi/2, \pi/2)$ this solution exists on the interval $(-\pi/2, \pi/2)$. ■

Newton's Law of Cooling

It has been shown experimentally that under certain conditions, the temperature of a body can be predicted by using Newton's law of cooling which states:

> "The rate of change of the temperature of a body is proportional to the difference between the temperature of the body and the temperature of the surrounding medium."

Hence, if $T(t)$ is the temperature of the body at time t and A is the temperature of the surrounding medium, then according to Newton's law of cooling, $T(t)$ satisfies the differential equation

$$\text{(40)} \qquad \frac{dT}{dt} = k(T - A) = kT - kA$$

where k is the constant of proportionality. Equation (40) is linear in the dependent variable T—that is, $T' = a(t)T + b(t)$ where $a(t) = k$ and $b(t) = -kA$. Since k and A are constants, $a(t)$ and $b(t)$ are constant functions which are defined on $(-\infty, \infty)$. Integrating, we find

$$T_1(t) = e^{\int^t a(s)\,ds} = e^{\int^t k\,ds} = e^{kt+C}.$$

Setting $C = 0$, substituting into $v(t)$, and integrating, yields

$$v(t) = \int^t \frac{b(s)}{T_1(s)}\,ds = \int^t \frac{-kA}{e^{ks}}\,ds = -kA\int^t e^{-ks}ds = Ae^{-kt} + D.$$

Setting $D = 0$ and substituting $T_1(t)$ and $v(t)$ into equation (36) where y has been replaced by T and x has been replaced by t, we find the solution of the linear differential equation (40) is

$$T(t) = e^{kt}(K + Ae^{-kt}) = Ke^{kt} + A \tag{41}$$

where K is an arbitrary constant.

Example 6 An Application of Newton's Law of Cooling A cup of coffee whose temperature is $190°F$ is poured in a room whose temperature is $65°F$. Two minutes later the temperature of the coffee is $175°F$. How long after the coffee is poured does it reach a temperature of $150°F$?

Solution

For this problem, A $= 65°F$ and the constants K and k of equation (41) must be chosen to satisfy the two conditions $T(0) = 190°F$ and $T(2) = 175°F$. Evaluating equation (41) at $t = 0$ and imposing the first condition, we find K must satisfy

$$T(0) = K + A \quad \text{or} \quad 190°F = K + 65°F.$$

Solving for K, we find

$$K = 190°F - 65°F = 125°F.$$

Substituting the value $125°F$ for K in equation (41), evaluating the resulting equation at $t = 2$, and imposing the second condition, we obtain the equation

$$T(2) = 125°Fe^{2k} + A \quad \text{or} \quad 175°F = 125°Fe^{2k} + 65°F.$$

Solving the right-hand equation for e^{2k}, we find

$$e^{2k} = \frac{175 - 65}{125} = \frac{110}{125} = \frac{22}{25}.$$

Hence,

$$e^k = \left(\frac{22}{25}\right)^{1/2}.$$

It is not necessary to determine the value of the constant of proportionality k explicitly, since the expression $e^{kt} = (e^k)^t$ appears in equation (41). Substituting the expression above for e^k in equation (41), we find the temperature of the coffee as a function of the time after it is poured is

$$T(t) = 125^\circ F \left(\frac{22}{25}\right)^{t/2} + 65^\circ F.$$

The coffee reaches the temperature of $150^\circ F$ when t satisfies the equation

$$150^\circ F = 125^\circ F \left(\frac{22}{25}\right)^{t/2} + 65^\circ F.$$

Solving for t, we find

$$t = \frac{2\ln\left(\frac{150-65}{125}\right)}{\ln\left(\frac{22}{25}\right)} = \frac{2\ln\left(\frac{85}{125}\right)}{\ln\left(\frac{22}{25}\right)} = 6.034 \text{ minutes. } \blacksquare$$

2.3.6 Exercises

In Exercises 1–10 find the general solution of the given linear differential equation.

1. $y' = 3y$
2. $y' = y + 1$
3. $y' = \dfrac{3y}{x} + x^3$
4. $y' = 2y + x^2 e^{2x}$
5. $\dfrac{di}{dt} + i - 2 = 0$
6. $\dfrac{di}{dt} + i = e^{-t}$
7. $dy = (\csc x + y\cot x)dx$
8. $dy = \csc x - y\cot x$
9. $\dfrac{dr}{d\theta} + \dfrac{4r}{\theta} = \theta$
10. $\dfrac{dr}{d\theta} + \dfrac{r}{\theta} = \dfrac{\sin\theta}{\theta}$

In Exercises 11–17 solve the given initial value problem by finding the general solution of the differential equation and then determining the value of the constant of integration which satisfies the initial condition. Also specify the interval on which each solution exists.

11. $y' = 4y + 1;\quad y(0) = 1$
12. $y' = xy + 2;\quad y(0) = 1$
13. $y' = y/x;\quad y(-1) = 2$
14. $y' = y/(x-1) + x^2;\quad y(0) = 1$
15. $y' = y/x + \sin x^2;\quad y(-1) = -1$

16. $y' = 2y/x + e^x; \quad y(1) = \dfrac{1}{2}$

17. $y' = (\cot x)y + \sin x; \quad y(\pi/2) = 0$

18. Let $y_1(x)$ be a solution of the homogeneous linear first-order differential equation

$$y' = a(x).$$

Show that $cy_1(x)$ is also a solution for any constant c.

19. a. Show that $y(x) = 0$ is *the* solution of the initial value problem

$$(0) \qquad y' = a(x)y; \quad y(c) = 0$$

b. Suppose that $y_1(x)$ and $y_2(x)$ are two different solutions of the nonhomogeneous linear initial value problem

$$(*) \qquad y' = a(x)y + b(x); \quad y(c) = d$$

Show that $y_3(x) = y_1(x) - y_2(x)$ satisfies the homogeneous linear initial value problem (0). Therefore, $y_3(x) \equiv 0$ and, consequently, $y_1(x) = y_2(x)$. This result proves that the solution of the linear initial value problem $(*)$ is "unique."

20. A thermometer reading $70°F$ is taken outside where the temperature is $10°F$. Five minutes later the thermometer reads $40°F$. How long after being taken outside is the thermometer reading within one-half a degree of the outside temperature?

21. A thermometer reading $80°F$ is taken outside. Five minutes later, the thermometer reads $60°F$. After another 5 minutes, the thermometer reads $50°F$. What is the temperature outside?

22. At 1:00 p.m., a thermometer reading $10°F$ is removed from a freezer and placed in a room whose temperature is $65°F$. At 1:05 p.m. the thermometer reads $25°F$. Later, the thermometer is placed back in the freezer. At 1:30 p.m., the thermometer reads $32°F$. When was the thermometer returned to the freezer, and what was the thermometer reading at that time?

23. A invited B for morning coffee. A poured two cups of coffee. B added enough cream to lower the temperature of his coffee $1°F$. After 5 minutes A added enough cream to his coffee to lower the temperature $1°F$ and both A and B began to drink their coffee. Who had the cooler coffee?

24. Various nonlinear first-order differential equations can be transformed into linear first-order differential equations by some change of variables. Perhaps most notable among such equations is **Bernoulli's equation**

$$(B) \qquad y' = P(x)y + Q(x)y^n$$

where n is a real constant. For $n = 0$ and $n = 1$, equation (B) is linear and the solution is immediate. Equation (B) is named in honor of Jacques Bernoulli (1654-1705), who is also called James or Jacques. After graduating from the University in Basel, Jacques traveled across Europe from 1676 to 1682, learning about the latest discoveries in mathematics and science. In 1683, he returned to the University in Basel and began teaching mechanics. In 1687, he was appointed professor of mathematics and remained in that position for the rest of his life. In 1696, Leibniz (1646-1716) showed that the change of variable $z = y^{1-n}$ reduces Bernoulli's equation is a linear first-order differential equation. Differentiating the transformation equation with respect to x, we obtain $z' = (1-n)y^{-n}y'$. Multiplying equation (B) by $(1-n)y^{-n}$ and performing the change of variables results in the following linear first-order differential equation

$$z' = (1-n)P(x)z + (1-n)Q(x).$$

Solve the following Bernoulli initial value problems and specify the interval on which each solution is valid.

a. $y' = 2x/y;\quad y(0) = 2$ b. $y' = -2y + y^2;\quad y(0) = 1$

c. $y' = -y/x + y^{1/2};\quad y(1) = 1$ d. $y' = y/x + x^2y^3;\quad y(-1) = -1$

In Exercises 25–30 use any technique to solve the given differential equation.

25. $x\,dx - y\,dy = 0$

26. $y\,dx - x\,dy = 0$

27. $(x^2 - y)\,dx + x\,dy = 0$

28. $xy(1-y)\,dx - 2\,dy = 0$

29. $x(1-y^3)\,dx - 3y^2\,dy = 0$

30. $y(2x-1)\,dx + x(x+1)\,dy = 0$

2.4 Numerical Solution

The oldest and simplest algorithm for generating a numerical approximation to a solution of a differential equation was developed by Leonhard Euler in 1768. Given a specific point (x_0, y_0) on the solution of the differential equation $y' = f(x, y)$, Euler wrote the equation for the tangent line to the solution through (x_0, y_0)—namely, $y = y_0 + f(x_0, y_0)(x - x_0)$. To obtain an approximation to the solution through (x_0, y_0) at x_1, Euler took a small step along the tangent line and arrived at the approximation $y_1 = y_0 + f(x_0, y_0)(x_1 - x_0)$ to the solution at x_1, $y(x_1)$. Continuing to generate points successively in this manner and by connecting the points (x_0, y_0), (x_1, y_1), $(x_2, y_2), \ldots$ in succession, Euler produced a polygonal path which approximated the solution. This first numerical algorithm for solving the initial value problem $y' = f(x, y)$; $y(x_0) = y_0$ is called **Euler's method** or, due to its particular geometric construction, the tangent line method. Euler's method is a single-step method. In single-step methods, only one solution value (x_0, y_0), is required to produce the next approximate solution value. On the other hand, multistep methods require two or more previous solution values to produce the next approximate solution value. In 1883, more than a century after Euler developed the first single-step method, the English mathematicians Francis Bashforth (1819-1912) and John Couch Adams (1819-1892) published an article on the theory of capillary action, which included multistep methods that were both explicit methods and implicit methods. In 1895, the German mathematician Carl David Tolmé Runge (1856-1927) wrote an article in which he developed two single-step methods. The second-order method was based on the midpoint rule while the third-order method was based on the trapezoidal rule. In an article that appeared in 1900, Karl Heun (1859-1929) improved Runge's results by increasing the order of the method to four. And in 1901, Martin Wilhelm Kutta (1867-1944) completed the derivation for the fourth-order methods by finding the complete set of eight equations that the coefficients must satisfy. He also specified the values for the coefficients of the classic fourth-order Runge–Kutta method and those of a fifth-order method.

Prior to 1900, most calculations were performed by hand with paper and pencil. Euler's method and Runge–Kutta methods are single-step methods. Euler's method is of order one and requires only one f function evaluation per step. The classic Runge–Kutta method is fourth-order and requires four f function evaluations per step. Adams–Bashforth, Adams–Moulton, and predictor-corrector methods require only two f function evaluations per step; however, since these methods are multistep methods, they require starting values obtained by some other method. By the 1930s, significant numerical integration techniques had been developed; however, their effective implementation was severely limited by the need to perform the computations by hand or with the aid of primitive mechanical calculators.

In the late 19th century and early 20th century, several commercially viable mechanical calculators capable of adding, subtracting, multiplying, and dividing were invented and manufactured. Electric motor driven calculators began to appear as early as 1900. These mechanical and electrical computing devices improved the speed and accuracy of generating numerical solutions of simple differential equations. In 1936, the German civil engineer Konrad Zuse (1910-1995) built the first mechanical binary computer, the Z1, in the living room of his parents' home. From 1942 to 1946, the first large scale, general-purpose electronic computer was designed and built by John W. Mauchly (1907-1980) and J. Presper Eckert (1919-1995) at the University of Pennsylvania. The computer was named ENIAC, which is an acronym for "Electronic Numerical Integrator and Computer." ENIAC, which used vacuum tube technology, was operated from 1946 to 1955. After many technological inventions such as the transistor and integrated circuitry, the first hand-held, battery-powered, pocket calculator capable of performing addition, subtraction, multiplication, and division was introduced by Texas Instruments in 1967. The first scientific pocket calculator, the HP-35, was produced in 1972 by Hewlett Packard.

In the 1960s and 1970s, several sophisticated computer programs were developed to solve differential equations numerically. Since then, significant advances in graphical display capabilities have occurred also. Consequently, at the present time there are many computer software packages available to generate numerical solutions of differential equations and to graphically display the results.

Most differential equations and initial value problems cannot be solved explicitly or implicitly; and, therefore, we must be satisfied with obtaining a numerical approximation to the solution. Thus, we need to know how to generate a numerical approximation to the solution of the initial value problem

$$(1) \qquad y' = f(x, y);\ y(x_0) = y_0.$$

In the differential equation in (1) replace y' by dy/dx, multiply by dx, and integrate both sides of the resulting equation from x_0 to x to obtain

$$\int_{x_0}^{x} dy = \int_{x_0}^{x} f(t, y(t))\, dt \quad \text{or} \quad y(x) - y(x_0) = \int_{x_0}^{x} f(t, y(t))\, dt.$$

Adding $y(x_0) = y_0$ to the last equation, we find the symbolic solution to the IVP (1) on $[x_0, x]$ to be

$$(2) \qquad y(x) = y_0 + \int_{x_0}^{x} f(t, y(t))\, dt.$$

When $f(x, y)$ in (1) is a function of the independent variable alone—that is, when the initial value problem is $y' = f(x);\ y(x_0) = y_0$—we can approximate the function $f(x)$ on the interval $[x_0, x_1]$, where x_1 is a specific point, by step functions or some polynomial in x, say $p_1(x)$, and then using this approximation integrate (2) over $[x_0, x_1]$ to obtain an approximation y_1 to the solution $\phi(x_1)$. Next, we approximate $f(x)$ on $[x_1, x_2]$ by some function $p_2(x)$ and integrate over $[x_1, x_2]$ to obtain $y_2 = y_1 + \int_{x_1}^{x_2} p_2(t)\, dt$ which is an approximation to the solution $\phi(x_2)$, and so on.

When $f(x, y)$ in (1) is a function of the dependent variable y, the value of the approximate solution y_1 at x_1 depends on the unknown solution $\phi(x)$ on the interval $[x_0, x_1]$ and the function $f(x, y)$ on the rectangle

$$R_1 = \{(x, y) \mid x_0 \le x \le x_1,\ y \in \{\phi(x) \mid x_0 \le x \le x_1\}\}.$$

Thus, we must approximate $\phi(x)$ on $[x_0, x_1]$ and $f(x, y)$ on R_1 in order to be able to integrate (2) over the interval $[x_0, x_1]$ and obtain an approximate solution y_1. In this case, additional approximate values $y_2, \ldots, y_n$ are obtained in a like manner.

For single-step methods, it is convenient to symbolize the numerical approximation to the exact solution $y = \phi(x)$ of the IVP (1) $y' = f(x, y);\ y(x_0) = y_0$ by the recursive formula

$$(3) \qquad y_{n+1} = y_n + \psi(x_n, y_n, h_n)$$

where $h_n = x_{n+1} - x_n$. The quantity h_n is called the **stepsize at** x_n and it can vary with each step. However, for computations performed by hand it is usually best to keep the stepsize constant—that is, set $h_n = h$, a constant, for $n = 0, 1, \ldots$.

If the function $y(x)$ has $m + 1$ continuous derivatives on an interval I containing x_0, then by **Taylor's formula with remainder**,

$$(4) \qquad \begin{aligned} y(x) = {} & y(x_n) + y^{(1)}(x_n)(x - x_n) + \frac{y^{(2)}(x_n)}{2}(x - x_n)^2 + \cdots \\ & + \frac{y^{(m)}(x_n)}{m!}(x - x_n)^m + \frac{y^{(m+1)}(\xi)}{(m+1)!}(x - x_n)^{m+1} \end{aligned}$$

where ξ is between x and x_n. Usually, one chooses m in equation (4) to be reasonably small and approximates $y(x_{n+1})$ by

$$(5) \quad y_{n+1} = y_n + f(x_n, y_n)h_n + \frac{f^{(1)}(x_n, y_n)}{2}h_n^2 + \cdots + \frac{f^{(m-1)}(x_n, y_n)}{m!}h_n^m.$$

The **discretization error**, **truncation error**, or **formula error** for this method is given by

$$(6) \qquad E_n = \frac{f^{(m)}(\xi, y(\xi))}{(m+1)!}h_n^{m+1}$$

where $\xi \in (x_n, x_{n+1})$.

If all calculations were performed with infinite precision, the discretization error would be the only error present. **Local discretization error** is the error that would be made in one step if the previous values were exact and there were no round-off error. Ignoring round-off error, **global discretization error** is the difference between the solution $\phi(x)$ of the IVP (1) and the numerical approximation at x_n—that is, the global discretization error is $e_n = y_n - \phi(x_n)$.

A derivation of the series that bears his name was published by the English mathematician Brook Taylor (1685-1731) in 1715. However, the Scottish mathematician James Gregory (1638-1675) seems to have discovered the series more than forty years before Taylor published it. And Johann Bernoulli had published a similar result in 1694. The series was published without any discussion of convergence and without giving the truncation error term—equation (6).

2.4.1 Euler's Method

Again consider the IVP (1) $y' = f(x,y);\; y(x_0) = y_0$. Substituting the initial condition $y(x_0) = y_0$ into the differential equation of (1), we can calculate $y'(x_0) = f(x_0, y(x_0)) = f(x_0, y_0)$, which is the slope of the tangent line of the exact solution $\phi(x)$ at $x = x_0$. If in equation

$$(2) \qquad y(x) = y_0 + \int_{x_0}^{x} f(t, y(t))\, dt$$

which we obtained from (1) earlier by integration, we approximate $y(t)$ by the constant function y_0 and $f(t, y(t))$ by the constant function $f(x_0, y_0)$ on the interval $[x_0, x_1]$, then we obtain the following approximation to the exact solution at x_1 :

$$y_1 = y_0 + \int_{x_0}^{x_1} f(x_0, y_0)\, dt = y_0 + f(x_0, y_0)(x_1 - x_0).$$

Knowing y_1 we can compute $f(x_1, y_1)$ from the differential equation in (1). The value $f(x_1, y_1)$ is an approximation to the slope of the tangent line of the exact solution $\phi(x)$ at $x = x_1$. Notice that $f(x_1, y_1)$ is only an approximation to $\phi'(x_1) = f(x_1, \phi(x_1))$, the slope of the tangent line of the exact solution $\phi(x)$ at $x = x_1$, since in general $y_1 \neq \phi(x_1)$. See Figure 2.11.

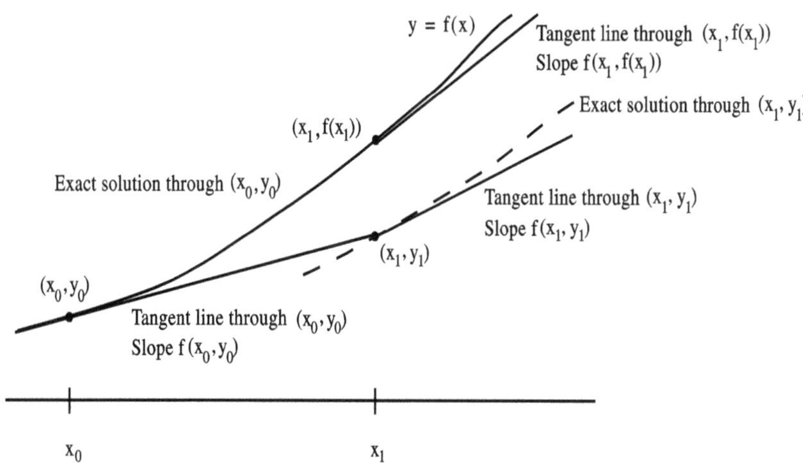

FIGURE 2.11
Euler's Method.

Just as we found that the IVP (1) is equivalent to the integral equation (2), we find the IVP $y' = f(x,y);\; y(x_1) = y_1$ is equivalent to the integral equation

$$y(x) = y_1 + \int_{x_1}^{x} f(t, y(t))\, dt.$$

Using the constant function y_1 to approximate $y(t)$ and the constant function $f(x_1, y_1)$ to approximate the integrand $f(t, y(t))$ on the interval $[x_1, x_2]$, we obtain the following approximation to the exact solution at x_2

$$y_2 = y_1 + \int_{x_1}^{x_2} f(x_1, y_1)\,dt = y_1 + f(x_1, y_1)(x_2 - x_1).$$

Proceeding in this manner, we find the following general recursion for this numerical approximation method to be

$$y_{n+1} = y_n + f(x_n, y_n)(x_{n+1} - x_n). \tag{7}$$

Equation (7) is known as **Euler's method** or the **tangent line method.**

We note that Euler's method is the Taylor series expansion method of order one, so the local discretization error is

$$E_n^E = \frac{1}{2!} f^{(1)}(\xi, y(\xi))h_n^2 = \frac{h_n^2}{2} y^{(2)}(\xi) \tag{8}$$

where $\xi \in (x_n, x_{n+1})$ and $h_n = x_{n+1} - x_n$. Equation (7) is a simple formula to use, especially for hand calculation; but, in general, it is not very accurate. The error of equation (8) is a local error—that is, the error per single step. As one moves away from the initial point x_0, the total error usually builds up. The total error at any point is composed of the accumulated local error and round-off error. It can be shown that if the exact solution of the IVP (1), $\phi(x)$, has a continuous second derivative on $[x_0, x_n]$ and if $|f_y(x, y)| \le L$ and $|y^{(2)}(x)| \le Y$ for $x \in [x_0, x_n]$, then the **global discretization error at** x_n, which is $y_n - \phi(x_n)$, is bounded by

$$|y_n - \phi(x_n)| \le \frac{hY}{2L}(e^{(x_n - x_0)L} - 1) \tag{9}$$

for a fixed stepsize h. Suppose we have reasonable estimates for L and Y on the interval $[x_0, x_n]$ and we wish to maintain a specified accuracy A throughout the interval $[x_0, x_n]$. Then we can require that

$$|y_n - \phi(x_n)| \le \frac{hY}{2L}(e^{(x_n - x_0)L} - 1) \le A.$$

Solving the right-most inequality for h, we usually obtain an underestimate of the stepsize to use throughout the interval $[x_0, x_n]$ in order to maintain the specified accuracy. It should be noted that the global discretization error bound in equation (9) is an upper bound and not a realistic error estimate. However, equation (9) does show that as $h \to 0$, the global discretization error goes to zero. Thus, neglecting round-off error, as $h \to 0$, the Euler approximation converges to the exact solution.

The Swiss mathematician Leonhard Euler (1707-1783) was one of the most prolific authors in the field of mathematics. He introduced the now common notations of e for the base of natural logarithms, π for the ratio of the circumference of a circle to its diameter, Σ for the sign for summation, i for the imaginary unit, and $\sin x$ and $\cos x$ for the trigonometric functions. Euler's famous formula $e^{i\theta} = \cos\theta + i\sin\theta$ expresses a relation between the exponential function and trigonometric functions. The special case $\theta = \pi$ provides the following relation between the numbers e, π, and i: $e^{i\pi} = -1$. Euler's contribution to differential equations includes the concept of an integrating factor, the method for solving linear differential equations with constant coefficients, the method of reduction of order, and power series solutions, to name a few.

Example 1 Euler's Approximation to the Solution of the Initial Value Problem: $\mathbf{y' = y + x;\ \ y(0) = 1}$

a. Find an approximate solution to the initial value problem

$$y' = y + x = f(x, y); \quad y(0) = 1 \tag{10}$$

on the interval $[0, 1]$ using Euler's method and constant stepsizes $h = .2$ and $h = .1$.

b. Use equation (9) to estimate the maximum global discretization error at $x = 1$ for constant stepsizes $h = .2$ and $h = .1$.

Solution

a. Table 2.1 contains Euler's approximation to the IVP (10) on the interval $[0, 1]$ obtained using a constant stepsize of $h = .2$. And Table 2.2 contains Euler's approximation to the IVP (10) on the interval $[0, 1]$ obtained using a constant stepsize of $h = .1$. All calculations were performed using six significant digits.

b. Taking the partial derivative of $f(x, y) = y + x$ with respect to y, we find $f_y = 1$, so $|f_y| \le 1 = L$ on $[0, 1]$. Differentiating the differential equation

TABLE 2.1
Euler's Approximation to the Initial Value Problem: (10) $\mathbf{y' = y + x;\ y(0) = 1}$ on $[\mathbf{0}, \mathbf{1}]$ with Stepsize $\mathbf{h = .2}$

n	x_n	y_n	$f(x_n, y_n) =$ $y_n + x_n$	$hf(x_n, y_n)$	$y_{n+1} =$ $y_n + hf(x_n, y_n)$
0	.0	1.0	1.0	.2	1.2
1	.2	1.2	1.4	.28	1.48
2	.4	1.48	1.88	.376	1.856
3	.6	1.856	2.456	.4912	2.3472
4	.8	2.3472	3.1472	.62944	2.97664
5	1.0	2.97664			

TABLE 2.2
Euler's Approximation to the Initial Value Problem: (10) $\mathbf{y}' = \mathbf{y} + \mathbf{x}$; $\mathbf{y}(\mathbf{0}) = \mathbf{1}$ on $[\mathbf{0}, \mathbf{1}]$ with Stepsize $\mathbf{h} = .\mathbf{1}$

n	x_n	y_n	$f(x_n, y_n) =$ $y_n + x_n$	$hf(x_n, y_n)$	$y_{n+1} =$ $y_n + hf(x_n, y_n)$
0	.0	1.0	1.0	.1	1.1
1	.1	1.1	1.2	.12	1.22
2	.2	1.22	1.42	.142	1.362
3	.3	1.362	1.662	.1662	1.5282
4	.4	1.5282	1.9282	.19282	1.72102
5	.5	1.72102	2.22102	.222102	1.94312
6	.6	1.94312	2.54312	.254312	2.19743
7	.7	2.19743	2.89743	.289743	2.48718
8	.8	2.48718	3.28718	.328718	2.81590
9	.9	2.81590	3.71590	.371590	3.18748
10	1.0	3.18748			

appearing in (10), $y' = y + x$, we get $y^{(2)} = y' + 1 = y + x + 1$. Examining the Euler numerical approximation values in Tables 2.1 and 2.2, we see that $|y| < 3.2$ for $x \in [0, 1]$. So, for $x \in [0, 1]$ we assume $|y| < 7$ which is approximately twice the largest y value appearing in Table 2.2. Using the triangle inequality, we find that $|y^{(2)}| < |y| + |x| + 1 < 9 = Y$ for $x \in [0, 1]$. Therefore, by equation (9) with $h = .2$ the maximum global discretization error satisfies

$$|y_5 - \phi(1)| \leq \frac{hY}{2L}(e^{(x_n - x_0)L} - 1) = \frac{(.2)(9)}{2}(e - 1) \approx 1.54645.$$

For $h = .1$ the maximum global discretization error is approximately .773227. Since the given differential equation is linear, we easily calculate the exact solution of the IVP (10) to be $\phi(x) = 2e^x - x - 1$. So the value of the exact solution at $x = 1$ is $\phi(1) = 2e - 2 \approx 3.43656$. Hence, the actual total error due to discretization and round-off at $x = 1$ for $h = .2$ is $3.43656 - 2.97664 = .45992$, which is about one-third of the estimated error of 1.54645. ■

2.4.2 Comments on Computing

Historically, the numerical solution of initial value problems originated in the 1700s with the work of Euler and Taylor. In the late 1800s and early 1900s, Runge, Heun, and Kutta developed their numerical integration techniques. All of these techniques were used to generate numerical solutions of initial value problems by hand—that is, with paper and pencil. There were no electronic calculators or computers available to compute the solutions. The numerical results which appear in Example 1 and Tables 2.1 and 2.2 were generated using relatively simple computer programs. If you know a scientific computing language, you should be able to write computer programs to produce similar output. The software which accompanies this text does not include programs to produce a numerical solution using the Taylor series method, Euler's method, the improved Euler's method, the modified Euler's method, or the fourth-order Runge–Kutta method. None of these methods is capable of producing an extremely accurate numerical solution over a "large" interval of the independent variable, because none of these methods is an adaptive method and because most of these methods are of low order. Computer algebra systems (CAS) such as Mathematica, MATLAB®, and MAPLE contain commands that allow you to generate a numerical solution using the methods listed above. For example, the following six MAPLE statements enable you to generate Euler's approximation to the IVP (10) $y' = y + x;\ y(0) = 1$ on the interval $[0, 1]$ with a constant stepsize of $h = .2$. When rounded to six significant digits, the values output by this program are the same values which appear in the columns labeled x_n and y_n of the table of values in Example 3.

```
with(DEtools):
de:=diff(y(x), x) = y(x) + x:
sol:=dsolve({de, y(0) = 1}, numeric, method=classical[foreuler], stepsize=
    0.2):
for i from 0 to 5 do
sol(.2 * i);
end do;
```

The first statement informs MAPLE to load the required software package DEtools. The second statement specifies the differential equation which is to be solved—$dy/dx = y + x$. In order to solve a different differential equation, the expression $y(x)+x$ must be replaced with an appropriate expression which conforms to MAPLE syntax. The third statement specifies the initial condition to use—in this case, $y(0) = 1$. To change the initial condition, change the values 0 and 1 accordingly. The specification "numeric" which appears in the third statement tells the computer to solve the initial value problem numerically. The third statement also indicates the integration is to be performed using the forward Euler's method (method=classical[foreuler]) and the stepsize is to be 0.2. The forward Euler's method, called foreuler in MAPLE, denotes Euler's method. Other methods we have discussed are specified in MAPLE as follows: heunform denotes the improved Euler's method, impoly denotes the modified Euler's method, and rk4 denotes the fourth-order Runge–Kutta method. The last three statements in the program form a "loop" which computes the values $x = .2*i$ at which the independent variable and its associated numerical solution value, sol$(.2 * i)$, are to be printed. In this example, the values printed will be (0, sol(0)), (.2, sol(.2)), ... , (1.0, sol(1.0)). If the initial condition is specified at $x = a$ and integration takes place over the interval $[a, b]$ where the stepsize h is chosen so that $n = (b - a)/h$ is an integer, then in the fourth statement above change 5 to n and in the fifth statement above change $.2 * i$ to $a + h * i$. This will cause $n + 1$ pairs of values $(x, \text{sol}(x))$ to be printed at $n + 1$ equally spaced points throughout the interval $[a, b]$. The spacing between the values will be h. When h is very small, instead of printing the solution every step, you may want to print the solution only every ten steps, or every one hundred steps, or every one thousand steps. This is accomplished by changing the value 5 in statement four and the expression $.2 * i$ in statement six appropriately.

2.4.3 Exercises

1. a. Compute an approximate solution to the initial value problem $y' = x^2 - y$; $y(0) = 1$ on the interval $[0, 1]$ using Euler's method and a constant stepsize of $h = .1$.

   b. Find an upper bound for the total discretization error at $x = 1$.

   c. How small must the stepsize be to ensure six decimal place accuracy per step?

   d. How small must the stepsize be to ensure six decimal place accuracy over the interval $[0, 1]$?

2. 2. Use Euler's method to generate a numerical solution to the initial value problem $y' = y/x + 2$; $y(1) = 1$ on the interval $[1, 2]$ with a stepsize of $h = .05$

2.4.4 Pitfalls of Numerical Methods

2.4.4.1 Comments on Computing

Most mathematical software packages which contain algorithms to solve ordinary differential equations include one or more routines which attempt to numerically approximate the solution of the initial value problem $y' = f(x, y)$; $y(c) = d$ on the interval $[a, b]$ where $a \leq c \leq b$. The best way for us to illustrate how to use these programs, the typical output of the programs, and some of the pitfalls that may occur when the programs are used is through the following set of examples. We suggest that you run these examples and compare your results with those given here. It is very possible that your results will not be exactly the same. The numerical results and graphical output presented in Examples 2, 3, 4, and 5 were generated using MAPLE software. We selected the default integration technique which is the fourth-order Runge–Kutta method with fixed stepsize. The stepsize is determined by the software. You should also use your favorite software package or artificial intelligence tools to solve the initial value problem in Example 4 and compare those results with the ones given in Example 4.

2.4.4.2 Examples

Example 2 Numerical Approximation and Graph of the IVP: $\mathbf{y' = x - y; y(0) = 2}$ Calculate and graph a numerical approximation of the solution to the initial value problem $y' = x - y;\ \ y(0) = 2$ on the interval $[-1, 4]$ using MAPLE.

Solution

The following four MAPLE statements produced the output shown in Figure 2.12.

```
with(DEtools):with(plots):
de:=diff(y(x), x)=x-y(x):
p:=DEplot(de, y(x), x=-1..4, y=0..10, {[y(0)=2]}, arrows=LINE, axes=
    BOXED):
display(p);
```

The second statement specifies the differential equation to be solved is $y' = x - y$. The third statement instructs MAPLE to graph the solution in the rectangle bounded by the lines $x = -1$, $x = 4$, $y = 0$, and $y = 10$. The third statement also specifies the initial condition is $y(0) = 2$, the direction field is to be displayed as lines (arrows=LINE), and the x and y axes are to appear as a box outside the rectangle (axes=BOXED).

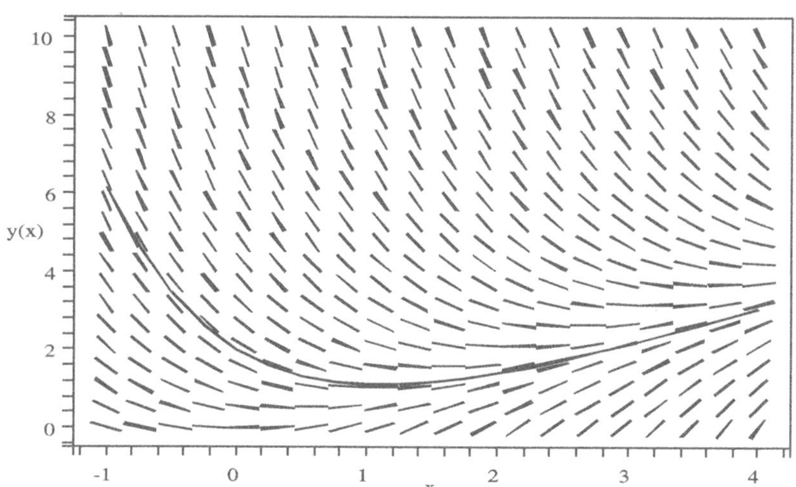

FIGURE 2.12
Direction Field and Graph of a Numerical Approximation of $y' = x$ - y; y(0) = 2.

We did not encounter any difficulties in generating the solution to the given initial value problem, because $y' = x - y$ is a linear differential equation—recall, $y' = f(x, y)$ is linear if and only if $f(x, y) = a(x)y + b(x)$—and because the functions $a(x) = -1$ and $b(x) = x$ are defined and continuous on the interval $[-1, 4]$. ■

The following example illustrates one type of result we might obtain in the event we attempt to produce a numerical approximation to a solution on an interval which is too large—that is, on an interval larger than the interval on which the solution exists.

Example 3 A Numerical Approximation Outside of the Interval of Existence Use the fundamental existence, uniqueness, and continuation theorems to analyze mathematically the initial value problem

$$(11) \qquad y' = \frac{y}{(x-1)(x+2)} + \frac{1}{x}; \qquad y(-1) = 2$$

on the interval $[-2.5, .5]$. Then use MAPLE to calculate and graph a numerical approximation of the solution to the IVP (11).

Solution

Mathematical Analysis

The differential equation of (11) is linear with $a(x) = 1/((x-1)(x+2))$ and $b(x) = 1/x$. The function $a(x)$ is not defined; and, therefore, not continuous at $x = -2$ and $x = 1$. The function $b(x)$ is not defined and not continuous at $x = 0$. Hence, the functions $a(x)$ and $b(x)$ are both continuous on the intervals $(-\infty, -2)$, $(-2, 0)$, $(0, 1)$, and $(1, \infty)$. Since the initial point $-1 \in (-2, 0)$, the IVP (11) has a unique solution on $(-2, 0)$ and this is the largest interval on which (11) has a solution.

Numerical Solution

A graph of the numerical approximation to the solution of the IVP (11) generated using MAPLE is shown in Figure 2.13.

This output was created using the following four Maple statements. These statements were obtained by modifying of the four statements in Example 2.

```
with(DEtools):with(plots):
de:=diff(y(x), x)= y(x)/((x-1)*(x-2))+1/x:
p:=DEplot(de, y(x), x=-2.5..0.5, {[y(-1)=2]}, arrows=NONE, axes=
    BOXED):
display(p);
```

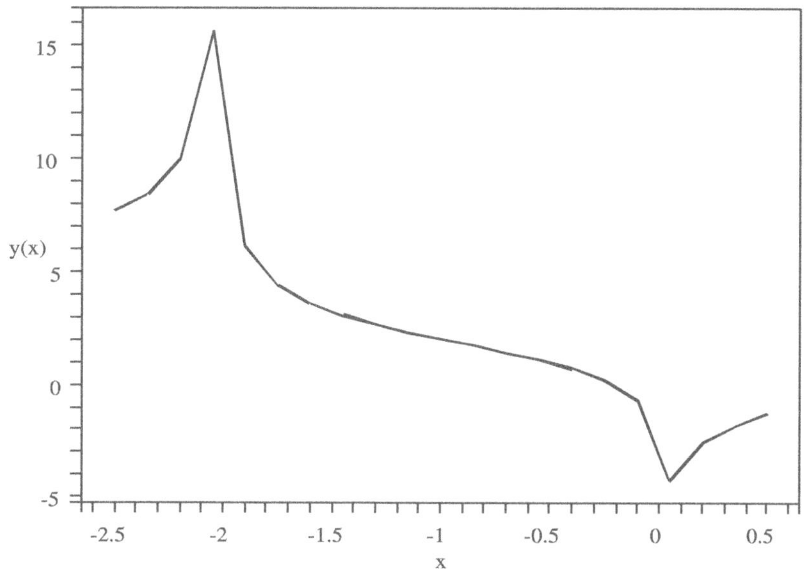

FIGURE 2.13
Numerical Approximation of the Solution to $y' = \dfrac{y}{(x-1)(x+2)} + \dfrac{1}{x}; y(-1) = 2$.

In the second statement, we changed the specification of the differential equation from $x - y(x)$ to $y(x)/((x-1)*(x-2)) + 1/x$. In the third statement, we replaced the range of x values $-1..4$ with the new range $-2.5..0.5$. We did not specify a range for the y values, so the program selected a range based on the computed solution. From Figure 2.13 it appears the minimum y value selected by the software was -5 and the maximum y value selected was 16. Also, in the third statement we changed the initial condition to $y(-1) = 2$ and we indicated we did not want a direction field displayed by replacing LINE by NONE.

Notice that the graph indicates the presence of vertical asymptotes in the solution near $x = -2$ and $x = 0$. ■

Example 3 illustrates that it is necessary for you to perform a thorough mathematical analysis for each initial value problem prior to computing a numerical solution. In this example, the computer generated a numerical approximation to the solution outside the interval of existence—that is, outside of the interval $(-2, 0)$. So you must analyze each initial value problem separately and determine what the fundamental theorems tell you about the problem with respect to existence, uniqueness, and continuation of the solution. When you use your computer software to numerically solve the IVP (11), what does the graph of the solution look like?

Example 4 Numerical Approximation and Graph of the IVP: $\mathbf{y' = -x/y;\quad y(-1) = 1}$ Mathematically analyze the initial value problem

$$y' = -x/y; \quad y(-1) = 1 \tag{12}$$

on the interval $[-2, 2]$ taking into account the fundamental existence, uniqueness, and continuation theorems. Then calculate and graph a numerical approximation of the solution to the IVP (12) using MAPLE.

Solution

Mathematical Analysis
Here $f(x, y) = -x/y$ and $\partial f/\partial y = x/y^2$ are defined and continuous on any finite rectangle which does not contain any point $(x, 0)$—that is, on any finite rectangle which does not contain a point of the x-axis, where $y = 0$. So by the fundamental theorems there exists a unique solution of the IVP (12) and this solution can be continued in a unique manner until either $x \to -\infty$, $x \to +\infty$, $y \to 0$ (since existence and uniqueness may no longer be guaranteed at $y = 0$), or $y \to +\infty$ (since the y coordinate of the initial condition, $y(-1) = 1$, is positive).

Numerical Solution
A graph of the numerical solution to the IVP (12) generated using MAPLE is shown in Figure 2.14. The following four MAPLE statements were used to produce the graph.

```
with(DEtools):with(plots):
de:=diff(y(x), x)= -x/y(x):
p:=DEplot(de,y(x),x=-2..2,{[y(-1)=1]},arrows=NONE,axes=BOXED,
    view=[-2.2..2.2,-2.2..2.2]):
display(p);
```

The second statement specifies the differential equation of (12). The third statement specifies the interval of integration with $x = -2..2$ and the initial condition with $y(-1) = -1$. We included the specification: view= $[-2.2..2.2, -2.2..2.2]$ because we wanted the horizontal axis of the graph to be slightly longer than the interval of integration and because we wanted to control the range of the y values instead of letting the software do so. The solution is "reasonably" accurate—but, of course, not exact—on the interval $[-1.4, 1.4]$ where the computed solution is positive. The solution of the IVP (12) is $y(x) = \sqrt{2 - x^2}$. (Verify this fact.) This solution exists only on the interval $(-\sqrt{2}, \sqrt{2})$. The graph of the solution is the upper half of a circle with center at the origin and radius $\sqrt{2}$.

The general solution of the differential equation of the IVP (12), $y' = -x/y$, is $x^2 + y^2 = k^2$, where k is an arbitrary constant. For $k \neq 0$ the graph of $x^2 + y^2 = k^2$ is a circle with center at the origin and radius $|k|$. Thus, the graph of the one-parameter family of curves $x^2 + y^2 = k^2$ is the set of all concentric circles with center at the origin. Look at Figure 2.14 again. Notice for $-2 < x < -1.4$ and for $1.4 < x < 2$ the computer-generated solution is attempting to approximate other members of the general solution of the differential equation.

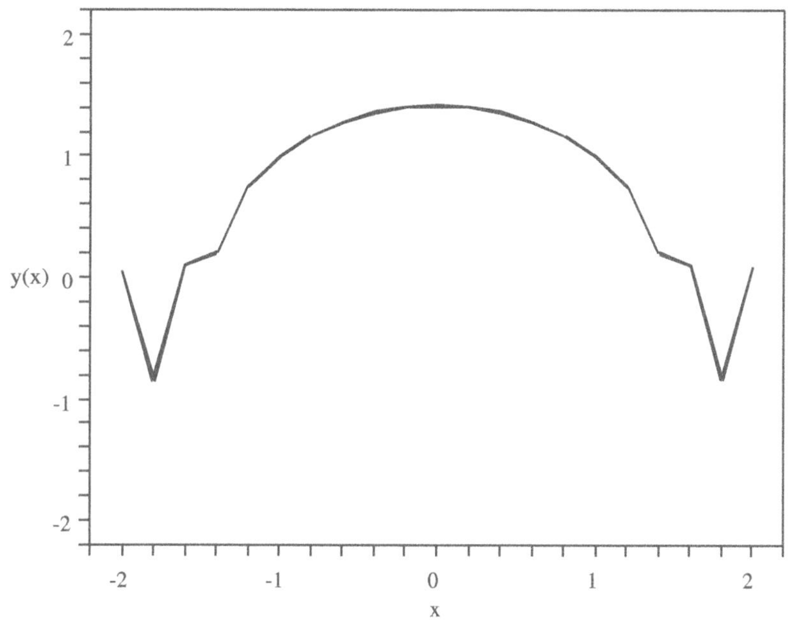

FIGURE 2.14
Graph of a Numerical Approximation of the Solution to $y' = -$ x /y; y (- 1) = 1. ■

Example 5 Numerical Approximation of the IVPs: $\mathbf{y' = \sqrt{xy}}$; $\mathbf{y(-1) = d}$ Mathematically analyze the following two initial value problems.

(13a) $\quad y' = \sqrt{xy}; \quad y(-1) = 0$

(13b) $\quad y' = \sqrt{xy}; \quad y(-1) = -1/9 \approx -.11111111$

Then calculate a numerical approximation of the solution to initial value problems on the interval $[-1, 1]$ using MAPLE.

Solution

Mathematical Analysis

The function $f(x, y) = \sqrt{xy}$ is defined and continuous in the first quadrant, in the third quadrant, on the x-axis (where $y = 0$), and on the y-axis (where $x = 0$). The function $\partial f / \partial y = \sqrt{x/y}$ is defined and continuous in the first quadrant, in the third quadrant, and on the y-axis; but it is not defined on the x-axis. Applying the fundamental theorems to the initial value problem

$$y' = \sqrt{xy}; \quad y(c) = d$$

we see that a solution exists as long as $(x, y(x))$ remains in the first or third quadrant or $y(x) = 0$. Furthermore, the solution is unique as long as $(x, y(x))$ remains in the first or third quadrant. Once the solution, $y(x)$, reaches the x-axis—that is, once $y(x) = 0$—the solution may no longer be unique.

a. The point $(-1, 0)$, which corresponds to the initial condition $y(-1) = 0$, is on the x-axis so the solution to the IVP (13a) exists, but the solution may not be unique. Notice that $y(x) \equiv 0$ is a solution of (13a). Verify that for all $r \leq -1$ and all $s \geq 0$

$$y_{rs}(x) = \begin{cases} -[(-x)^{3/2} - r]^2/9 & , \quad r \leq -1 \\ 0 & , \quad -1 < x < 0 \\ (x^{3/2} - s)^2/9 & , \quad 0 \leq s \end{cases}$$

is a solution of the IVP (13a). Hence (13a) has an infinite number of solutions on $(-\infty, \infty)$.

b. Since $(-1, -1/9)$ is in the third quadrant, the IVP (13b) has a unique solution until $x \to -\infty$, $x \to 0^-$, $y \to -\infty$, or $y \to 0^-$. On the interval $(-\infty, 0)$ the unique solution to the IVP (13b) is $y(x) = x^3/9$. (Verify this fact.) At $x = 0$, $y(x) = 0$ and the solution may no longer be unique. Show that for all $u \geq 0$

$$y_u(x) = \begin{cases} x^3/9 & , \quad x \leq 0 \\ 0 & , \quad 0 < x < u \\ (x^{3/2} - u)^2/9 & , \quad u \leq x \end{cases}$$

is a solution of the IVP (13b). Hence, (13b) has an infinite number of solutions on $(-\infty, \infty)$.

Numerical Solution

We generated numerical approximations to the IVPs (13a) and (13b) using MAPLE and another package.

a. The initial point $(-1, 0)$ is on the x-axis and as we saw in the mathematical analysis section, there are an infinite number of solutions to the IVP (13a). The other package generated the solution $y(x) \equiv 0$. The software gave us no indication that this is not a unique solution. MAPLE indicated floating point overflow no matter how small the interval of integration was chosen. This result would often lead one to assume the IVP (13a) has no solution.

b. In this case, the other package generated a numerical approximation on the interval $[-1, 1]$ which is a very good approximation of the solution $y(x) = x^3/9$ to the IVP (13b). However, the integration procedure gave us no indication that there are an infinite number of solutions of the IVP (13b) on the interval $[-1, 1]$ or that the solution is not unique once $y(x) = 0$. MAPLE generated a numerical approximation of the solution to the IVP (13b) on the interval $[-1, -.054]$ which is a very good approximation of the solution $y(x) = x^3/9$. The software gave us no indication that the solution could be extended further to the right, that there are an infinite number of solutions of the IVP (13b) on the interval $[-1, 1]$, or that the solution is not unique once $y(x) = 0$. ■

The previous examples were presented to show that computer programs which generate numerical approximations of solutions to initial value problems sometimes generate a solution where no solution exists, sometimes generate a single solution where there are multiple solutions, and sometimes do not produce a solution where there is one or more solutions. This is the typical behavior of such computer programs. It is not a flaw of the computer program. Hence, the previous examples should vividly illustrate to you the necessity of performing a thorough mathematical analysis for each individual initial value problem based on the fundamental theorems prior to generating any numerical approximation. Use your computer software numerical integration algorithm to solve the previous examples and compare your results to ours.

2.4.5 Exercises

Use the fundamental existence, uniqueness, and continuation theorems to mathematically analyze each of the following initial value problems. Then use a Computer Algebra System (CAS) such as MAPLE, *Mathematica*, or MATLAB to calculate and graph a numerical solution on the interval specified. Based on the analysis or graph specify, when possible, the subinterval on which the solution exists and the subinterval on which the solution is unique.

1. $y' = 1/(x - 1);\quad y(0) = 1$ on $[-2, 2]$

2. $y' = y + x;\quad y(0) = 0$ on $[-1, 1]$

3. $y' = y/x$ on $[-2, 2]$ a. $y(-1) = 1$ b. $y(-1) = -1$

4. $y' = y/(1 - x^2) + \sqrt{x}$ on $[-2, 2]$

   a. $y(.5) = 1$ b. $y(1) = 1$ c. $y(2) = 1$

5. $y' = y^2$ on $[-5, 5]$

   a. $y(-1) = 1$ b. $y(-1) = 0$ c. $y(1) = .5$

6. $y' = y^3$ on $[-2, 2]$

   a. $y(-1) = 1$ b. $y(-1) = 0$ c. $y(-1) = -1$

7. $y' = -3x^2/(2y)$ on $[-1, 1]$

   a. $y(-1) = 1$ b. $y(-1) = .5$ c. $y(-1) = 0$ d. $y(-1) = -1$

8. $y' = \sqrt{y}/x$ on $[-2, 2]$

   a. $y(-1) = 1$ b. $y(-1) = 0$ c. $y(-1) = -1$ d. $y(1) = 1$

9. $y' = 3xy^{1/3}$ on $[-1, 1]$
   [Hint: Input $y^{1/3}$ as sgn(y)*exp((log(abs(y)))/3)]

   a. $y(-1) = 1.5$ b. $y(-1) = 1$ c. $y(-1) = 0.5$
   d. $y(-1) = 0$ e. $y(-1) = -1$

10. $y' = \sqrt{(y + 2)(y - 1)}$ on $[-2, 2]$

    a. $y(0) = 0$ b. $y(0) = 1$ c. $y(0) = -3$

11. $y' = y/(y - x)$ on $[-1, 1]$

    a. $y(1) = 2$ b. $y(1) = 1$ c. $y(1) = 0$ d. $y(1) = -1$

12. $y' = xy/(x^2 + y^2)$ on $[-1, 1]$

    a. $y(0) = 1$ b. $y(0) = 0$ c. $y(0) = -1$

13. $y' = x\sqrt{1 - y^2}$ on $[-1, 1]$

    a. $y(0) = 1$ b. $y(0) = .9$ c. $y(0) = .5$ d. $y(0) = 0$

14. $y' = (-x + \sqrt{x^2 + 4y})/2$ on $[-1, 1]$

    a. $y(0) = 1$ b. $y(0) = 0$ c. $y(0) = -1$ d. $y(1) = -.2$
    e. $y(1) = -.25$

3

Applications of the Initial Value Problem $y' = f(x, y);\ y(c) = d$

In this chapter, we consider a variety of applications of the initial value problem $y' = f(x, y);\ y(c) = d$. First, in the section titled Calculus Revisited, we show that the solution to the particular initial value problem $y' = f(x);\ y(a) = 0$ is equivalent to the definite integral $\int_a^x f(t)\,dt$. Then, we show how to use computer software to calculate an approximation to the definite integral $\int_a^b f(x)\,dx$. This will allow us to solve problems from calculus numerically. In the sections titled Learning Theory Models, Population Models, Simple Epidemic Models, Falling Bodies, Mixture Problems, Curves of Pursuit, and Chemical Reactions, we examine some physical problems from a number of diverse disciplines which can be written as initial value problems and then solved using numerical integration software. Finally, we present a few additional applications in the Miscellaneous Exercises which appear at the end of this chapter.

3.1 Calculus Revisited

Many calculus problems involve computing a value for the definite integral $\int_a^b f(x)\,dx$. Examples of such problems include finding the area under a curve, finding the area between two curves, finding the length of an arc of a curve, finding the area of a surface generated by revolving a curve about an axis, finding the volume of a solid generated by revolving a region about an axis, and computing physical quantities such as work, force, pressure, moments, center of mass, and centroids.

A function $F(x)$ is called an **antiderivative of the function f(x) on an interval** $[a, b]$, if $F'(x) = f(x)$ for all $x \in [a, b]$. The fundamental theorem of integral calculus stated below tells us how to compute the value of the definite integral of $f(x)$ on $[a, b]$, $\int_a^b f(x)\,dx$, provided we can find an antiderivative of $f(x)$ on $[a, b]$.

DOI: 10.1201/9781003678014-3

The Fundamental Theorem of Integral Calculus

If $f(x)$ is continuous on the interval $[a, b]$ and if $F'(x) = f(x)$ for all x in $[a, b]$, then

$$\int_a^b f(x)\,dx = F(x)\Big|_a^b = F(b) - F(a).$$

Problems which appear in calculus texts are chosen very carefully so that required integrations can be performed—that is, so that the antiderivative of the integrand can be written as an elementary function. In this respect, problems which appear in calculus texts are somewhat artificial, because in practice the required integration usually cannot be performed explicitly. Look at the section on computing arc length in any calculus text and you will see that very few examples and Exercises are given. This is because it is difficult to choose many functions $y(x)$ such that the integral of $\sqrt{1 + [y'(x)]^2}$ can be written as an elementary function.

Unfortunately, **most functions f(x) which arise in practice do not have antiderivatives that can be written as elementary functions.** So we need to answer the following questions:

> "How can we rewrite any definite integral $\int_a^b f(x)\,dx$ as an initial value problem?"
>
> "How can we numerically solve the resulting initial value problem and thereby compute a value for the definite integral?"

If $f(x)$ is integrable on $[a, b]$, we can symbolically write the antiderivative of $f(x)$ on $[a, b]$ as

(1) $$F(x) = \int_a^x f(t)\,dt \quad \text{for } x \in [a, b].$$

Suppose $F(x)$ is differentiable on $[a, b]$. Differentiating (1) with respect to x, we find $F'(x) = f(x)$ for all $x \in [a, b]$. So $F(x)$ satisfies the differential equation $y' = f(x)$ on $[a, b]$. Evaluating $F(x)$ at $x = a$, we get $F(a) = 0$. Thus, the antiderivative $F(x)$ of $f(x)$ satisfies the initial value problem

(2) $$y' = f(x); \quad y(a) = 0$$

on the interval $[a, b]$ and the value of the definite integral of $f(x)$ on $[a, b]$ is

$$\int_a^b f(t)\,dt = y(b) = F(b).$$

Hence, the integral

$$y(x) = \int_a^x f(t)\,dt$$

is equivalent to $y(x)$ being the solution of the initial value problem

$$y' = f(x); \quad y(a) = 0$$

and

$$\int_a^b f(t)\,dt$$

is equivalent to $y(b)$, where $y(b)$ is the solution to the initial value problem $y' = f(x);\ y(a) = 0$ evaluated at b.

Example 1 A Continuous Integrand Whose Antiderivative Is Not an Elementary Function Compute $\int_1^2 \sin x^2\,dx$.

Solution

The function $f(x) = \sin x^2$ is defined and continuous on the interval $[1, 2]$. So $f(x)$ is integrable on $[1, 2]$, but its antiderivative $F(x)$ cannot be expressed as an elementary function. However, as we have just seen, the antiderivative $F(x)$ satisfies the initial value problem

(3) $$y' = \sin x^2; \quad y(1) = 0$$

and $F(2) = \int_1^2 \sin x^2\,dx$. We used a computer algebra system to solve the initial value problem (3) numerically on the interval $[1, 2]$. We found that $\int_1^2 \sin x^2\,dx$ is approximately equal to 0.4945103.

The two MAPLE statements

```
sol:=dsolve({diff(y(x), x) = sin(x ∧ 2), y(1) = 0} , numeric):
sol(2);
```

produces the value .494508245219554054 as its numerical approximation to $\int_1^2 \sin x^2\,dx$. This result was calculated using the default MAPLE numerical integration procedure Runge–Kutta–Fehlberg 4(5), RKF45.

Use your computer software to solve the initial value problem (3) on the interval $[1, 2]$. How does your result compare with the ones we obtained above? ■

In what follows we summarize some results and formulas from calculus for computing area, arc length, and volume when the curves involved are defined in rectangular coordinates, in polar coordinates, and parametrically.

Formulas Involving Curves Defined in Rectangular Coordinates

The Area Under a Curve

Let $y = f(x)$ be a continuous, nonnegative function ($f(x) \geq 0$) on the interval $[a, b]$. The area, A, of the region in the xy-plane bounded above by the curve

$y = f(x)$, bounded below by the x-axis ($y = 0$), and bounded by the vertical lines $x = a$ and $x = b$ is

$$A = \int_a^b f(x)\,dx.$$

The Area between Two Curves

Let $y = f(x)$ and $y = g(x)$ be continuous functions on the interval $[a, b]$ with the property that $f(x) \geq g(x)$ for all $x \in [a, b]$. The area bounded above by the curve $y = f(x)$, bounded below by the curve $y = g(x)$, and bounded by the vertical lines $x = a$ and $x = b$ is

$$A = \int_a^b [f(x) - g(x)]\,dx.$$

Arc Length

If $y = f(x)$ has a continuous first derivative, $f'(x)$, on the interval $[a, b]$, then the arc length of the curve $y = f(x)$ from a to b is

$$s = \int_a^b \sqrt{1 + [f'(x)]^2}\,dx.$$

Areas of Surfaces of Revolution

If $y = f(x)$ has a continuous first derivative, $f'(x)$, on the interval $[a, b]$, then

(1) the area of the surface generated by revolving the curve $y = f(x)$ from a to b about the x-axis is

$$S_x = 2\pi \int_a^b |f(x)|\sqrt{1 + [f'(x)]^2}\,dx$$

and

(2) the area of the surface generated by revolving about the y-axis the curve $y = f(x)$ from a to b where $0 \leq a \leq b$ is

$$S_y = 2\pi \int_a^b x\sqrt{1 + [f'(x)]^2}\,dx.$$

Volumes of Solids of Revolution

Let $f(x)$ and $g(x)$ be continuous functions on the interval $[a, b]$ with the property that $f(x) \geq g(x) \geq 0$ for all $x \in [a, b]$. The volume of the solid generated by revolving the region bounded by the curves $y = f(x)$, $y = g(x)$, $x = a$, and $x = b$

(1) about the x-axis is

$$V_x = \pi \int_a^b [f^2(x) - g^2(x)]\, dx$$

and

(2) about the y-axis for $0 \le a \le b$ is

$$V_y = 2\pi \int_a^b x[f(x) - g(x)]\, dx.$$

Formulas Involving Curves Defined in Polar Coordinates

Area

Let $r = f(\theta)$ be a continuous, nonnegative function on the interval $[\alpha, \beta]$ where $0 < \beta - \alpha \le 2\pi$. The area of the region bounded by the curves $r = f(\theta)$, $\theta = \alpha$, and $\theta = \beta$ is

$$A = \frac{1}{2} \int_\alpha^\beta f^2(\theta)\, d\theta.$$

The Area between Two Curves

Let $r = f(\theta)$ and $r = g(\theta)$ be continuous functions on the interval $[\alpha, \beta]$ where $0 < \beta - \alpha \le 2\pi$. And let f and g have the property that $0 \le g(\theta) \le f(\theta)$ for all θ in $[\alpha, \beta]$. The area of the region bounded by the curves $r = f(\theta)$, $r = g(\theta)$, $\theta = \alpha$, and $\theta = \beta$ is

$$A = \frac{1}{2} \int_\alpha^\beta [f^2(\theta) - g^2(\theta)]\, d\theta.$$

Arc Length

If $r = f(\theta)$ has a continuous first derivative on the interval $[\alpha, \beta]$, then the arc length of the curve $r = f(\theta)$ from α to β is

$$s = \int_\alpha^\beta \sqrt{r^2 + [df/d\theta]^2}\, d\theta.$$

Formulas Involving Curves Defined Parametrically

Area

If $y(x)$ is a nonnegative, continuous function of x on the interval $[a, b]$, if x and y are defined parametrically by

$$x = f(t), \quad y = g(t) \quad \text{for } t_1 \le t \le t_2$$

where $f(t_1) = a$ and $f(t_2) = b$, and if $f'(t)$ and $g(t)$ are both continuous on the interval $[t_1, t_2]$, then the definite integral of $y(x)$ on the interval $[a, b]$—the area under the curve $y(x)$ over $[a, b]$—is

$$\int_a^b y(x)\,dx = \int_{t_1}^{t_2} g(t)f'(t)\,dt.$$

Arc Length

If a curve is defined parametrically on the interval $[a, b]$ by $x = f(t)$, $y = g(t)$ for $t_1 \le t \le t_2$ where $f(t_1) = a$ and $f(t_2) = b$ and if $f'(t)$ and $g'(t)$ are continuous on the interval $[t_1, t_2]$, then the arc length of the curve over $[a, b]$ is

$$s = \int_{t_1}^{t_2} \sqrt{(dx/dt)^2 + (dy/dt)^2}\,dt = \int_{t_1}^{t_2} \sqrt{[f'(t)]^2 + [g'(t)]^2}\,dt.$$

Example 2 Numerical Calculation of Arc Length, Surface Area, and Volume

a. Find the arc length of the semi-circle $y(x) = \sqrt{9 - x^2}$ over the interval $[1, 2]$.

b. Find the area of the surface generated by revolving the given arc of the semi-circle over the interval $[1, 2]$ about the y-axis.

c. Find the volume of the solid generated by revolving the region bounded by the given arc of the semi-circle, the horizontal line $y = \sqrt{5}$, and the vertical line $x = 1$ about the y-axis.

Solution

a. Differentiating the equation for the semi-circle, we find

$$y'(x) = \frac{-x}{\sqrt{9 - x^2}}.$$

So the arc length is

$$s = \int_1^2 \sqrt{1 + (y')^2}\,dx = \int_1^2 \sqrt{1 + \frac{x^2}{9 - x^2}}\,dx$$

$$= \int_1^2 \sqrt{\frac{9 - x^2 + x^2}{9 - x^2}}\,dx = \int_1^2 \frac{3dx}{\sqrt{9 - x^2}}.$$

We numerically calculated an approximate value for this arc length using computing software by setting $f(x, y) = 3/\sqrt{9 - x^2}$, by inputting $[1, 2]$ for the interval of integration, and by inputting the initial condition $y(1) = 0$. The program output showed

$$\int_1^2 \frac{3dx}{\sqrt{9 - x^2}} \approx 1.169672.$$

Thus, the desired arc length, s, is approximately equal to 1.169672.

b. The surface area of the solid generated by revolving the given arc of the semi-circle about the y-axis is

$$S_y = 2\pi \int_1^2 x\sqrt{1 + (y')^2}\, dx = 6\pi \int_1^2 \frac{xdx}{\sqrt{9 - x^2}}.$$

We calculated an approximate value for the integral $\int_1^2 x\, dx/\sqrt{9 - x^2}$ using computing software by setting $f(x, y) = x/\sqrt{9 - x^2}$, by inputting $[1, 2]$ for the interval of integration, and inputting the initial condition $y(1) = 0$. From the program output, we found

$$\int_1^2 \frac{xdx}{\sqrt{9 - x^2}} \approx .5923588.$$

Hence, the surface area $S \approx 6\pi(.5923588) = 11.16570$.

c. The volume of the solid generated by revolving the given region about the y-axis is

$$V_y = 2\pi \int_1^2 x[y(x) - \sqrt{5}]\, dx = 2\pi \int_1^2 x[\sqrt{9 - x^2} - \sqrt{5}]\, dx.$$

Setting $f(x, y) = x[\sqrt{9 - x^2} - \sqrt{5}]$, inputting $[1, 2]$ as the interval of integration, and inputting the initial condition $y(1) = 0$, our computing software numerically computed

$$\int_1^2 x[\sqrt{9 - x^2} - \sqrt{5}]dx \approx 0.4615936.$$

So, the desired volume is $V_y \approx 2\pi(.4615936) = 2.900278$. ■

Graphs of polar equations of the form $r = a \pm b\cos\theta$ and $r = a \pm b\sin\theta$ are called *limaçons.* If $|a| < |b|$, the limaçon is a closed curve with two loops. (See Figure 3.1.) If $|a| = |b|$, the limaçon is a single heart-shaped loop and is, therefore, called a *cardioid.* (See Figure 3.2.) If $|a| > |b|$, the limaçon is a single, flattened, convex loop. (See Figure 3.3.)

Graphs of polar equations of the form $r = a\cos n\theta$ and $r = a\sin n\theta$ $(n \geq 2)$ are called *rose curves.* If n is odd, the rose curve has n loops (petals); whereas, if n is even the rose curve has $2n$ loops (petals). (See Figures 3.4 and 3.5.) Graphs of $r = a\cos\theta$ and $r = a\sin\theta$ are *circles.* And *lemniscates* are graphs of polar equations of the form $r^2 = a^2\cos 2\theta$ and $r^2 = a^2\sin 2\theta$. (See Figure 3.6.)

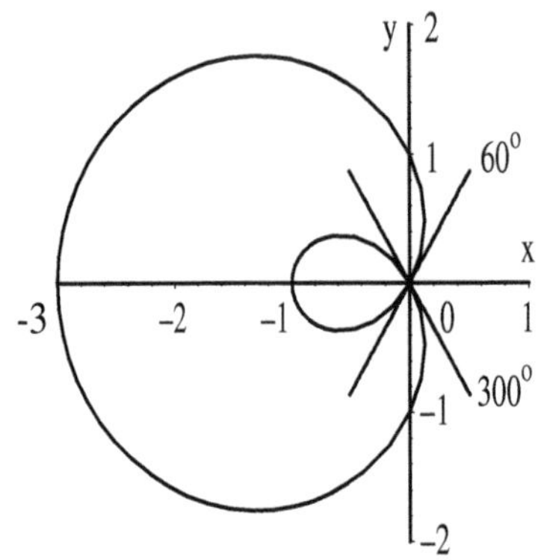

FIGURE 3.1
Limaçon $r = 1 - 2\cos\theta$.

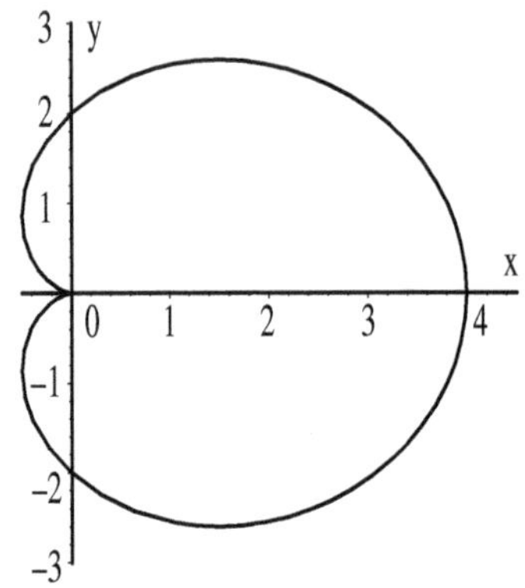

FIGURE 3.2
Cardioid $r = 2 + 2\cos\theta$.

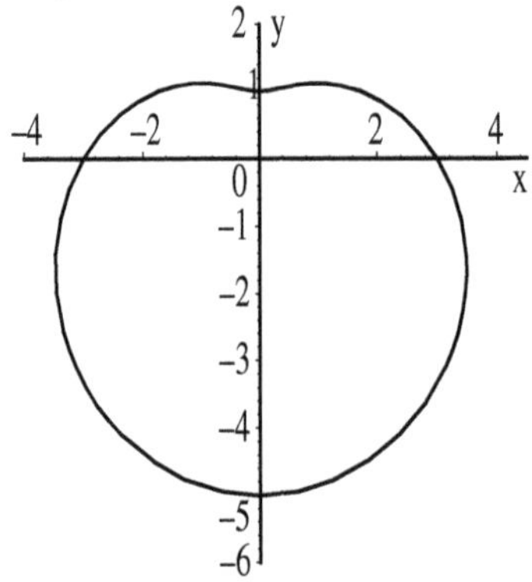

FIGURE 3.3
Limaçon $r = 3 - 2\sin\theta$.

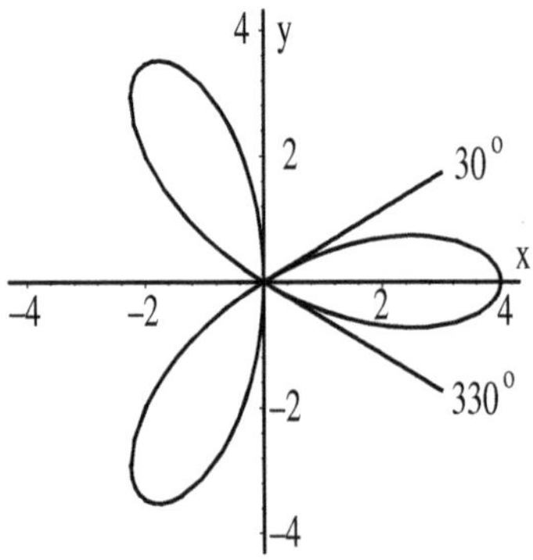

FIGURE 3.4
Rose Curve $r = 4\cos 3\theta$.

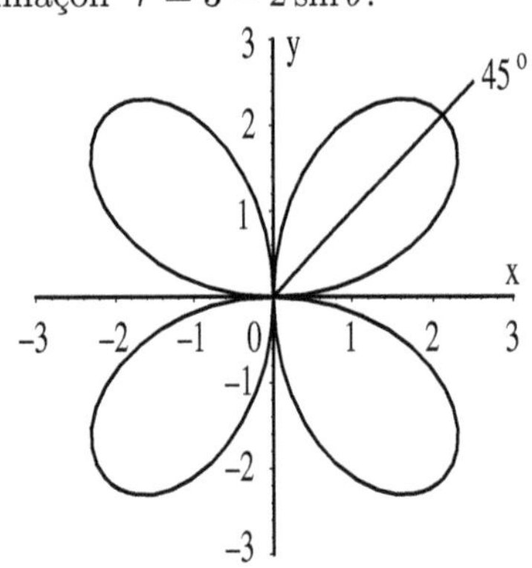

FIGURE 3.5
Rose Curve $r = 3\sin 2\theta$.

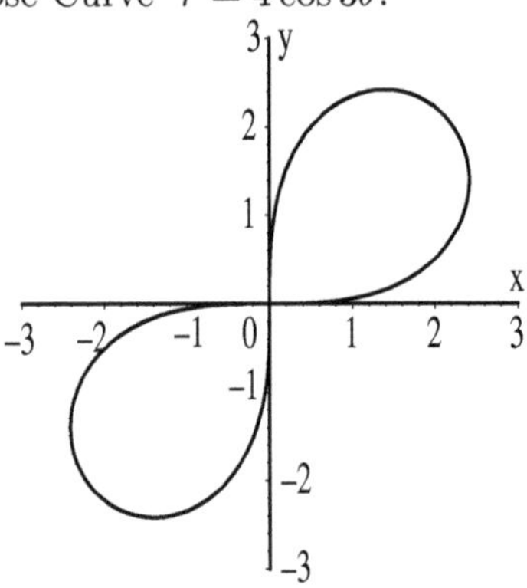

FIGURE 3.6
Lemniscate $r^2 = 9\sin 2\theta$.

3.1.1 Exercises

In the following Exercises approximate π by 3.141593 when necessary.

In Exercises 1–6 use your computer software to evaluate numerically the given definite integrals.

1. $\int_{-1}^{0} \sqrt{x^3+1}\,dx$
2. $\int_{-1}^{1} \sqrt{1+e^{-2x}}\,dx$
3. $\int_{1}^{3} e^{-x^2}\,dx$
4. $\int_{0}^{2} x^x\,dx$
5. $\int_{-1}^{3} \dfrac{e^x}{x}\,dx$
6. $\int_{0}^{\pi} \sqrt{1+\cos x}\,dx$

In Exercises 7–12 use your computer software to calculate the area under the given curve y = f(x) over the given interval [a, b].

7. $f(x) = 1/\ln x$ on $[2,3]$
8. $f(x) = e^x \ln x$ on $[.5,2]$
9. $f(x) = x\tan x$ on $[0,\pi/4]$
10. $f(x) = \ln(\tan x)$ on $[0,\pi/3]$
11. $f(x) = 1/\sqrt{1-x^3}$ on $[-1,0]$
12. $f(x) = \sqrt{1+\sin x}$ on $[0,\pi]$

In Exercises 13–16 numerically calculate the area of the region bounded by the given sets of curves.

13. $y = \sqrt{2+x^2}, \quad y = x^2, \quad x = 0$
14. $y = \sqrt{1+x}, \quad y = (x+1)^2$
15. $y = 1.5^{x^{1.5}}, \quad y = x^{2.3}, \quad x = 0, \quad x = 1$
16. $x = \sin y^2, \quad x = y/(y^3+1), \quad y = 0, \quad y = 1.2$

In Exercises 17–22 numerically compute the arc length of the given curve y = f(x) over the given interval [a, b].

17. $f(x) = x^3$ on $[0,1]$
18. $f(x) = 1/x$ on $[.5,1.5]$
19. $f(x) = 3x^2 + \dfrac{1}{6x^2}$ on $[1,2]$
20. $f(x) = \ln(1+e^x)$ on $[0,1.5]$
21. $f(x) = \dfrac{\sin x}{x}$ on $[.5,2]$
22. $f(x) = x\tan x$ on $[0,\pi/4]$

In Exercises 23–26 calculate numerically the area of the surface generated by revolving the given curve y = f(x) over the given interval [a, b] about (a) the x-axis and (b) the y-axis.

23. $f(x) = 2x^2$ on $[0,1]$
24. $f(x) = \sqrt{x}$ on $[1,2]$
25. $f(x) = \sin x$ on $[0,\pi]$
26. $f(x) = \dfrac{\sin x}{x}$ on $[1,2]$

27. Find the surface area and volume of the ellipsoid obtained by revolving the ellipse $x^2/16 + y^2/25 = 1$ (a) about the x-axis and (b) about the y-axis.

28. Find the surface area and volume of the hyperboloid obtained by revolving the portion of the hyperbola $y^2/25 - x^2/16 = 1$ between $x = -4$ and $x = 4$ (a) about the x-axis and (b) about the y-axis.

29. Find the surface area and volume of the torus obtained by revolving the circle $(x-1)^2 + (y-2)^2 = 1$ (a) about the x-axis and (b) about the y-axis.

In Exercises 30–34 numerically calculate the volume of the solid obtained by revolving the region bounded by the given set of curves (a) about the x-axis and (b) about the y-axis.

30. $y = \sin x^2, \quad y = 0, \quad x = 0, \quad x = \pi/4$

31. $y = \cos x, \quad y = \tan x, \quad x = 0$

32. $y = \ln x, \quad y = 0, \quad x = 1, \quad x = 2$

33. $y = 3^x, \quad y = 3x^3, \quad x = 0$

34. $x^{2/3} + y^{2/3} = 1$

35. The *spiral of Archimedes* is given by the equation $r = a\theta$ where a is a constant. For $a = 3$ find the area inside the spiral of Archimedes from $\theta = 0$ to $\theta = 2\pi$. For $a = 4$ find the arc length of the spiral of Archimedes from $\theta = 0$ to $\theta = \pi$.

36. Find the area inside the *logarithmic spiral* $r = 2e^{3\theta}$ from $\theta = 0$ to $\theta = \pi$. Find the arc length of this spiral.

37. Find the area and arc length of the cardioid $r = 2 + 2\cos\theta$. (See Figure 3.2.)

38. Find the area and arc length of the limaçon $r = 3 - 2\sin\theta$. (See Figure 3.3.)

39. Find the area and arc length of the rose curve $r = 4\cos 3\theta$. (See Figure 3.4.)

40. Find the area and arc length of the lemniscate $r^2 = 9\sin 2\theta$. (See Figure 3.6.)

41. Find the area between the two loops of the limaçon $r = 1 - 2\cos\theta$. Find the arc length of each loop of the limaçon. (See Figure 3.1.)

42. Find the area inside both lemniscates $r^2 = \cos 2\theta$ and $r^2 = \sin 2\theta$.

43. Find the area between the two limaçons $r = 5 + 3\cos\theta$ and $r = 2 - \sin\theta$.

44. About 1638, René Descartes (1596-1650) sent the equation $x^3 + y^3 = 3xy$ to Pierre Fermat (1601-1665) and challenged him to determine the tangent line to the curve at any point. (Can you determine the tangent line at any point?) The graph of the equation $x^3 + y^3 = 3xy$ is called the *folium of Descartes*. A sketch of the graph is shown in Figure 3.7. The equation may be rewritten in polar coordinates as $r = 3\sin\theta\cos\theta/(\cos^3\theta + \sin^3\theta)$. Find the area and arc length of the loop ("leaf") in the folium of Descartes.

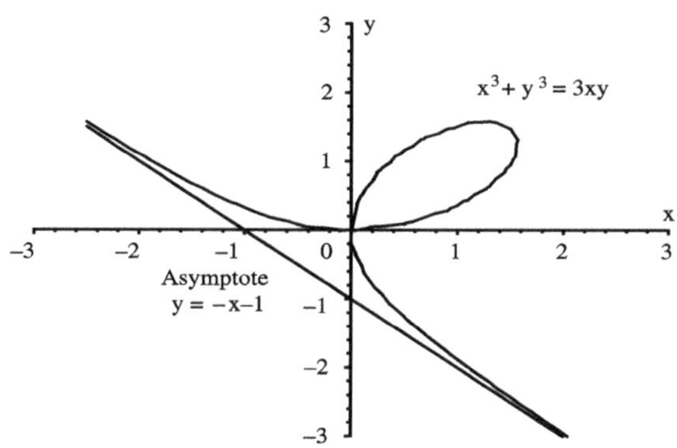

FIGURE 3.7
Graph of the Folium of Descartes.

45. Find the area and arc length of the ellipse $x = 4\cos t$, $y = 5\sin t$.

46. The curve traced out by a point P on the circumference of a circle of radius R as the circle rolls along the x-axis is called a *cycloid*. If when $t = 0$, the point P is at the origin, then the equation of the cycloid is

$$x = R(t - \sin t), \qquad y = R(1 - \cos t).$$

a. Find the area under one arch of the cycloid with $R = 2$.

b. Find the arc length of one arch of the cycloid with $R = 3$.

47. The curve traced out by a point P on the circumference of a circle of radius r which rolls around on the *inside* of a larger circle of radius R is called a *hypocycloid*. If the larger circle has equation $x^2 + y^2 = R^2$ and if when $t = 0$ the point P is at $(R, 0)$, then the equation of the hypocycloid is

$$x = (R - r)\cos t + r\cos\frac{(R - r)t}{r}, \qquad y = (R - r)\sin t - r\sin\frac{(R - r)t}{r}.$$

Find the area and arc length of the hypocycloid with $R = 3$ and $r = 1$.

48. The curve traced out by a point P on the circumference of a circle of radius r which rolls around the *outside* of a circle of radius R is called an *epicycloid.* If the larger circle has equation $x^2+y^2=R^2$ and if when $t=0$ the point P is at $(R,0)$, then the equation of the epicycloid is

$$x=(R+r)\cos t-r\cos\frac{(R+r)t}{r}, \qquad y=(R+r)\sin t-r\sin\frac{(R+r)t}{r}.$$

Find the area and arc length of the epicycloid with $R=3$ and $r=1$.

49. The *involute of a circle* is the curve traced out by a point P at the end of a string which is being unwound tautly from about a circle. If the equation of the circle is $x^2+y^2=R^2$ and if when $t=0$ the point P is at $(R,0)$, then the equation of the involute of the circle is

$$x=R(\cos t+t\sin t), \qquad y=R(\sin t-t\cos t).$$

Find the arc length of the involute of a circle with $R=2$ for $t=0$ to $t=2\pi$.

50. Find the arc length of the epitrochoid

$$x=3\cos 2\theta+4\cos\theta, \qquad y=3\sin 2\theta+4\sin\theta.$$

3.2 Learning Theory Models

Psychologists have studied the process of learning extensively. We will now derive a few simple models of the memorization process. Let A be the total amount of material to be memorized. Psychologists refer to the amount of material memorized at time t as the **attainment**. Let $y(t)$ denote the attainment. In the simplest model of memorization, it is assumed that the rate of change of attainment is proportional to the amount of material that remains to be memorized. Thus, in mathematical symbolism

$$y'(t)=k(A-y(t))$$

where $k>0$ is the constant of proportionality which indicates the natural learning ability of the particular subject (person or perhaps animal). We assume when the memorization process begins at $t=0$, the subject has not memorized any material. So, $y(0)=0$. Hence, for the simplest memorization model, we need to solve the initial value problem

$$y'=k(A-y); \quad y(0)=0. \tag{1}$$

Example 1 A Simple Learning Theory Model Solve the initial value problem (1) numerically and graph the "learning curve" (the solution to the IVP (1)) for a subject whose natural learning ability is $k = .06$ items/minute, if the number of things to be memorized is $A = 50$.

Solution

We ran our computing software by setting $f(x, y) = .06(50 - y)$ and setting the left endpoint of the interval of integration $a = 0$. We needed a reasonable value for the right endpoint of the interval of integration b. Since $0 \le y(t) \le 50$ and $y'(t) = .06(50 - y(t)) = 3 - .06y(t)$, $y'(t)$ satisfies the inequality $0 \le y'(t) \le 3$. Dividing the total number of items to be memorized, A, by the conservative estimate of the rate of learning, $y' = 1$, we decided to set $b = A/1 = 50$. At the start of the experiment, the subject has learned nothing, so the initial condition we used was $y(0) = 0$. A graph of the solution is shown in Figure 3.8. Observe that $y(t)$ increases rapidly for t near zero and $\lim_{t\to\infty} y(t) = 50 = A$. Thus, the largest amount of material is memorized at the beginning of the learning experience and in a relatively short period of time. As t increases and maximum attainment, A, is approached, the rate of attainment, y', becomes small. This phenomenon is called the "law of diminishing returns." Displaying our solution values on the monitor, we saw that 25 items were memorized in the first 11.6 minutes and a total of 37 items were memorized in 22.5 minutes. So it took almost as long to memorize items 26 through 37 (12 items) as it did to memorize the first 25 items. Thus, later in the learning experience it takes a longer period of time to learn the same number of items.

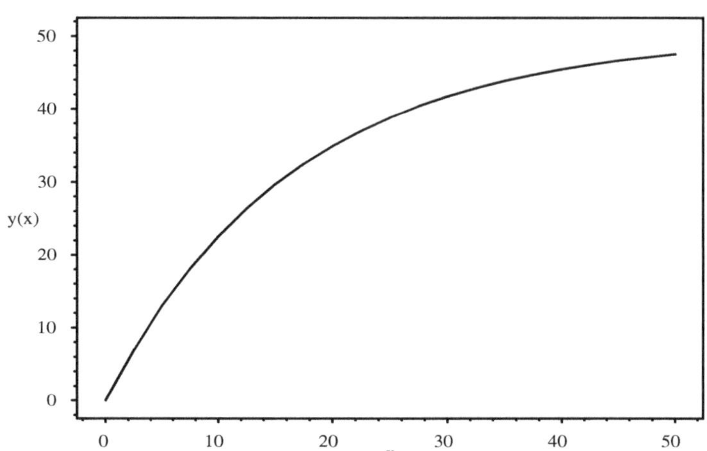

FIGURE 3.8
Graph of the "Learning Curve": $y' = 3 - .06y; y(0) = 0$. ■

3.2.1 Exercises

1. Find the general solution of the differential equation $y' = k(A - y)$ by any method.

2. Company XYZ tested two employees W_1 and W_2 to see who was better suited to perform a particular job based upon the IVP (1). For $i = 1, 2$ the explicit solution of the differential equation in the IVP (1) is $y_i'(t) = A_i + C_i e^{-k_i t}$. The employee who achieves the larger A_i value is better suited to perform the job. At the end of the first hour worker W_1 had performed 25 tasks and at the end of the second hour W_1 had performed 45 tasks. At the end of the first hour worker W_2 had performed 35 tasks and at the end of the second hour W_2 had performed 50 tasks. Which worker, W_1 or W_2, is better suited to perform the job?

3. When the learning process is lengthy, k is not a constant but a function which decreases with time because the subject becomes fatigued or distracted. Suppose $k(t) = k_0/(1 + .05t)$ where k_0 is a constant. Numerically solve and graph the learning curve for the initial value problem

   $$(2) \qquad y' = \frac{k_0(A - y)}{1 + .05t}; \qquad y(0) = 0$$

   with $k_0 = .06$ items/minute and $A = 50$ items on the interval $[0, 50]$. When has the subject memorized 25 items? 37 items? Compare these results with the corresponding answers for Example 1.

4. When a subject learns, he also forgets. In order to include the process of forgetting in our model of the memorization process, we assume the rate at which a subject forgets is proportional to the amount already learned. Thus, we wish to solve the following initial value problem

   $$(3) \qquad y' = k(t)(A - y) - By; \qquad y(0) = 0$$

   where B is a positive constant. The term $-By$ represents the process of forgetting and since $B > 0$ it produces a decrease in attainment. Numerically solve and graph the learning curve of the initial value problem

   $$y' = \frac{.06(50 - y)}{1 + .05t} - .02y; \qquad y(0) = 0$$

   on the interval $[0, 50]$. When does $y(t) = 25$? When does $y(t) = 37$? Compare your results with the results for Example 1 and Exercise 3. What is the maximum value of $y(t)$?

3.3 Population Models

A central figure in the history of population growth modeling is Thomas Robert Malthus (1766-1834). Malthus was the second of eight children of an English country gentleman. He graduated from Cambridge University and in 1788 was ordained as a minister in the Church of England. In the first edition of his essay on population, which was published in 1798, Malthus noted that the population of Europe doubled at regular intervals. Further research indicated that the rate of increase of the population was proportional to the present population. So, Malthus formulated what we presently call the **Malthusian population model:**

(1) $$\frac{dP}{dt} = kP \qquad \text{where } k > 0.$$

In this model, $P(t)$ represents the population at time t and the positive constant of proportionality k represents the "growth rate" per individual. If at some time t_0 we determine that the population has size P_0, then the solution at any time t of the differential equation (1) which satisfies the initial condition $P(t_0) = P_0$, is

(2) $$P(t) = P_0 e^{k(t-t_0)}.$$

Verify that (2) satisfies the initial value problem

(3) $$\frac{dP}{dt} = kP; \qquad P(t_0) = P_0$$

for any constant k. Observe that k need not be positive. If we let b represent the "birth rate" per individual and d represent the "death rate" per individual, then the constant $k = b - d$ represents the "growth rate" per individual of the population. A graph of the solution (2) of the initial value problem (3) for $k > 0$, $k = 0$, and $k < 0$ is shown in Figure 3.9. Observe for $k > 0$ the population grows exponentially and is unbounded—that is, as $t \to \infty$, $P(t) \to \infty$. For $k = 0$ the population maintains the constant value P_0. And for $k < 0$ the population decreases exponentially to zero.

For $k > 0$ the Malthusian population model indicates there is no limit to the size of the population. This is totally unrealistic. Nonetheless, for short intervals of time, exponential growth of a population is possible when the population has enough room in which to expand and an abundance of food and other natural resources to support its expansion. For example, the population of the United States from 1800 through 1860 can be modeled approximately by the function

(4) $$P(t) = P_0 e^{.03(t-1800)}$$

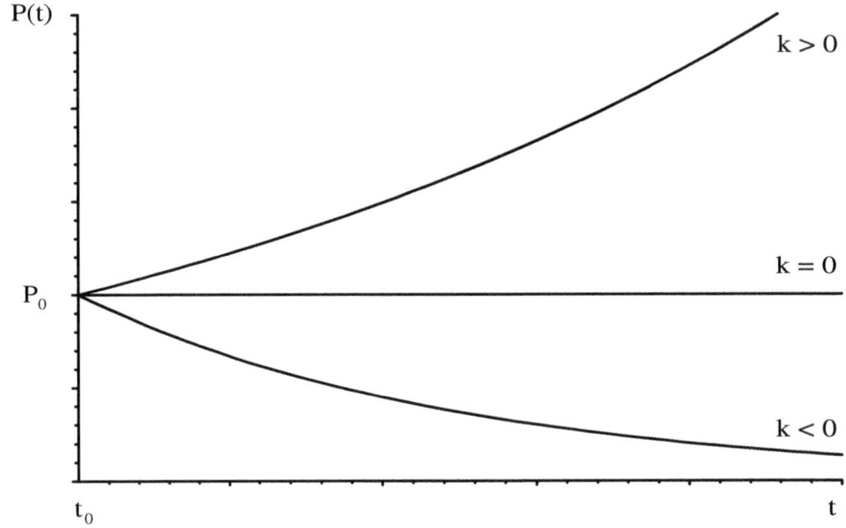

FIGURE 3.9
$P(t) = P_0 e^{k(t-t_0)}$ for $k > 0, k = 0$, and $k < 0$.

where $P_0 = 5.31$ million is the population of the United States in 1800. Of course, the function (4) is the solution of the Malthusian population model

$$\frac{dP}{dt} = .03P; \qquad P(1800) = 5.31 \text{ million.}$$

A comparison of the predicted population and the actual population of the United States from 1800 through 1900 is given in Table 3.1. As we can see from this table, the function (4) does not model the population growth of the United States from 1870 to 1900 very well. This is due to the fact that the Malthusian population model ignores important factors such as the change in the birth or death rate with time, wars, disease, immigration, emigration, and changes in the age structure of the population. (What happened in the United States between 1860 and 1870 which might account for the cessation of exponential growth?)

For $k < 0$ the Malthusian population model predicts that the population will decrease exponentially to zero. In 1946, the pesticide DDT began to be

TABLE 3.1
United States Population (in millions), 1800-1900

Year	1800	1810	1820	1830	1840	1850	1860	1870	1880	1890	1900
Actual	5.31	7.24	9.64	12.87	17.07	23.19	31.44	38.56	50.16	62.95	76.00
Predicted	5.31	7.17	9.68	13.06	17.63	23.80	32.12	43.36	58.53	79.01	106.65
Error (%)	0.00	-.97	.41	1.48	3.28	2.63	2.16	12.44	16.69	25.51	40.33

used extensively throughout the United States. The following year a decrease in the population of peregrine falcons was noted. By 1970, due to the use of DDT and other similar pesticides, the peregrine falcon was nearly extinct in the continental United States. This particular tragedy illustrates that it is currently possible for humans to alter the death rate of a species so that it becomes greater than the birth rate thereby producing a dramatic population decrease. Similar events may occur in the future if the delicate balance of environmental factors are not properly considered before various courses of action are pursued.

In his text of 1835, Lambert Quetelet (1796-1874) criticized Malthus and others who studied population growth for not establishing their results within a more mathematical framework. Quetelet advanced the theory that populations tend to grow geometrically, but the resistance to growth increases in proportion to the square of the velocity with which the population tends to increase. He drew an analogy between population growth and the motion of a body through a resisting medium; however, he presented no mathematical treatment of the problem. Pierre-François Verhulst (1804-1849) studied under Quetelet in Ghent. In memoirs published in 1838, 1845, and 1847, Verhulst developed his "logistic growth" model for populations. He was frustrated in his attempts to verify the model because no accurate census information was available at that time. Verhulst's population model lay dormant for approximately eighty years. It was independently rediscovered in the early 1920s by two American scientists, Raymond Pearl (1879-1940) and Lowell Reed (1886-1966). The **logistic law model** or **Verhulst–Pearl model** for population growth is

$$\frac{dP}{dt} = kP - \epsilon P^2$$

where k and ϵ are positive constants and ϵ is small relative to k. When the population P is small, the term ϵP^2 is very small compared to kP and so the population will grow at nearly an exponential rate. However, as the population becomes large, the term ϵP^2 will approach the term kP in size and the rate of population growth will approach zero. Using the techniques of separation of variables and partial fraction decomposition, it can be shown that the explicit solution of the initial value problem

$$\frac{dP}{dt} = kP - \epsilon P^2; \quad P(t_0) = P_0 \tag{5}$$

is

$$P(t) = \frac{kP_0}{\epsilon P_0 + (k - \epsilon P_0)e^{-k(t-t_0)}} \tag{6}$$

where P_0 is the size of the population at time t_0. Notice that as $t \to \infty$, $P(t) \to k/\epsilon$. So, regardless of the initial population size, the population ultimately approaches the limiting value of k/ϵ. Consequently, the constants k and ϵ are called the **vital coefficients** of a population and the constant

$K = k/\epsilon$ is called the **carrying capacity** of the population. The graph of equation (6) has an elongated S-shape and is called the **logistic curve**. See Figure 3.10.

Experts have estimated that the earth's human population has a vital coefficient $k = .029$. Given that the population of the world in 1960 was 3 billion people and the growth rate, $(dP/dt)/P$, was 1.8% per year, we can determine the vital coefficient ϵ in the following manner. Dividing the differential equation of (5) by P, we find that the growth rate expressed as a percentage satisfies

$$\frac{dP/dt}{P} = k - \epsilon P.$$

Substituting the estimated value for k and the known 1960 values given above for P and $(dP/dt)/P$, we find that the vital coefficient ϵ must satisfy

$$.018 = .029 - \epsilon(3 \times 10^9).$$

Solving this equation for ϵ yields $\epsilon = 3.667 \times 10^{-12}$. Using these vital coefficients, the logistic law model predicts a limiting value for the human population of the earth of

$$K = k/\epsilon = .029/(3.667 \times 10^{-12}) = 7.91 \times 10^9 \text{ people.}$$

In 2000, the size of the population reached 6.08×10^9. The graph displayed in Figure 3.10 is the logistic curve for the earth's human population obtained from equation (6) by setting $P_0 = 3$ billion, $t_0 = 1960$, $k = .029$, and $\epsilon = 3.667 \times 10^{-12}$.

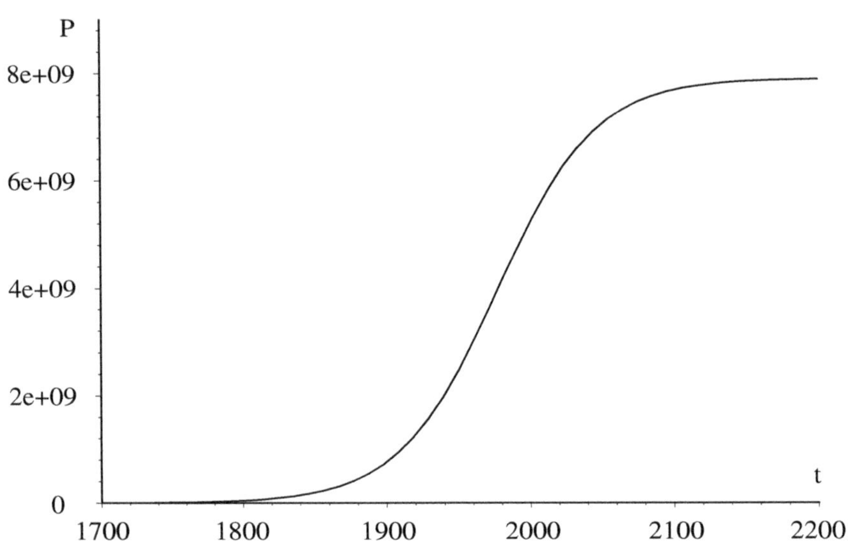

FIGURE 3.10
Graph of the "Logistic Curve" for the Earth's Human Population.

3.3.1 Exercises

1. Assume the human population of the earth obeys the Malthusian population model. In 1650 A.D. the earth's human population numbered 2.5×10^8. By 1950 A.D. the population had grown to 2.5×10^9. Write a general equation for the population of the earth. (Hint: Use the information given to determine the constants P_0, k, and t_0 in equation (2). No computer is needed.) Suppose further that the earth can support at most 2.5×10^{10} people. When will this limit be reached?

2. If a population is being harvested by a predator or dying due to a disease at a constant rate $H > 0$, then a possible model for the population based on a modification of the Malthusian model is

$$\frac{dP}{dt} = kP - H; \qquad P(0) = P_0.$$

   Suppose for a specific population $k = .1$, and the initial population size is 500. Use your computer software to compute and graph the population on the interval $[0, 10]$ for the following values of the harvesting constant:

   a. $H = 40$ b. $H = 50$ c. $H = 60$

   What do you think will happen to $P(t)$ as t increases in each case?

3. The logistic law model with $k = .03134$ and $\epsilon = 1.589 \times 10^{-10}$ pro- vides a model for the population of the United States. Based on this model, what is the limiting population of the United States? Assume that in 1800 the population of the United States was 5.31 million (5.31×10^6). Use your computer software to compute the population of the United States on the interval $[1800, 1900]$. Compare your results with the actual results given in Table 3.1. Graph the solution. (Hint: In the initial value problem (5) let $t_0 = 0$ correspond to the year 1800. So that $t = 10$ corresponds to 1810, $t = 20$ corresponds to 1820, and so forth. Thus, the interval of integration becomes $[0, 100]$ instead of $[1800, 1900]$. Also let the population be expressed in millions. That is, let $P(t) = p(t) \times 10^6$. Substituting for P in the differential equation of (5), we find $d(p \times 10^6)/dt = k(p \times 10^6) - \epsilon(p \times 10^6)^2$. Dividing by 10^6, we see that the population expressed in millions satisfies the differential equation $dp/dt = kp - \epsilon p^2(10^6)$ and the initial condition is $p(0) = 5.31$. Thus, in this example, you have translated the independent variable, t, and scaled the dependent variable, P.)

4. If a population which is known to grow according to the logistic law model begins to be harvested at a constant rate, H, at time t_0, then for $t > t_0$ the population satisfies the following initial value problem

$$\frac{dP}{dt} = kP - \epsilon P^2 - H; \qquad P(t_0) = P_0$$

where k, ϵ, and H are known positive constants, and P_0 is the population size at time t_0. Suppose when the population reaches the size $P_0 = 200$ a catfish farmer decides to harvest fish from his pond at a constant rate H. And suppose for his species of catfish the vital coefficients of the population are $k = .2$ and $\epsilon = .0004$. Numerically compute and graph the population of the catfish on the interval $[0, 25]$ for

a. $H = 10$ b. $H = 25$ c. $H = 50$.

What do you think will happen to the catfish population as time increases in each case? That is, what is $\lim_{t\to\infty} P(t)$ in each case?

5. Suppose the catfish farmer in Exercise 4 waits to start harvesting fish until the population size reaches 1000 instead of 200. Compute and graph the catfish population on the interval $[0, 25]$ for

   a. $H = 10$ b. $H = 25$ c. $H = 50$.

   What do you think will happen to the catfish population in each case now?

Several other population growth models have been proposed by various researchers. In the Exercises which follow we present some of these models.

6. The following population growth model, which is sometimes used in actuarial predictions, was proposed by Gompertz:

$$\frac{dP}{dt} = P(a - b\ln P); \qquad P(0) = P_0.$$

   Numerically compute and graph the solution to this initial value problem on the interval $[0, 25]$ for $P_0 = 75$, $a = .15$, and (i) $b = .03$ and (ii) $b = -.03$. What is $\lim_{t\to\infty} P(t)$ in each case?

7. The following two initial value problems are variations of the logistic law model. Compute and graph numerical solutions to these problems on the interval $[0, 25]$. However, before computing a solution see if you can guess whether the population will increase or decrease in each case.

   (a) $\dfrac{dP}{dt} = .2P\ln P - .02P^2; \qquad P(0) = 1$

   (b) $\dfrac{dP}{dt} = \dfrac{.2P}{\ln P} - .02P^2; \qquad P(0) = 2$

8. In 1963, F. E. Smith proposed the following model to explain the population growth of a species of water fleas under laboratory conditions.

$$\frac{dP}{dt} = \frac{.2P - .02P^2}{1 + .01P}; \qquad P(0) = 1.$$

Use your computer software to compute and graph the solution of this initial value problem on the interval $[0, 25]$. How does the solution behave as time increases? Does $\lim_{t\to\infty} P(t)$ exist? If so, what is it?

9. Another proposed modification of the logistic law model is

$$\frac{dP}{dt} = P(a - bP)(1 - \frac{c}{P}); \qquad P(0) = P_0$$

where $a, b, c > 0$. Compute and graph a numerical solution to this initial value problem on the interval $[0, 25]$ for $a = .3$, $b = .005$, $c = .02$, and $P_0 = 100$.

10. Compute and graph numerical solutions of the following two initial value problems on the interval $[0, 25]$. Before computing the solutions see if you can guess the behavior of the solution curves. Do they increase? Do they decrease? Do they increase and then decrease? Or, do they decrease and then increase? Does $\lim_{t\to\infty} P(t)$ exist? If so, what is it?

a. $\dfrac{dP}{dt} = .15P\ln(\dfrac{100}{P}); \qquad P(0) = 25$

b. $\dfrac{dP}{dt} = -.15P(1 - \dfrac{100}{P}); \qquad P(0) = 25$

3.4 Simple Epidemic Models

One of the greatest causes of human suffering and misery is the outbreak of various types of epidemics. Over the ages, numerous deaths have been caused by epidemics and frequently, a large proportion of a community has perished. Today well-developed countries are relatively free from the threat of death producing epidemics; however, epidemics of influenza still occur occasionally. Less well-developed countries in Africa and the Far East are still susceptible to lethal epidemics.

One of the earliest recorded accounts of an epidemic occurred in Athens from 430 B.C. to 428 B.C. The "Golden Age" of Athens coincided closely with the reign of Pericles, who rose to power in 469 B.C. The Peloponnesian War began in 431 B.C. and lasted nearly one quarter of a century. The war was essentially a conflict between the Greek city-states of Athens and Sparta. Athens had a strong navy and a weak army, while Sparta had a strong army and a weak navy. Pericles decided to bring the people from the areas surrounding Athens into the fortified city to protect his state from attack by land. Simultaneously, he had his navy attack the coastal areas of the Spartan state. This strategy worked well during the first year and it appeared Athens would soon win the war. However, overcrowding and unsanitary conditions produced an outbreak of a highly contagious disease. Victims of the disease usually died within six to eight days after being infected and all told between 30% and 60% of the population of Athens perished due to the disease. When the epidemic of 430 B.C. ended, Pericles sent his navy to capture the Spartan stronghold of Potidaea. The dreaded disease struck the crews while they were still at sea and forced them to return to Athens prior to accomplishing their mission. The epidemic swept through Athens again in 429 B.C. and in 428 B.C. when Pericles died from it. The war itself continued for several more years and finally the Spartans, aided by the epidemics, were able to defeat the Athenians.

In the 14th century an epidemic of the bubonic plague killed approximately one-fourth of the population of Europe, which was estimated to number 100 million. In 1520 a smallpox epidemic caused the death of half of the Aztec population of 3.5 million. An epidemic of measles on the island of Fiji in 1875 resulted in the death of 40,000 people out of a population of 150,000. From 1918 to 1921, there was a typhus epidemic in the Soviet Union which killed approximately 2.5 million people. In 1919 in a worldwide influenza epidemic, an estimated 20 million people perished from the disease or the complication of pneumonia.

In the 17th century John Graunt (1620-1674) and Sir William Petty (1623-1687) collected information on the incidence and location of epidemics. On April 30, 1760, Daniel Bernoulli presented a paper to the *Academie Royale des Sciences* in which a mathematical model was used for the first time to study the population dynamics of infectious disease. Bernoulli was investigating mortality due to smallpox and trying to assess the risks and advantages of preventive inoculation. In his mathematical model, Bernoulli formulated and solved a relevant differential equation. He evaluated the results in terms of the value of preventive inoculation. The modern mathematical theory of epidemics originated in the works of William Hamer and Sir Ronald Ross which appeared early in the 20th century.

In the simplest epidemic model, we assume the population size has the constant value of N. Thus, we assume there are no births and no immigration to increase the population size, and we assume there are no deaths and no emigration to decrease the population size. Since the time span of an epidemic is short (usually a few weeks or months) in comparison to the life span of a person, the assumption of a constant population size is fairly reasonable. Next, we assume the population is divided into two mutually exclusive sets: The **infectives** is the set of people who are infected with and capable of transmitting the disease. The **susceptibles** are the set of people who do not have the disease but may become infected later. We denote the number of infectives at time t by $I(t)$ and the number of susceptibles at time t by $S(t)$. Under the assumptions we have made

$$(1) \qquad I(t) + S(t) = N \quad \text{for all } t.$$

An assumption which was first made by William Hamer in 1906, and which has been included in every deterministic epidemic model ever since, is that the rate of change of the number of susceptibles is proportional to the product of the number of susceptibles and the number of infectives, which represents the rate of contact between susceptibles and infectives. Thus, it is assumed that

$$\frac{dS}{dt} = S'(t) = -\beta S(t)I(t) \quad \text{for all } t \tag{2}$$

where $\beta > 0$ is called the **infection rate**. Solving equation (1) for $I(t)$ and substituting into equation (2), we obtain the following differential equation for the number of susceptibles

$$S'(t) = -\beta S(t)(N - S(t)). \tag{3}$$

Differentiating (1), we see that

$$I'(t) + S'(t) = 0.$$

So the number of infectives satisfies the differential equation

$$I'(t) = -S'(t) = \beta S(t)I(t) = \beta(N - I(t))I(t) \tag{4}$$

since $S(t) = N - I(t)$ from (1). The differential equations (3) and (4) can both be solved explicitly by the technique of separation of variables, or one of these differential equations can be solved and the remaining function ($I(t)$ or $S(t)$) can be determined from equation (1). Observe that equation (4), $I' = \beta NI - \beta I^2$, has the same form as the logistic law model, $P' = kP - \epsilon P^2$, which we studied in the previous section.

Example 1 Solution of a Simple Epidemic Model with Only One Initial Infective Solve numerically and graph the solution of the differential equation

$$I'(t) = \beta(N - I(t))I(t) \tag{4}$$

for the number of infectives, $I(t)$, on the interval $[0, 10]$ if $\beta = .002$, $N = 1000$, and $I(0) = 1$. What is $\lim_{t\to\infty} I(t)$? What is $\lim_{t\to\infty} S(t)$?

Solution

We associated $I(t)$ with $y(x)$ and ran our computing software by setting $f(x, y) = .002(1000 - y)y$, by inputting the interval of integration as $[0, 10]$, and by inputting the initial condition $y(0) = 1$. A graph of the solution is shown in Figure 3.11. Notice that as t approaches 6, $I(t)$ approaches 1000. And since $I(t) + S(t) = N = 1000$, as t approaches 6, $S(t)$ approaches 0.

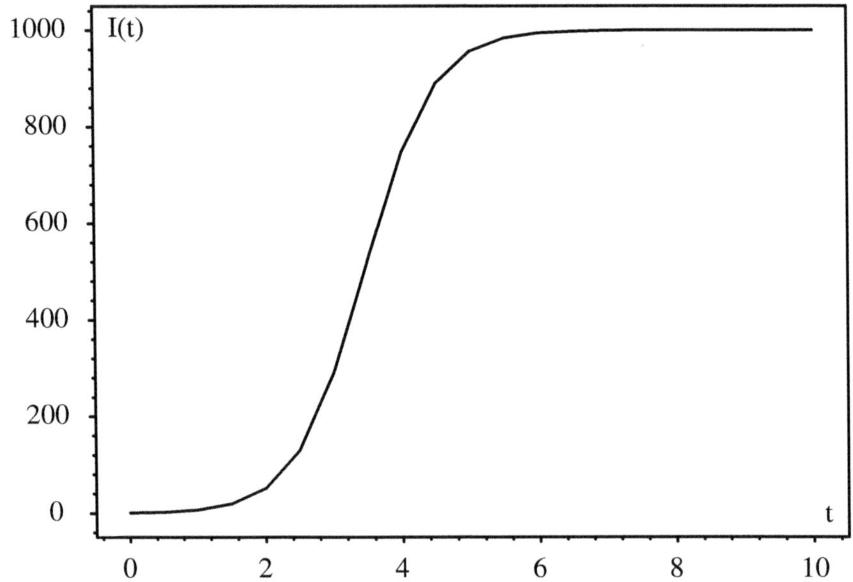

FIGURE 3.11
Numerical Approximation to the IVP: y' = .002(1000-y)y; y(0) = 1. ■

Example 1 illustrates a common characteristic of all epidemic models based solely on equations (1) and (2)—or equivalently on equation (3) or (4)—namely, regardless of how few infectives there are initially, all members of the population acquire the disease before the epidemic ends. This is a shortcoming of this particular model. By definition an epidemic ends when there are no new infectives over a certain period of time. When a real-life epidemic ends there are still many susceptibles in the population. Therefore, the model we have just studied must be modified in order to improve the results so they more closely reflect the results observed in real life. We will study such models in Section 10.6.

3.4.1 Exercises

1. When an epidemic is discovered, steps are normally taken to prevent its spread. Suppose health officials begin inoculations at a rate of $i'(t)$ per unit of time and this procedure is continued until the epidemic ends. As before, we let $I(t)$ denote the number of infectives at time t and $S(t)$ denote the number of susceptibles at time t. We denote the number of people who have been inoculated at time t by $i(t)$. Under these assumptions,

(5) $$I(t) + S(t) + i(t) = N \quad \text{for all } t.$$

In this case, it is assumed that

$$\frac{dS}{dt} = S'(t) = -\beta S(t)I(t) - i'(t) \quad \text{for all } t \tag{6}$$

where $\beta > 0$ is the infection rate. The first term on the right-hand side of equation (6) represents the rate of decrease in susceptibles due to contact between susceptibles and infectives while the second term represents the rate of decrease in susceptibles due to the inoculation of susceptibles. Differentiating equation (5) and solving the resulting equation for $I'(t)$, we find

$$I'(t) = -S'(t) - i'(t). \tag{7}$$

Substituting for $S'(t)$ from equation (6), we obtain

$$I'(t) = \beta S(t)I(t). \tag{8}$$

Solving equation (5) for $S(t)$ and substituting the result into equation (8), yields the following differential equation for the number of infectives

$$I'(t) = \beta(N - i(t) - I(t))I(t). \tag{9}$$

Recall that an epidemic ends at time t^* when $I(t^*) = 0$. In this case, when the epidemic ends, the entire population is divided into two mutually exclusive sets—the infected and the inoculated. Use your computer software to compute and graph the solution of the differential equation (9) on the interval $[0, 10]$, if $I(0) = 1$, $\beta = .002$, $N = 500$, the inoculation function $i(t) = \alpha t$ (this function indicates that health personnel are able to inoculate α units of people per unit of time), and (a) $\alpha = 30$ and (b) $\alpha = 50$. In each case, what is the duration of the epidemic, t^*, and what is the maximum number of infectives at that time, $I(t^*)$?

2. Suppose that the inoculation procedure (see Exercise 1) accelerates over time so that $i(t) = t + 30$. Then the differential equation for the number of infectives becomes

$$I'(t) = \beta[N - (t + 30)t - I(t)]I(t). \tag{10}$$

Compute and graph numerical solutions of the differential equation (10) on the interval $[0, 10]$, if $I(0) = 1$, $\beta = .002$, and $N = 500$. What is the duration of the epidemic, t^*, and the maximum number of infectives at that time, $I(t^*)$?

3.5 Falling Bodies

On April 28, 1686, Isaac Newton presented and dedicated to the Royal Society the first volume of a three-volume work which he chose to title *Philosophiea*

Naturalis Principia Mathematica (*Mathematical Principles of Natural Philosophy*). This work is almost always referred to simply as the *Principia*. The first volume deals mainly with various problems of motion under the idealized conditions of no friction and no resistance. In Book I of the *Principia*, Newton formulated his three laws of motion, and in Book III, he expanded upon them.

Newton's Laws of Motion

Newton's three laws of motion are as follows:

> **First Law of Motion:** "A body at rest tends to remain at rest, while a body in motion tends to remain in motion in a straight line with constant velocity unless acted upon by an external force."
>
> **Second Law of Motion:** "The rate of change of momentum of a body is proportional to the force acting on the body and is in the direction of the force."
>
> **Third Law of Motion:** "To every action there is an equal and opposite reaction."

If we let m denote the mass of a body and v denote its velocity, then the momentum of the body is mv. Hence, Newton's second law of motion may be expressed mathematically as

$$\frac{d(mv)}{dt} = kF \tag{1}$$

where k is a constant of proportionality, and F is the magnitude of the force acting on the body. If we make the assumption that the mass of the body is a constant and that the units of mass, velocity, and force are chosen so that $k = 1$, then Newton's second law becomes

$$m\frac{dv}{dt} = F. \tag{2}$$

Consider an origin located somewhere above the earth's surface. Let the positive axis of a one-dimensional coordinate system extend from the origin through the gravitational center of the earth. Suppose that at time $t = 0$ a body with mass m is initially located at y_0 and is traveling with a velocity v_0. See Figure 3.12. Assuming that the body is falling freely in a vacuum and that it is close enough to the earth's surface so that the only significant force acting on the body is the earth's gravitational attraction, then $F = mg$, where g is the gravitational attraction—approximately 32 ft/sec^2 in the English system of measurement and 9.8 m/sec^2 in the metric system. Substituting $F = mg$ in equation (2) and then dividing by $m \neq 0$, we find the velocity of the body satisfies the initial value problem $dv/dt = g$; $v(0) = v_0$. Integrating and satisfying the initial condition, we easily find the solution of this initial value

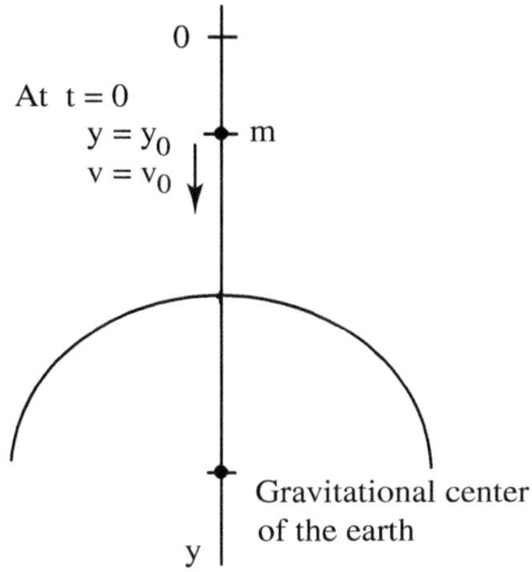

FIGURE 3.12
A Falling Body.

problem to be $v(t) = gt + v_0$. Notice that the velocity increases with time and as $t \to \infty$, $v(t) \to \infty$.

If we assume that the body is falling freely in air instead of in a vacuum, then we must make some assumption regarding the retarding force due to air resistance. Let us assume that the retarding force is proportional to the velocity of the body. Then the initial value problem for the velocity becomes $dv/dt = g - cv$; $v(0) = v_0$, where $c > 0$ is the constant of proportionality. Separating variables results in

$$\frac{dv}{g - cv} = dt.$$

Integrating, we find

$$\frac{-\ln|g - cv|}{c} = t + A,$$

where A is an arbitrary constant. Multiplying by $-c$ and exponentiating, we get

$$|g - cv| = e^{-c(t+A)} = e^{-cA}e^{-ct} = Be^{-ct}$$

where B is an arbitrary positive constant. By letting B be an arbitrary constant—positive, negative, or zero—we can remove the absolute value appearing in this equation. When $t = 0$, $v = v_0$, so $B = g - cv_0$ and therefore

$$g - cv = (g - cv_0)e^{-ct}.$$

Solving for the velocity v, we find

$$v(t) = \frac{g}{c}(1 - e^{-ct}) + v_0e^{-ct}.$$

Since $c > 0$, as t approaches ∞ the velocity $v(t)$ approaches the value g/c, the **terminal velocity**—the maximum velocity that the falling body can attain.

Example 1 Velocity of a Body Falling Freely in Air Assume the equation for the velocity of a body falling freely in air is

$$\frac{dv}{dt} = g - cv; \quad v(0) = 20 \text{ ft/sec},$$

where $g = 32$ ft/sec^2 and $c = .25$. Compute numerically and graph the solution to this initial value problem on the interval $[0, 10]$. What is the value of $v(10)$? How does this compare with the terminal velocity of $v_\infty = g/c = 128$ ft/sec?

Solution

Associating $v(t)$ with $y(x)$, we ran our computing software by setting $f(x, y) = 32 - .25y$, specifying the interval of integration to be $[0, 10]$, and specifying the initial condition to be $y(0) = 20$. A graph of the velocity as a function of time over the first 10 seconds is shown in Figure 3.13. From the table of values for the numerical solution, we found that $v(10) = 117.4931$ ft/sec, which is within 8.21% of the terminal velocity.

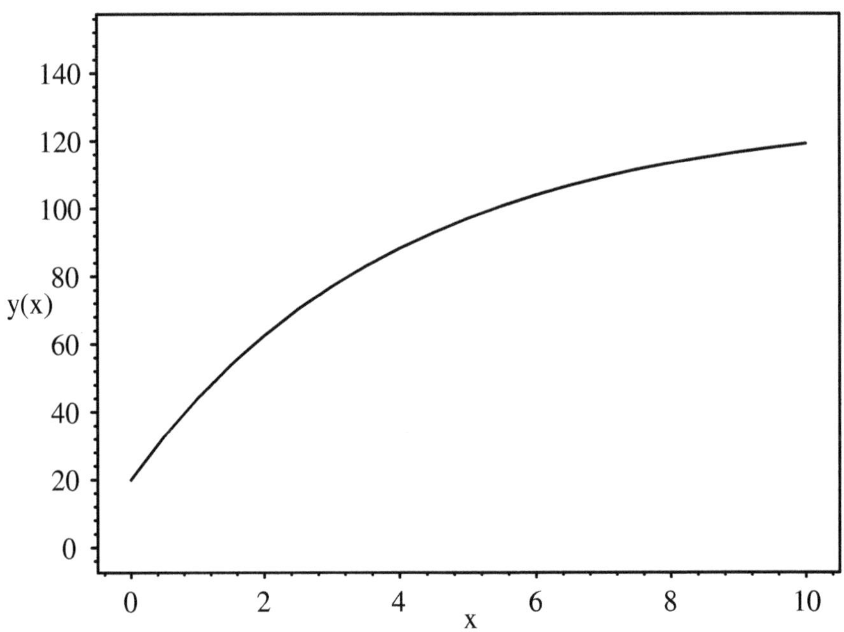

FIGURE 3.13
Numerical Approximation to the IVP: $y' = 32 - .25y$; y(0) = 20. ■

3.5.1 Exercises

1. Write the equation for the velocity of a falling body if the air resist- ance is proportional to the square of the velocity. Numerically compute and graph the velocity on the interval $[0, 10]$, if $g = 32$ ft/sec^2, if the constant of proportionality is $c = .25$, and if the initial velocity is $v(0) = 20$ ft/sec. Estimate the terminal velocity in this case.

2. Write the differential equation for the velocity of a falling body, if the air resistance is proportional to $v^{1.6}$. Numerically compute and graph the velocity on the interval $[0, 10]$, if $g = 32$ ft/sec^2, if the constant of proportionality is $c = .25$, and if the initial velocity is $v(0) = 20$ ft/sec. Estimate the terminal velocity.

3. Write the differential equation for the velocity of a falling body, if the air resistance is proportional to $\sqrt{v}$. Compute and graph the velocity on the interval $[0, 10]$, if $g = 32$ ft/sec^2, if the constant of proportionality is $c = .25$, and if the initial velocity is $v(0) = 20$ ft/sec. Is there a terminal velocity? If so, estimate its value.

4. A parachutist jumps from an airplane, falls freely for 10 seconds and then opens his parachute. Assume the parachutist's initial downward velocity was $v(0) = 0$ ft/sec, assume the air resistance is proportional to $v^{1.8}$, and assume the constant of proportionality without the parachute is .2 and with the parachute is 1.35. Compute the velocity of the parachutist 20 seconds after jumping from the airplane. (Hint: You need to solve two initial value problems. The solution of the first problem is the velocity of the parachutist for the first 10 seconds and the solution of the second problem is the velocity of the parachutist for the second 10 seconds. The final velocity of the first problem is the initial velocity (initial condition) for the second problem.)

3.6 Mixture Problems

Suppose at time $t = 0$ a quantity q_0 of a substance is present in a container. Also assume at time $t = 0$ a fluid containing a concentration $c_{in}(t)$ of the substance is allowed to enter the container at the rate $r_{in}(t)$ and that the mixture in the container is kept at a uniform concentration throughout by a mixing device. Furthermore, assume for $t \geq 0$ the mixture in the container with concentration $c_{out}(t)$ is allowed to escape at the rate $r_{out}(t)$. The problem is to determine the amount, $q(t)$, of substance in the container at any time. See Figure 3.14.

Since the rate of change of the amount of substance in the container, dq/dt, equals the rate at which the fluid enters the container times the concentration

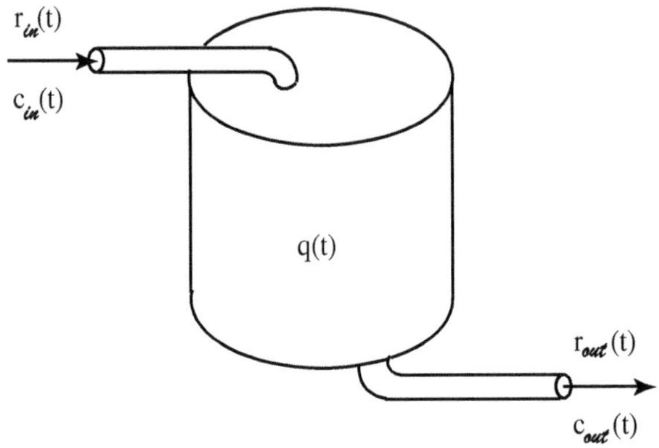

FIGURE 3.14
Diagram for a One Tank Mixture Problem.

of the substance in the entering fluid minus the rate at which the fluid leaves the container times the concentration of the substance in the container, $q(t)$ must satisfy the initial value problem

(1) $$\frac{dq}{dt} = r_{in}(t)c_{in}(t) - r_{out}(t)c_{out}(t); \quad q(0) = q_0.$$

Example 1 One Tank Mixture Problem A 200-gallon capacity tank initially contains a salt solution consisting of 20 lbs of salt dissolved in 165 gallons of water. A salt solution with concentration of 2 lbs/gal enters the tank at the rate of 8 gal/min, and the resulting uniform mixture leaves the tank at a rate of 3 gal/min. Compute and graph the amount of salt in the tank as a function of time and find the amount of salt in the tank at the time the tank starts to overflow.

Solution

Let $q(t)$ denote the number of pounds of salt in the tank at time t. Initially the tank contains 20 lbs of salt, so $q(0) = 20$ lbs. The rate at which the solution enters the tank is $r_{in}(t) = 8$ gal/min, and the concentration of the entering solution is $c_{in}(t) = 2$ lbs/gal. The rate at which the solution leaves the tank is $r_{out}(t) = 3$ gal/min. So the rate at which the tank is filling is $r(t) = r_{in}(t) - r_{out}(t) = 8$ gal/min -3 gal/min $= 5$ gal/min. Let $n(t)$ be the number of gallons of solution in the tank at time t. Since the tank initially contains 165 gallons and the rate of increase is 5 gal/min, $n(t) = 165 + 5t$. Hence, the concentration of the salt solution in the tank and flowing out of the tank at time t is $c_{out}(t) = q(t)/n(t) = q(t)/(165+5t)$. So the initial value problem which must be solved is

$$
\begin{aligned}
(2) \qquad \frac{dq}{dt} &= r_{in}(t)c_{in}(t) - r_{out}(t)c_{out}(t) \\
&= (8 \text{ gal/min})(2 \text{ lbs/gal}) - (3 \text{ gal/min})\frac{q(t)}{165+5t} \text{ (lbs/gal)} \\
&= 16 - \frac{3q(t)}{165+5t} \quad \text{(lbs/min)}; \\
q(0) &= 20 \text{ lbs.}
\end{aligned}
$$

The tank is full when $n(t) = 165 + 5t = 200$. Solving for t, we find $t = (200 - 165)/5 = 7$ minutes.

Associating $q(t)$ with $y(x)$, we used computing software to solve the initial value problem (2) on the interval $[0, 7]$ by setting $f(x, y) = 16 - 3y/(165+5x)$, by inputting the interval of integration as $[0, 7]$, and by inputting the initial condition $y(0) = 20$. A graph of the amount of salt in the tank is shown in Figure 3.15. From the table of values for the numerical solution, we find $q(7) = 123.7934$ lbs.

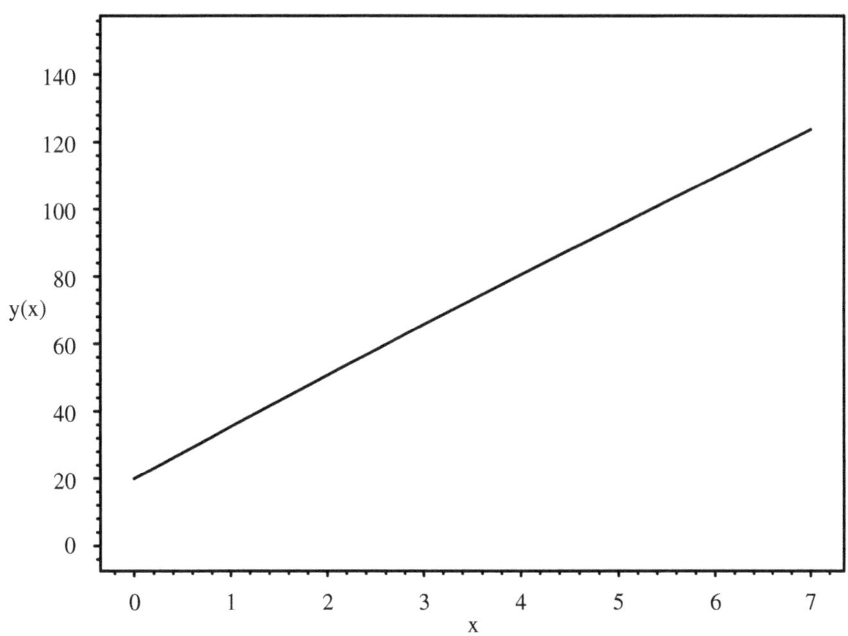

FIGURE 3.15
Numerical Approximation to the IVP: $y' = 16 - 3y/(165+5x)$; y(0) = 20. ■

Example 2 Two Tank Mixture Problem A 150-gallon tank is initially filled with a 60% dye solution. A 10% dye solution flows into the tank at a rate of 5 gal/min. The mixture flows out of the tank at the same rate into a 75 gallon tank which was initially filled with pure water. The mixture in the second tank flows out at the rate of 5 gal/min. Compute and graph the amount of dye in the second tank on the interval $[0, 125]$. When is the amount of dye in the second tank a maximum? What is the maximum amount of dye in the second tank?

Solution

Let $q_1(t)$ and $q_2(t)$ denote the number of gallons of dye in tanks 1 and 2 at time t, respectively. A diagram for this example is displayed in Figure 3.16. The initial number of gallons of dye in tank 1 at time $t = 0$ is $q_1(0) = 60\% \times 150$ gal $= 90$ gal. Since the second tank is initially filled with pure water, $q_2(0) = 0$ gal.

For tank 1, we have $r_{1\_in}(t) = r_{1\_out}(t) = 5$ gal/min, $c_{1\_in}(t) = 10\% = .1$, and $c_{1\_out}(t) = q_1(t)/150$. So the number of gallons of dye in tank 1 must satisfy the differential equation

$$\begin{aligned}\frac{dq_1}{dt} &= r_{1\_in}(t)c_{1\_in}(t) - r_{1\_out}(t)c_{1\_out}(t)\\ &= (5 \text{ gal/min})(.1) - (5 \text{ gal/min})\left(\frac{q_1}{150}\right)\\ &= (.5 - \frac{q_1}{30}) \quad (\text{gal/min}).\end{aligned}$$

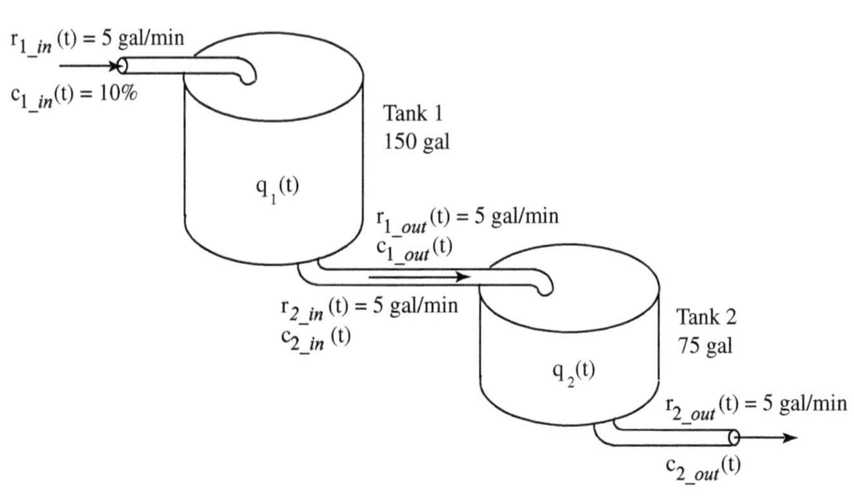

FIGURE 3.16
Diagram for a Two Tank Mixture Problem.

For tank 2, we have $r_{2\_in}(t) = r_{2\_out}(t) = 5$ gal/min, $c_{2\_in}(t) = q_1(t)/150$, and $c_{2\_out}(t) = q_2(t)/75$. So the number of gallons of dye in tank 2 must satisfy the differential equation

$$\begin{aligned}\frac{dq_2(t)}{dt} &= r_{2\_in}(t)c_{2\_in}(t) - r_{2\_out}(t)c_{2\_out}(t)\\ &= (5 \text{ gal/min})\left(\frac{q_1(t)}{150}\right) - (5 \text{ gal/min})\left(\frac{q_2(t)}{75}\right)\\ &= \left(\frac{q_1(t)}{30} - \frac{q_2(t)}{15}\right) \quad (\text{gal/min}).\end{aligned}$$

Hence, the number of gallons of dye in tanks 1 and 2 must simultaneously satisfy the two initial value problems

(3a) $$\frac{dq_1}{dt} = .5 - \frac{q_1}{30}; \qquad q_1(0) = 90$$

(3b) $$\frac{dq_2}{dt} = \frac{q_1}{30} - \frac{q_2}{15}; \qquad q_2(0) = 0. \blacksquare$$

Since the differential equation in (3a) involves the single variable q_1 and is linear, we can solve the initial value problem (3a) explicitly and substitute the result into (3b). The resulting initial value problem will involve the variable q_2 only. We will then use our computing software to solve this initial value problem numerically. (In Chapter 7, we will show how to solve the system of two, first-order, initial value problems consisting of (3a) and (3b) simultaneously—that is, we will show how to find q_1 and q_2 without first solving for q_1 explicitly and substituting into (3b).)

In Chapter 2, we proved that the general solution of the nonhomogeneous linear first-order differential equation $y' = a(t)y + b(t)$ where $a(t)$ and $b(t)$ are continuous on the interval I is

(4) $$y(t) = y_1(t)(K + v(t))$$

where K is an arbitrary constant and

$$y_1(t) = e^{\int^t a(x)\,dx} \qquad \text{and} \qquad v(t) = \int^t \frac{b(x)}{y_1(x)}dx.$$

Equation (3a) is a nonhomogeneous linear differential equation in which $y(t) = q_1(t)$, $a(t) = -1/30$, and $b(t) = .5$. Hence,

$$y_1(t) = e^{-\int^t dx/30} = e^{-t/30 + C_1}$$

where C_1 is a constant of integration. Since we need only one constant in the general solution—which is the constant K—we set $C_1 = 0$. Substituting the resulting expression for $y_1(t)$ into the equation for $v(t)$, yields

$$v(t) = \int^t \frac{b(x)}{y_1(x)}dx = \int^t \frac{.5}{e^{-x/30}}dx = .5\int^t e^{x/30}dx = 15e^{t/30} + C_2$$

where C_2 is a constant of integration. Again since we need only one constant in the general solution, we set $C_2 = 0$. Substituting our expressions for $y_1(t)$ and $v(t)$ with $C_1 = C_2 = 0$ into equation (4), we find the general solution for equation (3a) to be

$$q_1(t) = y_1(t)(K + v(t)) = e^{-t/30}(K + 15e^{t/30}) = Ke^{-t/30} + 15.$$

Since $q_1(0) = 90 = K + 15$, we have $K = 90 - 15 = 75$. Hence,

$$q_1(t) = 75e^{-t/30} + 15.$$

Substituting this expression for $q_1(t)$ into the differential equation of (3b), we find q_2 must satisfy the initial value problem

$$(5) \qquad \frac{dq_2}{dt} = 2.5e^{-t/30} + .5 - \frac{q_2}{15}; \quad q_2(0) = 0.$$

Associating $q_2(t)$ with $y(x)$, we used computing software to solve the initial value problem (5) on $[0, 125]$ by setting $f(x, y) = 2.5e^{-x/30} + .5 - y/15$, by inputting the interval of integration as $[0, 125]$, and by entering the initial condition $y(0) = 0$. A graph of the solution for the number of gallons of dye in tank 2 on the interval $[0, 125]$ is shown in Figure 3.17. From the graph we see that the maximum number of gallons of dye in the tank occurs when t is approximately 25 minutes. By searching the table of values of the numerical solution near $t = 25$ minutes, we find the maximum number of gallons of dye in the tank is 24.545 gallons and that the maximum occured when $t = 23.625$ minutes.

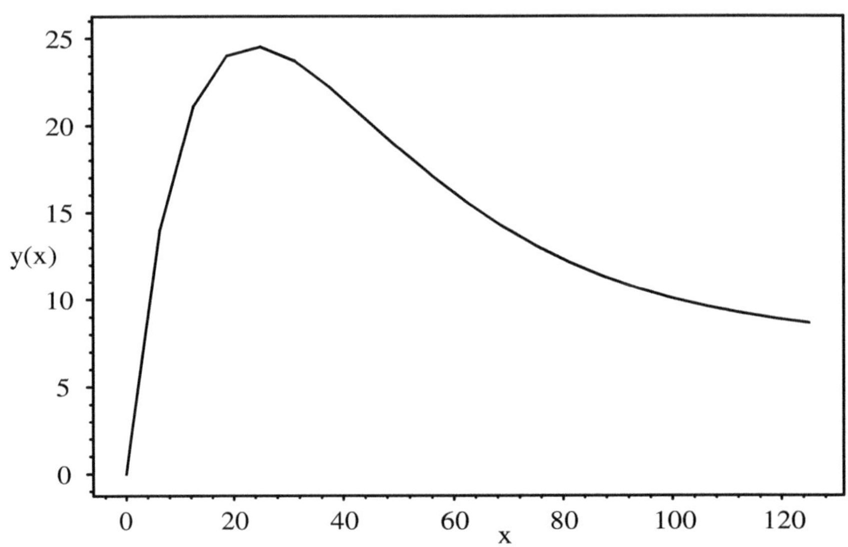

FIGURE 3.17
Numerical Approximation to the IVP: $y' = 2.5e^{-x/30} + .5 - y/15; y(0) = 0$.
■

3.6.1 Exercises

1. The bloodstream carries a drug to an organ at the rate of 1.5 in^3/sec and leaves the organ at the same rate. The organ contains 8 in^3 of blood and the concentration of the drug in the blood entering the organ is 0.5 g/in^3. If at time $t = 0$ there is no trace of the drug in the organ, and if we can assume that the organ behaves like a container in which the blood is uniformly mixed, compute and graph the amount of drug in the organ over the time interval $[0, 10]$. When is the amount of drug in the organ 2 g?

2. An office 12 ft by 10 ft and 8 ft high (the size of some mathematics professors' offices) initially contains air with no carbon monoxide. Beginning at time $t = 0$, cigarette smoke containing 3.5% carbon monoxide is blown into the office at a rate of .2 ft^3/min. Assuming the air in the room is uniformly mixed and leaves the room at the same rate, at what time does the amount of carbon monoxide in the air reach .015%? (Prolonged exposure to this concentration of carbon monoxide is dangerous.)

3. One of the major problems facing industrialized nations is water pol- lution. Rivers and lakes become polluted with various types of waste products which can kill plant and marine life. Once pollution of a river is stopped, the river will clean itself fairly rapidly. However, as this example will illustrate, large lakes require much longer to clean themselves by the natural process of clean water flowing into the lake and polluted water flowing out of the lake. If c_L is the concentration of pollution in a lake and V is the volume of the lake, then the total amount of pollutants in the lake is $Q = c_L V$. If r is the rate at which water enters and leaves a lake and if c_{in} is the concentration of pollutants entering the lake, then we have
$$\frac{dQ}{dt} = \frac{d(c_L V)}{dt} = rc_{in} - rc_L = r(c_{in} - c_L).$$
If we assume the volume of the lake is constant, then dividing by V, we find the concentration of pollution in the lake, c_L, satisfies the differential equation
$$\frac{dc_L}{dt} = \frac{r(c_{in} - c_L)}{V}.$$
Lake Michigan has a volume of 1180 mi^3 and the yearly flow rate is $r = 38$ mi^3/yr. Assuming at time $t = 0$ the concentration of pollutants in Lake Michigan is $c_L(0) = 0.4\%$, assuming the concentration of pollutants in the entering water is successfully reduced to 0.05%, and assuming the water in the lake is well mixed, how many years will it take to reduce the pollution concentration in Lake Michigan to 0.3%? 0.25%? 0.2%?

4. Let the subscript e denote Lake Erie and the subscript o denote Lake Ontario. The volumes of these lakes are $V_e = 116$ mi^3 and $V_o = 393$ mi^3.

Pollution enters Lake Ontario from Lake Erie, from rivers, and from water run off from the surrounding land. Suppose the rate of water entering Lake Ontario from Lake Erie is $r_e = 85$ mi^3/yr and the rate of water entering from non-Erie sources is $r = 14$ mi^3/yr. Assuming the rate at which water leaves Lake Ontario is $r_o = r_e + r = 99$ mi^3/yr, the amount of pollutants in Lake Ontario, $Q = c_o V_o$, satisfy the differential equation

$$\frac{dQ}{dt} = \frac{d(c_o V_o)}{dt} = r_e c_e + rc - r_o c_o$$

where c_e is the concentration of pollutants in Lake Erie and c is the concentration of pollutants in non-Erie sources. The first term on the right represents the amount of pollutants entering Ontario from Erie, the second term represents the amount of pollutants entering Ontario from non-Erie sources, and the third term represents the amount of pollutants leaving Ontario through outflow. Assuming the volume of Lake Ontario, V_o, is constant, we find, by dividing the previous equation by V_o, that the concentration of pollution in Lake Ontario, c_o, satisfies the differential equation

$$\frac{dc_o}{dt} = \frac{r_e c_e + rc - r_o c_o}{V_o}.$$

Compute and graph the concentration of pollution in Lake Ontario if $c_o(0) = 0.3\%$, $c_e(t) = 0.4\% e^{-t/10}$, and $c(t) = .08\%$. How many years will it take for the concentration of pollution in Lake Ontario to reach 0.2%? 0.1%?

3.7 Curves of Pursuit

Interesting problems result when one tries to determine the path that one object must take in order to pursue, and perhaps capture, a second object when either or both objects move according to specified constraints.

3.7.1 Exercises

Exercise 1. A boy attaches a stiff rod of length L to a toy boat. The boy places the boat at the edge of a rectangular pool at $(L, 0)$ and then moves to the corner of the pool at $(0, 0)$. As the boy walks along the other edge of the pool—the y-axis, the boat glides through the water after him. See Figure 3.18. The path followed by the boat is called a **tractrix** (Latin *tractum*, drag).

The line segment PQ, which represents the stiff rod of length L, is tangent to the path followed by the boat. So the slope of the line PQ is

$$\frac{dy}{dx} = -\frac{\sqrt{L^2 - x^2}}{x}. \tag{1}$$

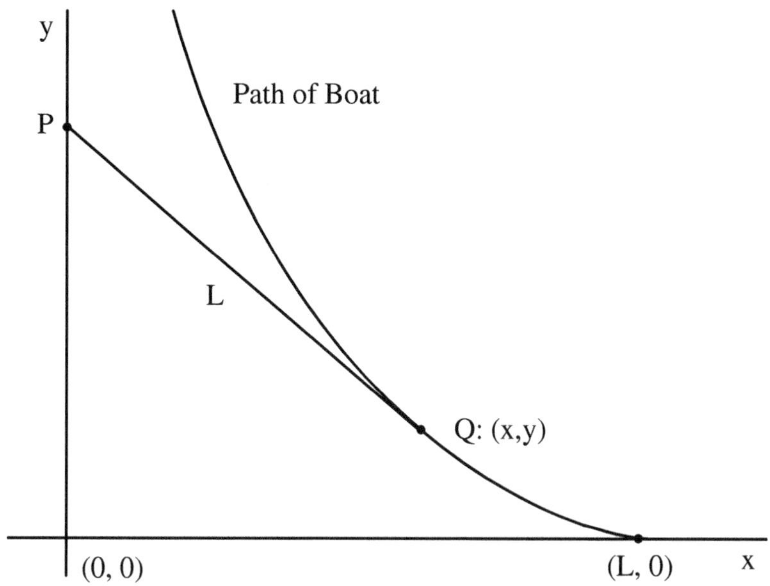

FIGURE 3.18
The Tractrix.

Graph the tractrix which corresponds to a rod of length $L = 10$ ft. (Hint: Solve the initial value problem consisting of the differential equation (1) and the initial condition $y(L) = 0$ on the interval $[.5, 10]$.)

Exercise 2. The y-axis and the line $x = W > 0$ represent the banks of a river. The river flows in the negative y-direction with speed s_r. A boat whose speed in still water is s_b is launched from the point $(W, 0)$. The boat is steered so that it is always headed toward the origin. See Figure 3.19.

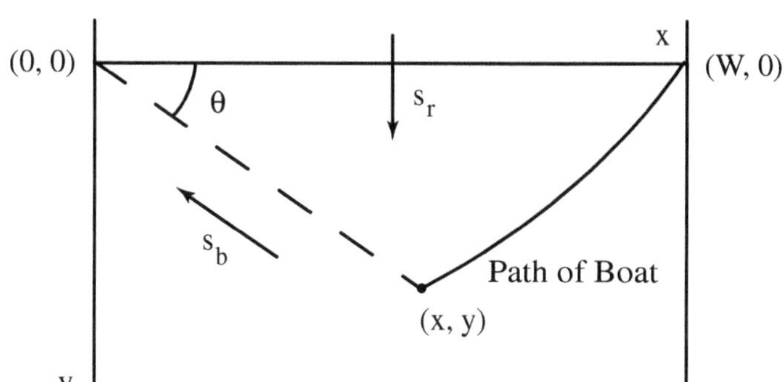

FIGURE 3.19
Path of a Boat in a River.

The components of the boat's velocity in the x-direction and y-direction are

$$\frac{dx}{dt} = -s_b \cos\theta \quad \text{and} \quad \frac{dy}{dt} = -s_r + s_b \sin\theta.$$

So

$$\frac{dy}{dx} = \frac{dy/dt}{dx/dt} = \frac{-s_r + s_b \sin\theta}{-s_b \cos\theta} = \frac{-s_r + s_b(-y/\sqrt{x^2+y^2})}{-s_b(x/\sqrt{x^2+y^2})}.$$

Simplifying, we get

$$\frac{dy}{dx} = \frac{s_r\sqrt{x^2+y^2} + s_b y}{s_b x}. \tag{2}$$

Graph the solutions of the initial value problems consisting of the differential equation (2) and the initial condition $y(W) = 0$ on the interval $[0, W]$ where $W = .5$ mi, $s_r = 3$ mi/hr, and (a) $s_b = 2$ mi/hr, (b) $s_b = 3$ mi/hr, and (c) $s_b = 4$ mi/hr. In each case, decide if the boat lands on the opposite bank. In those cases in which the boat does land, tell where it lands. How is the landing point related to s_b versus s_r?

Exercise 3. The y-axis and the line $x = W > 0$ represent the banks of a river. The river flows in the negative y-direction with speed s_r. At time $t = 0$ a man starts walking from the origin along the negative y-axis with speed s_m. At the same time, a boat whose speed in still water is s_b is launched from the point $(W, 0)$. The boat is steered so that it is always headed toward the man. Draw an appropriate figure for this problem, determine the differential equation satisfied by the path of the boat, compute and graph the path of the boat, if $W = .5$ mi, $s_r = 3$ mi/hr, $s_m = 1$ mi/hr, and (a) $s_b = 2$ mi/hr, (b) $s_b = 3$ mi/hr, and (c) $s_b = 4$ mi/hr. In each case, decide if the boat lands on the opposite bank. Tell where the boat lands, when it does land. (Hint: Use a "moving rectangular coordinate system" with the origin at the man.)

Exercise 4. A rabbit starts at $(0, a)$ and runs along the y-axis in the positive direction with a constant speed of s_r. A dog starts at $(b, 0)$ and pursues the rabbit with speed s_d. The dog runs so that he is always pointed toward the rabbit. See Figure 3.20.

At time t the rabbit will be at the point $(0, a + s_r t)$ and the dog will be at (x, y). Since the line between these two points is tangent to the path of the dog, the slope of the line is

$$\frac{dy}{dx} = \frac{y - a - s_r t}{x}.$$

Multiplying by x, we get

$$xy' = y - a - s_r t. \tag{3}$$

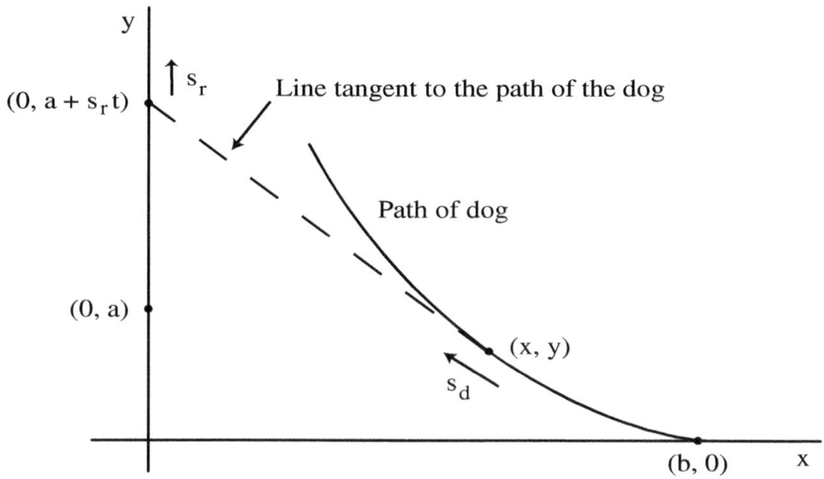

FIGURE 3.20
Path of a Dog Pursuing a Rabbit.

From calculus we know the arc length s of the path of the dog satisfies the differential equation

$$\frac{ds}{dt} = -\sqrt{1 + (dy/dx)^2}\frac{dx}{dt} = s_d.$$

The minus sign is due to the fact that s increases as x decreases. Solving for dt/dx, we have

$$(4) \qquad \frac{dt}{dx} = -\frac{\sqrt{1 + (y')^2}}{s_d}.$$

Differentiating (3) with respect to x, we get $xy'' + y' = y' - s_r dt/dx$. Solving this equation for dt/dx, we obtain

$$(5) \qquad \frac{dt}{dx} = \frac{-xy''}{s_r}.$$

Equating (4) and (5) results in

$$(6) \qquad \frac{xy''}{s_r} = \frac{\sqrt{1 + (y')^2}}{s_d}.$$

Let $p = y'$. Then $p' = y''$. When $t = 0$, $(x, y) = (b, 0)$, and $p = -a/b$. Substituting for p and p' in equation (6) and solving for p', we see that p satisfies the initial value problem

$$p' = \frac{s_r\sqrt{1+p^2}}{xs_d}; \qquad p(b) = \frac{-a}{b}.$$

Separating variables and letting $R = s_r/s_d$, we find

$$\frac{dp}{\sqrt{1+p^2}} = \frac{R\,dx}{x}.$$

Integrating, we get

$$\ln|p + \sqrt{1+p^2}| = R\ln|x| + C$$

where C is the constant of integration. Exponentiating, we obtain

$$p + \sqrt{1+p^2} = Kx^R. \tag{7}$$

Subtracting p and squaring, we find

$$1 + p^2 = K^2x^{2R} - 2Kpx^R + p^2.$$

Eliminating p^2 and solving for p, yields

$$p = y' = \frac{K^2x^{2R} - 1}{2Kx^R}. \tag{8}$$

The constant K must be chosen to satisfy the initial condition $p(b) = -a/b$. Returning to equation (7), solving for K, and setting $x = b$ and $p = -a/b$, we see that

$$K = (-a/b + \sqrt{1 + a^2/b^2})/b^R = (-a + \sqrt{a^2 + b^2})/b^{R+1}. \tag{9}$$

The path of the dog is the solution to the initial value problem consisting of the differential equation (8) where K is given by equation (9) and the initial condition $y(b) = 0$. Compute and graph the path of the dog if $a = 200$ ft, $b = 100$ ft, $s_r = 50$ ft/sec, and (a) $s_d = 40$ ft/sec, (b) $s_d = 50$ ft/sec, and (c) $s_d = 75$ ft/sec. In which case(s) does the dog capture the rabbit? When and where does the dog catch the rabbit?

3.8 Chemical Reactions

Chemical equations indicate how molecules of substances combine or decompose to produce other substances. For example, the chemical equation

$$2H_2 + O_2 \rightarrow 2H_2O$$

states that 2 molecules of hydrogen, H_2, combine with 1 molecule of oxygen, O_2, to form 2 molecules of water, H_2O. A general chemical equation has the form

$$aA + bB + cC + \cdots \to pP + qQ + \cdots$$

where $a, b, c, \ldots, p, q, \ldots$ are positive integers. This equation indicates that a molecules of the substance A, b molecules of the substance B, c molecules of the substance C, ... react to form p molecules of the substance P, q molecules of the substance Q,.... The rate at which a substance is formed is called the **velocity of the reaction**. In many cases, the law of mass action stated below applies in determining the velocity of reaction.

The Law of Mass Action

> If the temperature is kept constant, the velocity of a chemical reaction is proportional to the product of the concentrations of the reacting substances.

Recall from elementary chemistry that a **mole** is the number of grams of a substance which is equal to the molecular weight of the substance. For example, 1 mole of hydrogen, H_2, has a weight of 2.016 grams, since the atomic weight of the element hydrogen, H, is 1.008 grams; 1 mole of oxygen, O_2, has a weight of 32 grams, since the atomic weight of the element oxygen, O, is 16 grams; and 1 mole of water, H_2O, has a weight of $2(1.008) + 16 = 18.016$ grams.

If we let $C_A(t), C_B(t), C_C(t), \ldots$ represent the concentrations of substances $A, B, C, \ldots$ at time t expressed in moles per liter and if y is the number of moles per liter which have reacted at time t, then the law of mass action states that the velocity of reaction, dy/dt, satisfies the differential equation

$$\frac{dy}{dt} = k(C_A)^a(C_B)^b(C_C)^c \cdots \tag{1}$$

where the constant of proportionality k is called the **velocity constant**. The sum of the exponents $a + b + c + \cdots$ is called the **order of the reaction**.

Example 1 An Order 2 Chemical Reaction Consider the order 2 reaction

$$A + B \to P$$

Suppose two reactants, A and B, are combined in solution and have an initial concentration $C_A(0) = 7$ moles/liter and $C_B(0) = 3$ moles/liter. Assume the law of mass action applies and the velocity constant is $k = 2$ liters/(mole·sec).

a. Numerically solve equation (1) for the number of moles per liter which have reacted at time t, $y(t)$. (In this case, $a = b = 1$.)

b. What is the limiting concentration of the product P? (That is, what is $\lim_{t\to\infty} y(t)$?)

c. How long does it take to produce one-half of the limiting concentra- tion? 90% of the limiting concentration?

d. What is the concentration of substances A and B at time t? That is, what is $C_A(t)$ and $C_B(t)$?

e. What is $\lim_{t\to\infty} C_A(t)$ and $\lim_{t\to\infty} C_B(t)$?

Solution

Let $y(t)$ denote the number of moles/liter of the substance P at time t. The concentration of substance A at time t is $C_A(t) = C_A(0) - y(t) = 7 - y(t)$ and the concentration of substance B at time t is $C_B(t) = C_B(0) - y(t) = 3 - y(t)$. Since the concentrations C_A and C_B are positive and can only decrease as the reaction proceeds, $0 \leq C_A(t) \leq 7$ and $0 \leq C_B(t) \leq 3$. Thus, we must have $0 \leq 7 - y \leq 7$ and $0 \leq 3 - y \leq 3$. So we must have both $0 \leq y \leq 7$ and $0 \leq y \leq 3$ which implies $0 \leq y \leq 3$. According to the law of mass action

$$\frac{dy}{dt} = kC_A(t)C_B(t) = 2(7 - y)(3 - y).$$

At time $t = 0$, we have $y(t) = 0$, since no reaction has taken place. We ran our computing software by setting $f(x, y) = 2(7 - y)(3 - y)$, by inputting the interval of integration as $[0, 1]$, and by inputting the initial condition $y(0) = 0$.

a. A graph of $y(t)$ on $[0, 1]$ is displayed in Figure 3.21.

b. As noted above $0 \leq y(t) \leq 3$ and from Figure 3.21 or the corresponding table of values for the numerical solution, we see that $\lim_{t\to\infty} y(t) = 3$.

c. From the table of values for the numerical solution we see that $y(t) = 50\% \times 3 = 1.5$ moles/liter when t is approximately .06 seconds, and we also see that $y(t) = 90\% \times 3 = 2.7$ moles/liter when t is approximately .24 seconds.

d. As stated earlier $C_A(t) = 7 - y(t)$ and $C_B(t) = 3 - y(t)$.

e. Since $\lim_{t\to\infty} y(t) = 3$, we easily compute

$$\lim_{t\to\infty} C_A(t) = C_A(0) - \lim_{t\to\infty} y(t) = 7 - 3 = 4 \text{ moles/liter}$$

and

$$\lim_{t\to\infty} C_B(t) = C_B(0) - \lim_{t\to\infty} y(t) = 3 - 3 = 0 \text{ moles/liter.}$$

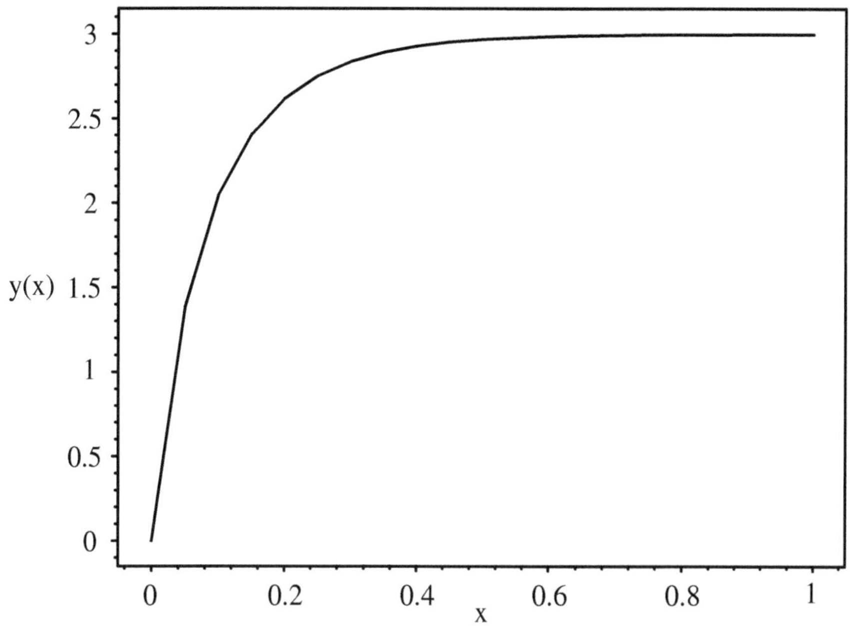

FIGURE 3.21
Numerical Approximation to the IVP: $y' = 2(7 - y)(3 - y)$; y(0) = 0. ■

3.8.1 Exercises

1. Assume that the law of mass action applies to the chemical reaction $A + B \to P$. Suppose the initial concentrations are $C_A(0) = 6$ moles/liter and $C_B(0) = 2$ moles/liter and that the velocity constant is $k = 1$ liter/(mole·sec).

   a. Find and graph the number of moles per liter, $y(t)$, which have reacted at time t.

   b. What is the limiting concentration of the product P?

   c. What is the limiting concentration of the reactant B?

2. The chemical reaction $A+B \to P$ satisfies the law of mass action. Suppose $C_A(0) = C_B(0) = 5$ moles/liter and $k = 1.5$ liters/(mole·sec).

   a. Find and graph the number of moles per liter, $y(t)$, which have reacted at time t.

   b. What is $\lim_{t\to\infty} y(t)$? $\lim_{t\to\infty} C_A(t)$? $\lim_{t\to\infty} C_B(t)$?

3. The chemical reaction $A + B + C \to P$ satisfies the law of mass action. If $y(t)$ is the number of moles per liter which have reacted at time t;

if $C_A(t)$, $C_B(t)$, and $C_C(t)$ are the concentrations of substances A, B, and C respectively at time t; if $C_A(0)$, $C_B(0)$, and $C_C(0)$ are the initial concentrations; and if k is the velocity constant; then $y(t)$ satisfies the initial value problem

$$y' = kC_A(t)C_B(t)C_C(t) = k(C_A(0) - y)(C_B(0) - y)(C_C(0) - y); \; y(0) = 0.$$

Suppose $k = .5$ liter/(mole·sec), $C_A(0) = 8$ moles/liter, $C_B(0) = 5$ moles/liter, and $C_C(0) = 3$ moles/liter.

a. Find and graph $y(t)$ on the interval $[0, 5]$.

b. Find $\lim_{t\to\infty} y(t)$.

c. When is $y(t) = .95 \times \lim_{t\to\infty} y(t)$?

4. The chemical reaction $A + B + C \to P$ obeys the law of mass action. Find and graph $y(t)$ on $[0, 5]$, if $C_A(0) = C_B(0) = C_C(0) = 6$ moles/liter and $k = .3$ liter/(mole·sec). (See Exercise 3.)

5. The chemical reaction $2A + B \to 2P + Q + R$ obeys the law of mass action. Let $y(t)$ be the number of moles per liter of reactant B which have reacted at time t, if $C_A(t)$ and $C_B(t)$ are the concentrations of substances A and B at time t, if $C_A(0)$ and $C_B(0)$ are the initial concentrations, and if k is the velocity constant, then $y(t)$ satisfies the initial value problem

$$y' = k[C_A(t)]^2 C_B(t) = k[C_A(0) - 2y]^2[C_B(0) - y]; \quad y(0) = 0.$$

Suppose $k = 1.5$ liter/(mole·sec), $C_A(0) = 9$ moles/liter, and $C_B(0) = 5$ moles/liter.

a. Find and graph $y(t)$ on the interval $[0, 5]$.

b. Find $\lim_{t\to\infty} y(t)$.

c. What is the limiting concentrations of the reactants A and B?

3.9 Miscellaneous Exercises

1. The error function is usually defined as

$$\operatorname{erf}(x) = \frac{2}{\sqrt{\pi}} \int_0^x e^{-t^2}\, dt.$$

This definition of the error function as an integral is equivalent to the initial value problem

$$y' = \frac{2e^{-x^2}}{\sqrt{\pi}}; \quad y(0) = 0.$$

Solve this initial value problem on the interval $[0, 5]$ and graph the solution.

2. Suppose the pollution index expressed in parts per million over a 24 hour period satisfies the differential equation

$$\frac{dP}{dt} = \frac{.34 - .04t}{\sqrt{50 + 25t - t^2}}.$$

At 12 midnight ($t = 0$) the pollution index is found to be .43 parts per million. Compute and graph the pollution index for the next 24 hours. At what time is the pollution index the highest? Lowest?

3. Let T be the absolute temperature of a body and let A be the absolute temperature of the surrounding medium. According to **Stefan's law of radiation** the rate of change of the temperature of the body is proportional to the difference between the fourth power of the temperature of the body and the fourth power of the temperature of the surrounding medium. Thus, according to Stefan's law $dT/dt = -k(T^4 - A^4)$. Compute and graph on the interval $[0, 20]$ the temperature T of a body whose initial temperature is $3000°K$, whose constant of proportionality is $k = 5 \times 10^{-12}/((°K)^3$-min), and which is surrounded by a medium with constant temperature $A = 500°K$. ($°K$ denotes degrees Kelvin. On the Kelvin temperature scale, $0°K$ corresponds to absolute zero, which is $-273.15°C$, and one degree Kelvin is the same size as one degree Celsius.)

4. When a periodic voltage $E(t) = E_0 \sin \omega t$ is suddenly applied to a coil wound around an iron core, the magnetic flux, $y(t)$, satisfies the initial value problem

$$y' = -Ay - By^3 + E_0 \sin \omega t; \quad y(0) = 0$$

where A, B, E_0, and ω are positive constants. Solve this initial value problem and graph the solution on the interval $[-.5, .5]$ for $A = 2$, $B = 3$, $E_0 = 2$, and $\omega = 4$.

5. A simple model for the spread of a rumor assumes there is a fixed population size, N, and that each person who has heard the rumor tells the rumor to m people each day. Some of them will already have heard the rumor. Let $H(t)$ be the number of people who have heard the rumor at time t. Each day each person who has heard the rumor will tell it to $m(N - H(t))/N$ people who have not already heard the rumor, so $H(t)$ satisfies the differential equation $H'(t) = H(t)m(N - H(t))/N$. Assume $N = 500$, $m = 3$, and $H(0) = 1$. Compute and graph $H(t)$ on the interval $[0, 10]$. When is $H(t) = 250$? What is $\lim_{t\to\infty} H(t)$?

6. Compute and graph on the interval $[1, 4]$ the curve which lies above the x-axis and has the property that the length of the arc joining any two points is equal to the area under the arc.

7. In 1913, L. Michaelis and M. Menton developed the following simple model for chemical enzyme kinetics: $dy/dt = -y/(y+1)$; $y(0) = 1$. Solve and graph the solution to this initial value problem on the interval $[0, 5]$.

8. **The Cycloid** The curve traced out by a point P on the circumference of a circle of radius R as the circle rolls along the x-axis is called a *cycloid.* If when $\theta = 0$, the point P is at the origin, then the parametric equation of the cycloid is

$$x = R(\theta - \sin\theta), \qquad y = R(1 - \cos\theta)$$

where θ is the angle, measured in radians, through which the rolling circle has rotated. A graph of a cycloid is displayed in Figure 3.22. Galileo Galilei (1564-1642) coined the term "cycloid" and was the first to conduct a meaningful study of the curve.

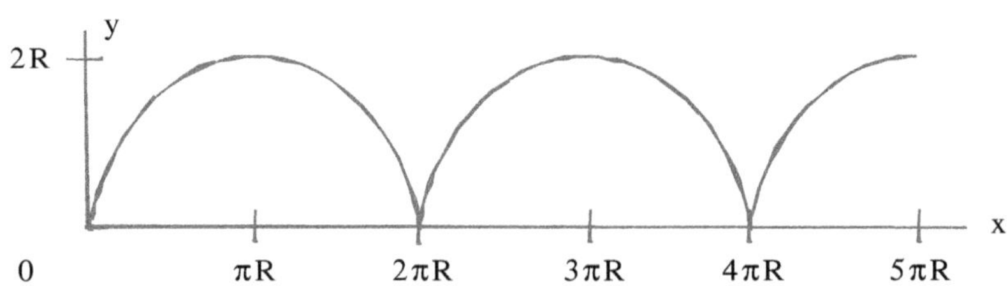

FIGURE 3.22
A Graph of the Cycloid.

a. The following theorem is due to the great English architect, astronomer, and mathematician Christopher Wren (1632-1723). "The length of one arch ($0 \le \theta \le 2\pi$) of a cycloid is equal to eight times the radius of the generating circle." Prove Wren's theorem.

b. The Italian physicist and mathematician Evangelista Torricelli (1608-1647) invented the barometer and was a student of Galileo. The following theorem is due to Torricelli. "The area under one arch of a cycloid is three times the area of the generating circle." Prove Torricelli's theorem.

9. **The Brachistochrone Problem** (From the Greek *brakhisto khrónos* meaning "shortest time.") In June, 1696, in *Acta Eruditorum* Johann Bernoulli (1667-1748) posed the following challenge problem to other mathematicians:

"Given two points A and B in a vertical plane, what is the curve traced out by a point acted on only by gravity, which starts at A and reaches B in the shortest time."

Earlier, in 1638, Galileo in his *Discourse on Two New Sciences* had tried to solve a similar problem of finding the path of fastest descent from a point to a wall. Galileo erroneously concluded that the arc of a circle was the fastest path. However, he had warned earlier in his *Discourse* of the possibility of fallacies in his results due to the need for a "higher science"—perhaps, he meant calculus.

Of course, Johann Bernoulli already knew how to solve the brachistochrone problem before he posed the problem to the mathematical community at large. Initially, he wanted to impose a six-month time limit on solving the problem. However, Leibniz persuaded him to lengthen the time for solution, so that foreign mathematicians would have a better opportunity to solve the problem. Bernoulli and Leibniz were deliberately taunting Newton to solve the problem. According to one Newtonian scholar, when Newton returned home one evening at about 4 p.m. after working at the Royal Mint in the Tower of London on the great coinaging, he found a letter from Johann Bernoulli. Newton stayed up all night in order to solve the problem and mailed the solution by the next post. By way of comparison, it took Bernoulli two weeks to solve the problem. In January 1697, the Royal Society published Newton's solution anonymously in the *Philosophical Transactions of the Royal Society.* In May 1697, *Acta Eruditorum* published solutions to the brachistochrone problem by Johann Bernoulli, by Jacob Bernoulli (Johann's older brother), by Leibniz, and a Latin translation of Newton's solution. It was not until 1988 that de L'Hôspital's solution was published.

In Section 3.5, Falling Bodies, we found that a body of mass m initially located at y_0, traveling with a velocity v_0, and close enough to the earth's surface so that the only significant force acting on the body is the earth's gravitational attraction, g, satisfies the initial value problem $dv/dt = g;\ v(0) = v_0$. (See Figure 3.12.) We found the solution of this initial value problem is $v(t) = gt + v_0$. When the initial velocity is zero—that is, when $v(0) = 0$ the solution is (a) $v(t) = gt$. Since, by definition, the velocity $v = dy/dt$ and since the mass has initial position $y(0) = y_0$, the position satisfies the initial value problem $dy/dt = gt;\ y(0) = y_0$. The solution of this initial value problem is $y(t) = gt^2/2 + y_0$. When the initial velocity is zero, that is, $y(0) = 0$ the solution is (b) $y(t) = gt^2/2$. Solving equation (a) for t, substituting the result into equation (b), and solving the result for v, we find that any mass which is located at the origin with initial velocity zero, satisfies the equation (c) $v = \sqrt{2gy}$. Equation (c) is the equation for the velocity of a body falling from the origin with initial velocity zero in terms of its position.

Now consider the following problem from optics. Suppose a ray of light travels from point A to point P with velocity v_1, then enters a less dense medium, and travels from P to B with a larger velocity v_2. From Figure 3.23, we see that the total time required for the light to travel from A to

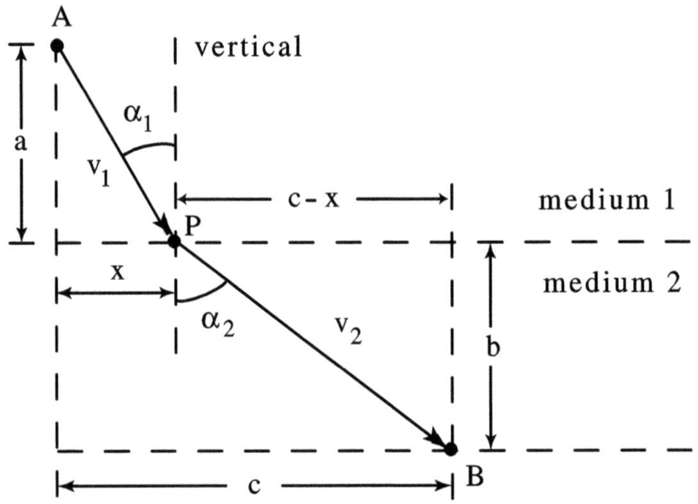

FIGURE 3.23
Snell's Law of Reflection.

P to B is

$$T = \frac{\sqrt{a^2 + x^2}}{v_1} + \frac{\sqrt{b^2 + (c - x)^2}}{v_2}.$$

In 1657, Pierre de Fermat (1601-1665) discovered the **principle of least time**, which states that light is able to select its path from A through P to B so as to minimize the time of travel T. Calculating dT/dx and setting the result equal to zero, we find

$$\frac{x}{v_1\sqrt{a^2 + x^2}} = \frac{c - x}{v_2\sqrt{b^2 + (c - x)^2}} v_2$$

or

$$\frac{\sin \alpha_1}{v_1} = \frac{\sin \alpha_2}{v_2}.$$

This expression is known as *Snell's Law of Reflection.* The Dutch astronomer and mathematician Willebrord Snellius (1580-1626) discovered Snell's Law experimentally in 1621 and stated it in the form $\sin \alpha_1 / \sin \alpha_2$ is a constant. In general, at any point (x, y) on the brachistochrone (d) $\sin \alpha / v = k$, where k is a constant.

In Figure 3.24 suppose a point mass is capable of selecting a path from the origin $A : (0, 0)$ to point B in the least possible time.

From the geometry of Figure 3.24 and trigonometry, we find that

$$\text{(e)} \qquad \sin \alpha = \cos \beta = \frac{1}{\sec \beta} = \frac{1}{1 + \tan^2 \beta} = \frac{1}{\sqrt{1 + (y')^2}}.$$

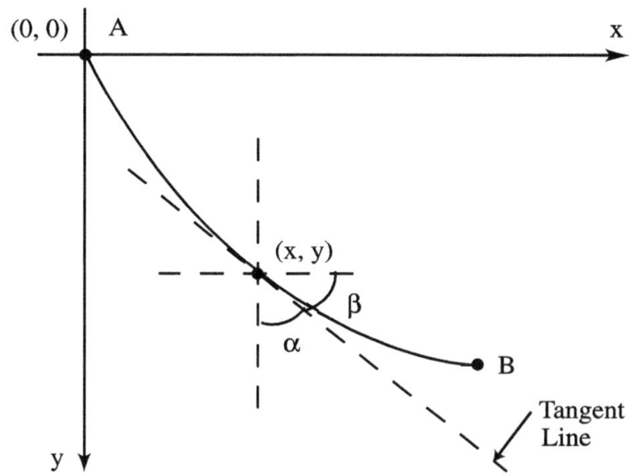

FIGURE 3.24
A Brachistochrone Curve.

From equations (c), (d), and (e) we see that the brachistochrone satisfies the differential equation

(f) $$y(1 + (y')^2) = k^2/(2g) = K$$

where K is a positive constant. Verify (1) that the cycloid

$$x = R(\theta - \sin\theta), \qquad y = R(1 - \cos\theta)$$

satisfies the differential equation (f), (2) that the cycloid passes through the origin, and (3) that R can be determined so that the cycloid passes through $B : (x_B, y_B)$ for any positive values x_B and y_B.

4

N-th Order Linear Differential Equations

As we noted earlier, ordinary differential equations are divided into two distinct classes—linear equations and nonlinear equations. In Chapters 2 and 3, we studied a few differential equations which can be solved explicitly in terms of elementary functions or which can be written as formulas involving quadratures. In particular, we found that the solution of the first-order linear differential equation $y' = a(x)y + b(x)$ can be written symbolically as $y(x) = y_1(x)(K + v(x))$ where K is an arbitrary constant,

$$y_1(x) = e^{\int^x a(t)\,dt} \quad \text{and} \quad v(x) = \int^x \frac{b(t)}{y_1(t)}\,dt.$$

Terms regarding higher-order linear differential equations are defined as follows. An **n-th order linear differential equation** is any differential equation of the form

$$(1) \quad a_n(x)y^{(n)}(x) + a_{n-1}(x)y^{(n-1)}(x) + \cdots + a_1(x)y^{(1)}(x) + a_0(x)y(x) = b(x)$$

where $a_n(x)$ is not identically zero. The functions $a_n(x), a_{n-1}(x), \ldots, a_1(x)$, and $a_0(x)$, which are all functions of the independent variable x alone, are called the **coefficient functions** of (1). If $b(x) \equiv 0$, then the linear differential equation (1) is said to be **homogeneous.** Whereas, if $b(x)$ is not the zero function, then (1) is said to be **nonhomogeneous.**

Observe that equation (1) is linear in y and its derivatives. The differential equation $x^2y^{(3)} - 2e^xy^{(2)} + (\cos x)y^{(1)} + 7y = \tan 4x$ is a third-order nonhomogeneous linear differential equation and $4y^{(2)} - 3xy = 0$ is a second-order homogeneous linear differential equation. The second-order differential equation $y^{(2)} + yy^{(1)} - 2y = 0$ is not linear because of the term $yy^{(1)}$. The differential equation $(y^{(2)})^3 + (\sin x)y = 3e^x$ is not linear because of the term $(y^{(2)})^3$.

In general, for $n \geq 2$ the n-th order linear differential equation (1) cannot be solved explicitly in terms of elementary functions or written as a formula involving quadratures as it can in the case when $n = 1$. Nonetheless, many physical phenomena such as mechanical systems and electrical circuits can be modeled by n-th order linear differential equations. Consequently, n-th order linear differential equations are important in the study of physics, engineering, and other applied sciences. Linear differential equations are "first order"

DOI: 10.1201/9781003678014-4

("lowest order") mathematical approximations to a wide variety of physical problems. In Section 4.1, we discuss the basic theory for n-th order linear differential equations. We state conditions which ensure the existence and uniqueness of solutions to (1). When (1) is homogeneous, we prove a superposition theorem which tells us how to combine solutions to obtain other, more general, solutions. We define the concept of linear independence for a set of functions and prove when (1) is homogeneous that there are n linearly independent solutions and show how to write the general solution in terms of those linearly independent solutions. Finally, we show how to write the general solution of (1) when it is nonhomogeneous.

In Section 4.2, we present a brief history of the search for methods to find roots of polynomial equations. Next, in Section 4.3, we show how to find the general solution of an n-th order homogeneous linear differential equation with constant coefficients by calculating the roots of an n-th degree polynomial equation. Then in Section 4.4, we indicate how to find the general solution of a nonhomogeneous linear differential equation with constant coefficients using the method of undetermined coefficients. In Chapter 5, we present the Laplace transform method for solving nonhomogeneous linear differential equations with constant coefficients. In Chapter 6, we present several applications whose solution ultimately requires the solution of some n-th order linear differential equation with constant coefficients. Later in Chapter 7, which concerns the solution of systems of n first-order differential equations, we show how to solve a general n-th order linear differential equation by rewriting it as a system of n first-order equations.

4.1 Basic Theory

An **initial value problem** for an n-th order linear differential equation consists of solving the differential equation

$$(1)\quad a_n(x)y^{(n)}(x) + a_{n-1}(x)y^{(n-1)}(x) + \cdots + a_1(x)y^{(1)}(x) + a_0(x)y(x) = b(x)$$

subject to a set of n constraints, called **initial conditions,** of the form

$$(2)\qquad y(x_0) = c_1,\ y^{(1)}(x_0) = c_2,\ \ldots,\ y^{(n-1)}(x_0) = c_n$$

where $c_1, c_2, \ldots, c_n$ are any specified constants and x_0 is some specified point.

In Section 2.2, we stated and proved an existence and uniqueness theorem for the first-order linear initial value problem: $y' = f(x, y);\ y(c) = d$. The following theorem, which we state without proof, generalizes this theorem to n-th order linear initial value problems.

An Existence and Uniqueness Theorem for an n-th Order Linear Initial Value Problem:

If the functions $a_n(x)$, $a_{n-1}(x)$, ..., $a_1(x)$, $a_0(x)$, and $b(x)$ are all continuous on an interval I and if $a_n(x) \neq 0$ for any x in I, then there exists a unique solution on I to the initial value problem consisting of the linear n-th order differential equation

$$(1)\quad a_n(x)y^{(n)}(x) + a_{n-1}(x)y^{(n-1)}(x) + \cdots + a_1(x)y^{(1)}(x) + a_0(x)y(x) = b(x)$$

and the initial conditions

$$(2)\qquad y(x_0) = c_1,\ y^{(1)}(x_0) = c_2,\ \ldots,\ y^{(n-1)}(x_0) = c_n$$

where $c_1, c_2, \ldots, c_n$ are any specified constants and x_0 is some point in the interval I.

Observe that the existence and uniqueness theorem guarantees that the zero function, $y(x) \equiv 0$, is the unique solution of the initial value problem

$$a_n(x)y^{(n)}(x) + a_{n-1}(x)y^{(n-1)}(x) + \cdots + a_1(x)y^{(1)}(x) + a_0(x)y(x) = 0;$$

$$y(x_0) = 0,\ y^{(1)}(x_0) = 0,\ \ldots,\ y^{(n-1)}(x_0) = 0$$

on any interval on which the functions $a_n(x), a_{n-1}(x), \ldots, a_1(x), a_0(x)$ are all continuous and $a_n(x) \neq 0$. We will need to use this fact later.

Verify that the functions $y_1(x) = x^2$ and $y_2(x) = x^3$ are two distinct solutions on $(-\infty, \infty)$ of the initial value problem

$$(3)\qquad x^2y'' - 4xy' + 6y = 0; \quad y(0) = 0, \quad y'(0) = 0.$$

The fact that the IVP (3) has two distinct solutions does not violate the existence and uniqueness theorem, since the IVP (3) does not satisfy all of the hypotheses of the existence and uniqueness theorem. Notice that the initial conditions are specified at $x_0 = 0$ and when $x = 0$ the coefficient of y'', the leading coefficient of the differential equation, is $a_2(x) = x^2 = 0$. That is, the hypothesis of the theorem $a_2(0) \neq 0$ is not satisfied on any interval I which contains the point $x = 0$—which is where the initial conditions are specified. This example illustrates that if a single hypothesis of the existence and uniqueness theorem fails to be satisfied, then nothing can be concluded about the existence of the solution of an initial value problem or about the uniqueness of a solution when there is one. In general, when a single hypothesis of the existence and uniqueness theorem is not satisfied, then the initial value problem may have no solution, it may have multiple solutions, or it may have a single, unique solution.

Example 1 Analysis of a Differential Equation and Verification of the Solution of an Initial Value Problem Analyze the differential equation

$$(4)\qquad y'' - 3y' + 2y = 4x^2$$

and verify that

(5) $$y(x) = 2e^x - 3e^{2x} + 2x^2 + 6x + 7$$

is the unique solution on $(-\infty, \infty)$ of the initial value problem

(6) $$y'' - 3y' + 2y = 4x^2; \quad y(0) = 6, \quad y'(0) = 2.$$

Solution

The DE (4) is linear and nonhomogeneous. The functions $a_2(x) = 1$, $a_1(x) = -3$, $a_0(x) = 2$, and $b(x) = 4x^2$ are all defined and continuous on $(-\infty, \infty)$ and $a_2(x) \neq 0$ on $(-\infty, \infty)$. Therefore, by the existence and uniqueness theorem there exists a unique solution to the IVP (6) on the interval $(-\infty, \infty)$ which satisfies $y(x_0) = c_1$ and $y'(x_0) = c_2$ for any choice of the constants x_0, c_1, and c_2.

The function (5) $y(x) = 2e^x - 3e^{2x} + 2x^2 + 6x + 7$ is defined and continuous on $(-\infty, \infty)$. Differentiating twice, we find

$$y'(x) = 2e^x - 6e^{2x} + 4x + 6$$

and

$$y''(x) = 2e^x - 12e^{2x} + 4.$$

Substituting these expressions for y, y', and y'' into the DE (4), we see that

$$\begin{aligned}
&y'' - 3y' + 2y \\
&= (2e^x - 12e^{2x} + 4) - 3(2e^x - 6e^{2x} + 4x + 6) + 2(2e^x - 3e^{2x} + 2x^2 + 6x + 7) \\
&= (2 - 6 + 4)e^x + (-12 + 18 - 6)e^{2x} + (4 - 18 + 14) + (-12 + 12)x + 4x^2 \\
&= 4x^2.
\end{aligned}$$

Thus, since y' and y'', as well as y, are all defined on $(-\infty, \infty)$, $y(x)$ is a solution of the DE (4) on $(-\infty, \infty)$. Evaluating y and y' at $x = 0$, we find

$$y(0) = 2 - 3 + 7 = 6 \quad \text{and} \quad y'(0) = 2 - 6 + 6 = 2.$$

Thus, the initial conditions specified in the IVP (6) are satisfied. Hence, $y(x)$ is the unique solution of the IVP (6) on the interval $(-\infty, \infty)$. ■

Example 2 Analysis of a Differential Equation and Verification of the Solution of an Initial Value Problem Analyze the differential equation

(7) $$x^2y'' + xy' - 4y = 0$$

and verify that

(8) $$y(x) = 2x^2 + \frac{1}{x^2}$$

is the unique solution on $(0, \infty)$ of the initial value problem

(9) $$x^2y'' + xy' - 4y = 0; \quad y(1) = 3, \quad y'(1) = 2.$$

Solution

The DE (7) is linear and homogeneous. The functions $a_2(x) = x^2$, $a_1(x) = x$, $a_0(x) = -4$, and $b(x) = 0$ are all defined and continuous on $(-\infty, \infty)$ and $a_2(x) \neq 0$ for $x \neq 0$. Since the initial conditions of the IVP (9) are specified at $x_0 = 1 \in (0, \infty)$, by the existence and uniqueness theorem there exists a unique solution to the IVP (9) in the interval $(0, \infty)$. Differentiating (8) $y(x) = 2x^2 + \dfrac{1}{x^2}$ twice, we get

$$y'(x) = 4x - \frac{2}{x^3}$$

and

$$y''(x) = 4 + \frac{6}{x^4}.$$

Substituting these expressions for y, y', and y'' into the DE (7), we find

$$x^2y'' + xy' - 4y = x^2(4 + \frac{6}{x^4}) + x(4x - \frac{2}{x^3}) - 4(2x^2 + \frac{1}{x^2})$$

$$= 4x^2 + \frac{6}{x^2} + 4x^2 - \frac{2}{x^2} - 8x^2 - \frac{4}{x^2}$$

$$= 0, \quad \text{provided } x \neq 0.$$

Since y, y', and y'' are defined for $x \neq 0$, $y(x)$ is a solution of the DE (7) on $(-\infty, 0)$ and $(0, \infty)$. Evaluating y and y' at $x = 1$, yields $y(1) = 2+1 = 3$ and $y'(1) = 4 - 2 = 2$. Hence, $y(x) = 2x^2 + \dfrac{1}{x^2}$ is the unique solution of the IVP (9) on the interval $(0, \infty)$. ■

Suppose that $y_1(x)$ and $y_2(x)$ are both solutions of the same differential equation and suppose that c_1 and c_2 are two arbitrary constants, then the sum $c_1y_1(x) + c_2y_2(x)$ is called a **linear combination** of the two solutions. The following example illustrates that a linear combination of two solutions of a homogeneous linear differential equation is also a solution.

Example 3 A Linear Combination of Two Solutions of a Homogeneous Linear Differential Equation is a Solution

a. Show that $y_1(x) = e^{-2x}$ and $y_2(x) = e^x$ are solutions on $(-\infty, \infty)$ of the homogeneous linear differential equation

$$y'' + y' - 2y = 0. \tag{10}$$

b. Show that the linear combination $y(x) = c_1y_1(x)+c_2y_2(x) = c_1e^{-2x}+c_2e^x$, where c_1 and c_2 are arbitrary constants, is also a solution to the DE (10) on $(-\infty, \infty)$.

Solution

1. Differentiating $y_1 = e^{-2x}$ twice, we find $y_1' = -2e^{-2x}$ and $y_1'' = 4e^{-2x}$. Substitution into (10), yields

$$y'' + y' - 2y = (4e^{-2x}) + (-2e^{-2x}) - 2(e^{-2x}) = (4 - 2 - 2)e^{-2x} = 0.$$

Since y_1, y_1', and y_1'' are defined on $(-\infty, \infty)$, $y_1 = e^{-2x}$ is a solution of the DE (10) on $(-\infty, \infty)$. Likewise, differentiating $y_2 = e^x$ twice, we find $y_2' = e^x$ and $y_2'' = e^x$. Substitution into equation (10) yields

$$y'' + y' - 2y = (e^x) + (e^x) - 2(e^x) = (1 + 1 - 2)e^x = 0.$$

Since y_2, y_2', and y_2'' are defined on $(-\infty, \infty)$, $y_2 = e^x$ is a solution of the DE (10) on $(-\infty, \infty)$.

2. Differentiating the linear combination $y = c_1y_1 + c_2y_2 = c_1e^{-2x} + c_2e^x$ twice, yields

$$y' = c_1y_1' + c_2y_2' = -2c_1e^{-2x} + c_2e^x$$

and

$$y'' = c_1y_1'' + c_2y_2'' = 4c_1e^{-2x} + c_2e^x.$$

Substituting into (10), we find

$$\begin{aligned} y'' + y' - 2y &\\ &= (4c_1e^{-2x} + c_2e^x) + (-2c_1e^{-2x} + c_2e^x) - 2(c_1e^{-2x} + c_2e^x) \\ &= c_1(4 - 2 - 2)e^{-2x} + c_2(1 + 1 - 2)e^x \\ &= 0e^{-2x} + 0e^x = 0. \end{aligned}$$

Since y_1, y_1', y_1'', y_2, y_2', and y_2'' are defined and continuous on $(-\infty, \infty)$, $y(x) = c_1y_1(x) + c_2y_2(x) = c_1e^{-2x} + c_2e^x$ is defined and continuous on $(-\infty, \infty)$. And since the linear combination $y(x) = c_1y_1(x) + c_2y_2(x)$ satisfies the DE (10) on $(-\infty, \infty)$ for arbitrary constants c_1 and c_2, $y(x)$ is a solution of (10) on $(-\infty, \infty)$. ■

The following superposition theorem generalizes the result of Example 3 for homogeneous linear differential equations.

Superposition Theorem

If $y_1(x), y_2(x), \ldots, y_k(x)$ are solutions on the interval I of the homogeneous linear differential equation

$$(11)\quad a_n(x)y^{(n)}(x) + a_{n-1}(x)y^{(n-1)}(x) + \cdots + a_1(x)y^{(1)}(x) + a_0(x)y(x) = 0,$$

then the linear combination $y(x) = c_1y_1(x) + c_2y_2(x) + \cdots + c_ky_k(x)$ where $c_1, c_2, \ldots, c_k$ are arbitrary constants is a solution of (11) on I.

Proof: We will prove this theorem for the special case in which $k = 2$ and $n = 3$. Then it will be obvious how to prove the theorem in general. We assume that $y_1(x)$ and $y_2(x)$ are solutions on an interval I of the third-order, homogeneous linear differential equation

$$a_3(x)y''' + a_2(x)y'' + a_1(x)y' + a_0(x)y = 0. \tag{12}$$

Since $y_1(x)$ is a solution of (12) on I,

$$a_3(x)y_1''' + a_2(x)y_1'' + a_1(x)y_1' + a_0(x)y_1 = 0 \quad \text{for all } x \in I$$

and since $y_2(x)$ is a solution of (12) on I,

$$a_3(x)y_2''' + a_2(x)y_2'' + a_1(x)y_2' + a_0(x)y_2 = 0 \quad \text{for all } x \in I.$$

Differentiating $y(x) = c_1y_1(x) + c_2y_2(x)$ three times, we find

$$y'(x) = c_1y_1'(x) + c_2y_2'(x),$$

$$y''(x) = c_1y_1''(x) + c_2y_2''(x),$$

and

$$y'''(x) = c_1y_1'''(x) + c_2y_2'''(x).$$

Substituting into the left-hand side of (12) for y, y', y'', and y''', we see that

$$\begin{aligned}
a_3(x)&y''' + a_2(x)y'' + a_1(x)y' + a_0(x)y \\
&= a_3(x)[c_1y_1'''(x) + c_2y_2'''(x)] + a_2(x)[c_1y_1''(x) + c_2y_2''(x)] \\
&\quad + a_1(x)[c_1y_1'(x) + c_2y_2'(x)] + a_0(x)[c_1y_1(x) + c_2y_2(x)] \\
&= c_1[a_3(x)y_1''' + a_2(x)y_1'' + a_1(x)y_1' + a_0(x)y_1] \\
&\quad + c_2[a_3(x)y_2''' + a_2(x)y_2'' + a_1(x)y_2' + a_0(x)y_2] \\
&= c_10 + c_20 = 0 \qquad \text{for all } x \in I.
\end{aligned}$$

■

For example, $y_1(x) = e^{2x}$ and $y_2(x) = \sin x$ are both solutions of the third-order homogeneous linear differential equation

$$y''' - 2y'' + y' - 2y = 0 \tag{13}$$

on the interval $(-\infty, \infty)$. Consequently, $y(x) = c_1e^{2x} + c_2\sin x$ is a solution to (13) on $(-\infty, \infty)$. In particular, $z(x) = 3e^{2x} - 5\sin x$ is a solution of (13) on $(-\infty, \infty)$.

The superposition theorem states that any linear combination of solutions of an n-th order homogeneous linear differential equation is a solution of the same differential equation. A useful corollary of the superposition theorem is the following: If $y(x)$ is any solution of an n-th order homogeneous linear differential equation, then $cy(x)$ is also a solution for any arbitrary constant *c*. The superposition theorem only applies to homogeneous linear differential equations. It does not apply to nonhomogeneous linear equations or nonlinear equations.

Let $\{y_1(x), y_2(x), \ldots, y_m(x)\}$ be a set of functions defined on an interval I. The set is **linearly dependent on the interval** I if there exist constants $c_1, c_2, \ldots, c_m$ not all zero such that

$$c_1y_1(x) + c_2y_2(x) + \cdots + c_my_m(x) = 0$$

for all $x \in I$. Otherwise, the set is **linearly independent on the interval** I.

That is, the set of functions $\{y_1(x), y_2(x), \ldots, y_m(x)\}$ is linearly independent on the interval I if and only if $c_1y_1(x) + c_2y_2(x) + \cdots + c_my_m(x) = 0$ for all x in I implies $c_1 = c_2 = \cdots = c_m = 0$. In other words, the set of functions is linearly independent on the interval I if and only if the only way the linear combination $c_1y_1(x) + c_2y_2(x) + \cdots + c_my_m(x)$ can be identically zero on the interval I is for $c_1 = c_2 = \cdots = c_m = 0$.

For example, the following sets of functions are linearly independent on the interval $I = (-\infty, \infty)$—that is, on the entire real line: $\{1,\ x,\ 3x^2,\ x^5,\ -6x^{10}\}$, $\{e^{2x},\ xe^{2x},\ e^{3x},\ 5x\}$, and $\{\sin x,\ \cos 3x,\ e^{-x},\ xe^x \sin 2x\}$.

The set $\left\{\dfrac{1}{x^2}, \dfrac{1}{x-1}, \dfrac{1}{x+1}\right\}$ is not linearly independent on $(-\infty, \infty)$. This set is linearly independent on each of the intervals $(-\infty, -1)$, $(-1, 0)$, $(0, 1)$, and $(1, \infty)$. The set is not linearly independent on any interval which contains -1, 0, or 1, since some function in the set is not defined at -1, 0, or 1.

The set $\{2,\ x,\ 3 - 4x\}$ is linearly dependent on every interval I, since $(-3) \cdot 2 + (8) \cdot x + (2) \cdot (3 - 4x) = 0$ for all real x.

The set $\{x,\ |x|\}$ is linearly dependent on the interval $(0, 3)$, since $(-1) \cdot x + (1) \cdot |x| = 0$ for all x in $(0, 3)$. But the set $\{x,\ |x|\}$ is linearly independent on the interval $(-3, 3)$, since $c_1x + c_2|x| = 0$ for $x = 1$ implies $c_1 + c_2 = 0$ and since $c_1x + c_2|x| = 0$ for $x = -1$ implies $-c_1 + c_2 = 0$. The simultaneous solution of $c_1 + c_2 = 0$ and $-c_1 + c_2 = 0$ is $c_1 = c_2 = 0$. This example illustrates that the set of functions $\{y_1(x), y_2(x), \ldots, y_m(x)\}$ can be shown to be linearly independent on an interval I by evaluating the linear combination $c_1y_1(x) + c_2y_2(x) + \cdots + c_my_m(x) = 0$ at m distinct points $x_1,\ x_2, \ldots, x_m$ in the interval I and showing that the only solution of the m simultaneous equations

$$\begin{array}{ccccccccc}
c_1y_1(x_1) & + & c_2y_2(x_1) & + & \cdots & + & c_my_m(x_1) & = & 0 \\
c_1y_1(x_2) & + & c_2y_2(x_2) & + & \cdots & + & c_my_m(x_2) & = & 0 \\
\vdots & \vdots & \vdots & \vdots & \vdots & \vdots & \vdots & \vdots & \vdots \\
c_1y_1(x_m) & + & c_2y_2(x_m) & + & \cdots & + & c_my_m(x_m) & = & 0
\end{array}$$

is $c_1 = c_2 = \cdots = c_m = 0$.

In order to prove that a set of functions $\{y_1(x), y_2(x), \ldots, y_m(x)\}$ is linearly dependent on an interval I, we must find explicit constants $c_1, c_2, \ldots, c_m$ not all zero such that $c_1y_1(x) + c_2y_2(x) + \cdots + c_my_m(x) = 0$ for all $x \in I$. On the other hand, to show that a set of functions $\{y_1(x), y_2(x), \ldots, y_m(x)\}$ is linearly independent on an interval I, we must prove that we cannot find constants $c_1, c_2, \ldots, c_m$ not all zero such that $c_1y_1(x) + c_2y_2(x) + \cdots + c_my_m(x) = 0$ for all $x \in I$. This is usually very difficult to do directly. However, when $y_1(x), y_2(x), \ldots, y_m(x)$ are all solutions of the same homogeneous linear differential equation, then, as we shall soon discover, it is fairly easy to check for linear dependence or linear independence.

Let the functions $y_1(x), y_2(x), \ldots, y_m(x)$ all be differentiable at least $m-1$ times for all x in some interval I. The **Wronskian** of $y_1, y_2, \ldots, y_m$ on I is the determinant

$$W(y_1, y_2, \ldots, y_m, x) = \begin{vmatrix} y_1 & y_2 & \cdots & y_m \\ y_1' & y_2' & \cdots & y_m' \\ \vdots & \vdots & & \vdots \\ y_1^{(m-1)} & y_2^{(m-1)} & \cdots & y_m^{(m-1)} \end{vmatrix}.$$

The Wronskian is named in honor of Józef Maria Hoëné Wronski (1778-1853), who was born in Poland, studied mathematics and philosophy in Germany, and lived much of his life in France. For $m = 2$ the Wronskian of y_1 and y_2 is

$$W(y_1, y_2, x) = \begin{vmatrix} y_1 & y_2 \\ y_1' & y_2' \end{vmatrix} = y_1y_2' - y_1'y_2.$$

And for $m = 3$ the Wronskian of y_1, y_2, and y_3 is

$$W(y_1, y_2, y_3, x) = \begin{vmatrix} y_1 & y_2 & y_3 \\ y_1' & y_2' & y_3' \\ y_1'' & y_2'' & y_3'' \end{vmatrix}$$
$$= y_1y_2'y_3'' + y_2y_3'y_1'' + y_3y_1'y_2'' - y_1''y_2'y_3 - y_2''y_3'y_1 - y_3''y_1'y_2.$$

Theorem 4.1 Let the functions $y_1(x), y_2(x), \ldots, y_m(x)$ all be differentiable at least $m-1$ times for all x in some interval I. If the functions $y_1, y_2, \ldots, y_m$ are linearly dependent on the interval I, then for all $x \in I$, the Wronskian $W(y_1, y_2, \ldots, y_m, x) = 0$.

Proof: Since by hypothesis the set of functions $\{y_1(x), y_2(x), \ldots, y_m(x)\}$ is assumed to be linearly dependent on the interval I, there exist constants $c_1, c_2, \ldots, c_m$ not all zero such that

$$c_1y_1(x) + c_2y_2(x) + \cdots + c_my_m(x) = 0 \quad \text{for all} \quad x \in I.$$

Differentiating this equation $m-1$ times, we find for all $x \in I$

$$\begin{array}{ccccccccc} c_1y_1^{(1)}(x) & + & c_2y_2^{(1)}(x) & + & \cdots & + & c_my_m^{(1)}(x) & = & 0 \\ \vdots & \vdots & \vdots & \vdots & & \vdots & \vdots & \vdots & \vdots \\ c_1y_1^{(m-1)}(x) & + & c_2y_2^{(m-1)}(x) & + & \cdots & + & c_my_m^{(m-1)}(x) & = & 0. \end{array}$$

This system of m equations in the m unknowns $c_1, c_2, \ldots, c_m$ may be written in matrix-vector notation as

$$(14) \qquad \begin{pmatrix} y_1 & y_2 & \cdots & y_m \\ y_1' & y_2' & \cdots & y_m' \\ \vdots & \vdots & & \vdots \\ y_1^{(m-1)} & y_2^{(m-1)} & \cdots & y_m^{(m-1)} \end{pmatrix} \begin{pmatrix} c_1 \\ c_2 \\ \vdots \\ c_m \end{pmatrix} = \begin{pmatrix} 0 \\ 0 \\ \vdots \\ 0 \end{pmatrix}$$

(If you are unfamiliar with matrix-vector notation, see Section 8.1.) Recall from linear algebra the theorem which states: "A homogeneous system of m equation in m unknowns has a nonzero solution if and only if the determinant of the coefficient matrix is zero." Since we have assumed that not all of the unknowns (the c_is) are zero, equation (14) has a nonzero solution and, therefore, the determinant of the coefficient matrix is zero—that is, $W(y_1, y_2, \ldots y_m, x) = 0$ for all $x \in I$. ■

The functions $y_1 = \sin 2x$ and $y_2 = \sin x \cos x$ are linearly dependent on the interval $(-\infty, \infty)$, since $1 \cdot y_1 - 2 \cdot y_2 = 1 \cdot \sin 2x - 2 \cdot \sin x \cos x = 0$ for all $x \in (-\infty, \infty)$. Hence, by Theorem 4.1, we must have $W(y_1, y_2, x) = 0$ for all $x \in (-\infty, \infty)$. Calculating this Wronskian, we find for all $x \in (-\infty, \infty)$

$$\begin{aligned} W(y_1, y_2, x) &= \begin{vmatrix} \sin 2x & \sin x \cos x \\ 2\cos 2x & -\sin^2 x + \cos^2 x \end{vmatrix} \\ &= \sin 2x(-\sin^2 x + \cos x^2) - 2\cos 2x(\sin x \cos x) \\ &= (2 \sin x \cos x)(\cos 2x) - 2 \sin x \cos x \cos 2x = 0. \end{aligned}$$

From the study of logic, we know that if the statement "A implies B" is true, then the statement "not B implies not A" is also true. That is, if "A implies B" is a theorem, then the contrapositive of the statement, which is "not B implies not A" is also a theorem. The contrapositive of Theorem 4.1 is the following theorem.

Theorem 4.2 Let the functions $y_1(x), y_2(x), \ldots, y_m(x)$ all be differentiable at least $m-1$ times for all x in some interval I. If for some $x \in I$ the Wronskian $W(y_1, y_2, \ldots, y_m, x) \neq 0$, then the functions $y_1, y_2, \ldots, y_m$ are linearly independent on the interval I.

Example 4 Verification of the Linear Independence of Two Differentiable Functions

Show that e^x and e^{2x} are linearly independent on the interval $(-\infty, \infty)$.

Solution

The functions e^x and e^{2x} are differentiable on $(-\infty, \infty)$ and their Wronskian is

$$W(e^x, e^{2x}, x) = \begin{vmatrix} e^x & e^{2x} \\ e^x & 2e^{2x} \end{vmatrix} = 2e^{3x} - e^{3x} = e^{3x} \neq 0 \quad \text{for all } x \in (-\infty, \infty).$$

Therefore, by Theorem 4.2, the functions e^x and e^{2x} are linearly independent on $(-\infty, \infty)$. ■

Example 5 Verification of the Linear Independence of Three Differentiable Functions Show that the functions $1, x$, and x^2 are linearly independent on the interval $(-\infty, \infty)$.

Solution

The functions $1, x$, and x^2 are at least twice differentiable on $(-\infty, \infty)$ and their Wronskian is

$$W(1, x, x^2, x) = \begin{vmatrix} 1 & x & x^2 \\ 0 & 1 & 2x \\ 0 & 0 & 2 \end{vmatrix} = 2 \neq 0 \quad \text{for all } x \in (-\infty, \infty).$$

Therefore, by Theorem 4.2, the functions $1, x$, and x^2 are linearly independent on $(-\infty, \infty)$. ■

If the Wronskian of a set of functions is zero at every point in an interval, that does not imply the set of functions is linearly dependent on the interval. Consider the functions $y_1(x) = x^3$ and $y_2(x) = |x|^3$ on the interval $(-2, 2)$. Differentiating, we find $y_1'(x) = 3x^2$ and

$$y_2'(x) = \begin{cases} -3x^2, & x < 0 \\ 0, & x = 0 \\ 3x^2, & x > 0 \end{cases}$$

(Verify that $y_2'(0)$ exists and has the value zero.) Computing the Wronskian of $\{x^3, |x|^3\}$ on $(-2, 0)$, we find

$$W(x^3, |x|^3, x) = \begin{vmatrix} x^3 & -x^3 \\ 3x^2 & -3x^2 \end{vmatrix} = -3x^5 + 3x^5 = 0.$$

At 0, we get

$$W(x^3, |x|^3, 0) = \begin{vmatrix} 0 & 0 \\ 0 & 0 \end{vmatrix} = 0.$$

And on $(0, 2)$, we see

$$W(x^3, |x|^3, x) = \begin{vmatrix} x^3 & x^3 \\ 3x^2 & 3x^2 \end{vmatrix} = 3x^5 - 3x^5 = 0.$$

Hence, $W(x^3, |x|^3, x) = 0$ for every $x \in (-2, 2)$. Now assume there exist constants c_1 and c_2 such that $c_1 y_1(x) + c_2 y_2(x) = c_1 x^3 + c_2 |x|^3 = 0$ for all $x \in (-2, 2)$. For $x = -1$, we must have $-c_1 + c_2 = 0$ and for $x = 1$, we must have $c_1 + c_2 = 0$. Simultaneously solving these two equations in c_1 and c_2, we find $c_1 = c_2 = 0$, which shows that the set $\{x^3, |x|^3\}$ is linearly independent on $(-2, 2)$. Hence, $\{x^3, |x|^3\}$ is a set of functions which is linearly independent on $(-2, 2)$ and whose Wronskian is identically zero on $(-2, 2)$.

We now prove the following important theorem regarding the relationship of linearly independent solutions of n-th order homogeneous linear differential equations and the Wronskian of the solutions.

Theorem 4.3 Let $a_n(x)$, $a_{n-1}(x)$, $\ldots$, $a_1(x)$, $a_0(x)$ be continuous on the interval I and let $a_n(x) \neq 0$ for all $x \in I$. The functions $y_1(x)$, $y_2(x)$, $\ldots$, $y_n(x)$ are linearly independent solutions on the interval I of the n-th order homogeneous linear differential equation

$$(15) \quad a_n(x) y^{(n)}(x) + a_{n-1}(x) y^{(n-1)}(x) + \cdots + a_1(x) y^{(1)}(x) + a_0(x) y(x) = 0,$$

if and only if $y_1(x), y_2(x), \ldots, y_n(x)$ are solutions of (15) on I and the Wronskian $W(y_1, y_2, \ldots, y_n, x) \neq 0$ for some $x \in I$.

Proof: The "if" portion of this theorem is a special case of Theorem 4.2 in which the functions $y_1(x), y_2(x), \ldots, y_n(x)$ are all known to be solutions of the same n-th order homogeneous linear differential equation.

We prove the "only if" portion of this theorem by contradiction. Thus, we assume that the functions $y_1(x), y_2(x), \ldots, y_n(x)$ are linearly independent on the interval I and that the Wronskian $W(y_1, y_2, \ldots, y_n, x) = 0$ for all $x \in I$. Choose any $a \in I$. By the theorem from linear algebra stated earlier, since $W(y_1, y_2, \ldots, y_n, a) = 0$ there exists a nonzero solution to the following linear homogeneous system of n equations in the n unknowns $c_1, c_2, \ldots, c_n$

$$(16) \quad \begin{array}{ccccccccc} c_1 y_1(a) & + & c_2 y_2(a) & + & \cdots & + & c_n y_n(a) & = & 0 \\ c_1 y_1^{(1)}(a) & + & c_2 y_2^{(1)}(a) & + & \cdots & + & c_n y_n^{(1)}(a) & = & 0 \\ \vdots & \vdots & \vdots & \vdots & \vdots & \vdots & \vdots & \vdots & \vdots \\ c_1 y_1^{(n-1)}(a) & + & c_2 y_2^{(n-1)}(a) & + & \cdots & + & c_n y_n^{(n-1)}(a) & = & 0 \end{array}$$

Let the nonzero solution be denoted by $k_1, k_2, \ldots, k_n$ and consider the linear combination

$$(17) \qquad y(x) = k_1 y_1(x) + k_2 y_2(x) + \cdots + k_n y_n(x).$$

Since $y_1(x), y_2(x), \ldots, y_n(x)$ are solutions of (15) on I, by the superposition theorem $y(x)$ is a solution of (15) on I. Differentiating (17) $n - 1$ times and then evaluating (17) and each derivative at $x = a$, we find from equations (16) that $y(x)$ satisfies the conditions

$$(18) \qquad y(a) = 0, \; y^{(1)}(a) = 0, \ldots, \; y^{(n-1)}(a) = 0.$$

That is, $y(x)$ satisfies the initial value problem consisting of the differential equation (15) and the initial conditions (18). But by the existence and uniqueness theorem, the unique solution of this initial value problem is the zero function. Hence,

$$y(x) = k_1 y_1(x) + k_2 y_2(x) + \cdots + k_n y_n(x) = 0 \quad \text{for all } x \in I.$$

That is, the functions $y_1(x), y_2(x), \ldots, y_n(x)$ are linearly dependent on the interval I, which is a contradiction. ■

In effect, Theorem 4.3 says, as illustrated by the proof, "If $y_1(x)$, $y_2(x), \ldots$, $y_n(x)$ are solutions on the interval I of the n-th order homogeneous linear differential equation (15), then either

(1) $W(y_1, y_2, \ldots, y_n, x) = 0$ for all $x \in I$ and the solutions are linearly dependent on I

or

(2) $W(y_1, y_2, \ldots, y_n, x) \neq 0$ for all $x \in I$ and the solutions are linearly independent on I."

That is, if $y_1(x), y_2(x), \ldots, y_n(x)$ are solutions on an interval I of the same n-th order homogeneous linear differential equation, it is not possible for their Wronskian to be zero at one point in I and to be nonzero at another point in I. Hence, to check a set of n solutions to (15) on an interval I to see if they are linearly dependent on I or linearly independent on I, all we need to do is to evaluate the Wronskian of the solutions at some convenient point in I and see if it is zero or not.

Example 6 Determination of Linear Dependence or Linear Independence The functions e^x, xe^x, and x^2e^x are solutions on $(-\infty, \infty)$ of the third-order homogeneous linear differential equation

$$y^{(3)} - 3y^{(2)} + 3y^{(1)} - y = 0.$$

Determine if they are linearly dependent or linearly independent on $(-\infty, \infty)$.

Solution

By definition

$$W(e^x, xe^x, x^2e^x, x) = \begin{vmatrix} e^x & xe^x & x^2e^x \\ e^x & (x+1)e^x & (x^2+2x)e^x \\ e^x & (x+2)e^x & (x^2+4x+2)e^x \end{vmatrix}.$$

Computing this Wronskian directly is tedious at best. However, evaluating the Wronskian at $x = 0 \in (-\infty, \infty)$ and computing, we find easily that

$$W(e^x, xe^x, x^2e^x, 0) = \begin{vmatrix} 1 & 0 & 0 \\ 1 & 1 & 0 \\ 1 & 2 & 2 \end{vmatrix} = 2 \neq 0.$$

So the functions e^x, xe^x, and x^2e^x are linearly independent on $(-\infty, \infty)$ by Theorem 4.3. ■

Example 7 Determination of Linear Dependence or Linear Independence The functions e^x, e^{-x}, and $\sinh x$ are solutions on $(-\infty, \infty)$ of the third-order homogeneous linear differential equation

$$y^{(3)} + y^{(2)} - y^{(1)} - y = 0.$$

Determine if they are linearly dependent or linearly independent on $(-\infty, \infty)$.

Solution

By definition

$$W(e^x, e^{-x}, \sinh x, x) = \begin{vmatrix} e^x & e^{-x} & \sinh x \\ e^x & -e^{-x} & \cosh x \\ e^x & e^{-x} & \sinh x \end{vmatrix}.$$

Evaluating the Wronskian at $x = 0$ and computing, we find

$$W(e^x, e^{-x}, \sinh x, 0) = \begin{vmatrix} 1 & 1 & 0 \\ 1 & -1 & 1 \\ 1 & 1 & 0 \end{vmatrix} = 0.$$

So by Theorem 4.3 the functions e^x, e^{-x}, and $\sinh x$ are linearly dependent on $(-\infty, \infty)$. ■

The following existence theorem proves that there are at least n linearly independent solutions on the interval I to the differential equation

$$(19) \quad a_n(x)y^{(n)}(x) + a_{n-1}(x)y^{(n-1)}(x) + \cdots + a_1(x)y^{(1)}(x) + a_0(x)y(x) = 0$$

where $a_n(x), a_{n-1}(x), \ldots, a_1(x), a_0(x)$ are all continuous on the interval I and $a_n(x) \neq 0$ for all $x \in I$.

Existence of N Linearly Independent Solutions to N-th Order Homogeneous Linear Differential Equations

Theorem 4.4 Let $a_n(x)$, $a_{n-1}(x)$, $\ldots$, $a_1(x)$, $a_0(x)$ be continuous on the interval I and let $a_n(x) \neq 0$ for all $x \in I$. There exist n linearly independent solutions on I of the n-th order homogeneous linear differential equation

$$(19) \quad a_n(x)y^{(n)}(x) + a_{n-1}(x)y^{(n-1)}(x) + \cdots + a_1(x)y^{(1)}(x) + a_0(x)y(x) = 0.$$

Proof: Let $x_0 \in I$. By the existence and uniqueness theorem, there exist n unique solutions on I, say $y_1(x), y_2(x), \ldots, y_n(x)$, of the n initial value problems consisting of the DE (19) and the following n sets of initial conditions

$$(20.1) \quad y_1(x_0) = 1, \quad y_1^{(1)}(x_0) = 0, \quad y_1^{(2)}(x_0) = 0, \ \ldots, \quad y_1^{(n-1)}(x_0) = 0$$

$$(20.2) \quad y_2(x_0) = 0, \quad y_2^{(1)}(x_0) = 1, \quad y_2^{(2)}(x_0) = 0, \ \ldots, \quad y_2^{(n-1)}(x_0) = 0$$

$$\vdots \qquad \vdots \qquad \vdots \qquad \vdots$$

$$(20.\text{n}) \quad y_n(x_0) = 0, \quad y_n^{(1)}(x_0) = 0, \quad y_n^{(2)}(x_0) = 0, \ \ldots, \quad y_n^{(n-1)}(x_0) = 1$$

To prove that the n solutions $y_1(x), y_2(x), \ldots, y_n(x)$ of the specified n initial value problems are linearly independent solutions of the DE (19) on the interval I, we examine their Wronskian evaluated at x_0 and find

$$W(y_1, y_2, \ldots, y_n, x_0) = \begin{vmatrix} y_1(x_0) & y_2(x_0) & \cdots & y_n(x_0) \\ y_1'(x_0) & y_2'(x_0) & \cdots & y_n'(x_0) \\ \vdots & \vdots & \ddots & \vdots \\ y_1^{(n-1)}(x_0) & y_2^{(n-1)}(x_0) & \cdots & y_n^{(n-1)}(x_0) \end{vmatrix}$$

$$= \begin{vmatrix} 1 & 0 & \cdots & 0 \\ 0 & 1 & \cdots & 0 \\ \vdots & \vdots & \ddots & \vdots \\ 0 & 0 & \cdots & 1 \end{vmatrix} = 1 \neq 0.$$

Hence, by Theorem 4.3 we know that the functions $y_1(x), y_2(x), \ldots, y_n(x)$ are linearly independent solutions of the DE (19) on the interval I. ■

By Theorem 4.4 there are at least n linearly independent solutions to the DE (19) on the interval I. The following representation theorem shows that there are at most n linearly independent solutions of (19) on I and it provides a representation for every solution of the DE (19) in terms of any set of n linearly independent solutions. The following theorem does not say that there is only one set of n linearly independent solutions—the set specified in the proof of Theorem 4.4—but that the maximum number of members in any solution set which is linearly independent on I is n.

A Representation Theorem for N-th Order Homogeneous Linear Differential Equations

Let $a_n(x), a_{n-1}(x), \ldots, a_1(x), a_0(x)$ be continuous on the interval I and let $a_n(x) \neq 0$ for all $x \in I$. If $y_1(x), y_2(x), \ldots, y_n(x)$ are linearly independent solutions on I of the n-th order homogeneous linear differential equation

$$(21) \quad a_n(x)y^{(n)}(x) + a_{n-1}(x)y^{(n-1)}(x) + \cdots + a_1(x)y^{(1)}(x) + a_0(x)y(x) = 0,$$

and if $y(x)$ is any solution of (21) on I, then

$$y(x) = c_1y_1(x) + c_2y_2(x) + \cdots + c_ny_n(x)$$

for suitably chosen constants c_i.

Proof: Let $y(x)$ be any solution of the DE (21) on the interval I and let x_0 be any point in I. Consider the function $z(x) = c_1y_1(x) + c_2y_2(x) + \cdots + c_ny_n(x)$ where the c_i are arbitrary constants. By the superposition theorem, $z(x)$ is a solution of the DE (21) on I. If we can choose the c_i so that

$$(22) \qquad z(x_0) = y(x_0),\ z^{(1)}(x_0) = y^{(1)}(x_0),\ \ldots,\ z^{(n-1)}(x_0) = y^{(n-1)}(x_0),$$

then $z(x) \equiv y(x)$ for all $x \in I$, since by the existence and uniqueness theorem there is only one solution of the DE (21) which satisfies the initial conditions (22). Differentiating the function $z(x)$ successively $n-1$ times and then evaluating $z(x)$ and its $n-1$ derivatives at x_0, we obtain the following system of n equations in the n unknowns c_i:

$$\begin{array}{ccccccccc}
c_1y_1(x_0) & + & c_2y_2(x_0) & + & \cdots & + & c_ny_n(x_0) & = & y(x_0) \\
c_1y_1^{(1)}(x_0) & + & c_2y_2^{(1)}(x_0) & + & \cdots & + & c_ny_n^{(1)}(x_0) & = & y^{(1)}(x_0) \\
\vdots & \vdots & \vdots & \vdots & & \vdots & \vdots & \vdots & \vdots \\
c_1y_1^{(n-1)}(x_0) & + & c_2y_2^{(n-1)}(x_0) & + & \cdots & + & c_ny_n^{(n-1)}(x_0) & = & y^{(n-1)}(x_0)
\end{array}$$

Or, in matrix-vector notation,

$$(23) \qquad \begin{pmatrix} y_1(x_0) & y_2(x_0) & \cdots & y_n(x_0) \\ y_1'(x_0) & y_2'(x_0) & \cdots & y_n'(x_0) \\ \vdots & \vdots & & \vdots \\ y_1^{(n-1)}(x_0) & y_2^{(n-1)}(x_0) & \cdots & y_n^{(n-1)}(x_0) \end{pmatrix} \begin{pmatrix} c_1 \\ c_2 \\ \vdots \\ c_n \end{pmatrix} = \begin{pmatrix} y(x_0) \\ y^{(1)}(x_0) \\ \vdots \\ y^{(n-1)}(x_0) \end{pmatrix}$$

From linear algebra, we know that this system of equations has a unique solution if and only if the determinant of the square matrix in (23) is nonzero. Of course, the determinant of the square matrix in (23) is the Wronskian of $y_1, y_2, \ldots, y_n$ evaluated at x_0. Since the functions $y_1(x), y_2(x), \ldots, y_n(x)$ are linearly independent on I, their Wronskian evaluated at x_0 is nonzero. Therefore, there exists a unique solution $(c_1, c_2, \ldots, c_n)^T$ to (23) and consequently $y(x) \equiv z(x) = c_1y_1(x) + c_2y_2(x) + \cdots + c_ny_n(x)$ on the interval I. ■

If $y_1(x), y_2(x), \ldots, y_n(x)$ are n linearly independent solutions of the n-th order homogeneous linear differential equation (21) on an interval I, then the **general solution** of the DE (21) on I is

$$y(x) = c_1y_1(x) + c_2y_2(x) + \cdots + c_ny_n(x)$$

where the c_i are arbitrary constants.

Example 8 General Solution of a Third-Order, Homogeneous Linear Differential Equation Given that the functions e^x, xe^x, and x^2e^x are linearly independent solutions on $(-\infty, \infty)$ of the third-order homogeneous linear differential equation $y^{(3)} - 3y^{(2)} + 3y^{(1)} - y = 0$, write the general solution.

Solution

The general solution of the given differential equation on the interval $(-\infty, \infty)$ is $y(x) = c_1e^x + c_2xe^x + c_3x^2e^x$ where c_1, c_2, and c_3 are arbitrary constants. ■

The **general n-th order nonhomogeneous linear differential equation** has the form

$$(24) \quad a_n(x)y^{(n)}(x)+a_{n-1}(x)y^{(n-1)}(x)+\cdots+a_1(x)y^{(1)}(x)+a_0(x)y(x) = b(x)$$

where $a_n(x) \neq 0$ in some interval I and $b(x) \neq 0$ for some $x \in I$. The **associated homogeneous linear differential equation** is

$$(25) \quad a_n(x)y^{(n)}(x) + a_{n-1}(x)y^{(n-1)}(x) + \cdots + a_1(x)y^{(1)}(x) + a_0(x)y(x) = 0.$$

Any function $y_p(x)$ which satisfies the nonhomogeneous linear DE (24) and which contains no arbitrary constant is called a **particular solution** of the nonhomogeneous equation. For example, $y_p(x) = 2$ is a particular solution of the nonhomogeneous linear differential equation $y''+4y = 8$, since $y_p''+4y_p = 0+4(2) = 8$ and y_p contains no arbitrary constant.

Example 9 Verification of a Particular Solution Show that $y(x) = x^2 - 2x$ is a particular solution of the nonhomogeneous linear differential equation

$$(26) \qquad y^{(3)} - 3y^{(2)} + 3y^{(1)} - y = -x^2 + 8x - 12.$$

Solution

Differentiating $y(x) = x^2-2x$ three times, we find $y^{(1)}(x) = 2x-2$, $y^{(2)}(x) = 2$, and $y^{(3)} = 0$. Substituting for $y(x)$ and its derivatives in the DE (26), we see that

$$\begin{aligned} y^{(3)} - 3y^{(2)} + 3y^{(1)} - y &= 0 - 3(2) + 3(2x - 2) - (x^2 - 2x) \\ &= -6 + 6x - 6 - x^2 + 2x = -x^2 + 8x - 12. \end{aligned}$$

Since $y(x)$ satisfies the DE (26) and contains no arbitrary constant, $y(x)$ is a particular solution of (26). ■

A Representation Theorem for N-th Order Nonhomogeneous Linear Differential Equations

If $y_p(x)$ is any particular solution on the interval I of the nonhomogeneous linear differential equation

$$(24)\quad a_n(x)y^{(n)}(x)+a_{n-1}(x)y^{(n-1)}(x)+\cdots+a_1(x)y^{(1)}(x)+a_0(x)y(x)=b(x),$$

and if $y_1(x), y_2(x), \ldots, y_n(x)$ are n linearly independent solutions on I of the associated homogeneous equation

$$(25)\quad a_n(x)y^{(n)}(x)+a_{n-1}(x)y^{(n-1)}(x)+\cdots+a_1(x)y^{(1)}(x)+a_0(x)y(x)=0,$$

then every solution of the DE (24) on the interval I has the form

$$y(x)=c_1y_1(x)+c_2y_2(x)+\cdots+c_ny_n(x)+y_p(x)$$

where $c_1, c_2, \ldots, c_n$ are suitably chosen constants.

Proof: Let $y_c(x)=c_1y_1(x)+c_2y_2(x)+\cdots+c_ny_n(x)$ where the c_i are arbitrary constants and let $z(x)$ be any solution of the nonhomogeneous linear DE (24) on I. In order to prove this theorem, we must show that it is possible to choose the c_i so that $z(x)=y_c(x)+y_p(x)$. Since $z(x)$ and $y_p(x)$ are both solutions on the interval I of the nonhomogeneous DE (24), $w(x)=z(x)-y_p(x)$ is a solution on I of the associated homogeneous DE (25). By the representation theorem for n-th order homogeneous linear differential equations, there exist constants $c_1, c_2, \ldots, c_n$ such that

$$w(x)=z(x)-y_p(x)=c_1y_1(x)+c_2y_2(x)+\cdots+c_ny_n(x).$$

Hence, $z(x)=y_c(x)+y_p(x)$ for suitably chosen constants $c_1, c_2, \ldots, c_n$. ■

Let $y_1(x), y_2(x), \ldots, y_n(x)$ be n linearly independent solutions on I of the homogeneous DE (25) associated with the nonhomogeneous DE (24). The linear combination $y_c(x)=c_1y_1(x)+c_2y_2(x)+\cdots+c_ny_n(x)$, where $c_1, c_2, \ldots, c_n$ are arbitrary constants, is called the **complementary solution** of the nonhomogeneous DE (24). Observe that the complementary solution is the general solution of the associated homogeneous equation (25). The **general solution** of the nonhomogeneous DE (24) is $y(x)=y_c(x)+y_p(x)$, where $y_c(x)$ is the complementary solution and $y_p(x)$ is any particular solution.

Example 10 General Solution of a Third-Order, Nonhomogeneous Linear Differential Equation Write the general solution of the nonhomogeneous linear differential equation

$$(26)\qquad y^{(3)}-3y^{(2)}+3y^{(1)}-y=-x^2+8x-12.$$

Solution

In Example 9, we showed that a particular solution of the DE (26) is $y_p(x) = x^2 - 2x$. And in Example 8, we showed that the general solution of the associated homogeneous equation $y^{(3)} - 3y^{(2)} + 3y^{(1)} - y = 0$ is $y_c(x) = c_1e^x + c_2xe^x + c_3x^2e^x$, where c_1, c_2, and c_3 are arbitrary constants. Hence, the general solution of the nonhomogeneous DE (26) is

$$y(x) = y_c(x) + y_p(x) = c_1e^x + c_2xe^x + c_3x^2e^x + x^2 - 2x,$$

where c_1, c_2, and c_3 are arbitrary constants. ■

In order to solve a nonhomogeneous, linear differential equation, we need to find a complementary solution—which is a linear combination of n linearly independent solutions of the associated homogeneous equation—and a particular solution. In this chapter, we will show how to solve a nonhomogeneous linear differential equation on the interval $(-\infty, \infty)$ when the functions $a_n(x), a_{n-1}(x), \ldots, a_0(x)$ are all constants. In Chapter 7, we will show how to write the general n-th order nonhomogeneous linear differential equation (1) as a system of n first-order differential equations. Then we will be able to generate a numerical solution to the initial value problem consisting of the initial conditions (2), and the linear nonhomogeneous differential equation (1) for arbitrary functions $a_n(x)$, $a_{n-1}(x), \ldots, a_0(x)$, $b(x)$.

The following is a summary of the results we obtained in this section.

A. If the functions $a_n(x), a_{n-1}(x), \ldots, a_1(x)$, and $a_0(x)$ are all continuous on the interval I and if $a_n(x) \neq 0$ for any x in I, then **for the homogeneous n-th order linear differential equation**

$$a_n(x)y^{(n)}(x) + a_{n-1}(x)y^{(n-1)}(x) + \cdots + a_1(x)y^{(1)}(x) + a_0(x)y(x) = 0$$

i. there exists a set containing exactly n linearly independent solutions on the interval I, say $\{y_1(x), y_2(x), \ldots, y_n(x)\}$,

ii. the general solution of the homogeneous linear differential equation on I is

$$y_c(x) = c_1y_1(x) + c_2y_2(x) + \cdots + c_ny_n(x)$$

where $c_1, c_2, \ldots, c_n$ are arbitrary constants, and

iii. there exists a unique solution of the homogeneous linear differential equation which satisfies the initial conditions

$$y(x_0) = k_1, \quad y^{(1)}(x_0) = k_2, \ \ldots, \quad y^{(n-1)}(x_0) = k_n$$

where $k_1,\ k_2,\ \ldots,\ k_n$ are specified constants and x_0 is some point in I.

B. If the functions $a_n(x), a_{n-1}(x), \ldots, a_1(x), a_0(x)$, and $b(x)$ are all continuous on the interval I, if $a_n(x) \neq 0$ for any x in I, and if $b(x) \neq 0$ for some

x in I, then **for the nonhomogeneous n-th order linear differential equation**

$$a_n(x)y^{(n)}(x) + a_{n-1}y^{(n-1)}(x) + \cdots + a_1(x)y^{(1)}(x) + a_0(x)y(x) = b(x)$$

i. there exists a particular solution of the nonhomogeneous differential equation on I, say $y_p(x)$,

ii. the general solution of the nonhomogeneous differential equation on I is

$$y(x) = y_c(x) + y_p(x)$$

where $y_c(x)$ is the general solution of the associated homogeneous equation, and

iii. there exists a unique solution to nonhomogeneous differential equation which satisfies the initial conditions

$$y(x_0) = k_1, \quad y^{(1)}(x_0) = k_2, \ldots, \quad y^{(n-1)}(x_0) = k_n$$

where $k_1, k_2, \ldots, k_n$ are specified constants and x_0 is some point in I.

4.1.1 Exercises

In Exercises 1–6 determine the largest interval on which the existence and uniqueness theorem guarantees the existence of a unique solution for the given initial value problem.

1. $3y'' - 2y' + 4y = x, \quad y(-1) = 2, \quad y'(-1) = 3$

2. $xy''' + xy' = 4; \quad y(1) = 0, \quad y'(1) = 1, \quad y''(1) = -1$

3. $x(x-3)y'' + 3y' = x^2; \quad y(1) = 0, \quad y'(1) = 1$

4. $x(x-3)y'' + 3y' = x^2; \quad y(5) = 0, \quad y'(5) = 1$

5. $\sqrt{1-x}\, y'' - 4y = \sin x; \quad y(-2) = 3, \quad y'(-2) = -1$

6. $(x^2 - 4)y'' + (\ln x)y = xe^x; \quad y(1) = 1, \quad y'(1) = 2$

7. Verify that e^x and e^{-x} are both solutions of the differential equation $y'' - y = 0$. Why are $\sinh x = (e^x - e^{-x})/2$ and $\cosh x = (e^x + e^{-x})/2$ also solutions?

8. Verify that the complex valued functions e^{ix} and e^{-ix} are solutions of the differential equation $y'' + y = 0$. Why are $\sin x = (e^{ix} - e^{-ix})/2i$ and $\cos x = (e^{ix} + e^{-ix})/2$ also solutions?

9. Verify that x, x^{-2}, and $c_1x + c_2x^{-2}$ where c_1 and c_2 are arbitrary constants are solutions of the differential equation $x^2y'' + 2xy' - 2y = 0$ for $x > 0$.

10. Verify that the functions 1 and x^2 are linearly independent on $(-\infty, \infty)$ and are solutions of the differential equation $2yy'' - (y')^2 = 0$. Is the linear combination $y(x) = c_1 + c_2x^2$ a solution of this differential equation for arbitrary constants c_1 and c_2? Why does this not violate the superposition theorem?

11. Show that the following sets of functions are linearly dependent on $(-\infty, \infty)$ by finding constants $c_1, c_2, \ldots, c_n$ not all zero such that $c_1y_1 + c_2y_2 + \cdots + c_ny_n = 0$.

    a. $\{2x, 3x\}$
    b. $\{1, x, 3x - 4\}$
    c. $\{1, \sin^2 x, \cos^2 x\}$
    d. $\{x, e^x, xe^x, (x + 2)e^x\}$

12. Use the Wronskian to show that the following sets of functions are linearly independent on $(-\infty, \infty)$.

    a. $\{\sin x, \cos x\}$
    b. $\{\sin x, \sin 2x\}$
    c. $\{1, x, x^2\}$
    d. $\{e^x, e^{2x}, xe^{2x}\}$

13. a. Verify that $y_1 = e^x$ and $y_2 = e^{-x}$ are linearly independent solutions on $(-\infty, \infty)$ of the homogeneous linear differential equation

    $$y'' - y = 0. \tag{27}$$

    b. Write the general solution of (27).

    c. Find the solution which satisfies the initial conditions $y(0) = 0$, $y'(0) = 1$.

14. a. Verify that $y_1 = 1$, $y_2 = \sin x$ and $y_3 = \cos x$ are linearly independent solutions on $(-\infty, \infty)$ of the homogeneous linear differential equation

    $$y''' + y' = 0. \tag{28}$$

    b. Write the general solution of (28).

    c. Find the solution which satisfies the initial conditions $y(0) = 1$, $y'(0) = 0$, $y''(0) = -1$.

15. a. Verify that $y_1 = x$ and $y_2 = x \ln x$ are linearly independent solutions on $(0, \infty)$ of the homogeneous linear differential equation

    $$x^2y'' - xy' + y = 0. \tag{29}$$

    b. Write the general solution of (29).

    c. Find the solution which satisfies the initial conditions $y(1) = 2$, $y'(1) = -1$.

16. a. Verify that $y_p = 8$ is a particular solution on $(-\infty, \infty)$ of the nonhomogeneous linear differential equation

$$y'' - 4y = -32. \tag{30}$$

b. Write the associated homogeneous equation.

c. Verify that $y_1 = e^{2x}$ and $y_2 = e^{-2x}$ are linearly independent solutions on $(-\infty, \infty)$ of the associated homogeneous equation.

d. Write the complementary solution for (30).

e. Write the general solution of (30).

f. Find the solution of (30) which satisfies the initial conditions $y(0) = -9$, $y'(0) = 6$.

17. a. Verify that $y_p = 3x + 2$ is a particular solution on $(-\infty, \infty)$ of the nonhomogeneous linear differential equation

$$y'' + 9y = 27x + 18. \tag{31}$$

b. Write the associated homogeneous equation.

c. Verify that $y_1 = \sin 3x$ and $y_2 = \cos 3x$ are linearly independent solutions on $(-\infty, \infty)$ of the associated homogeneous equation.

d. Write the complementary solution for (31).

e. Write the general solution of (31).

f. Find the solution of (31) which satisfies the initial conditions $y(0) = 23$, $y'(0) = 21$.

18. a. Verify that $y_p = x + 1/x$ is a particular solution on $(0, \infty)$ of the nonhomogeneous, linear differential equation

$$x^2y'' + xy' - 4y = -3x - 3/x. \tag{32}$$

b. Write the associated homogeneous equation.

c. Verify that $y_1 = x^2$ and $y_2 = 1/x^2$ are linearly independent solutions on $(0, \infty)$ of the associated homogeneous equation.

d. Write the complementary solution for (32).

e. Write the general solution of (32).

f. Find the solution of (32) which satisfies the initial conditions $y(1) = 3$, $y'(1) = -6$.

4.2 Roots of Polynomials

Previously you found roots of polynomials in order to solve word problems in algebra, to aid in the graphing of polynomials, and to find critical points, relative maxima, and relative minima of polynomials in calculus. In the next section, we will see how roots of polynomials enter into the solution of differential equations. Furthermore, as you continue to study mathematics, you will encounter a multitude of other occasions on which you will need to find the roots of a polynomial.

A **polynomial of degree n in one variable** x is a function of the form

$$p(x) = a_n x^n + a_{n-1} x^{n-1} + \cdots + a_1 x + a_0$$

where n is a positive integer; a_n, $a_{n-1}, \ldots$, a_1, a_0 are complex numbers; and $a_n \neq 0$. A **root of a polynomial** is any complex number r such that $p(r) = 0$.

As we shall see shortly, the history of solving the polynomial equation $p(x) = 0$—that is, of finding the roots of the polynomial $p(x)$—is divided into two distinct searches. One search consisted of trying to find an explicit formula for the roots, while the other search consisted of developing techniques for approximating roots.

In order to truly appreciate the algebra of old, we need to realize that the symbolism we currently use is approximately four hundred years old—some symbolism being much more recent. The first printed occurrence of the $+$ and $-$ sign, for example, appeared in an arithmetic book published by Johann Widman in 1489. However, the symbols were not used to represent the operations of addition and subtraction but merely to indicate excess and deficiency. The signs $+$ and $-$ were used to represent operations by the Dutch mathematician Giel Vander Hoecke in 1514 and probably by others somewhat earlier. The equal sign, $=$, appeared in 1557 in *The Whetstone of Witte* by Robert Recorde. Our inequality symbols of $<$ and $>$ are due to Thomas Harriot in 1631. In 1637, the French mathematician René Descartes introduced our custom of using letters early in our alphabet to denote constants and letters late in our alphabet to represent unknowns. Descartes also introduced our system of exponents—x^2 for $x \cdot x$, x^3 for $x \cdot x \cdot x$, etc.

In 1842, G. H. F. Nesselmann divided the development of algebra into three stages. The first stage was rhetorical algebra. In rhetorical algebra, problems were posed and solved using pure prose—no abbreviations or symbolism was employed. This stage lasted from antiquity until the time of Diophantus of Alexandria or approximately 250 A.D. The second stage was syncopated algebra. In this stage, abbreviations were used for some of the quantities and operations which occurred most frequently. The final stage is called symbolic algebra. In this stage, the solution of a problem is obtained by purely algebraic manipulation of mathematical shorthand.

In addition to the changing aspect of algebra, we should also be aware that different types of numerals and number representation systems were used and that the various sets of numbers—natural, rational, real, and complex—were themselves in the process of being developed. The ancient Babylonians, Chinese, Egyptians, Greeks, and Romans each had their own distinct numerals and systems of representing numbers. The Babylonians used wedge-shaped characters and base 60. The Egyptians used hieroglyphics. The Greek alphabetic numeral system was derived from the initial letters of the number name. And the Romans also used a letter-based numeral system with which most of us are familiar. These systems of numeration did not represent numbers compactly nor were they easy systems in which to develop calculating algorithms. The Hindu-Arabic numeral system—the system which we, and the majority of the world, currently use to represent numbers and for computing—was invented by the Hindus and preserved and transmitted to the Western World by the Arabs. The earliest existing examples of our present numeral system appeared in the inscriptions of King Asoka who ruled most of India in the third century B.C. Early examples do not contain a symbol for zero or the idea of positional notation. Exactly when these crucial ideas were introduced in India is not certain. But it is clear that they were introduced prior to 800 A.D. It was not until the 13th century that our current computing patterns used in conjunction with the Hindu-Arabic numeral system reached its present form.

The development of the various sets of numbers proceeded as follows: First, man invented the natural numbers for counting. Then, he invented the positive rational numbers for the purpose of measurement. Next, the positive real numbers were devised to accommodate irrational numbers such as $\sqrt{2}$. Later, negative numbers were accepted and finally the complex numbers were devised to incorporate imaginary numbers.

Until the beginning of written history, the natural numbers seem to have served the purpose of mankind adequately. The idea of unit fractions—that is, a fraction with numerator one—arose early in both Babylon and Egypt. About 2000 B.C. the Babylonians conceived the idea of fractions with numerators greater than one. However, no acceptable treatment of fractions which predates the Egyptian Ahmes Papyrus of approximately 1550 B.C. has yet been discovered. The followers of Pythagoras (c. 540 B.C.) knew and demonstrated the incommensurability of the diagonal and the side of a square. That is, the Pythagoreans knew that $\sqrt{2}$ was irrational. About 375 B.C. Theaetetus developed a general theory of quadratic irrationals. The first mention of negative numbers other than as subtrahends appears in the *Arithmetica* of Diophantus in about 275 A.D., where the equation $4x + 20 = 4$ is called absurd since the solution is $x = -4$. In India, negative numbers were viewed as distinct entities by at least 628. In his *Ars Magna* (1545), Cardan first accepted negative numbers as roots of polynomials. Also, Cardan was the first person to use the square root of a negative number in computations. He demonstrated that $x = 5 + \sqrt{-15}$ and $x = 5 - \sqrt{-15}$ were both solutions of the equation $x^2 + 40 = 10x$. In 1637, Descartes coined the terms real and imaginary. In 1748, Euler introduced the use of i for $\sqrt{-1}$ and in 1832, Gauss gave the name complex numbers to the quantities $a + bi$.

Prior to 1545, mathematicians recognized only positive real numbers as roots of polynomials. Therefore, to solve a polynomial equation before 1545 meant to find the positive real roots. The linear equation $ax = b$, where a and b were both positive, could be solved both algebraically and geometrically by early civilizations. The ancient Babylonians (c. 2000 B.C.) knew how to solve the quadratic equation $ax^2 + bx + c = 0$ algebraically by both the method of completing the square and by substituting into the quadratic formula. They could also solve certain special cubic equations. The Persian poet, astronomer, and mathematician Omar Khayyam (c. 1044-1123) was able to geometrically solve every type of cubic equation for its positive roots. It was not until almost five hundred years later that the algebraic solution of the cubic equation was accomplished. Shortly thereafter, the algebraic solution of the quartic equation was achieved also.

About 1515, Scipio del Ferro discovered the algebraic solution of the cubic equation of the form $x^3 + bx = c$. He did not publish his result but revealed his secret to his pupil Antonio Maria Fior (Florido). In 1530, Zuanne de Tonini da Coi sent the following two problems to Niccolo Fontana to be solved:

$$x^3 + 3x^2 = 5 \quad \text{and} \quad x^3 + 6x^2 + 8x = 1000.$$

Fontana was also known as Tartaglia (the stammerer) because of a saber wound he received when he was only 13 years old at the hands of the French in the 1512 massacre at Brescia. In 1535, Tartaglia claimed to have algebraically solved the cubic equation of the form $x^3 + ax^2 = c$. Florido, who believed that Tartaglia was merely boasting, challenged Tartaglia to a public contest. Each contestant was to submit to the other the same number of cubic equations to be solved within a given period of time. Tartaglia accepted the challenge. He knew he could defeat his opponent if he submitted only problems of the type which he could solve, namely, $x^3 + ax^2 = c$, and if he could also solve problems of the type which Florido could solve, namely, $x^3 + bx = c$. Tartaglia exerted himself, and a few days before the scheduled contest he discovered how to solve problems of the type Florido could solve. Knowing how to solve both types of cubic equations, whereas his opponent only knew how to solve one type of cubic equation, Tartaglia triumphed completely. However, Tartaglia did not publish his method of solution. In 1539, Girolamo Cardano (Cardan), an unscrupulous man who practiced medicine in Milan, wrote Tartaglia requesting a meeting. At the meeting Tartaglia, upon having pledged Cardan to secrecy, revealed his method of solution—at first in cryptic verse and later in full detail. Cardan admits receiving the solution from Tartaglia but denies receiving any explanation of the method. In 1545, Cardan published his *Ars Magna* in which he reduced the general cubic equation $x^3 + px^2 + qx = r$ to the form $x^3 + bx = c$ and then solved the latter equation. Tartaglia protested vehemently. But Cardan was ably defended by one of his students, Ludovico

Ferrari, who claimed Cardan received the solution from Ferro through a third party and that Tartaglia, himself, was guilty of plagiarizing Ferro's solution.

In 1540, da Coi proposed a problem to Cardan which required the solution of the quartic equation $x^4 + 6x^2 + 36 = 60x$. Cardan was unable to solve this problem, but passed it on to Ferrari. Ferrari solved the problem and in the process showed how to reduce the solution of all quartic equations of the form $x^4 + px^2 + qx + r = 0$ to the solution of a cubic equation. In effect, Ferrari solved the general quartic polynomial $x^4 + ax^3 + bx^2 + cx + d = 0$, since it reduces to $x^4 + px^2 + qx + r = 0$ by means of a simple linear transformation. Cardan included Ferrari's solution of the quartic equation in his *Ars Magna*. Thus, the *Ars Magna* contains the first published solution of both the general cubic equation and the general quartic equation although neither was the work of the author!

In 1803, in 1805, and again in 1813, the Italian physician Paola Ruffini published inconclusive proofs that the roots of the fifth and higher order polynomials cannot be written in terms of the coefficients of the polynomial by means of radicals. This fact was later successfully proven by the Norwegian mathematician Niels Henrik Abel in 1824.

Sixteenth century Italian mathematicians assumed every polynomial with rational coefficients had a root. Late in the century, they were aware that a quadratic polynomial has two roots, a cubic polynomial has three roots, and a quartic polynomial has four roots. Peter Roth seems to be the first writer to explicitly state the fundamental theorem of algebra in his *Arithmetica philosophica* published in 1608. The fundamental theorem of algebra states that an n-th degree polynomial has n roots. D'Alembert attempted to prove this theorem in 1746. Euler (1749) and Lagrange also attempted to prove the theorem. The first rigorous proof is due to Gauss in 1799. A simpler proof was given in 1849.

The second search for roots of polynomials, and in retrospect the more fruitful search, consisted of developing techniques for approximating the values of the roots. As we mentioned earlier, the ancient Babylonians knew in about 2000 B.C. how to solve a quadratic equation algebraically using the quadratic formula. On a practical level, since the quadratic formula involves extracting a square root, the Babylonians found it necessary to devise a method for computing a square root. The following scheme for computing the square root of a positive real number is due to the Babylonians.

Let a be a positive real number, and let $x_1 > 0$ be a first approximation (guess) of the value of $\sqrt{a}$. Either

$$(1) \quad x_1 = \sqrt{a} \qquad \text{or} \qquad (2) \quad x_1 < \sqrt{a} \qquad \text{or} \qquad (3) \quad x_1 > \sqrt{a}$$

If (1) $x_1 = \sqrt{a}$, we are done.
If (2) $x_1 < \sqrt{a}$, then $\sqrt{a}x_1 < a$ and $\sqrt{a} < a/x_1$.
If (3) $x_1 > \sqrt{a}$, then $\sqrt{a}x_1 > a$ and $\sqrt{a} > a/x_1$.

Case (2) states if x_1 is less than $\sqrt{a}$, then a/x_1 is greater than $\sqrt{a}$. While case (3) states if x_1 is greater than $\sqrt{a}$, then a/x_1 is less than $\sqrt{a}$. In either case, both x_1 and a/x_1 are approximations of $\sqrt{a}$. One is an under estimate and the other is an over estimate. Notice that the product of x_1 and a/x_1 is a. Let x_2 denote the average of x_1 and a/x_1. That is, let

$$x_2 = (x_1 + a/x_1)/2.$$

Since x_1 and a/x_1 are estimates of $\sqrt{a}$, their average x_2 is also an estimate of $\sqrt{a}$. As before, either $x_2 = \sqrt{a}$, or $x_2 < \sqrt{a}$, or $x_2 > \sqrt{a}$. And if $x_2 \neq \sqrt{a}$, then

$$x_3 = (x_2 + a/x_2)/2$$

will be a new approximation of $\sqrt{a}$. Hence, we obtain the following iteration procedure for calculating an approximate value for $\sqrt{a}$: Make an initial guess $x_1 > 0$ which approximates $\sqrt{a}$. Then successively compute

$$x_{n+1} = (x_n + a/x_n)/2$$

for $n = 1, 2, 3, \ldots$ until x_{n+1} is as good an approximation to $\sqrt{a}$ as desired. Neither we nor the Babylonians proved that for any initial guess $x_1 > 0$, as $n \to +\infty$, $x_n \to \sqrt{a}$; but it does.

We should also note that the ancient Babylonians knew how to compute approximate roots of certain cubic polynomials.

In the third century A.D., Chinese mathematicians developed a method known as *fan fa* for computing approximate roots of polynomials. In 1303, Chu Shih-chieh computed roots of polynomials up to degree 14 using the method of *fan fa*. Unaware that the method of *fan fa* had been invented in China and in use for nearly 15 centuries, the English mathematician W. G. Horner published the equivalent of this method in 1819. In the West the method is called "Horner's method."

After inventing the calculus, Isaac Newton used the derivative of a function in a procedure which he invented in 1669 to iteratively calculate approximate roots of the function. As far as we know, Newton used his method, Newton's method, to find only the positive root of just one polynomial, $x^3 - 2x - 5$. Joseph Raphson simplified and improved Newton's method and published a new version of the method in 1690. It is the Newton–Raphson method which is currently often used to find roots of equations. Unfortunately, 19th century textbook writers ignored Raphson's contribution and the Newton–Raphson method is commonly called Newton's method today.

The Newton–Raphson method for finding a root of a differentiable function f is developed as follows. Let x_1 be an initial guess for the unknown value of a root of f. We approximate the graph of $y = f(x)$ at $(x_1, f(x_1))$ by the tangent line. See Figure 4.1. The equation of the tangent line at $(x_1, f(x_1))$—that is, the equation of the line through the point $(x_1, f(x_1))$ with slope $f'(x_1)$—is

$$y - f(x_1) = f'(x_1)(x - x_1).$$

Provided $f'(x_1) \neq 0$, the tangent line intersects the x-axis at some point, say, $(x_2, 0)$. Substituting $x = x_2$ and $y = 0$ into the above equation, we find x_2 satisfies

$$-f(x_1) = f'(x_1)(x_2 - x_1).$$

Or solving for x_2, we get

$$x_2 = x_1 - f(x_1)/f'(x_1).$$

The value x_2 is a new approximation to a root of f. The process is repeated.

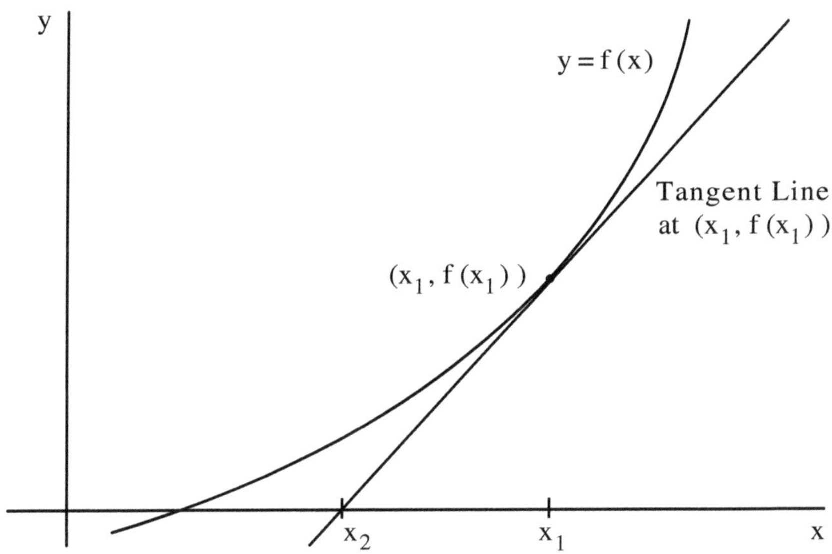

FIGURE 4.1
The Newton–Raphson Method.

Thus, Newton–Raphson's method for finding a root of a differentiable function f is as follows:

1. Make an initial guess, x_1, of a root.
2. Then provided $f'(x_n) \neq 0$ iteratively compute

$$x_{n+1} = x_n - f(x_n)/f'(x_n)$$

   until x_{n+1} is as good an approximation of a root as desired or until a specified number of iterations have been computed without the partial sequence $x_1, x_2, \ldots, x_{n+1}$ appearing to converge to a root.
3. If the partial sequence does not appear to be converging, then the process can be started over again with a new guess for x_1.

Consider the polynomial equation $f(x) = x^2 - a = 0$ where $a > 0$. Differentiating we get $f'(x) = 2x$. So the iterative portion of the Newton–Raphson method for this function is

$$\begin{aligned} x_{n+1} &= x_n - f(x_n)/f'(x_n) = x_n - (x_n^2 - a)/(2x_n) \\ &= (x_n^2 + a)/(2x_n) = (x_n + a/x_n)/2. \end{aligned}$$

This is exactly the recursion of the ancient Babylonian technique for computing a square root.

Now, consider the general n-th degree polynomial

$$p(x) = a_n x^n + a_{n-1}x^{n-1} + a_{n-2}x^{n-2} + \cdots + a_2x^2 + a_1x + a_0$$

where $a_n \neq 0$. For any given value of x, computing $p(x)$ as written above requires n additions and $2n - 1$ multiplications—one multiplication each to successively compute $x^2 = x \cdot x$, $x^3 = x^2 \cdot x$, ..., $x^n = x^{n-1} \cdot x$ and one multiplication each for $a_k \cdot x^k$ for $k = 1, 2, \ldots, n$. The polynomial $p(x)$ may be rewritten in the nested multiplication or Horner form

$$p(x) = ((\ldots((a_nx + a_{n-1})x + a_{n-2})x + \cdots + a_2)x + a_1)x + a_0.$$

In this form, n additions are still required but only n multiplications are required. (The computation of the value in the innermost set of parentheses requires one multiplication and one addition; the computation of the value in the next innermost set of parentheses also requires one multiplication and one addition, and so on.) This is a considerable savings in computing time—almost a 50% savings—since the operation of multiplication requires much more computing time than the operation of addition.

For any given value t, we can recursively compute a new list of coefficients b_{n-1}, b_{n-2}, ..., b_1, b_0, R as follows:

$$(1) \qquad \begin{aligned} b_{n-1} &= a_n \\ b_{n-2} &= b_{n-1}t + a_{n-1} \\ &\vdots \\ b_1 &= b_2t + a_2 \\ b_0 &= b_1t + a_1 \\ R &= b_0t + a_0 \end{aligned}$$

The coefficient $b_{n-2} = b_{n-1}t + a_{n-1} = a_nt + a_{n-1}$ is the number one gets by computing the value of the expression in the innermost set of parentheses of $p(t)$, the coefficient $b_{n-3} = b_{n-2}t + a_{n-2} = (a_nt + a_{n-1})t + a_{n-2}$ is the number one gets by computing the value of the expression in the first two innermost set of parentheses of $p(t)$, and so on. So $R = p(t)$.

Solving the set of equations (1) for the a's, we get

$$\begin{aligned}
a_n &= b_{n-1} \\
a_{n-1} &= b_{n-2} - b_{n-1}t \\
&\vdots \\
a_2 &= b_1 - b_2 t \\
a_1 &= b_0 - b_1 t \\
a_0 &= R - b_0 t
\end{aligned}$$

Substituting in $p(x)$ and rearranging algebraically, we find for any t

$$\begin{aligned}
p(x) &= a_n x^n + a_{n-1}x^{n-1} + \cdots + a_2 x^2 + a_1 x + a_0 \\
&= b_{n-1}x^n + (b_{n-2} - b_{n-1}t)x^{n-1} + \cdots + (b_1 - b_2 t)x^2 + (b_0 - b_1 t)x + \\
&\quad (R - b_0 t) \\
&= b_{n-1}x^{n-1}(x-t) + b_{n-2}x^{n-2}(x-t) + \cdots + b_1 x(x-t) + b_0(x-t) + R.
\end{aligned}$$

Defining $q(x) = b_{n-1}x^{n-1} + b_{n-2}x^{n-2} + \cdots + b_1 x + b_0$, we see that $p(x) = (x-t)q(x)+R$. That is, when $p(x)$ is divided by $x-t$, the quotient is $q(x)$ and the remainder is R. If t is a root of $p(x)$, then $R = 0$ and $p(x) = (x-t)q(x)$. Additional roots of $p(x)$ can then be found by finding roots of $q(x)$. The process of computing the coefficients $b_{n-1}, b_{n-2}, \ldots, b_0$ of the polynomial $q(x)$—which is of degree $n-1$—from the coefficients $a_n, a_{n-1}, \ldots, a_0$ when t is a root of $p(x)$ is called **deflation**.

Horner's method—which is actually the much more ancient Chinese method, *fan fa*—for finding the roots of a polynomial consists of applying the Newton–Raphson method to polynomials. To apply this method we must be able to evaluate both $p(x)$ and $p'(x)$ at any point t. Since $p(x) = (x-t)q(x)+R$, $p'(x) = (x-t)q'(x) + q(x)$ and therefore $p'(t) = q(t)$. Now $p(t) = R$ can be evaluated using equation (1) and $p'(t) = q(t)$ can be calculated from the analogous set of equations for q, namely

$$\begin{aligned}
c_{n-2} &= b_{n-1} \\
c_{n-3} &= c_{n-2}t + b_{n-2} \\
&\vdots \\
c_0 &= c_1 t + b_1 \\
S &= c_0 t + b_0 = q(t) = p'(t).
\end{aligned} \tag{2}$$

So, Horner's method for computing a root of $p(x)$ is as follows:

1. Make an initial guess for the value of a root, t_1.
2. Use equations (1) and (2) to compute $R = p(t_n)$ and $S = p'(t_n)$ for $n = 1, 2, 3, \ldots$.
3. Provided $S \neq 0$, use Newton–Raphson's method to compute
$$t_{n+1} = t_n - p(t_n)/p'(t_n) = t_n - R/S.$$
4. If for some n, R is as near zero as desired, then take a root of $p(x)$ to be t_n. If n becomes large, but R is not as near zero as desired, start over with a different value for t_1.

4.2.1 Comments on Computing

We used a computing program for the computing tasks discussed in this section. This program computes all the roots of a polynomial with complex coefficients of degree less than or equal to 10. It uses the Newton–Raphson method, Horner's method, and deflation in conjunction with other techniques to approximate the roots of a polynomial. The next two examples illustrate the typical output of our computing program. *You should compare the results you obtain using your software with these results.* The following two MAPLE statements also solve numerically the polynomial equation (3) appearing in Example 1 below.

polyeqn:= $x \wedge 7 + x \wedge 6 + 12 * x \wedge 5 - 28 * x \wedge 4 - 733 * x \wedge 3 +$

$$1011 * x \wedge 2 - 1784 * x - 38480 = 0:$$

factor(polyeqn,complex);

Example 1 Calculation of the Roots of a Polynomial Find the roots of the polynomial equation

$$x^7 + x^6 + 12x^5 - 28x^4 - 733x^3 + 1011x^2 - 1784x - 38480 = 0 \qquad (3)$$

Solution

We entered the value 7 into our computing program for the degree of the polynomial equation to be solved. Then we input the coefficients of the polynomial—namely, $a_7 = 1$, $a_6 = 1$, $a_5 = 12$, $a_4 = -28$, $a_3 = -733$, $a_2 = 1011$, $a_1 = -1784$, and $a_0 = -38480$. The roots of the polynomial as calculated by our program exercises appear in Figure 4.2.

One zero is computed to be $2.000000 + 3.000000i$. The actual zero of the polynomial corresponding to this computed zero is $2 + 3i$. So, in this case, the approximation is excellent. The second and third zeros computed were $-4.000000 - 1.153824 \times 10^{-11}i$ and $-4.000000 + 1.153809 \times 10^{-11}i$. The actual zeros of the polynomial corresponding to these roots are -4 and -4.

THE ZEROS OF THIS POLYNOMIAL ARE

ZERO	REAL PART	IMAGINARY PART
1	2.000000E+00	3.000000E+00
2	-4.000000E+00	-1.153824E-11
3	-4.000000E+00	1.153809E-11
4	2.000000E+00	-3.000000E+00
5	-1.000000E+00	6.000000E+00
6	5.000000E+00	2.419410E-16
7	-1.000000E+00	-6.000000E+00

FIGURE 4.2
Zeros of the Polynomial Equation (3).

This is the worst approximation of a zero for this polynomial, but it is not unacceptable. One thing you must learn is how to interpret results from a computer. When a polynomial has repeated real roots—such as, -4, -4—small, erroneous imaginary parts are often computed. Since 1.15×10^{-11} is small compared to 4, we can easily surmise that the actual zeros may be -4 and -4. The fourth zero is $2 - 3i$ and the fifth is $-1 + 6i$. We conclude that the sixth zero of the polynomial is probably 5 instead of the computed value of $5 + 2.419410 \times 10^{-16}i$. The seventh zero is $-1 - 6i$. ■

Example 2 Calculation of the Roots of a Polynomial Find the roots of the polynomial equation

(4) $$x^3 + (-7 - 3i)x^2 + (10 + 15i)x + 8 - 12i = 0.$$

Solution

We entered 3 for the degree of the polynomial equation to be solved into our computing program. Then we entered the values for the coefficients. The roots of this polynomial as calculated by our program are displayed in Figure 4.3. The values shown should be interpreted as i, $3 + 2i$, and 4.

THE ZEROS OF THIS POLYNOMIAL ARE

ZERO	REAL PART	IMAGINARY PART
1	-7.414635E-17	1.000000E+00
2	3.000000E+00	2.000000E+00
3	4.000000E+00	-4.440892E-16

FIGURE 4.3
Zeros of the Polynomial Equation (4). ■

4.2.2 Exercises

In the following exercises use computer routines available to you to calculate the roots of appropriate polynomial equations.

1. In 1225, Leonardo Fibonacci showed in his text *Flos* that the equation

$$x^3 + 2x^2 + 10x = 20$$

has no solution of the form $a+\sqrt{b}$ where a and b are rational numbers. Then he obtained the following root by some undisclosed numerical method.

$$x = 1^o 22' 7'' 42^{iii} 33^{iv} 4^v 40^{vi} \approx 1.3688081075$$

Find all solutions of the given cubic equation and compare with the above solution.

2. Find the roots of the following equations which da Coi sent to Tartaglia for solution in 1530.

$$x^3 + 3x^2 = 5 \quad \text{and} \quad x^3 + 6x^2 + 8x = 1000$$

3. Find the roots of the following quartic equation which da Coi sent to Cardan for solution in 1540.

$$x^4 + 6x^2 + 36 = 60x$$

4. In his *Ars Magna* of 1545, Cardan demonstrated that $x = 5 \pm \sqrt{-15}$ were the solutions of $x^2 + 40 = 10x$. Compare values you obtain with the values obtained by Cardan.

5. Listed below are some cubic equations which Cardan solved. Compare the values you obtain with the values obtained by Cardan.

Equation	**Roots**
$x^3 + 10x = 6x^2 + 4$	$2,\ 2 \pm \sqrt{2}$
$x^3 + 21x = 9x^2 + 5$	$5,\ 2 \pm \sqrt{3}$
$x^3 + 26x = 12x^2 + 12$	$2,\ 5 \pm \sqrt{19}$

6. In 1567, Nicolas Petri of Deventer calculated $1 + \sqrt{2}$ to be the positive root of the quartic equation

$$x^4 + 6x^3 = 6x^2 + 30x + 11$$

and neglected the negative roots. Find all the roots of this equation and compare with Petri's single positive root.

7. About 1600, the French lawyer, politician, and part-time mathemati- cian, François Viète introduced a numerical method similar to, but more cumbersome than, Newton's method for approximating roots of polynomials. Perhaps Viète's method inspired Newton to invent his method of 1669. Find the roots of the following equations considered by Viète.

$$x^5 - 5x^3 + 500x = 7905504$$

$$x^6 - 15x^4 + 85x^3 - 225x^2 + 274x = 120$$

8. Evidently, the only equation which Newton, himself, ever solved using Newton's method was $x^3 - 2x - 5 = 0$. Newton solved this problem in 1669. Of course, every numerical technique devised since that time has been used to solve this equation. Solve this cubic equation.

9. Find the roots of the quintic polynomial

$$p(x) = x^5 - (13.999 + 5i)x^4 + (74.99 + 55.998i)x^3 - (159.959 + 260.982i)x^2 +$$

$$(1.95 + 463.934i)x + (150 - 199.95i).$$

4.3 Homogeneous Linear Equations with Constant Coefficients

Many practical and important physical phenomena can be modeled by n-th order linear differential equations with constant coefficients. We will now show how a simple pendulum can be modeled by a second-order homogeneous linear differential equation with constant coefficients.

Simple Pendulum A simple pendulum consists of a rigid straight wire of negligible mass and length L with a bob of mass m attached to one end. The other end of the wire is attached to a fixed support. The pendulum is free to move in a vertical plane. Let θ be the angle the wire makes with the vertical—the equilibrium position of the system. We will choose θ to be positive if the wire is to the right of vertical and negative if the wire is to the left of vertical. See Figure 4.4.

If we neglect resistance due to friction and the medium (usually air) in which the system is operating, then there are only two forces acting on the mass m: F_W, the tension in the wire, which acts along the wire and toward the support; and $F_g = mg$, the force of gravity, which acts vertically downward. The force of gravity may be represented as two forces: one that acts parallel to F_W but in the opposite direction, F_N; and one that acts perpendicular to F_W, F_T. From Figure 4.4 we see that

$$F_N = mg\cos\theta = -F_W \quad \text{and} \quad F_T = mg\sin\theta.$$

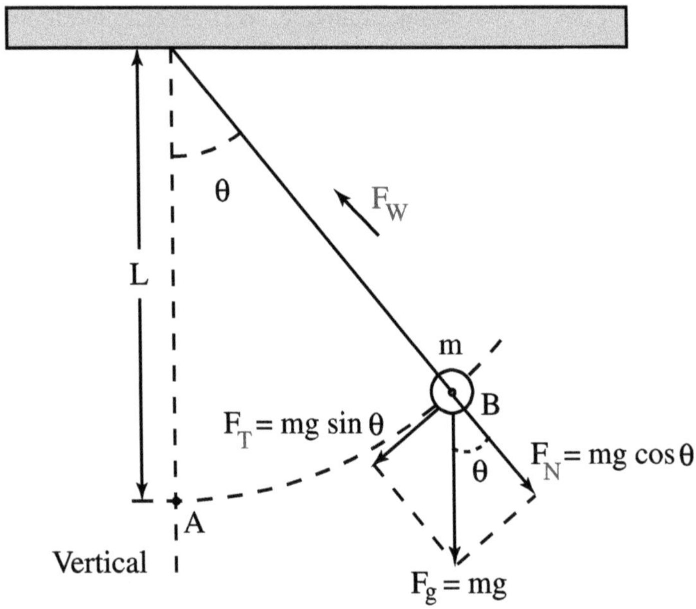

FIGURE 4.4
A Simple Pendulum.

The net force that tends to restore the system to equilibrium is F_T. If we let s represent the arc length AB, then applying Newton's second law of motion ($F = ma$), we get

$$(1) \qquad F_T = mg\sin\theta = -m\frac{d^2s}{dt^2}$$

because $-d^2s/dt^2$ is the acceleration along the arc AB. Since $s = L\theta$,

$$(2) \qquad \frac{d^2s}{dt^2} = L\frac{d^2\theta}{dt^2}.$$

Substituting (2) into (1), we obtain

$$g\sin\theta = -L\frac{d^2\theta}{dt^2}$$

or

$$(3) \qquad L\frac{d^2\theta}{dt^2} + g\sin\theta = 0.$$

This differential equation is nonlinear due to the function $\sin\theta$. The explicit solution of the DE (3) involves elliptic integrals.

The Maclaurin expansion for $\sin\theta$ is

$$\sin\theta = \theta - \frac{\theta^3}{3!} + \frac{\theta^5}{5!} - \cdots .$$

For θ small, $\sin\theta$ is approximately equal to θ (which is written mathematically as $\sin\theta \approx \theta$). Replacing $\sin\theta$ in equation (3) by θ, we obtain the following **linearized** differential equation

$$(4) \qquad L\frac{d^2\theta}{dt^2} + g\theta = 0$$

which approximates the motion of the simple pendulum for θ small.

We could try to solve the DE (4) by trial-and-error. That is, we could try to guess the form of the solution of (4) which contains one unknown constant A, differentiate the guessed solution twice, substitute into (4), and see if it is possible to determine A. For instance, if we guess the solution of (4) has the form $\theta = t^A$. Then differentiating twice, we get

$$\frac{d\theta}{dt} = At^{A-1} \qquad \text{and} \qquad \frac{d^2\theta}{dt^2} = A(A-1)t^{A-2}.$$

Substituting into (4), we see that A must satisfy

$$LA(A-1)t^{A-2} + gt^A = 0.$$

There is no constant value A which satisfies this equation for all values of t. Consequently, there is no solution of the DE (4) of the form $\theta = t^A$. Guessing $\theta = A\sin t$ or $\theta = \sin(t + A)$ produces no solution to the DE (4) either. (You may want to verify this statement.) Next, we seek a solution of (4) of the form $\theta = \sin At$. Differentiation yields

$$\frac{d\theta}{dt} = A\cos At \qquad \text{and} \qquad \frac{d^2\theta}{dt^2} = -A^2\sin At.$$

Substituting into the DE (4), we find A must be chosen to satisfy

$$-LA^2\sin At + g\sin At = 0 \qquad \text{or} \qquad (-LA^2 + g)\sin At = 0.$$

Hence,

$$-LA^2 + g = 0 \qquad \text{or} \qquad \sin At = 0.$$

Thus,

$$A = \pm\sqrt{g/L} \qquad \text{or} \qquad A = 0.$$

Notice that the choice $A = 0$ yields the zero solution, $\theta = \sin 0 \equiv 0$, to the DE (4). The choice $A = \sqrt{g/L}$ produces the particular solution $\theta_1 = \sin\sqrt{g/L}t$. A second linearly independent solution $\theta_2 = \cos\sqrt{g/L}t$ is obtained by guessing a solution of (4) of the form $\theta = \cos Bt$ and discovering $B = \sqrt{g/L}$. (Use the Wronskian to prove that the functions $\sin Ct$ and $\cos Ct$ where C

is an arbitrary constant are linearly independent functions on the interval $(-\infty, \infty)$.) Since $\theta_1 = \sin\sqrt{g/L}t$ and $\theta_2 = \cos\sqrt{g/L}t$ are linearly independent on $(-\infty, \infty)$, the general solution of the DE (4) is $\theta(t) = c_1 \sin\sqrt{g/L}t + c_2 \cos\sqrt{g/L}t$ where c_1 and c_2 are arbitrary constants.

Now let us consider the n-th order homogeneous linear differential equation

$$(5) \qquad a_n y^{(n)}(x) + a_{n-1} y^{(n-1)}(x) + \cdots + a_1 y^{(1)}(x) + a_0 y(x) = 0$$

with constant coefficients $a_n, a_{n-1}, \ldots, a_1, a_0$ where $a_n \neq 0$. By the representation theorem for n-th order homogeneous linear differential equations, since the constants $a_n, a_{n-1}, \ldots, a_1, a_0$ are all continuous functions on the interval $(-\infty, \infty)$ and $a_n \neq 0$, there exists a set containing exactly n linearly independent solutions of (5) on $(-\infty, \infty)$. Our immediate problem, then, is to determine a set of n linearly independent solutions to (5) on $(-\infty, \infty)$.

Both Daniel Bernoulli and Leonhard Euler knew how to solve second-order (n=2) homogeneous linear differential equations with constant coefficients prior to 1740. Euler was the first to publish his results in 1743. Following his approach, we suppose $y = e^{rx}$, where r is an unknown constant (real or complex), is a solution of (5). Successively differentiating, we find

$$y^{(1)} = re^{rx}, \quad y^{(2)} = r^2 e^{rx}, \ldots, \quad y^{(n-1)} = r^{n-1} e^{rx}, \quad y^{(n)} = r^n e^{rx}.$$

Substituting into (5), we get

$$a_n r^n e^{rx} + a_{n-1} r^{n-1} e^{rx} + \cdots + a_1 r e^{rx} + a_0 e^{rx} = 0.$$

Factoring e^{rx} from each term, leads to

$$e^{rx}(a_n r^n + a_{n-1} r^{n-1} + \cdots + a_1 r + a_0) = 0.$$

Since $e^{rx} \neq 0$ for any x and any constant r, the function $y = e^{rx}$ is a solution to the differential equation (5) if and only if r satisfies the polynomial equation

$$(6) \qquad p(r) = a_n r^n + a_{n-1} r^{n-1} + \cdots + a_1 r + a_0 = 0.$$

That is, e^{rx} is a solution of (5) if and only if r is a root of $p(r)$. Equation (6) is called the **auxiliary equation** associated with the differential equation (5).

Distinct Real Roots If the roots $r_1, r_2, \ldots, r_n$ of the auxiliary equation (6) are all real and no two roots are equal, then the functions $y_1(x) = e^{r_1 x}$, $y_2(x) = e^{r_2 x}, \ldots, y_n(x) = e^{r_n x}$ form a linearly independent set of real-valued solutions to (5) on the interval $(-\infty, \infty)$ and the general solution of (5) on $(-\infty, \infty)$ is $y(x) = c_1 y_1(x) + c_2 y_2(x) + \cdots + c_n y_n(x)$ where the c_i's are arbitrary constants. It is clear from the discussion above that the functions $y_i(x) = e^{r_i x}$ are all solutions of (5) on $(-\infty, \infty)$. All we need to do is verify that they are linearly independent on $(-\infty, \infty)$. We may do so by showing that their Wronskian is nonzero at some $x_0 \in (-\infty, \infty)$. By definition

$$W(y_1, y_2, \ldots, y_n, x) = \begin{vmatrix} e^{r_1 x} & e^{r_2 x} & \cdots & e^{r_n x} \\ r_1 e^{r_1 x} & r_2 e^{r_2 x} & \cdots & r_n e^{r_n x} \\ \vdots & \vdots & & \vdots \\ r_1^{n-1} e^{r_1 x} & r_2^{n-1} e^{r_2 x} & \cdots & r_n^{n-1} e^{r_n x} \end{vmatrix}.$$

A judicious choice for x in the Wronskian above is $x = 0$. Making this choice, we find

$$W(y_1, y_2, \ldots, y_n, 0) = \begin{vmatrix} 1 & 1 & \cdots & 1 \\ r_1 & r_2 & \cdots & r_n \\ \vdots & \vdots & & \vdots \\ r_1^{n-1} & r_2^{n-1} & \cdots & r_n^{n-1} \end{vmatrix} =$$

$$[(r_2-r_1)(r_3-r_1)\cdots(r_n-r_1)]\cdot[(r_3-r_2)(r_4-r_2)\cdots(r_n-r_2)]\cdots[(r_{n-1}-r_n)] \neq 0,$$

since the roots are distinct. The last determinant in the calculation above is known as the **Vandermonde determinant** and its value is well known.

For example, to find the general solution of the third-order homogeneous linear differential equation

$$(7) \qquad 2y''' - y'' - 2y' + y = 0,$$

we write the auxiliary equation

$$p(r) = 2r^3 - r^2 - 2r + 1 = 0.$$

Factoring, we find

$$p(r) = (r + 1)(r - 1)(2r - 1) = 0.$$

So the roots of the auxiliary equation are -1, 1, and $\frac{1}{2}$, and three linearly independent solutions of the DE (7) are

$$y_1(x) = e^{-x}, \quad y_2(x) = e^{x}, \quad \text{and} \quad y_3(x) = e^{x/2}.$$

Consequently, the general solution of the DE (7) is

$$y(x) = c_1 e^{-x} + c_2 e^{x} + c_3 e^{x/2}$$

where c_1, c_2, and c_3 are arbitrary constants.

Repeated Real Roots Consider the differential equation

$$(8) \qquad y''' - 6y'' + 12y' - 8y = 0.$$

The auxiliary equation is

$$p(r) = r^3 - 6r^2 + 12r - 8 = 0.$$

Factoring, we find

$$p(r) = (r - 2)^3 = 0.$$

The roots of this auxiliary equation are $r_1 = r_2 = r_3 = 2$. Thus, one solution of the DE (8) is $y_1(x) = e^{r_1 x} = e^{2x}$. However, $y_2(x) = e^{r_2 x} = e^{2x} = y_1(x)$ and $y_3(x) = e^{r_3 x} = e^{2x} = y_1(x)$. That is, y_2 and y_3 are identical to y_1.

Consequently, we have only one solution to the DE (8). So when some roots of the auxiliary equation are real and equal, the technique of the previous paragraph will not suffice.

A root s of the auxiliary equation (6), $p(r) = 0$, is called a **root of multiplicity k**, if the unique factorization of the polynomial $p(r)$ contains the factor $(r - s)$ exactly k times. The function $y(x) = e^{sx}$ will be one member of the linearly independent solution set; however, for $k \geq 2$ we must find $k - 1$ other linearly independent solutions corresponding to the root s. To find the additional linearly independent solutions, we assume that there are $k - 1$ solutions of the form $y_{m+1} = x^m e^{sx}$, where $1 \leq m \leq k - 1$. We then verify that $y_{m+1}(x)$ satisfies the differential equation and that the set $\{y_1(x), y_2(x), \ldots, y_k(x)\}$ is linearly independent on the interval $(-\infty, \infty)$. The functions y_{m+1} associated with the repeated root s are linearly independent, because the functions 1, x, x^2, ... , x^{k-1} are linearly independent. The functions y_{m+1} are also linearly independent of all solutions associated with other distinct roots of the auxiliary equation. This is always the case, but we will not present a proof.

Returning to our example, we will verify that $y_2(x) = xe^{2x}$ and $y_3(x) = x^2e^{2x}$ are solutions of the DE (8) and that $y_1(x) = e^{2x}$, $y_2(x) = xe^{2x}$, and $y_3(x) = x^2e^{2x}$ are linearly independent on $(-\infty, \infty)$. Calculating the first, second, and third derivatives of y_2 and y_3, we find

$$\begin{aligned} y_2' &= (1 + 2x)e^{2x} & \text{and} \qquad y_3' &= (2x + 2x^2)e^{2x} \\ y_2'' &= (4 + 4x)e^{2x} & \text{and} \qquad y_3'' &= (2 + 8x + 4x^2)e^{2x} \\ y_2''' &= (12 + 8x)e^{2x} & \text{and} \qquad y_3''' &= (12 + 24x + 8x^2)e^{2x}. \end{aligned}$$

Substitution of y_2 into (8), yields

$$y_2''' - 6y_2'' + 12y_2' - 8y_2 = [(12+8x) - 6(4+4x) + 12(1+2x) - 8x]e^{2x} = 0 \cdot e^{2x} = 0.$$

Hence, y_2 is a solution of the DE (8). Substitution of y_3 into (8), yields

$$\begin{aligned} y_3''' - 6y_3'' + 12y_3' - 8y_3 &= [(12+24x+8x^2) - 6(2+8x+4x^2) + 12(2x+2x^2) - 8x^2]e^{2x} \\ &= 0 \cdot e^{2x} = 0. \end{aligned}$$

Thus, y_3 is a solution of the DE (8). To verify that y_1, y_2, and y_3 are linearly independent on $(-\infty, \infty)$, we examine their Wronskian at $x = 0$.

$$W(y_1, y_2, y_3, 0) = \begin{vmatrix} y_1(0) & y_2(0) & y_3(0) \\ y_1'(0) & y_2'(0) & y_3'(0) \\ y_1''(0) & y_2''(0) & y_3''(0) \end{vmatrix} = \begin{vmatrix} 1 & 0 & 0 \\ 2 & 1 & 0 \\ 4 & 4 & 2 \end{vmatrix} = 2 \neq 0.$$

Consequently, the set $\{y_1(x), y_2(x), y_3(x)\}$ is linearly independent on the interval $(-\infty, \infty)$ and the general solution of the DE (8) on $(-\infty, \infty)$ is

$$y(x) = c_1e^{2x} + c_2xe^{2x} + c_3x^2e^{2x}$$

where c_1, c_2, and c_3 are arbitrary constants.

Example 1 Solution of a Homogeneous Linear Differential Equation with Constant Coefficients Find the general solution of the fourth-order homogeneous linear differential equation

$$y^{(4)} - 2y^{(2)} + y = 0. \tag{9}$$

Solution

The auxiliary equation corresponding to the DE (9) is

$$p(r) = r^4 - 2r^2 + 1 = 0.$$

Factoring, we find

$$p(r) = (x + 1)^2(x - 1)^2 = 0.$$

So the roots of the auxiliary equation are -1, -1, 1, and 1. Hence, the general solution of the DE (9) on the interval $(-\infty, \infty)$ is

$$y(x) = c_1e^{-x} + c_2xe^{-x} + c_3e^x + c_4xe^x$$

where c_1, c_2, c_3, and c_4 are arbitrary constants. ■

Example 2 Solution of a Homogeneous Linear Differential Equation with Constant Coefficients Find the general solution of the tenth-order homogeneous linear differential equation.

$$y^{(10)} + 3y^{(9)} - 6y^{(8)} - 22y^{(7)} - 3y^{(6)} + 39y^{(5)} + 40y^{(4)} + 12y^{(3)} = 0.$$

Solution

The auxiliary equation corresponding to this differential equation is

$$r^{10} + 3r^9 - 6r^8 - 22r^7 - 3r^6 + 39r^5 + 40r^4 + 12r^3 = 0. \tag{10}$$

Using the computing software, we find the roots of this polynomial are -3, -1, -1, -1, -1, 0, 0, 0, 2, and 2. That is, -3 is a root of multiplicity one, -1 is a root of multiplicity four, 0 is a root of multiplicity three, and 2 is a root of multiplicity two. Corresponding to the root -3, we have the solution

$$y_1 = e^{-3x}.$$

Corresponding to the root -1 of multiplicity four, we have the four linearly independent solutions

$$y_2 = e^{-x}, \quad y_3 = xe^{-x}, \quad y_4 = x^2e^{-x}, \quad \text{and} \quad y_5 = x^3e^{-x}.$$

Corresponding to the root 0 of multiplicity three, we have the three linearly independent solutions

$$y_6 = e^{0x} = 1, \quad y_7 = xe^{0x} = x, \quad \text{and} \quad y_8 = x^2e^{0x} = x^2.$$

And corresponding to the root 2 of multiplicity two, we have the two linearly independent solutions

$$y_9 = e^{2x} \quad \text{and} \quad y_{10} = xe^{2x}.$$

The set $\{y_1, y_2, \ldots, y_{10}\}$ contains ten linearly independent solutions on the interval $(-\infty, \infty)$ of the DE (10). Therefore, the general solution of the DE (10) is

$$\begin{aligned} y &= c_1y_1 + c_2y_2 + c_3y_3 + \cdots + c_{10}y_{10} \\ &= c_1e^{-3x} + c_2e^{-x} + c_3xe^{-x} + c_4x^2e^{-x} + c_5x^3e^{-x} + c_6 1 + c_7x + c_8x^2 \\ &\quad + c_9e^{2x} + c_{10}xe^{2x} \end{aligned}$$

where $c_1, c_2, \ldots, c_{10}$ are arbitrary constants. ■

Complex Roots Recall from calculus that the Taylor series expansions about $x = 0$ for e^x, $\cos x$, and $\sin x$ are

$$e^x = \sum_{n=0}^{\infty} \frac{x^n}{n!} = 1 + x + \frac{x^2}{2!} + \frac{x^3}{3!} + \frac{x^4}{4!} + \cdots$$

$$\cos x = \sum_{n=0}^{\infty} \frac{(-1)^n x^{2n}}{(2n)!} = 1 - \frac{x^2}{2!} + \frac{x^4}{4!} - \frac{x^6}{6!} + \cdots$$

$$\sin x = \sum_{n=0}^{\infty} \frac{(-1)^n x^{2n+1}}{(2n+1)!} = x - \frac{x^3}{3!} + \frac{x^5}{5!} - \frac{x^7}{7!} + \cdots.$$

Each of these series converges for all x. Substituting ix for x in the expansion for e^x, using the fact that $i^2 = -1$, $i^3 = -i$, $i^4 = 1$, $i^5 = i$, $i^6 = -1$, etc., and rearranging, we find

$$\begin{aligned} e^{ix} &= \sum_{n=0}^{\infty} \frac{(ix)^n}{n!} \\ &= 1 + ix + \frac{i^2x^2}{2!} + \frac{i^3x^3}{3!} + \frac{i^4x^4}{4!} + \frac{i^5x^5}{5!} + \frac{i^6x^6}{6!} + \frac{i^7x^7}{7!} + \cdots \\ &= (1 - \frac{x^2}{2!} + \frac{x^4}{4!} - \frac{x^6}{6!} + \cdots) + i(x - \frac{x^3}{3!} + \frac{x^5}{5!} - \frac{x^7}{7!} + \cdots) \\ &= \cos x + i \sin x. \end{aligned}$$

The identity $e^{ix} = \cos x + i \sin x$ is known as **Euler's formula**. If $\alpha + i\beta$ is a root of multiplicity one of the auxiliary equation associated with some linear differential equation with constant (real or complex) coefficients, then a solution of the differential equation corresponding to the root $\alpha + i\beta$ is the complex function

$$y = e^{(\alpha+i\beta)x} = e^{\alpha x}e^{i\beta x} = e^{\alpha x}(\cos\beta x + i\sin\beta x).$$

Suppose the complex number $\alpha + i\beta$ where $\beta \neq 0$ and its complex conjugate $\alpha - i\beta$ are both roots of the auxiliary equation associated with a homogeneous linear differential equation. Two linearly independent, complex solutions corresponding to these two roots are

$$\begin{aligned} y_1 &= e^{(\alpha+i\beta)x} = e^{\alpha x}e^{i\beta x} = e^{\alpha x}(\cos\beta x + i\sin\beta x) \\ y_2 &= e^{(\alpha-i\beta)x} = e^{\alpha x}e^{-i\beta x} = e^{\alpha x}(\cos(-\beta x) + i\sin(-\beta x)) \\ &= e^{\alpha x}(\cos\beta x - i\sin\beta x). \end{aligned}$$

Hence, the general solution of the differential equation will include the linear combination

$$\begin{aligned} c_1y_1 + c_2y_2 &= c_1e^{\alpha x}(\cos\beta x + i\sin\beta x) + c_2e^{\alpha x}(\cos\beta x - i\sin\beta x) \\ &= e^{\alpha x}[(c_1 + c_2)\cos\beta x + i(c_1 - c_2)\sin\beta x] \end{aligned}$$

where c_1 and c_2 are arbitrary complex constants. Choosing $c_1 = c_2 = 1/2$, we see that $y_3 = e^{\alpha x}\cos\beta x$ is a real solution to the differential equation (since any particular linear combination of solutions is also a solution). And choosing $c_1 = -c_2 = 1/(2i)$, we see that $y_4 = e^{\alpha x}\sin\beta x$ is also a real solution of the differential equation. So two real linearly independent solutions on $(-\infty, \infty)$ corresponding to the complex conjugate pair of roots $\alpha + i\beta$ and $\alpha - i\beta$ are $y_3 = e^{\alpha x}\cos\beta x$ and $y_4 = e^{\alpha x}\sin\beta x$. (Can you prove y_3 and y_4 are linearly independent? Hint: Calculate their Wronskian.) Therefore, in the general solution the linear combination of complex solutions $c_1y_1 + c_2y_2$ can be replaced by the linear combination of real solutions $k_1y_3 + k_2y_4$ where k_1 and k_2 are arbitrary real constants.

Furthermore, if the complex number $\alpha + i\beta$ where $\beta \neq 0$, and its complex conjugate $\alpha - i\beta$ are both roots of multiplicity $k > 0$ of the auxiliary equation associated with a homogeneous linear differential equation, then $2k$ linearly independent, real solutions corresponding to these two roots are

$$\begin{array}{rclrcl} y_1 &=& e^{\alpha x}\cos\beta x, & y_2 &=& e^{\alpha x}\sin\beta x, \\ y_3 &=& xe^{\alpha x}\cos\beta x, & y_4 &=& xe^{\alpha x}\sin\beta x, \\ &\vdots& & &\vdots& \\ y_{2k-1} &=& x^{k-1}e^{\alpha x}\cos\beta x, & y_{2k} &=& x^{k-1}e^{\alpha x}\sin\beta x \end{array}$$

and the general solution of the differential equation will include the linear combination $c_1y_1 + c_2y_2 + \cdots + c_{2k-1}y_{2k-1} + c_{2k}y_{2k}$. This result is important since the **conjugate root theorem** states that if the coefficients of an n-th degree polynomial are all real, then complex roots occur in conjugate pairs. So if the coefficients of an n-th order linear differential equation are all real constants, then complex roots of the associated auxiliary equation will always occur in conjugate pairs, all n linearly independent solutions may be written as real solutions, and the general solution can be written as a linear combination of real solutions. It should be noted that if any of the constant coefficients $a_n, a_{n-1}, \ldots, a_1, a_0$ of the differential equation (hence, auxiliary equation) is complex, then it is no longer true that the complex conjugate of a complex root will also be a root. In this case, the corresponding general solution will contain some complex components.

Example 3 Real General Solution of a Homogeneous Linear Differential Equation Find the real general solution of the fifth-order linear homogeneous differential equation

$$y^{(5)} - 11y^{(4)} + 50y^{(3)} - 94y^{(2)} + 13y^{(1)} + 169y = 0.$$

Solution

Using the computing software, we find the roots of the associated auxiliary equation

$$r^5 - 11r^4 + 50r^3 - 94r^2 + 13r^1 + 169 = 0$$

are -1, $3+2i$, $3+2i$, $3-2i$, and $3-2i$. Corresponding to the real root -1, is the real solution

$$y_1 = e^{-x}.$$

Corresponding to the double complex root $3+2i$ and the double complex conjugate root $3-2i$, are the four linearly independent real solutions

$$y_2 = e^{3x}\cos 2x, \quad y_3 = e^{3x}\sin 2x, \quad y_4 = xe^{3x}\cos 2x, \quad \text{and} \quad y_5 = xe^{3x}\sin 2x.$$

So the real, general solution to the given differential equation is

$$\begin{aligned} y &= c_1e^{-x} + c_2e^{3x}\cos 2x + c_3e^{3x}\sin 2x + c_4xe^{3x}\cos 2x + c_5xe^{3x}\sin 2x \\ &= c_1e^{-x} + (c_2 + c_4x)e^{3x}\cos 2x + (c_3 + c_5x)e^{3x}\sin 2x. \ \blacksquare \end{aligned}$$

4.3.1 Exercises

In Exercises 1–20 determine the real general solution of the given homogeneous linear differential equations.

1. $y'' - 4y = 0$
2. $y'' - 3y' + 2y = 0$
3. $y'' - 4y' = 0$
4. $y'' - 13y' + 42y = 0$
5. $y'' + 2y' + 2y = 0$

6. $y'' + 4y = 0$
7. $4y^{(2)} + 4y^{(1)} - 3y = 0$
8. $y^{(3)} - y^{(2)} + y^{(1)} - y = 0$
9. $y^{(3)} - 5y^{(2)} + 3y^{(1)} + 9y = 0$
10. $y^{(3)} - 3y^{(2)} + 3y^{(1)} - y = 0$
11. $y^{(3)} - 4y^{(2)} + y^{(1)} + 26y = 0$
12. $y^{(3)} - 4y^{(2)} + 6y^{(1)} - 4y = 0$
13. $y^{(4)} - 16y = 0$
14. $y^{(4)} + 16y = 0$
15. $y^{(4)} - 4y^{(3)} + 8y^{(2)} - 8y^{(1)} + 4y = 0$
16. $y^{(4)} - 8y^{(1)} = 0$
17. $36y^{(4)} - 12y^{(3)} - 11y^{(2)} + 2y^{(1)} + y = 0$
18. $y^{(5)} - 3y^{(4)} + 3y^{(3)} - 3y^{(2)} + 2y^{(1)} = 0$
19. $y^{(5)} - y^{(4)} + y^{(3)} + 35y^{(2)} + 16y^{(1)} - 52y = 0$
20. $y^{(8)} + 8y^{(4)} + 16y = 0$

In Exercises 21–25 assume α and $\beta \neq \alpha$ are real constants.

21. Find the real general solution of $y'' + \alpha y = 0$ on $(-\infty, \infty)$ for a. $\alpha > 0$, b. $\alpha = 0$, and c. $\alpha < 0$.

22. Find the real general solution of $y'' + \alpha y' = 0$ on $(-\infty, \infty)$ for a. $\alpha \neq 0$ and b. $\alpha = 0$.

23. Find the real general solution of $y'' - 2\alpha y' + \alpha^2 y = 0$ on $(-\infty, \infty)$ for $\alpha \neq 0$.

24. Find the real general solution of $y'' - (\alpha + \beta)y' + \alpha\beta y = 0$ on $(-\infty, \infty)$.

25. Find the real general solution of $y'' - 2\alpha y' + (\alpha^2 + \beta^2)y = 0$ on $(-\infty, \infty)$.

26. Let α be a real constant and consider the differential equation (*) $y'' + 2\alpha y' + 1 = 0$.

    a. Write the auxiliary equation associated with (*).

    b. Use the quadratic formula to find the roots of the auxiliary equa- tion.

    c. Find the values of α which produce two equal real roots of the auxiliary equation and write the general solution of (*) in each case.

d. Find the intervals in which α produces two unequal real roots and write the general solution of (*) in these intervals.

e. Find the interval in which α produces complex conjugate roots and write the general solution of (*) for α in this interval.

When c is a complex constant, $\dfrac{de^{cx}}{dx} = ce^{cx}$. Consequently, the complex function $y(x) = e^{cx}$ will be a solution of a homogeneous linear differential equation with complex coefficients if and only if c is a root of the associated auxiliary equation, just as is the case when the coefficients are all real.

In Exercises 27–28 write the auxiliary equation, find its roots, and then write the complex general solution of the given homogeneous linear differential equation with constant complex coefficients.

27. $y^{(3)} - (3+4i)y^{(2)} - (4-12i)y^{(1)} + 12y = 0$

28. $y^{(4)} - (3+i)y^{(3)} + (4+3i)y^{(2)} = 0$

29. Consider the complex initial value problem (†) $y' - iy = 0;\ y(0) = 1$.

a. Show that $y_1(x) = e^{ix}$ satisfies the IVP (†) for all x in $(-\infty, \infty)$.

b. Show that $y_2(x) = \cos x + i \sin x$ satisfies the IVP (†) for all x in $(-\infty, \infty)$.

c. Assuming the existence and uniqueness theorem stated in sec- tion 4.1 applies to the IVP (†) what can you conclude about $y_1(x) = e^{ix}$ and $y_2(x) = \cos x + i \sin x$ for all x in $(-\infty, \infty)$. (Note: This exercise provides a different proof of Euler's formula.)

4.4 Nonhomogeneous Linear Equations with Constant Coefficients

Now let us consider the n-th order nonhomogeneous linear differential equation

$$(1) \qquad a_n y^{(n)}(x) + a_{n-1} y^{(n-1)}(x) + \cdots + a_1 y^{(1)}(x) + a_0 y(x) = b(x)$$

with constant coefficients $a_n, a_{n-1}, \ldots, a_1, a_0$ where $a_n \neq 0$ and $b(x)$ is not the zero function. Since the constants $a_n, a_{n-1}, \ldots, a_1, a_0$ are all continuous functions on the interval $(-\infty, \infty)$, by the representation theorem for n-th order nonhomogeneous linear differential equations proven in Section 4.1 there exists a particular solution, $y_p(x)$, of (1) on any interval I on which $b(x)$ is continuous and the general solution of (1) is $y(x) = y_c(x) + y_p(x)$ where $y_c(x)$ is the complementary solution of the associated homogeneous linear equation

$$(2) \qquad a_n y^{(n)}(x) + a_{n-1} y^{(n-1)}(x) + \cdots + a_1 y^{(1)}(x) + a_0 y(x) = 0.$$

We have seen how to solve (2) and find the complementary solution, $y_c(x)$, previously. So the problem of solving (1) reduces to one of finding the particular solution $y_p(x)$. When the nonhomogeneity $b(x)$ is a polynomial function, an exponential function, a sine or cosine function, or a finite sum or product of such functions, then the method of undetermined coefficients may be used to find a particular solution $y_p(x)$. That is, when $b(x)$ is a function which consists entirely of terms of the form $x^p, x^p e^{\alpha x}, x^p e^{ax} \sin bx$, or $x^p e^{ax} \cos bx$ where p is a nonnegative integer, α is a constant (perhaps complex), and a and b are real constants, then the method of undetermined coefficients may be used. When $b(x)$ is not of this form, then one may try to find a particular solution by the Laplace transform method which we will discuss in Chapter 5.

The Method of Undetermined Coefficients The method of undetermined coefficients for solving the nonhomogeneous linear differential equation with constant coefficients (1) consists of (i) solving the associated homogeneous equation (2), (ii) judiciously guessing the form of the particular solution, $y_p(x)$, of the nonhomogeneous equation (1) with the coefficients left unspecified—hence, the name **undetermined coefficients,** (iii) differentiating the assumed form of particular solution n times and substituting the particular solution and its derivatives into (2), and (iv) determining, if possible, specific values for the unspecified coefficients. If all the coefficients are determined, then we have guessed the correct form of the particular solution. If some mathematical anomaly occurs along the way, we have guessed the wrong form of the particular solution and we must guess again or we have made an error in the calculations. The following three simple examples illustrate this method.

Example 1 Solution of a Nonhomogeneous Linear Differential Equation with Constant Coefficients Solve the nonhomogeneous linear differential equation

(3) $$y'' - 3y' + 2y = 5.$$

Solution

The associated homogeneous equation $y'' - 3y' + 2y = 0$ has auxiliary equation $p(r) = r^2 - 3r + 2 = 0$. Since the two roots of this equation are 1 and 2, the complementary solution of equation (3) is

$$y_c(x) = c_1 e^x + c_2 e^{2x}.$$

Since the nonhomogeneity $b(x) = 5$, a constant, we guess that the particular solution has the form $y_p(x) = A$, where A is an unspecified constant. Differentiating $y_p(x) = A$ twice, we find $y_p'(x) = 0$ and $y_p''(x) = 0$. Substituting $y_p(x)$ and its derivatives into equation (3), we find A must satisfy $2A = 5$. Hence, $A = 5/2$ and a particular solution of (3) is $y_p(x) = 5/2$. Therefore, the general solution of (3) is

$$y(x) = c_1 e^x + c_2 e^{2x} + \frac{5}{2}. \blacksquare$$

Example 2 Solution of a Nonhomogeneous Linear Differential Equation with Constant Coefficients Solve the nonhomogeneous linear differential equation

$$(4) \qquad y'' - 3y' + 2y = 4e^{3x}.$$

Solution

As in Example 1, the auxiliary equation of the associated homogeneous equation $y'' - 3y' + 2y = 0$ has roots 1 and 2. So the complementary solution of equation (4) is

$$y_c(x) = c_1 e^x + c_2 e^{2x}.$$

Since the nonhomogeneity $b(x) = 4e^{3x}$, we guess that the particular solution has the form $y_p(x) = Ae^{3x}$, where A is an unspecified constant. Differentiating $y_p(x) = Ae^{3x}$ twice, we find $y_p'(x) = 3Ae^{3x}$ and $y_p''(x) = 9Ae^{3x}$. Substituting $y_p(x)$ and its derivatives into equation (4), we find A must satisfy

$$9Ae^{3x} - 3(3Ae^{3x}) + 2(Ae^{3x}) = 4e^{3x} \quad \text{or} \quad 2Ae^{3x} = 4e^{3x}.$$

Hence, $A = 2$ and a particular solution of (4) is $y_p(x) = 2e^{3x}$. Therefore, the general solution of (4) is

$$y(x) = c_1 e^x + c_2 e^{2x} + 2e^{3x}. \blacksquare$$

Example 3 Wrong Initial Guess of the Form of the Particular Solution Solve the nonhomogeneous linear differential equation

$$(5) \qquad y'' - 3y' + 2y = 4e^{2x}.$$

Solution

Again, the auxiliary equation of the associated homogeneous linear equation, $y'' - 3y' + 2y = 0$, has roots 1 and 2. So the complementary solution of equation (5) is

$$y_c(x) = c_1 e^x + c_2 e^{2x}.$$

Since the nonhomogeneity $b(x) = 4e^{2x}$, we guess that the particular solution has the form $y_p(x) = Ae^{2x}$, where A is an unspecified constant. Differentiating $y_p(x) = Ae^{2x}$ twice, we find $y_p'(x) = 2Ae^{2x}$ and $y_p''(x) = 4Ae^{2x}$. Substituting $y_p(x)$ and its derivatives into equation (5), we find A must satisfy

$$4Ae^{2x} - 3(2Ae^{2x}) + 2(Ae^{2x}) = 4e^{2x} \quad \text{or} \quad 0 = 4e^{2x}.$$

Since the last equation above is clearly false, there is no particular solution of (5) of the form we have assumed—namely, $y_p(x) = Ae^{2x}$. We guessed the wrong form for a particular solution of the DE (5), and the resulting false equation is telling us we made the wrong guess. Examining the complementary solution, $y_c(x) = c_1e^x + c_2e^{2x}$, we see that the term Ae^{2x} is already present in the complementary solution, appearing as c_2e^{2x}. So, of course, Ae^{2x} is a solution of the associated homogeneous equation (5) and produces the value zero when substituted into the left-hand side of the nonhomogeneous equation (6).

We need to guess a different form for a particular solution to the DE (5). But, what should we guess? Since 2 is a single root of the auxiliary equation $p(r) = r^2 - 3r + 2 = 0$, the function $y(x) = Ae^{2x}$ is one solution of the associated homogeneous linear differential equation $y'' - 3y' + 2y = 0$. If 2 were a double root of the auxiliary equation, then two linearly independent solutions of the associated homogeneous equation would be $y_1(x) = Ae^{2x}$ and $y_2(x) = Bxe^{2x}$. Since 2 is a single root and it is not a double root of the auxiliary equation, our second guess for the form of a particular solution is $y_p(x) = Bxe^{2x}$, where B is an unspecified constant which is to be determined. Differentiating $y_p(x) = Bxe^{2x}$ twice, we find

$$y_p'(x) = B(2x+1)e^{2x} \quad \text{and} \quad y_p''(x) = B(4x+4)e^{2x}.$$

Substituting $y_p(x)$ and its derivatives into equation (5), we find B must satisfy

$$B(4x+4)e^{2x} - 3(B(2x+1)e^{2x}) + 2(Bxe^{2x}) = 4e^{2x} \quad \text{or} \quad Be^{2x} = 4e^{2x}.$$

So $B = 4$, and a particular solution of (5) is $y_p(x) = 4xe^{2x}$. Therefore, the general solution of (5) is $y(x) = c_1e^x + c_2e^{2x} + 4xe^{2x}$. ■

From the previous three examples, we have discovered that the form of the particular solution of a nonhomogeneous differential equation with constant coefficients depends upon the roots of the auxiliary equation of the associated homogeneous differential equation as well as the nonhomogeneity itself. The following three cases delineate this relationship more explicitly.

Form of the Particular Solution of a Nonhomogeneous Linear Differential Equation with Constant Coefficients Based on the Form of the Nonhomogeneity

Case 1. If $r = 0$ is a root of multiplicity k of the auxiliary equation of the associated homogeneous linear differential equation (2) where $k \geq 0$ (Here $k = 0$ corresponds to $r = 0$ not being a root of the auxiliary equation.) and if $b(x)$ is a polynomial of degree m—that is, if

(6) $$b(x) = b_mx^m + b_{m-1}x^{m-1} + \cdots + b_1x + b_0$$

where b_m, b_{m-1}, $\ldots$, b_1, and b_0 are constants and $b_m \neq 0$, then there is a particular solution of the nonhomogeneous linear differential equation (1) of the form

$$y_p = x^k(A_mx^m + A_{m-1}x^{m-1} + \cdots + A_1x + A_0)$$

where A_m, A_{m-1}, ..., A_1, and A_0 are unknown constants which are to be determined so that y_p satisfies (1). Determination of the actual values of these constants requires n differentiations of y_p, substitution of y_p and its n derivatives into (1), and then the solution of a resulting system of linear equations in the $m+1$ unknowns A_m, A_{m-1}, ..., A_1, A_0.

Case 2. If $r = \alpha$ is a root of multiplicity k ($k \geq 0$) of the auxiliary equation of the associated homogeneous linear differential equation (2) and if $b(x)$ has the form

$$(7) \qquad b(x) = e^{\alpha x}(b_m x^m + b_{m-1}x^{m-1} + \cdots + b_1 x + b_0)$$

where α, b_m, b_{m-1}, ..., b_1, b_0 are constants and $b_m \neq 0$, then there is a particular solution of the nonhomogeneous linear differential equation (1) of the form

$$y_p = x^k e^{\alpha x}(A_m x^m + A_{m-1}x^{m-1} + \cdots + A_1 x + A_0)$$

where A_m, A_{m-1}, ..., A_1, and A_0 are unknown constants to be determined.

Case 3. If $r = a+bi$ and $r = a-bi$ are roots of multiplicity k ($k \geq 0$) of the auxiliary equation of the associated homogeneous linear differential equation (2) and if $b(x)$ has the form

$$(8) \qquad b(x) = e^{ax}(b_m x^m + b_{m-1}x^{m-1} + \cdots + b_1 x + b_0)\sin bx$$

or the form

$$(9) \qquad b(x) = e^{ax}(b_m x^m + b_{m-1}x^{m-1} + \cdots + b_1 x + b_0)\cos bx,$$

then there is a particular solution of (1) of the form

$$y_p = x^k e^{ax}[(A_m x^m + A_{m-1}x^{m-1} + \cdots + A_1 x + A_0)\cos bx +$$
$$(B_m x^m + B_{m-1}x^{m-1} + \cdots + B_1 x + B_0)\sin bx]$$

where A_m, A_{m-1}, ..., A_1, A_0, B_m, B_{m-1}, ..., B_1, and B_0 are unknown constants which are to be determined. ■

When $b(x)$ is a sum of terms of the form (6), (7), (8), or (9)—that is, when $b(x) = b_1(x) + b_2(x) + \cdots + b_s(x)$ where $b_1(x)$, $b_2(x)$, ..., $b_s(x)$ all have one of the forms (6), (7), (8), or (9)—it is usually easier, first, to calculate s particular solutions, $y_{p_j}(x)$, $1 \leq j \leq s$, to the corresponding s separate nonhomogeneous linear differential equations

$$a_n y^{(n)}(x) + a_{n-1}y^{(n-1)}(x) + \cdots + a_1 y^{(1)}(x) + a_0 y(x) = b_j(x).$$

And then add the s particular solutions to obtain $y_p(x) = y_{p_1}(x) + y_{p_2}(x) + \cdots + y_{p_s}(x)$ which is a particular solution of the original differential equation

$$a_n y^{(n)}(x) + a_{n-1}y^{(n-1)}(x) + \cdots + a_1 y^{(1)}(x) + a_0 y(x) =$$
$$b(x) = b_1(x) + b_2(x) + \cdots + b_s(x).$$

The following example illustrates this procedure.

Example 4 General Solution of a Nonhomogeneous Differential Equation by Superposition Find the general solution of

(10) $$y'' + 3y' = 4x^2 - 2e^{-3x}.$$

Solution

The associated homogeneous equation is $y'' + 3y' = 0$. And the auxiliary equation $r^2 + 3r = 0$ has roots 0 and -3. So the complementary solution of the DE (10) is

$$y_c(x) = c_1 + c_2 e^{-3x}.$$

Next, we find two particular solutions $y_{p_1}(x)$ and $y_{p_2}(x)$ to the two nonhomogeneous differential equations:

(11) $$y'' + 3y' = 4x^2 = b_1(x)$$

(12) $$y'' + 3y' = -2e^{-3x} = b_2(x).$$

Since $b_1(x) = 4x^2$ is of the form (6) and $r = 0$ is a root of the auxiliary equation of order $k = 1$, we seek a particular solution $y_{p_1}(x)$ of the DE (11) of the form

$$y_{p_1}(x) = x^1(A_2x^2 + A_1x + A_0) = A_2x^3 + A_1x^2 + A_0x.$$

Differentiating, we find $y'_{p_1}(x) = 3A_2x^2 + 2A_1x + A_0$ and $y''_{p_1}(x) = 6A_2x + 2A_1$. Substituting into the DE (11), we see A_2, A_1, and A_0 must satisfy

$$(6A_2x + 2A_1) + 3(3A_2x^2 + 2A_1x + A_0) = 4x^2$$

or

$$9A_2x^2 + (6A_2 + 6A_1)x + (2A_1 + 3A_0) = 4x^2.$$

Equating coefficients of x^2, x, and $x^0 = 1$, we see A_2, A_1, and A_0 must satisfy the following system of three linear equations simultaneously.

$$\begin{aligned} 9A_2 &= 4 \\ 6A_2 + 6A_1 &= 0 \\ 2A_1 + 3A_0 &= 0. \end{aligned}$$

Solving this system, we find $A_2 = \dfrac{4}{9}$, $A_1 = \dfrac{-4}{9}$, and $A_0 = \dfrac{8}{27}$. Consequently,

$$y_{p_1}(x) = \frac{4}{9}x^3 - \frac{4}{9}x^2 + \frac{8}{27}x.$$

Since $b_2(x) = -2e^{-3x}$ is of the form (7) and $r = -3$ is a root of the auxiliary equation of order $k = 1$, a particular solution $y_{p_2}(x)$ of (12) will have the form

$$y_{p_2}(x) = x^1e^{-3x}(C) = Cxe^{-3x}.$$

Differentiating, we get $y'_{p_2}(x) = C(-3x+1)e^{-3x}$ and $y''_{p_2}(x) = C(9x-6)e^{-3x}$. Substituting into the DE (12), we find C must satisfy

$$C(9x-6)e^{-3x} + 3C(-3x+1)e^{-3x} = -2e^{-3x} \quad \text{or} \quad -3Ce^{-3x} = -2e^{-3x}.$$

Hence, $C = \frac{2}{3}$ and $y_{p_2}(x) = \frac{2}{3}xe^{-3x}$.

Adding the particular solutions $y_{p_1}(x)$ and $y_{p_2}(x)$ of the two nonhomogeneous linear differential equations (11) and (12), we find that a particular solution of the given nonhomogeneous differential equation (10) is

$$y_p(x) = y_{p_1}(x) + y_{p_2}(x) = \frac{4}{9}x^3 - \frac{4}{9}x^2 + \frac{8}{27}x + \frac{2}{3}xe^{-3x}$$

and, consequently, the general solution of (10) is

$$y(x) = y_c + y_p = c_1 + c_2e^{-3x} + \frac{4}{9}x^3 - \frac{4}{9}x^2 + \frac{8}{27}x + \frac{2}{3}xe^{-3x}. \blacksquare$$

4.4.1 Exercises

Determine the general solution of the following nonhomogeneous linear differential equations.

1. $y'' + 4y = \sin x$
2. $y'' + y = x^3$
3. $y'' + 4y = \cos 2x$
4. $y'' - y = e^x$
5. $y'' + 2y' + y = 3e^x - x + 1$
6. $y^{(3)} - y^{(1)} = x$
7. $y^{(3)} - y^{(2)} + y^{(1)} - y = 4\sin x$
8. $y^{(4)} - y = e^{-x}$
9. $y^{(4)} - 6y^{(3)} + 13y^{(2)} - 12y^{(1)} + 4y = 2e^x - 4e^{2x}$
10. $y^{(4)} + 4y^{(2)} = 24x^2 - 6x + 14 + 32\cos 2x$
11. $y^{(4)} + 2y^{(2)} + y = 3 + \cos 2x$
12. $y^{(4)} - 3y^{(3)} + 3y^{(2)} - y^{(1)} = 6x - 20 - 120x^2e^x$
13. $y^{(3)} - 6y^{(2)} + 21y^{(1)} - 26y = 36e^{2x}\sin 3x$
14. $y^{(3)} + y^{(2)} - y^{(1)} - y = (2x^2 + 4x + 8)\cos x + (6x^2 + 8x + 12)\sin x$
15. $y^{(6)} - 12y^{(5)} + 63y^{(4)} - 184y^{(3)} + 315y^{(2)} - 300y^{(1)} + 125y = e^x(48\cos x + 96\sin x)$

4.5 Initial Value Problems

The **initial value problem for n-th order homogeneous linear differential equations with constant coefficients** consists of solving the differential equation

$$(1) \qquad a_n y^{(n)}(x) + a_{n-1}y^{(n-1)}(x) + \cdots + a_1 y^{(1)}(x) + a_0 y(x) = 0$$

where a_n, a_{n-1}, ... , a_1, and a_0 are constants and where $a_n \neq 0$ subject to the initial conditions

$$(2) \qquad y(x_0) = k_1, \quad y^{(1)}(x_0) = k_2, \ldots, \; y^{(n-1)}(x_0) = k_n$$

where k_1, k_2, ... , k_n are constants.

By the summary theorem stated at the end of Section 4.1, this initial value problem has a unique solution on the interval $(-\infty, \infty)$. One way to solve the initial value problem consisting of equations (1) and (2) is to find the general solution of (1), using a root-finding routine when necessary. The general solution of (1) will include n arbitrary constants, c_1, c_2, ... , c_n. The value of these constants must be determined so that equations (2) are satisfied. This requires the solution of a system of n linear equations in the n unknowns, c_1, c_2, ... , c_n.

The **initial value problem for n-th order nonhomogeneous linear differential equations with constant coefficients** consists of solving the differential equation

$$(3) \qquad a_n y^{(n)}(x) + a_{n-1}y^{(n-1)}(x) + \cdots + a_1 y^{(1)}(x) + a_0 y(x) = b(x)$$

where a_n, a_{n-1}, ... , a_1, and a_0 are constants and where $a_n \neq 0$ subject to the initial conditions

$$(4) \qquad y(x_0) = k_1, \quad y^{(1)}(x_0) = k_2, \; \ldots, \; y^{(n-1)}(x_0) = k_n.$$

This initial value problem has a unique solution on the largest interval containing x_0 on which the function $b(x)$ is continuous. When $b(x)$ is a linear combination of terms of the form (6), (7), (8), and (9) of Section 4.4, the solution of the initial value problem (3)-(4) can be found by calculating the complementary solution, y_c, which will include n arbitrary constants c_1, c_2, ... , c_n; by calculating a particular solution, y_p, using the method of undetermined coefficients; and then by determining values for the constants c_1, c_2, ... , c_n such that $y = y_c + y_p$ satisfies the equations of (4)—this requires the solution of a system of n linear equations in n unknowns. In this instance, you must remember that it is the coefficients of the general solution $y = y_c + y_p$ and not the coefficients of the complementary solution y_c which must be chosen to satisfy the initial conditions (4).

A second method for solving the initial value problems (1)-(2) and (3)-(4) as well as the general initial value problem consisting of the differential equation $y^{(n)}(x) = f(x, y(x), y^{(1)}(x), \ldots, y^{(n-1)}(x))$ and the initial conditions $y(x_0) = k_1$, $y^{(1)}(x_0) = k_2$, $\ldots$, $y^{(n-1)}(x_0) = k_n$ is to rewrite these n-th order differential equations as a system of n first-order differential equations and then use the techniques described in Chapter 7 to solve the corresponding system initial value problem. When an explicit equation for the solution is not required, the second method of solution is the simplest method to use. This method will produce the solution of the initial value problem as a set of ordered pairs (a function) on any finite interval about x_0 in which certain conditions on f, f_y, $f_{y^{(1)}}$, $\ldots$, $f_{y^{(n-1)}}$ are satisfied.

Example 1 Solution of a Nonhomogeneous Initial Value Problem
Solve the initial value problem

$$(5) \qquad y'' + 3y' = 4x^2 - 2e^{-3x}; \quad y(0) = 1, \quad y'(0) = 0.$$

Solution

Since the differential equation of this problem is equation (10) of Section 4.4, we know the general solution is $y(x) = y_c(x) + y_p(x)$ where $y_c(x)$ and $y_p(x)$ are as calculated in Example 4 of Section 4.4. That is,

$$y(x) = c_1 + c_2 e^{-3x} + \frac{4}{9}x^3 - \frac{4}{9}x^2 + \frac{8}{27}x + \frac{2}{3}xe^{-3x}.$$

Differentiating, we find

$$y'(x) = -3c_2 e^{-x} + \frac{4}{3}x^2 - \frac{8}{9}x + \frac{8}{27} + \frac{2}{3}(-3x + 1)e^{-3x}.$$

The constants c_1 and c_2 must be chosen so that the initial conditions $y(0) = 1$ and $y'(0) = 0$ are satisfied. Hence, c_1 and c_2 must be chosen to satisfy the following system of linear equations.

$$c_1 + c_2 = 1 \qquad \text{(from the condition } y(0) = 1)$$

$$-3c_2 + \frac{8}{27} + \frac{2}{3} = 0 \qquad \text{(from the condition } y'(0) = 0)$$

Solving this system, we find $c_1 = \dfrac{55}{81}$ and $c_2 = \dfrac{26}{81}$. Therefore, the solution of the IVP (5) is

$$y(x) = \frac{55}{81} + \frac{26}{81}e^{-3x} + \frac{4}{9}x^3 - \frac{4}{9}x^2 + \frac{8}{27}x + \frac{2}{3}xe^{-3x}. \ \blacksquare$$

4.5.1 Exercises

Solve the following initial value problems.

1. $y'' - y' - 6y = 0; \quad y(0) = 1, \quad y'(0) = -2$
2. $y'' + y = 0; \quad y(1) = 1, \quad y'(1) = 0$
3. $y^{(3)} - 3y^{(2)} - 4y^{(1)} + 12y = 0; \quad y(0) = 1, \quad y^{(1)}(0) = 5, \quad y^{(2)}(0) = -1$
4. $y^{(3)} - 3y^{(1)} - 2y = 0; \quad y(0) = 0, \quad y^{(1)}(0) = 9, \quad y^{(2)}(0) = 0$
5. $y^{(4)} + 3y^{(3)} + 2y^{(2)} = 0; \quad y(0) = 0, \quad y^{(1)}(0) = 4, \quad y^{(2)}(0) = -6,$ $y^{(3)}(0) = 14$
6. $y^{(4)} - 2y^{(3)} + 2y^{(1)} - y = 0; \quad y(0) = 1, \quad y^{(1)}(0) = -1, \quad y^{(2)}(0) = -3,$ $y^{(3)}(0) = 3$
7. $y'' + y = 10e^{2x}; \quad y(0) = 0, \quad y'(0) = 0$
8. $y'' - 4y = 2 - 8x; \quad y(0) = 0, \quad y'(0) = 5$
9. $y^{(3)} - y^{(2)} + y^{(1)} - y = 2e^{x}; \quad y(0) = 1, \quad y^{(1)}(0) = 3, \quad y^{(2)}(0) = -3$
10. $y^{(4)} + 2y^{(2)} + y = 3x + 4; \quad y(0) = 0, \quad y^{(1)}(0) = 0, \quad y^{(2)}(0) = 1,$ $y^{(3)}(0) = 1$

5

The Laplace Transform Method

In Sections 4.3 and 4.4, we showed how to solve homogeneous and nonhomogeneous linear differential equations with constant coefficients, and in Section 4.5, we showed how to solve initial value problems in which the differential equations were homogeneous or nonhomogeneous linear differential equations with constant coefficients. The technique consisted of finding the general solution of the differential equation and then choosing the constants in the general solution to satisfy the specified initial conditions. In this chapter, we present the Laplace transform method for solving homogeneous and nonhomogeneous linear differential equations with constant coefficients and their corresponding initial value problems. We begin by examining the Laplace transform and its properties.

5.1 The Laplace Transform and Its Properties

The Laplace transform is named in honor of the French mathematician and astronomer Pierre Simon Marquis de Laplace (1749-1827). Laplace studied the integral appearing in the definition of the transform in 1779 in conjunction with his research on probability. However, most of the results and techniques presented in this chapter were developed by the English electrical engineer Oliver Heaviside (1850-1925) more than a century later. The Laplace transform is defined as follows.

Let $f(x)$ be a function defined on the interval $[0, +\infty)$. The **Laplace transform** of $f(x)$ is

$$\mathcal{L}[f(x)] = \int_0^\infty f(x)e^{-sx}\,dx = F(s)$$

provided the improper integral exists for s sufficiently large.

The Laplace transform is an operator which assigns to one function, $f(x)$, another function, $F(s)$. It can be shown that if the improper integral appearing in the definition converges for some fixed value s_0, then it will converge for all $s > s_0$. For some functions, the Laplace transform exists for all real s. While for other functions the Laplace transform does not exist for any real s.

DOI: 10.1201/9781003678014-5

Recall from calculus that the improper integral

$$\int_0^\infty g(x)\,dx \quad \text{is defined to be the limit} \quad \lim_{B\to+\infty}\int_0^B g(x)\,dx,$$

provided the limit exists. When the limit exists, we say the integral exists and its value is the limit. When the limit does not exist, we say the integral does not exist. Consider the improper integral $\int_0^\infty e^{-sx}dx$. For $s = 0$, the function $e^{-sx} = 1$ and the improper integral diverges, since

$$\int_0^\infty 1\,dx = \lim_{B\to+\infty}\int_0^B 1\,dx = \lim_{B\to+\infty} B = +\infty.$$

For $s \neq 0$,

$$\int_0^\infty e^{-sx}\,dx = \lim_{B\to+\infty}\int_0^B e^{-sx}\,dx = \lim_{B\to+\infty}\left(\frac{e^{-sx}}{-s}\bigg|_0^B\right)$$

$$= \lim_{B\to+\infty}\frac{-1}{s}(e^{-sB}-1) = \begin{cases} \dfrac{1}{s} &, \quad \text{if } s>0 \\ \text{diverges} &, \quad \text{if } s<0 \end{cases}$$

Instead of writing the limit $\lim_{B\to+\infty}(\ )|_0^B$ over and over again, we will denote this limit simply by $(\ )|_0^\infty$. Writing the integrand of the integral under consideration as $1e^{-sx}$, we see that the integral is the Laplace transform of the function $f(x) = 1$. Thus, we have shown that

$$\mathcal{L}[1] = \int_0^\infty 1e^{-sx}\,dx = \frac{-1}{s}(e^{-sx}-1)\bigg|_0^\infty = \frac{1}{s}, \quad \text{provided } s>0.$$

Example 1 Find the Laplace transform of $\mathbf{f(x) = e^{ax}}$

Solution

$$\mathcal{L}[e^{ax}] = \int_0^\infty e^{ax}e^{-sx}\,dx = \int_0^\infty e^{-(s-a)x}\,dx = \frac{-1}{s-a}e^{-(s-a)x}\bigg|_0^\infty = \frac{1}{s-a}$$

provided $s > a = s_0$. If $s \leq a$, then $\mathcal{L}[e^{ax}]$ does not exist. ■

Example 2 Calculate the Laplace transform of $\mathbf{f(x) = x^n}$, where n is a positive integer.

Solution

By definition

$$\mathcal{L}[x^n] = \int_0^\infty x^n e^{-sx}\,dx.$$

Recall from calculus the following formula for integration by parts

$$\int_a^b u\,dv = uv\Big|_a^b - \int_a^b v\,du.$$

Letting $u = x^n$ and $dv = e^{-sx}dx$ and differentiating and integrating, we find $du = nx^{n-1}dx$ and $v = -\frac{1}{s}e^{-sx}$. Integration by parts, yields

(1)

$$\mathcal{L}[x^n] = \int_0^\infty x^n e^{-sx}\,dx = -\frac{x^n}{s}e^{-sx}\Big|_0^\infty + \frac{n}{s}\int_0^\infty x^{n-1}e^{-sx}\,dx = \frac{n}{s}\mathcal{L}[x^{n-1}]$$

provided $s > 0$.

Letting $n = 1$, we find

$$\mathcal{L}[x] = \frac{1}{s}\mathcal{L}[1] = \frac{1}{s^2} \quad \text{for} \quad s > 0.$$

If $n > 1$, we repeatedly use the result of equation (1) to obtain

$$\mathcal{L}[x^n] = \frac{n}{s}\mathcal{L}[x^{n-1}] = \frac{n(n-1)}{s^2}\mathcal{L}[x^{n-2}] = \cdots = \frac{n!}{s^n}\mathcal{L}[1] = \frac{n!}{s^{n+1}} \quad \text{for } s > 0. \blacksquare$$

Example 3 Find $\mathcal{L}[\sin bx] = \int_0^\infty (\sin bx)\, e^{-sx}\, dx$.

Solution

We use integration by parts. This time, we let $u = \sin bx$ and $dv = e^{-sx}dx$. Differentiation and integration yield $du = b\cos bx$ and $v = -\frac{1}{s}e^{-sx}$. Consequently,

(2)

$$\mathcal{L}[\sin bx] = \frac{-1}{s}(\sin bx)e^{-sx}\Big|_0^\infty + \frac{b}{s}\int_0^\infty (\cos bx)e^{-sx}\,dx = \frac{b}{s}\int_0^\infty (\cos bx)e^{-sx}\,dx$$

provided $s > 0$. Now, we use integration by parts a second time by letting $u = \cos bx$ and $dv = e^{-sx}dx$. Then $du = -b\sin bx$ and $v = -\frac{1}{s}e^{-sx}$. Integrating the integral appearing on the right-hand side of (2) by parts, we find

(3)

$$\mathcal{L}[\sin bx] = \frac{b}{s}\left\{\frac{-1}{s}(\cos bx)e^{-sx}\Big|_0^\infty - \frac{b}{s}\int_0^\infty (\sin bx)e^{-sx}\,dx\right\} = \frac{b}{s^2} - \frac{b^2}{s^2}\mathcal{L}[\sin bx]$$

provided $s > 0$. Solving equation (3) algebraically for $\mathcal{L}[\sin bx]$, we obtain

$$\mathcal{L}[\sin bx] = \frac{b/s^2}{1 + b^2/s^2} = \frac{b}{s^2 + b^2} \quad \text{for} \quad s > 0. \blacksquare$$

In the previous three examples, we used the definition of the Laplace transform to calculate the transform of three different functions—e^{ax}, x^n, and $\sin bx$. In elementary calculus, we learned the definition of the derivative of a function, calculated the derivative of a few functions from the definition, and then learned rules for the differentiation of the sum, difference, product, and quotient of two functions. These rules allowed us to differentiate a variety of functions without explicitly using the definition of the derivative. The operation of differentiating a function transforms the function $f(x)$ into the function $f'(x)$. If the operation of differentiation is denoted by D, then the transformation can be written as $D[f(x)] = f'(x)$. The function $f'(x)$ is the transform of $f(x)$ under the transformation D. Thus, for example, $D[x^3] = 3x^2$. Another transformation we encountered in calculus was integration. The operation of integration transforms the function $f(x)$ into the function $F(x) = \int_0^x f(t)\,dt$. If we let I denote integration, then this transformation can be written as $I[f(x)] = \int_0^x f(t)\,dt = F(x)$. For instance, $I[x^3] = x^4/4$.

An operator T is **linear** if for every pair of functions $f(x)$ and $g(x)$ and for every pair of constants c_1 and c_2,

$$T[c_1 f(x) + c_2 g(x)] = c_1 T[f(x)] + c_2 T[g(x)].$$

Differentiation and integration are both linear operators. That is, for any two differentiable functions $f(x)$ and $g(x)$ and any two constants c_1 and c_2

$$D[c_1 f(x) + c_2 g(x)] = c_1 D[f(x)] + c_2 D[g(x)].$$

Likewise, for any two integrable functions $f(x)$ and $g(x)$ and any two constants c_1 and c_2

$$I[c_1 f(x) + c_2 g(x)] = c_1 I[f(x)] + c_2 I[g(x)].$$

We now prove that the **Laplace transform is a linear operator**. That is, we prove that if $f_1(x)$ and $f_2(x)$ are functions which have Laplace transforms for $s > s_1$ and $s > s_2$, respectively, and if c_1 and c_2 are constants, then

$$\mathcal{L}[c_1 f_1(x) + c_2 f_2(x)] = c_1 \mathcal{L}[f_1(x)] + c_2 \mathcal{L}[f_2(x)] \quad \text{for } s > \max(s_1, s_2).$$

Proof: Let $s > \max(s_1, s_2)$. Then, by definition,

$$\begin{aligned}\mathcal{L}[c_1 f_1(x) + c_2 f_2(x)] &= \int_0^\infty \{c_1 f_1(x) + c_2 f_2(x)\} e^{-sx}\,dx \\ &= c_1 \int_0^\infty f_1(x) e^{-sx}\,dx + c_2 \int_0^\infty f_2(x) e^{-sx}\,dx \\ &= c_1 \mathcal{L}[f_1(x)] + c_2 \mathcal{L}[f_2(x)].\end{aligned}$$

■

Example 4 Calculate $\mathcal{L}[\mathbf{4x^2{+}3}]$.

Solution

Using the linearity property of the Laplace transform, we find

$$\mathcal{L}[4x^2+3] = 4\mathcal{L}[x^2] + 3\mathcal{L}[1] = 4\frac{2}{s^3} + 3\frac{1}{s} = \frac{8}{s^3} + \frac{3}{s} \quad \text{for} \quad s > 0. \ \blacksquare$$

Example 5 Calculate $\mathcal{L}[\mathbf{sinh\, bx}]$.

Solution

Since $\sinh bx = \frac{1}{2}e^{bx} - \frac{1}{2}e^{-bx}$, we find, using the linearity property of the Laplace transform, that

$$\mathcal{L}[\sinh bx] = \mathcal{L}[\frac{1}{2}e^{bx} - \frac{1}{2}e^{-bx}] = \frac{1}{2}\mathcal{L}[e^{bx}] - \frac{1}{2}\mathcal{L}[e^{-bx}]$$

$$= \frac{1}{2}\frac{1}{s-b} - \frac{1}{2}\frac{1}{s+b} = \frac{b}{s^2-b^2} \quad \text{for} \quad s > |b|. \ \blacksquare$$

Another property of Laplace transforms which is useful in calculating the transform of some functions is the following property which is called the **Translation Property of the Laplace Transform.**

If $\mathcal{L}[f(x)] = F(s)$ for $s > s_0$, then $\mathcal{L}[e^{ax}f(x)] = F(s-a)$ for $s > s_0 + a$.

Proof: By definition and hypothesis,

$$\mathcal{L}[f(x)] = \int_0^\infty f(x)e^{-sx}\,dx = F(s) \quad \text{for} \quad s > s_0.$$

Hence,

$$\mathcal{L}[e^{ax}f(x)] = \int_0^\infty e^{ax}f(x)e^{-sx}\,dx$$

$$= \int_0^\infty f(x)e^{-(s-a)x}\,dx = F(s-a) \quad \text{for} \quad s-a > s_0.$$

That is,

$$\mathcal{L}[e^{ax}f(x)] = F(s-a) \quad \text{for} \quad s > s_0 + a.$$

$\blacksquare$

As the following example illustrates, this property allows us to calculate easily the Laplace transform of the function $e^{ax}f(x)$, if we already know the transform of $f(x)$.

Example 6 Calculate $\mathcal{L}[\mathrm{x^n e^{ax}}]$ where n is a positive integer.

Solution

Earlier, we found

$$\mathcal{L}[x^n] = \frac{n!}{s^{n+1}} = F(s) \quad \text{for} \quad s > 0.$$

Using the translation property, we now find

$$\mathcal{L}[x^n e^{ax}] = F(s-a) = \frac{n!}{(s-a)^{n+1}} \quad \text{for} \quad s > a. \ \blacksquare$$

A function $f(x)$ is **piecewise continuous on a finite interval [a,b]** if and only if

(i) $f(x)$ is continuous on $[a,b]$ except at a finite number of points,

(ii) the limits

$$f(a^+) = \lim_{x \to a^+} f(x) \quad \text{and} \quad f(b^-) = \lim_{x \to b^-} f(x)$$

both exist and are finite, and

(iii) if $c \in (a,b)$ is a point of discontinuity of $f(x)$, then the following limits exist and are finite:

$$f(c^-) = \lim_{x \to c^-} f(x) \quad \text{and} \quad f(c^+) = \lim_{x \to c^+} f(x).$$

When the limits in (iii) are equal, f is said to have a **removable discontinuity at** c.

When the limits in (iii) are unequal, f is said to have a **jump discontinuity at** c.

If $f(x)$ is piecewise continuous on a finite interval $[a,b]$ and is continuous except possibly at the points $a = a_1 < a_2 < \cdots < a_n = b$, then f is integrable on $[a,b]$ and

$$\int_a^b f(x)\,dx = \int_{a_1}^{a_2} f(x)\,dx + \int_{a_2}^{a_3} f(x)\,dx + \cdots + \int_{a_{n-1}}^{a_n} f(x)\,dx.$$

The function graphed in Figure 5.1 is piecewise continuous on $[a,b]$. It has a removable discontinuity at c_1 and jump discontinuities at c_2 and c_3.

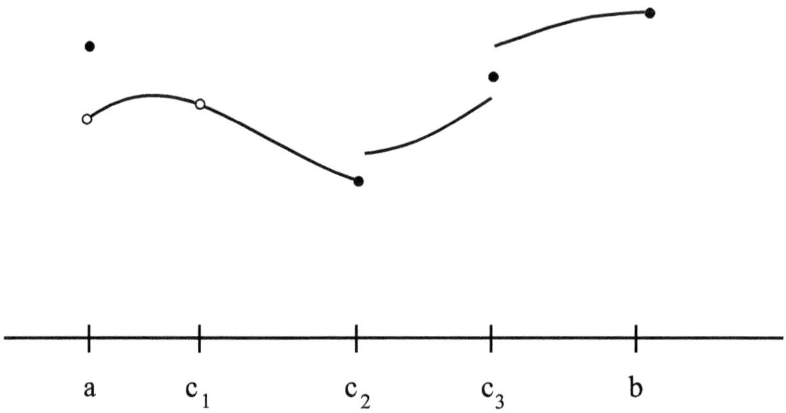

FIGURE 5.1
Graph of a Piecewise Continuous Function.

Example 7 Compute the Laplace transform of the piecewise continuous function

$$f(x) = \begin{cases} x, & 0 \le x < 2 \\ 3, & 2 \le x \end{cases}$$

Solution

By definition,

$$(4) \qquad \mathcal{L}[f(x)] = \int_0^\infty f(x)e^{-sx}\,dx = \int_0^2 xe^{-sx}\,dx + \int_2^\infty 3e^{-sx}\,dx.$$

To calculate the first integral on the right-hand side of equation (4), we use integrations by parts. Letting $u = x$ and $dv = e^{-sx}$ and differentiating and integrating, we find $du = dx$ and $v = -(1/s)e^{-sx}$. Substituting these expressions into the integration by parts formula, yields

$$(5) \quad \int_0^2 xe^{-sx}\,dx = \frac{-x}{s}e^{-sx}\Big|_0^2 + \frac{1}{s}\int_0^2 e^{-sx}\,dx = \frac{-2}{s}e^{-2s} - \left\{\frac{1}{s^2}e^{-sx}\Big|_0^2\right\}$$

$$= \frac{-2}{s}e^{-2s} - \frac{1}{s^2}[e^{-2s} - 1].$$

Evaluating the second integral on the right-hand side of equation (4), we find

$$(6) \qquad \int_2^\infty 3e^{-sx}\,dx = \frac{-3}{s}e^{-sx}\Big|_2^\infty = \frac{3}{s}e^{-2s} \qquad \text{for } s > 0.$$

Substituting the results from equations (5) and (6) into (4), we get

$$\mathcal{L}[f(x)] = \frac{-2}{s}e^{-2s} - \frac{1}{s^2}[e^{-2s} - 1] + \frac{3}{s}e^{-2s} = \frac{1}{s}e^{-2s} - \frac{1}{s^2}[e^{-2s} - 1] \quad \text{for } s > 0. \blacksquare$$

Not all functions have a Laplace transform as the following example shows.

Example 8 Show that $\mathcal{L}[e^{x^2}] = \int_0^\infty e^{x^2} e^{-sx}\,dx$ does not exist.

Solution

Clearly the integral does not exist if $s \leq 0$, since the integrand is positive. Suppose the integral does exist for some $s > 0$. Then

$$\text{(7)} \qquad \mathcal{L}[e^{x^2}] = \int_0^\infty e^{x^2} e^{-sx}\,dx = \int_0^{2s} e^{x(x-s)}\,dx + \int_{2s}^\infty e^{x(x-s)}\,dx.$$

The first integral on the right-hand side of equation (7) is positive, since the integrand is positive for all real x and s. For $x \geq 2s$, we have $x - s \geq s$ and $e^{x(x-s)} \geq e^{sx}$. Therefore, the second integral on the right-hand side of (7) satisfies the inequality

$$\int_{2s}^\infty e^{x(x-s)}\,dx \geq \int_{2s}^\infty e^{sx}\,dx = \infty.$$

Thus, for any $s > 0$, we have

$$\mathcal{L}[e^{x^2}] = \int_0^{2s} e^{x(x-s)}\,dx + \int_{2s}^\infty e^{x(x-s)}\,dx \geq \int_{2s}^\infty e^{sx}\,dx = \infty.$$

Consequently, $\mathcal{L}[e^{x^2}]$ does not exist. $\blacksquare$

For any fixed positive value of s, the factor e^{-sx}, which appears in the integrand of the definition of the Laplace transform, is a "damping factor"—a factor which decreases to zero as x increases. Provided the function $f(x)$ does not "grow too rapidly" as x increases, we expect the defining integral to converge and, therefore, the Laplace transform to exist. Classes of functions which do not "grow too rapidly" are said to be of exponential order a, or simply of exponential order. Such functions are defined as follows.

A function $f(x)$ is of **exponential order a as x $\to +\infty$** if and only if there exist positive constants M and x_0 and a constant a such that

$$|f(x)| < Me^{ax} \quad \text{for} \quad x \geq x_0.$$

A function which is of exponential order $a > 0$ as $x \to +\infty$ may become infinite as $x \to +\infty$, but it may not become infinite more rapidly than Me^{ax}. It follows from this definition that all bounded functions are of exponential order 0 as $x \to +\infty$. Also if $a < b$ and $f(x)$ is of exponential order a as $x \to +\infty$, then $f(x)$ is of exponential order b as $x \to +\infty$, since $a < b$ implies $Me^{ax} < Me^{bx}$. It should be noted that there are functions which are not of exponential order a as $x \to +\infty$ for any a. For instance, the function e^{x^2} is not of exponential order a as $x \to +\infty$ for any a. As a matter of fact, it can be shown that for any positive constants a and M, no matter how large, there exists an x_0—which depends upon M and a—such that $e^{x^2} > Me^{ax}$ for all $x > x_0$.

The following theorem provides sufficient conditions for the existence of a Laplace transform of a function $f(x)$. These conditions are not necessary conditions, as we shall show in the example following the theorem.

Theorem 5.1 An Existence Theorem for the Laplace Transform If $f(x)$ is piecewise continuous on $[0, b]$ for all finite $b > 0$ and if $f(x)$ is of exponential order a as $x \to +\infty$, then the Laplace transform of $f(x)$ exists for $s > a$.

Proof: Since $f(x)$ is assumed to be of exponential order a as $x \to +\infty$, there exist positive constants M and x_0 such that $|f(x)| < Me^{ax}$ for $x > x_0$. We rewrite the Laplace transform of $f(x)$ as follows:

$$(8) \qquad \mathcal{L}[f(x)] = \int_0^\infty f(x)e^{-sx}\,dx = \int_0^{x_0} f(x)e^{-sx}\,dx + \int_{x_0}^\infty f(x)e^{-sx}\,dx.$$

The first integral on the right-hand side of equation (8) exists, since $f(x)$ is piecewise continuous on $[0, x_0]$, which implies $f(x)e^{-sx}$ is piecewise continuous on $[0, x_0]$, because e^{-sx} is continuous on $[0, x_0]$. Since $f(x)e^{-sx}$ is piecewise continuous on $[0, x_0]$, it is integrable on $[0, x_0]$. The absolute value of the second integral on the right-hand side of equation (8) satisfies

$$\begin{aligned}\left|\int_{x_0}^\infty f(x)e^{-sx}\,dx\right| &\le \int_{x_0}^\infty |f(x)|e^{-sx}\,dx < M\int_{x_0}^\infty e^{ax}e^{-sx}\,dx \\ &= \frac{M}{a-s}e^{(a-s)x}\bigg|_{x_0}^\infty = \frac{-M}{a-s}e^{(a-s)x_0} \quad \text{provided} \quad s > a.\end{aligned}$$

Since the improper integral $\int_{x_0}^\infty e^{ax}e^{-sx}\,dx$ converges for $s > a$, the improper integral $\int_{x_0}^\infty |f(x)|e^{-sx}\,dx$ converges for $s > a$ by the comparison test for improper integrals. Since $\int_{x_0}^\infty |f(x)|e^{-sx}\,dx$ converges for $s > a$, the integral $\int_{x_0}^\infty f(x)e^{-sx}\,dx$ converges for $s > a$. Thus, the second integral on the right-hand side of equation (8) converges for $s > a$. Since both improper integrals on the right-hand side of (8) converge for $s > a$, the Laplace transform of $f(x)$, $\mathcal{L}[f(x)]$, exists for $s > a$. ■

The function $f(x) = 1/\sqrt{x}$ is of exponential order 0 as $x \to +\infty$, but it is not piecewise continuous on $[0, b]$ for any $b > 0$ since $\lim_{x\to 0^+} f(x) = +\infty$. Thus, $f(x) = 1/\sqrt{x}$ does not satisfy the first hypothesis of Theorem 5.1;

however, the Laplace transform exists, as the following calculations show. By definition

$$\mathcal{L}\left[\frac{1}{\sqrt{x}}\right] = \int_0^\infty \frac{e^{-sx}}{\sqrt{x}}\, dx.$$

Making the change of variable $t = sx$, we find

$$\mathcal{L}\left[\frac{1}{\sqrt{x}}\right] = \frac{1}{\sqrt{s}} \int_0^\infty \frac{e^{-t}}{\sqrt{t}}\, dt = \sqrt{\frac{\pi}{s}}.$$

The value of the definite integral on the right-hand side of the last equation was obtained from a table of integrals.

Early in calculus, you discovered that many functions have the same derivative. Letting D denote the differential operator as we did earlier, we find that $D[x^3] = D[x^3 + 1] = D[x^3 + C] = 3x^2$, where C is an arbitrary constant. Just as many functions have the same derivative, many functions have the same Laplace transform. At the beginning of this section, we calculated the Laplace transform of the function $f(x) = 1$ and found $\mathcal{L}[f(x)] = 1/s = F(s)$ provided $s > 0$. The Laplace transform of the piecewise continuous function

$$g(x) = \begin{cases} 1, & x \neq 2 \\ 3, & x = 2 \end{cases}$$

is

$$\mathcal{L}[g(x)] = \int_0^\infty 1e^{-sx}\, dx = \int_0^2 1e^{-sx}\, dx + \int_2^\infty 1e^{-sx}\, dx = \frac{1}{s} = F(s),$$

provided $s > a$. The function $g(x)$ could have been chosen to differ from $f(x)$ at any finite set of values of x, or even at an infinite set of values such as the set $\{1, 2, 3, \ldots\}$. In calculus, you learned if $D[f(x)] = D[g(x)]$ on some interval $[a, b]$, then $f(x) = g(x) + C$ on $[a, b]$ for some constant C. The following theorem, which we state without proof, is an analogous theorem for Laplace transforms.

Theorem 5.2 If $f(x)$ and $g(x)$ are defined and piecewise continuous on $[0, b]$ for all finite $b > 0$ and of exponential order a as $x \to +\infty$ and if $\mathcal{L}[f(x)] = \mathcal{L}[g(x)]$, then $f(x) = g(x)$ at all points $x \in [0, \infty)$ where $f(x)$ and $g(x)$ are both continuous.

A consequence of this theorem is that a given Laplace transform, $F(s)$, cannot have more than one continuous function $f(x)$ defined on $[0, \infty)$ which transforms into $F(s)$. Although there may be many functions $g(x)$ which transform into $F(s)$, there is only one continuous function defined on $[0, \infty)$ which transforms into $F(s)$. Therefore, by defining **the inverse Laplace transform** as follows, there is a unique inverse Laplace transform.

If there exists a continuous function $f(x)$ defined on the interval $[0, \infty)$ such that $\mathcal{L}[f(x)] = F(s)$, then $f(x)$ is called the **inverse Laplace transform** of $F(s)$ and we write $\mathcal{L}^{-1}(F(s)) = f(x)$.

Consider the Laplace transform $H(s) = c_1F(s) + c_2G(s)$ where c_1 and c_2 are arbitrary constants. Assume $f(x)$ and $g(x)$ are continuous functions on $[0, \infty)$ which are of exponential order a as $x \to +\infty$, and assume $f(x)$ and $g(x)$ have Laplace transforms $F(s) = \mathcal{L}[f(x)]$ and $G(s) = \mathcal{L}[g(x)]$. Since the Laplace transform is a linear operator,

$$(9) \qquad \mathcal{L}[c_1f(x) + c_2g(x)] = c_1\mathcal{L}[f(x)] + c_2\mathcal{L}[g(x)] = c_1F(s) + c_2G(s).$$

Taking the inverse Laplace transform of (9), we find

$$\mathcal{L}^{-1}[c_1F(s) + c_2G(s)] = c_1f(x) + c_2g(x) = c_1\mathcal{L}^{-1}[F(s)] + c_2\mathcal{L}^{-1}[G(s)].$$

Thus, we have shown that **the inverse Laplace transform is a linear operator.**

Just as there are functions $f(x)$ which do not have Laplace transforms (recall $f(x) = e^{x^2}$ does not have a Laplace transform), there are many functions $F(s)$ which do not have inverse Laplace transforms. For example, $F(s) = 1$ is a function which, by the definition above, does not have an inverse Laplace transform, since there is no continuous function $f(x)$ defined on the interval $[0, \infty)$ such that $\mathcal{L}[f(x)] = 1$. However, in Section 5.5, we define and discuss the Dirac delta function, $\delta(x)$. This "function" is not a function on $[0, +\infty)$ in the classical sense. Moreover, it is not continuous on $[0, \infty)$. Nonetheless, in distribution theory, the delta function has the property that $\mathcal{L}[\delta(x)] = 1$. Hence, $\delta(x)$ is called the inverse Laplace transform of $F(s) = 1$.

Table 5.1 contains several functions, $f(x)$, and their corresponding Laplace transform, $F(s)$. The left column of Table 5.1 contains functions, $f(x)$, which are continuous on $[0, \infty)$ and the corresponding entry in the right column contains their Laplace transform, $F(s)$. Since each function $f(x)$ is an inverse Laplace transform of the function $F(s)$ appearing in the corresponding right column, the left column is labelled $f(x) = \mathcal{L}^{-1}(F(s))$.

In integral calculus, you learned how to integrate rational functions by using partial fraction decomposition. In order to find inverse Laplace transforms of rational functions efficiently using a table of Laplace transforms, we need to know a variation of partial fraction decomposition.

Partial Fraction Expansion for Computing Inverse Laplace Transforms

Let $F(s) = P(s)/Q(s)$ where $P(s)$ and $Q(s)$ are polynomials with real coefficients, where $P(s)$ and $Q(s)$ have no common factor, and where the degree of $P(s)$ is less than the degree of $Q(s)$.

TABLE 5.1
Laplace Transforms and Inverse Laplace Transforms

$f(x) = \mathcal{L}^{-1}(F(s))$	$F(s) = \mathcal{L}[f(x)]$	
e^{ax}	$\frac{1}{s-a}$	for $s > a$
1	$\frac{1}{s}$	for $s > 0$
x^n, n a positive integer	$\frac{n!}{s^{n+1}}$	for $s > 0$
$\sin bx$	$\frac{b}{s^2+b^2}$	for $s > 0$
$\cos bx$	$\frac{s}{s^2+b^2}$	for $s > 0$
$\sinh bx$	$\frac{b}{s^2-b^2}$	for $s > \|b\|$
$\cosh bx$	$\frac{s}{s^2-b^2}$	for $s > \|b\|$
$x^n e^{ax}$, n a positive integer	$\frac{n!}{(s-a)^{n+1}}$	for $s > a$
$e^{ax}\sin bx$	$\frac{b}{(s-a)^2+b^2}$	for $s > a$
$e^{ax}\cos bx$	$\frac{s-a}{(s-a)^2+b^2}$	for $s > a$
$e^{ax}\sinh bx$	$\frac{b}{(s-a)^2-b^2}$	for $s > \|b\| + a$
$e^{ax}\cosh bx$	$\frac{s-a}{(s-a)^2-b^2}$	for $s > \|b\| + a$

A. For Linear Factors

When $s-r$ is a factor of $Q(s)$ exactly n times, the part of the partial fraction expansion for $P(s)/Q(s)$ corresponding to the term $(s-r)^n$ is

$$\frac{A_1}{s-r} + \frac{A_2}{(s-r)^2} + \cdots + \frac{A_n}{(s-r)^n}$$

where A_1, A_2, ... , A_n are real constants which must be determined.

B. For Irreducible Quadratic Factors

Let $(s-a)^2+b^2$ be a quadratic factor of $Q(s)$ which cannot be factored into linear factors with real coefficients. When $(s-a)^2+b^2$ is a factor of $Q(s)$ exactly n times, the part of the partial fraction expansion for $P(s)/Q(s)$ corresponding to the term $(s-a)^2+b^2$ is

$$\frac{B_1(s-a)+C_1 b}{(s-a)^2+b^2} + \frac{B_2(s-a)+C_2 b}{((s-a)^2+b^2)^2} + \cdots + \frac{B_n(s-a)+C_n b}{((s-a)^2+b^2)^n}$$

where B_1, B_2, ... , B_n and C_1, C_2, ... , C_n are real constants which must be determined.

Example 9 For $\mathbf{F}(\mathbf{s}) = \dfrac{\mathbf{2}}{\mathbf{s(s+1)}}$, calculate $\mathcal{L}^{-1}[\mathbf{F}(\mathbf{s})]$.

Solution

The denominator of $F(s)$, which is $Q(s) = s(s+1)$, has two linear factors of multiplicity 1, so the partial fraction expansion for $F(s)$ has the form

$$\text{(10)} \qquad F(s) = \frac{2}{s(s+1)} = \frac{A}{s} + \frac{B}{s+1}$$

where A and B are constants to be determined. Multiplying equation (10) by $s(s+1)$, we see A and B must be chosen to satisfy

$$\text{(11)} \qquad 2 = A(s+1) + Bs.$$

Setting $s=0$ in equation (11), we find $A=2$. And setting $s=-1$ in equation (11), we see $B=-2$. Hence,

$$\mathcal{L}[f(x)] = \frac{2}{s(s+1)} = \frac{2}{s} - \frac{2}{s+1} = 2\left(\frac{1}{s}\right) - 2\left(\frac{1}{s+1}\right).$$

Since

$$\mathcal{L}[1] = \frac{1}{s} \quad \text{and} \quad \mathcal{L}[e^{-x}] = \frac{1}{s+1},$$

$$\text{(12)} \qquad \mathcal{L}[f(x)] = 2\mathcal{L}[1] - 2\mathcal{L}[e^{-x}] = \mathcal{L}[2-2e^{-x}].$$

Taking the inverse Laplace transform of equation (12), we find $f(x) = 2-2e^{-x}$. ■

Example 10 For $\mathbf{F(s)} = \dfrac{\mathbf{-2s^3 + 3s^2 + 37s - 55}}{\mathbf{(s-4)^2(s^2 - 4s+13)}}$**, calculate** $\mathcal{L}^{-1}[\mathbf{F(s)}]$.

Solution

The quadratic factor $s^2 - 4s + 13$ appearing in the denominator of $F(s)$ is irreducible, since its discriminant $(-4)^2 - 4(1)(13) = 16 - 52 = -36 < 0$. Completing the square, we can rewrite this quadratic factor as

$$s^2 - 4s + 13 = (s^2 - 4s + 4) + 9 = (s-2)^2 + 3^2 = (s-a)^2 + b^2.$$

Since the linear factor $(s-4)$ appears to the second power in the denominator of $F(s)$ and the quadratic factor appears to the first power, the partial fraction expansion for $F(s)$ has the form

$$(13) \qquad \frac{-2s^3 + 3s^2 + 37s - 55}{(s-4)^2(s^2 - 4s + 13)} = \frac{A}{s-4} + \frac{B}{(s-4)^2} + \frac{C(s-2) + 3D}{(s-2)^2 + 3^2}.$$

Multiplication of (13) by $(s-4)^2(s^2 - 4s + 13)$, yields

$$(14) \quad -2s^3 + 3s^2 + 37s - 55 =$$
$$A(s-4)(s^2 - 4s + 13) + B(s^2 - 4s + 13) + C(s-2)(s-4)^2 + 3D(s-4)^2.$$

Setting $s = 4$ in (14), we obtain

$$-2(4)^3 + 3(4)^2 + 37(4) - 55 = B(4^2 - 4(4) + 13) \quad \text{or} \quad 13 = 13B.$$

So B $= 1$. Substituting $B = 1$ into (14), and then subtracting $s^2 - 4s + 13$ from both sides of the resulting equation, yields

$$(15) \quad -2s^3+2s^2+41s-68 = A(s-4)(s^2-4s+13)+C(s-2)(s-4)^2+3D(s-4)^2.$$

Letting $s = 0$ in (15), results in $\quad -68 = -52A - 32C + 48D.$
Letting $s = 2$ in (15), results in $\quad 6 = -18A + 12D.$
And letting $s = 1$ in (15), results in $\quad -27 = -30A - 9C + 27D.$

Solving the last three equations simultaneously, we find $A = -3$, $C = 1$, and $D = -4$. Hence,

$$F(s) = -3\left(\frac{1}{s-4}\right) + \frac{1}{(s-4)^2} + \frac{s-2}{(x-2)^2 + 3^2} - 4\left(\frac{3}{(x-2)^2 + 3^2}\right).$$

Since the inverse Laplace transform is a linear operator

$$\mathcal{L}^{-1}[F(s)] = -3\mathcal{L}^{-1}\left[\frac{1}{s-4}\right] + \mathcal{L}^{-1}\left[\frac{1}{(s-4)^2}\right]$$
$$+ \mathcal{L}^{-1}\left[\frac{s-2}{(x-2)^2 + 3^2}\right] - 4\mathcal{L}^{-1}\left[\frac{3}{(x-2)^2 + 3^2}\right]$$

$$= -3e^{4x} + xe^{4x} + e^{2x}\cos 3x - 4e^{2x}\sin 3x. \blacksquare$$

As we have discovered, manually calculating the Laplace transform of a function from the definition can involve using various integration techniques such as integration by parts, substitution, and so forth, or it can involve looking up definite integrals in a table of integrals. If we have a table of Laplace transforms available, then manually computing the Laplace transform can require using the linearity property or translation property of Laplace transforms. To manually calculate the inverse Laplace transform often requires using partial fraction expansion and the use of a table of Laplace transforms.

5.1.1 Comments on Computing

s Algorithms for calculating the Laplace transform and the inverse Laplace transform are often included in computer algebra systems (CAS). What the user needs to know in order to use such a CAS is the command to use to invoke the Laplace transform or inverse Laplace transform, the required arguments of the command, and the syntax for entering the function $f(x)$ or $F(s)$. A CAS will not show the computations used to arrive at the answer, it will just provide the answer. Thus, if a user specifies the proper syntax to request a CAS to calculate the Laplace transform of $f(x) = x^2 \sin x$ the CAS will simply respond

$$2\frac{-1+3s^2}{(1+s^2)^3}.$$

And if the user specifies the proper syntax to request a CAS to calculate the inverse Laplace transform of $F(s) = \dfrac{2}{s(s+1)}$, the CAS will respond $2 - 2e^{-x}$.

5.1.2 Exercises

In Exercises 1–6 manually calculate the Laplace transform from its definition. If you have a CAS available which calculates the Laplace transform, also use the CAS to calculate the Laplace transforms of Exercises 1–6 and compare those answers to the ones you obtained by hand.

1. Let a and b be real constants.

   a. Calculate $\mathcal{L}[x \sin bx]$.

   b. Use the translation property to calculate $\mathcal{L}[xe^{ax} \sin bx]$.

2. Let a and b be real constants.

   a. Calculate $\mathcal{L}[x \cos bx]$.

   b. Use the translation property to calculate $\mathcal{L}[xe^{ax} \cos bx]$.

3. Find $\mathcal{L}[f(x)]$ for

$$f(x) = \begin{cases} 0, & 0 \le x < 3 \\ 1, & 3 \le x \end{cases}$$

4. Find $\mathcal{L}[g(x)]$ for

$$g(x) = \begin{cases} 1 - x, & 0 \le x < 1 \\ x - 1, & 1 \le x \end{cases}$$

5. Find $\mathcal{L}[h(x)]$ for

$$h(x) = \begin{cases} x, & 0 \le x < 2 \\ -x + 4, & 2 \le x < 4 \\ 0, & 4 \le x \end{cases}$$

6. Find $\mathcal{L}[k(x)]$ for

$$k(x) = \begin{cases} 0, & 0 \le x < 1 \\ 1, & 1 \le x < 2 \\ 0, & 2 \le x \end{cases}$$

7. Use Table 5.1 and the linearity property of Laplace transforms to find the following transforms.

   a. $\mathcal{L}[5]$ b. $\mathcal{L}[e]$

   c. $\mathcal{L}[3x - 2]$ d. $\mathcal{L}[e^{-x}(2x + 1)]$

   e. $\mathcal{L}[e^{2x} \sin 3x - 2 \cos x]$ f. $\mathcal{L}[e^{3x+2}]$

5.1.2.1 Comments on Computing

The following two MAPLE statements may be used to calculate the Laplace transform of $f(x) = x^2 \sin x$.

```
with(inttrans):
F:=laplace(x ∧ 2 * sin(x), x, s);
```

The output displayed by MAPLE is

$$F := 2\frac{-1 + 3s^2}{(1 + s^2)^3}$$

The first statement above, with(inttrans):, instructs the computer to load software for calculating the Laplace transform.

The following four statements calculate the Laplace transform of the piecewise continuous function of Exercise 4.

```
with(inttrans):
g:=piecewise(0 <= x and x < 1, 1 − x, 1 <= x, x − 1):
H:=convert(g, Heaviside):
G:=laplace(H, x, s);
```

The third statement converts the representation of the function g to the proper format for use with the laplace command. The output is

$$G := \frac{1}{s} - \frac{1}{s^2} + \frac{2e^{(-s)}}{x^2}$$

8. Show that if $f(x)$ and $g(x)$ are both of exponential order a as $x \to +\infty$, then $f(x) - g(x)$ is also of exponential order a as $x \to +\infty$.

9. For each of the following functions $F(s)$ find a function $f(x)$ such that $\mathcal{L}[f(x)] = F(s)$. That is, for each given function $F(s)$ find an inverse Laplace transform $\mathcal{L}^{-1}[F(s)]$. If you have a CAS available which calculates the inverse Laplace transform, also use the CAS to calculate the inverse Laplace transforms of each of the given functions $F(s)$ and compare those answers to the ones you obtained by hand.

a. $\dfrac{3}{s^3}$ b. $\dfrac{4}{(s+2)^2}$ c. $\dfrac{-2s}{s^2+3}$

d. $\dfrac{1}{s^2(s+1)}$ e. $\dfrac{s-1}{s^2-2s+5}$ f. $\dfrac{2}{s^2-2s+5}$

g. $\dfrac{-4}{s(s^2+1)}$ h. $\dfrac{2s+5}{s^2+2s+2}$ i. $\dfrac{1}{s^2} + \dfrac{2}{s^2-1}$

j. $\dfrac{3s}{s^2-4s+3}$ k. $\dfrac{1}{s^2-4s+9}$ l. $\dfrac{s}{s^2-4s+9}$

5.1.2.2 Comments on Computing

The following three MAPLE statements compute the inverse Laplace transform of

$$F(s) = \frac{2}{s(s+1)} = \frac{2}{s} - \frac{2}{s+1}$$

in two different ways.

```
with(inttrans):
f:=invlaplace(2/(s * (s + 1)), s, x);
f:=invlaplace(2/s - 2/(s + 1), s, x);
```

The output of the second statement is

$$\text{f:=}\ 4e^{(-1/2\,x)}\sinh(\frac{1}{2}x)$$

while the output of the third statement is

$$\text{f:=}\ 2 - 2e^{(-x)}$$

Notice the two results are equal but are expressed differently.

5.2 Using the Laplace Transform and Its Inverse to Solve Initial Value Problems

The Laplace transform method for solving n-th order linear initial value problems in which the differential equation is homogeneous or nonhomogeneous is a three-step process. First, the differential equation is transformed by the Laplace transform into an algebraic equation in s and $\mathcal{L}[y(x)]$—the Laplace transform of the solution of the initial value problem. Next, the unknown in the algebraic equation $\mathcal{L}[y(x)]$ is solved for by algebraic manipulation. And finally, the inverse Laplace transformation is applied to obtain the solution of the initial value problem. The Laplace transform method immediately yields the solution of n-th order linear homogeneous and nonhomogeneous differential equations and n-th order linear initial value problems in which the differential equation is homogeneous or nonhomogeneous. That is, one does not have to (1) find the general solution of the associated homogeneous differential equation, (2) find a particular solution to the nonhomogeneous differential equation, and (3) add these solutions to get the general solution. In addition, if the problem is a linear initial value problem, the initial conditions are incorporated in the transforming equations. Hence, one does not have to find the general solution and then choose the constants to satisfy the specified initial equations.

Let us formally calculate the Laplace transform of the derivative of the function $y(x)$. By definition,

$$\mathcal{L}[y'(x)] = \int_0^\infty y'(x)e^{-sx}\,dx. \tag{1}$$

Letting $u = e^{-sx}$ and $dv = y'(x)\,dx$, and differentiating and integrating, we find $du = -se^{-sx}dx$ and $v = y(x)$. Then applying the integration by parts formula to equation (1), yields

$$\mathcal{L}[y'(x)] = \int_0^\infty y'(x)e^{-sx}\,dx = y(x)e^{-sx}\Big|_0^\infty + s\int_0^\infty y(x)e^{-sx}\,dx.$$

If $y(x)$ is of exponential order a as $x \to +\infty$, then for $s > a$, $y(x)e^{-sx} \to 0$ as $x \to +\infty$ and, therefore,

$$\mathcal{L}[y'(x)] = -y(0) + s\mathcal{L}[y(x)] \qquad \text{for} \quad s > a.$$

Next, we formally calculate the Laplace transform of the second derivative of the function $y(x)$. Again we use integration by parts. This time we set $u = e^{-sx}$ and $dv = y''(x)\,dx$, and differentiating and integrating, we find $du = -se^{-sx}dx$ and $v = y'(x)$. Substituting these expressions into the formula for integration by parts, yields

$$\mathcal{L}[y''(x)] = \int_0^\infty y''(x)e^{-sx}\,dx = y'(x)e^{-sx}\Big|_0^\infty + s\int_0^\infty y'(x)e^{-sx}\,dx.$$

If $y'(x)$ is of exponential order a as $x \to +\infty$, then for $s > a$, $y'(x)e^{-sx} \to 0$ as $x \to +\infty$ and, therefore,

$$\mathcal{L}[y''(x)] = -y'(0) - sy(0) + s^2\mathcal{L}[y(x)] \qquad \text{for} \quad s > a.$$

By induction, we obtain the following **general formula for the Laplace transform of the n-th derivative of the function y(x).**

Let $y(x)$, $y^{(1)}(x)$, ... , $y^{(n-1)}(x)$ be continuous on $[0, \infty)$ and let $y^{(n)}(x)$ be piecewise continuous on $[0, \infty)$. Furthermore, let $y(x)$, $y^{(1)}(x)$, ... , $y^{(n)}(x)$ be of exponential order a. Then, for $s > a$

$$\text{(2)} \qquad \mathcal{L}[y^{(n)}(x)] = -y^{(n-1)}(0) - sy^{(n-2)}(0) - \cdots - s^{n-1}y(0) + s^n\mathcal{L}[y(x)].$$

Since the solutions of homogeneous linear differential equations with constant coefficients and all of their derivatives are continuous and of exponential order a as $x \to +\infty$ for some constant a, equation (2) is valid for the all n-th order linear homogeneous differential equations with constant coefficients. If, in addition, the nonhomogeneity $b(x)$ of a nonhomogeneous linear differential equation with constant coefficients is of exponential order a as $x \to +\infty$, then $b(x)$ has a Laplace transform and equation (2) is valid for that nonhomogeneous differential equation.

The following example illustrates how to use the Laplace transform method to obtain the general solution of a second-order linear homogeneous differential equation with constant coefficients.

Example 1 Use the Laplace transform method to solve the homogeneous linear differential equation

$$\mathbf{y'' + 4y = 0.}$$

Solution

Apply the Laplace Transform to the Differential Equation

Taking the Laplace transform of given differential equation, we find

$$\mathcal{L}[y'' + 4y] = \mathcal{L}[0].$$

Using the fact that $\mathcal{L}[0] = 0$ and the linearity property of the Laplace transform, yields

$$\mathcal{L}[y''(x)] + 4\mathcal{L}[y(x)] = 0.$$

Replacing $\mathcal{L}[y''(x)]$ by the expression we obtain from equation (2) with $n = 2$, results in

$$-y'(0) - sy(0) + s^2\mathcal{L}[y(x)] + 4\mathcal{L}[y(x)] = 0$$

or

$$-y'(0) - sy(0) + (s^2 + 4)\mathcal{L}[y(x)] = 0. \tag{3}$$

Solve the Algebraic Equation for $\mathcal{L}[\mathbf{y}(\mathbf{x})]$

Since specific initial conditions are not given, we let $A = y(0)$ and $B = y'(0)$, where A and B are arbitrary real constants. Then solving equation (3) for $\mathcal{L}[y(x)]$, we obtain

$$\mathcal{L}[y(x)] = \frac{B + As}{s^2 + 4}. \tag{4}$$

Apply the Inverse Laplace Transform

Applying the inverse Laplace transform to (4) and using the linearity of the inverse Laplace transform, results in

$$\begin{aligned} y(x) = \mathcal{L}^{-1}\left[\frac{B + As}{s^2 + 4}\right] &= B\mathcal{L}^{-1}\left[\frac{1}{s^2 + 4}\right] + A\mathcal{L}^{-1}\left[\frac{s}{s^2 + 4}\right] \\ &= \frac{B}{2}\sin 2x + A\cos 2x. \end{aligned}$$

Hence, the general solution of the differential equation $y'' + 4y = 0$ is

$$y(x) = C\sin 2x + A\cos 2x,$$

where A and $C = B/2$ are arbitrary real constants. ■

The following example shows how to use the Laplace transform method to obtain the general solution of a second-order linear nonhomogeneous differential equation with constant coefficients.

Example 2 Find the general solution of the nonhomogeneous linear differential equation

$$\mathbf{y'' + y' - 2y = x^2 - 1}$$

using the Laplace transform method.

Solution

Apply the Laplace Transform to the Differential Equation

Successively taking the Laplace transform of the given equation, using the linearity property of the Laplace transform, replacing $\mathcal{L}[y''(x)]$ by the expression we obtain from equation (2) with $n = 2$, and replacing $\mathcal{L}[y'(x)]$ by the expression we obtain from equation (2) with $n = 1$, results in

$$\mathcal{L}[y'' + y' - 2y] = \mathcal{L}[x^2 - 1]$$

$$\mathcal{L}[y''(x)] + \mathcal{L}[y'(x)] - 2\mathcal{L}[y(x)] = \mathcal{L}[x^2] - \mathcal{L}[1]$$

$$(5)\quad (-y'(0) - sy(0) + s^2\mathcal{L}[y(x)]) + (-y(0) + s\mathcal{L}[y(x)]) - 2\mathcal{L}[y(x)] = \frac{2}{s^3} - \frac{1}{s}.$$

Solve the Algebraic Equation for $\mathcal{L}[\mathbf{y}(\mathbf{x})]$

Letting $A = y(0)$ and $B = y'(0)$ and solving for $\mathcal{L}[y(x)]$, yields

$$(6)\qquad \mathcal{L}[y(x)] = \frac{A(s+1) + B + \dfrac{2}{s^3} - \dfrac{1}{s}}{s^2 + s - 2}$$

$$= \frac{A(s+1) + B}{(s+2)(s-1)} + \frac{2}{s^3(s+2)(s-1)} - \frac{1}{s(s+2)(s-1)}.$$

Expanding the right-hand side of this equation using partial fraction expansion and combining like terms, we find

$$(7)\qquad \mathcal{L}[y(x)] = \frac{c_1}{s+2} + \frac{c_2}{s-1} - \frac{1}{4s} - \frac{1}{2s^2} - \frac{1}{s^3}$$

where $c_1 = (4A - 4B - 1)/12$ and $c_2 = (2A + B + 1)/3$ are arbitrary real constants, since A and B are arbitrary real constants.

Apply the Inverse Laplace Transform

Applying the inverse Laplace transform to (7) and using the linearity property of the inverse Laplace transform, yields

$$(8)\qquad y(x) = c_1\mathcal{L}^{-1}\left[\frac{1}{s+2}\right] + c_2\mathcal{L}^{-1}\left[\frac{1}{s-1}\right] - \frac{1}{4}\mathcal{L}^{-1}\left[\frac{1}{s}\right]$$

$$-\frac{1}{2}\mathcal{L}^{-1}\left[\frac{1}{s^2}\right] - \mathcal{L}^{-1}\left[\frac{1}{s^3}\right]$$

$$= c_1e^{-2x} + c_2e^x - \frac{1}{4} - \frac{1}{2}x - \frac{1}{2}x^2.$$

Hence, the general solution of the differential equation $y'' + y' - 2y = x^2 - 1$ is

$$y(x) = c_1e^{-2x} + c_2e^x - \frac{1}{4} - \frac{1}{2}x - \frac{1}{2}x^2$$

where c_1 and c_2 are arbitrary real constants. ■

The following example illustrates how to use the Laplace transform method to solve an initial value problem.

Example 3 Find the solution of the initial value problem

$$\mathbf{y'' - 4y' + 5y = 2e^x - \sin x; \quad y(0) = 1, \quad y'(0) = -1}$$

using the Laplace transform method.

Solution

Apply the Laplace Transform to the Differential Equation

The Laplace transform of the given differential equation is
(9)

$$(-y'(0) - sy(0) + s^2\mathcal{L}[y(x)]) - 4(-y(0) + s\mathcal{L}[y(x)]) + 5\mathcal{L}[y(x)] = \frac{2}{s-1} - \frac{1}{s^2+1}.$$

Solve the Algebraic Equation for $\mathcal{L}[\mathbf{y(x)}]$

Substituting the given initial conditions into (9) and solving the resulting equation for $\mathcal{L}[y(x)]$, we find

$$\mathcal{L}[y(x)] = \frac{s - 5 + \frac{2}{s-1} - \frac{1}{s^2+1}}{s^2 - 4s + 5} = \frac{s^4 - 6s^3 + 8s^2 - 7s + 8}{(s-1)(s^2+1)(s^2-4s+5)}.$$

Using partial fraction expansion, results in

$$(10)\quad \mathcal{L}[y(x)] = \frac{1}{s-1} - \frac{1}{8}\frac{1}{s^2+1} - \frac{1}{8}\frac{s}{s^2+1} - \frac{17}{8}\frac{1}{(s-2)^2+1} + \frac{1}{8}\frac{(s-2)}{(s-2)^2+1}.$$

Apply the Inverse Laplace Transform

Applying the inverse Laplace transform to (10) and using linearity of the inverse Laplace transform, yields

$$y(x) = \mathcal{L}^{-1}\left[\frac{1}{s-1}\right] - \frac{1}{8}\mathcal{L}^{-1}\left[\frac{1}{s^2+1}\right] - \frac{1}{8}\mathcal{L}^{-1}\left[\frac{s}{s^2+1}\right]$$

$$- \frac{17}{8}\mathcal{L}^{-1}\left[\frac{1}{(s-2)^2+1}\right] + \frac{1}{8}\mathcal{L}^{-1}\left[\frac{s-2}{(s-2)^2+1}\right]$$

$$= e^x - \frac{1}{8}\sin x - \frac{1}{8}\cos x - \frac{17}{8}e^{2x}\sin x + \frac{1}{8}e^{2x}\cos x.$$

Hence, the solution of the given initial value problem is

$$y(x) = e^x - \frac{1}{8}\sin x - \frac{1}{8}\cos x - \frac{17}{8}e^{2x}\sin x + \frac{1}{8}e^{2x}\cos x. \blacksquare$$

5.2.1 Comments on Computing

Manually using the Laplace transform method to solve a linear differential equation with constant coefficients or an initial value problem which includes such a differential equation is a three step process. First, one calculates the Laplace transform of the differential equation and substitutes for the initial conditions, if any are specified. Next, one uses partial fraction expansion, a table of Laplace transforms and their inverses, and some algebraic rearrangement, when necessary, so that the linearity property of the Laplace transform can be used. Finally, the inverse Laplace transform is determined. Some computer algebra systems (CAS) include one or more algorithms for using the Laplace transform method to solve linear differential equations and initial value problems. In one method of solution, the user follows the three steps listed above. That is, the user instructs the CAS to calculate the Laplace transform of the given differential equation and specifies the initial conditions, if specific values are given. Next, the user has the CAS perform a partial fraction expansion. And then, the user has the CAS calculate the inverse Laplace transform to obtain the required solution. In the second method, the user enters the differential equation and initial conditions, if any, and instructs the CAS to solve the differential equation or initial value problem using the Laplace transform method. In this case, the CAS will not show the step-by-step computations used to arrive at the answer, it will simply provide the answer, if it can. Thus, if a user specifies the proper syntax requesting a CAS to use the Laplace transform method to solve the initial value problem $y'' + 4y = 0$; $y(0) = 1$, $y'(0) = -1$, then the CAS will simply respond $y(x) = -\frac{1}{2}\sin 2x + \cos 2x$.

5.2.2 Exercises

In Exercises 1–7 use the Laplace transform method to calculate the general solution of the given differential equation manually. If you have a CAS available which contains algorithms for using the Laplace transform method, use them to calculate the solutions of Exercises 1–7 and compare those answers to the ones you obtained by hand.

1. $y' - y = 0$

2. $y'' - 2y' + 5y = 0$

3. $y' + 2y = 4$

4. $y'' - 9y = 2\sin 3x$

5. $y'' + 9y = 2\sin 3x$

6. $y'' + y' - 2y = xe^x - 3x^2$

7. $y^{(4)} - 2y^{(3)} + y^{(2)} = xe^x - 3x^2$

In Exercises 8–14 use the Laplace transform method to calculate the solution of the given initial value problem manually. If you have a CAS available which contains algorithms for using the Laplace transform method to solve initial value problems, use them to calculate the solutions of Exercises 8–14 and compare those answers to the ones you obtained by hand.

8. $y' = e^x; \quad y(0) = -1$

9. $y' - y = 2e^x; \quad y(0) = 1$

10. $y'' - 9y = x + 2; \quad y(0) = -1, \quad y'(0) = 1$

11. $y'' + 9y = x + 2; \quad y(0) = -1, \quad y'(0) = 1$

12. $y'' - y' + 6y = -2\sin 3x; \quad y(0) = 0, \quad y'(0) = -1$

13. $y'' - 2y' + 2y = 1 - x^2; \quad y(0) = 1, \quad y'(0) = 0$

14. $y''' + 3y'' + 2y' = x + \cos x; \quad y(0) = 1, \quad y'(0) = -1, \quad y''(0) = 2$

5.2.2.1 Comments on Computing

The following nine MAPLE statements calculate and print the general solution of the nonhomogeneous differential equation $y'' + y' - 2y = x^2 - 1$ which appears in Example 2.

```
with(inttrans):
alias(L(y(x)) =laplace(y(x), x, s)):
alias(A = y(0)):
alias(B = D(y)(0)):
DE2:=diff(y(x), x$2)+diff(y(x), x) − 2 * y(x) = x ∧ 2 − 1;
laplace(DE2, x, s);
L(y(x))=solve(%, L(y(x)));
convert(%, parfrac, s);
invlaplace(%, s, x);
```

Unless you give MAPLE instructions to the contrary, it uses its own notation for items. For example, in the second statement above, we use the alias command to tell MAPLE to print "$L(y(x))$" where it would normally print "laplace$(y(x), x, s)$." Statement three instructs MAPLE to print "A" instead of "$y(0)$" and statement four instructs it to print "B" instead of "$D(y)(0)$," which is the MAPLE representation for $y'(0)$. The fifth statement defines the differential equation. Notice that "diff$(y(x), x\$2)$" is MAPLE's notation for

$\frac{d^2y}{dx^2} = y''(x)$ and "diff$(y(x), x)$" is the notation for $\frac{dy}{dx} = y'(x)$. The sixth statement instructs the computer to calculate and print the Laplace transform of the differential equation DE2. The printed output is equivalent to equation (5) of Example 2. Since in MAPLE the % sign refers to the previous statement, the seventh statement instructs the computer to solve the result (output) of the previous statement for $L(y(x))$. The printed output of the seventh statement is equivalent to equation (6) of Example 2. The eighth statement instructs MAPLE to perform partial fraction expansion on the result of the seventh statement. The output of the eighth statement is equivalent to equation (7). The last statement causes the computer to calculate the inverse Laplace transform of the output of the eighth statement. The output of the final statement is equivalent to equation (8)—the general solution of the given differential equation.

You can find the general solution to other second-order differential equations by changing the definition of the differential equation, DE2, in the fifth statement. To solve higher-order differential equations, you must make additions as well. Furthermore, you can solve the initial value problem of Example 3 by changing the third statement to $y(0) := 1$:, by changing the fourth statement to $D(y)(0) := -1$:, and by changing the fifth statement to DE:= diff$(y(x), x\$2) - 4*$diff$(y(x), x) + 5 * y(x) = 2*exp(x)-sin(x)$;.

The following three MAPLE statements calculate and print the general solution of the differential equation, DE2, using the Laplace transform method without displaying the results of any intermediate steps.

```
with(inttrans):
DE2:=diff(y(x),x$2)+diff(y(x),x) - 2 * y(x) = x ^ 2 - 1;
dsolve({DE2, y(0) = A, D(y)(0) = B}, y(x), method=laplace);
```

5.3 Convolution and the Laplace Transform

One can often write a function $H(s)$ as the product of two functions $F(s)$ and $G(s)$ in such a way that both $F(s)$ and $G(s)$ are the Laplace transforms of known functions, say $f(x)$ and $g(x)$, respectively. That is, one sometimes encounters the situation in which $H(s) = F(s)G(s)$, where $F(s) = \mathcal{L}[f(x)]$ and $G(s) = \mathcal{L}[g(x)]$. Hence, $H(s) = \mathcal{L}[f(x)]\mathcal{L}[g(x)]$. Momentarily, we might expect that $H(s) = \mathcal{L}[f(x)]\mathcal{L}[g(x)] = \mathcal{L}[f(x)g(x)]$. Stated verbally, we might anticipate that the Laplace transform distributes over the multiplication of two functions. However, we know from calculus that, in general, the integral of the product of two functions is not equal to the product of the integrals of the two functions. And since the Laplace transform is defined in terms of an integral, we should not expect the Laplace transform to distribute over the multiplication of functions. The following example illustrates that, in general,

$$\mathcal{L}[f(x)]\mathcal{L}[g(x)] \neq \mathcal{L}[f(x)g(x)].$$

Let $f(x) = x$ and let $g(x) = e^x$. Then

$$\mathcal{L}[f(x)]\mathcal{L}[g(x)] = \mathcal{L}[x]\mathcal{L}[e^x] = \frac{1}{s^2}\frac{1}{s-1}$$

and

$$\mathcal{L}[f(x)g(x)] = \mathcal{L}[xe^x] = \frac{1}{(s-1)^2}.$$

Clearly,

$$\mathcal{L}[f(x)]\mathcal{L}[g(x)] \neq \mathcal{L}[f(x)g(x)] \quad \text{for any real } s.$$

Two questions should immediately come to mind: "Is there any function $h(x)$ such that $\mathcal{L}[h(x)] = H(s) = \mathcal{L}[f(x)]\mathcal{L}[g(x)]$?" and "If so, how is $h(x)$ related to $f(x)$ and $g(x)$?" We shall now answer these two questions.

The **convolution of f(x) and g(x)** is defined by the equation

(1) $$f(x) * g(x) = \int_0^x f(x-\xi)g(\xi)\,d\xi.$$

Making the change of variable $\eta = x - \xi$ in the integral appearing in (1), we see that

$$\begin{aligned} f(x) * g(x) &= \int_0^x f(x-\xi)g(\xi)\,d\xi = -\int_x^0 f(\eta)g(x-\eta)\,d\eta \\ &= \int_0^x g(x-\eta)f(\eta)\,d\eta = g(x) * f(x). \end{aligned}$$

Hence, we have shown that the convolution operator is commutative. Indeed, the convolution operator has many of the same properties as ordinary multiplication. For instance,

$$\begin{aligned} f(x) * (g_1(x) + g_2(x)) &= f(x) * g_1(x) + f(x) * g_2(x), \\ f(x) * (g(x) * h(x)) &= (f(x) * g(x)) * h(x), \end{aligned}$$

and

$$f(x) * 0 = 0.$$

Consequently, the convolution operator may be thought of as a "generalized multiplication" operator. However, the convolution operator does not have some of the properties of ordinary multiplication. For example, it is not true for all functions $f(x)$ that $f(x) * 1 = f(x)$.

Suppose that $f(x)$ and $g(x)$ both have a Laplace transform for $s > a$. That is, suppose

$$\mathcal{L}[f(x)] = \int_0^\infty f(x)e^{-sx}\,dx \quad \text{and} \quad \mathcal{L}[g(x)] = \int_0^\infty g(x)e^{-sx}\,dx$$

both exist for $s > a$. By definition,

$$\mathcal{L}[f(x) * g(x)] = \int_0^\infty \left[\int_0^x f(x-\xi)g(\xi)\,d\xi\right] e^{-sx}\,dx \tag{2}$$

provided both integrals exist. The domain of integration, which is the region above the positive x-axis and below the half-line $\xi = x,\ x \geq 0$, and the order of integration is shown in Figure 5.2a. Assuming that the order of integration can be interchanged, we find

$$\mathcal{L}[f(x) * g(x)] = \int_0^\infty \left[\int_\xi^\infty f(x-\xi)e^{-sx}\,dx\right] g(\xi)\,d\xi. \tag{3}$$

See Figure 5.2b. In the innermost integral in equation (3), we make the change of variable $\eta = x - \xi$ and thereby obtain

$$\begin{aligned}\mathcal{L}[f(x) * g(x)] &= \int_0^\infty \left[\int_0^\infty f(\eta)e^{-s(\xi+\eta)}\,d\eta\right] g(\xi)\,d\xi \\ &= \int_0^\infty f(\eta)e^{-s\eta}\,d\eta \int_0^\infty g(\xi)e^{-s\xi}\,d\xi \\ &= \mathcal{L}[f(x)]\mathcal{L}[g(x)].\end{aligned}$$

Thus, we have proven the following **convolution theorem**.

If $\mathcal{L}[f(x)]$ and $\mathcal{L}[g(x)]$ both exist for $s > a$, then

$$\mathcal{L}[f(x)]\mathcal{L}[g(x)] = \mathcal{L}[f(x) * g(x)] \quad \text{for } s > a.$$

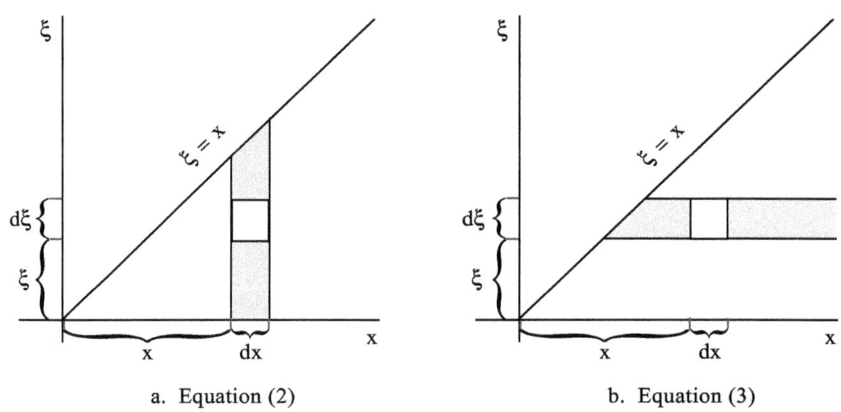

FIGURE 5.2
Domain and Order of Integration.

Example 1 Find a function h(x) whose Laplace transform is

$$\mathbf{H(s)} = \frac{\mathbf{1}}{\mathbf{s^2(s+1)}}.$$

Solution

We can rewrite $H(s)$ as the product

$$H(s) = \frac{1}{s^2}\frac{1}{s+1} = F(s)G(s),$$

where $F(s) = \dfrac{1}{s^2} = \mathcal{L}[x]$ and $G(s) = \dfrac{1}{s+1} = \mathcal{L}[e^{-x}]$. By the convolution theorem and the commutative property of the convolution operator, we have

$$h(x) = x * e^{-x} = \int_0^x (x-\xi)e^{-\xi}\, d\xi$$

and

$$h(x) = e^{-x} * x = \int_0^x e^{-(x-\xi)}\xi\, d\xi.$$

Evaluating the second integral, which is slightly simpler, we find

$$h(x) = e^{-x}\int_0^x \xi e^{\xi}\, d\xi = e^{-x}\{[(\xi-1)e^{\xi}]\Big|_0^x\} = e^{-x}[(x-1)e^x+1] = x-1+e^{-x}. \blacksquare$$

The next example shows how to solve Example 1 using partial fraction expansion.

Example 2 Use partial fraction expansion to find a function h(x) whose Laplace transform is

$$\mathbf{H(s)} = \frac{\mathbf{1}}{\mathbf{s^2(s+1)}}.$$

Solution

By the partial fraction expansion, there exist constants A, B, and C such that

$$\frac{1}{s^2(s+1)} = \frac{A}{s^2} + \frac{B}{s} + \frac{C}{(s+1)}.$$

Multiplying by $s^2(s+1)$, we obtain

$$(4) \qquad 1 = A(s+1) + Bs(s+1) + Cs^2.$$

Setting $s = 0$ in (4), we find $1 = A$.

Setting $s = -1$ in (4), we find $1 = C$.

Then setting $s = 1, A = 1$, and $C = 1$ in (4), we see that B must satisfy $1 = 2 + 2B + 1$. Hence, $B = -1$ and

$$\mathcal{L}[h(x)] = H(s) = \frac{1}{s^2(s+1)} = \frac{1}{s^2} + \frac{-1}{s} + \frac{1}{(s+1)}.$$

Taking the inverse Laplace transform, we obtain

$$h(x) = \mathcal{L}^{-1}\left[\frac{1}{s^2}\right] + \mathcal{L}^{-1}\left[\frac{-1}{s}\right] + \mathcal{L}^{-1}\left[\frac{1}{(s+1)}\right] = x - 1 + e^{-x}. \blacksquare$$

Examples 1 and 2 illustrate that using the convolution to find an inverse Laplace transform is often simpler than using partial fraction expansion.

Example 3 Use the Laplace transform method and the convolution theorem to solve the initial value problem

$$\mathbf{y' + y = x; \quad y(0) = 0.}$$

Solution

Taking the Laplace transform of the given differential equation, we obtain

$$\begin{aligned} \mathcal{L}[y' + y] &= \mathcal{L}[x] \\ \mathcal{L}[y'(x)] + \mathcal{L}[y(x)] &= \mathcal{L}[x] \\ -y(0) + s\mathcal{L}[y(x)] + \mathcal{L}[y(x)] &= \frac{1}{s^2}. \end{aligned}$$

Imposing the initial condition $y(0) = 0$ and combining like terms, we get

$$(s+1)\mathcal{L}[y(x)] = \frac{1}{s^2}.$$

Solving for $\mathcal{L}[y(x)]$, we find the Laplace transform of the solution, $y(x)$, of the given initial value problem satisfies

$$\mathcal{L}[y(x)] = \frac{1}{s^2(s+1)}.$$

Recalling the results from Examples 1 and 2, we see that the solution of the given initial value problem is

$$y(x) = x - 1 + e^{-x}. \blacksquare$$

5.3.1 Exercises

For each of the following functions H(s), use the Laplace transform information from Table 5.1 and the convolution theorem to find a function h(x) such that $\mathcal{L}[\mathbf{h(x)}] = \mathbf{H(s)}$. You may use a Computer Algebra System to solve these exercises by defining the convolution operator as in equation (1).

1. $\dfrac{1}{s(s^2+9)}$

2. $\dfrac{1}{(s+1)(s-2)}$

3. $\dfrac{1}{(s+1)(s-2)^2}$

4. $\dfrac{1}{s(s^2-2s+5)}$

5. $\dfrac{s}{(s-1)(s^2+4)}$

6. $\dfrac{1}{s^2(s^2-4)}$

5.3.1.1 Comments on Computer Software

In Example 1, we calculated the convolution $e^{-x} * x$ and found $e^{-x} * x = x - 1 + e^{-x}$. The following MAPLE statement calculates and prints the convolution of $x * e^{-x}$.

```
int((x-xi)*exp(-xi),xi=0..x);
```

As expected, the output is $e^{-x} + x - 1$.

In Exercises 7–14 use the Laplace transform method and the convolution theorem to find the solution to the given initial value problem.

7. $y' - 2y = 6; \quad y(0) = 2$
8. $y' + y = e^x; \quad y(0) = \dfrac{5}{2}$
9. $y'' + 9y = 1; \quad y(0) = 0, \quad y'(0) = 0$ (Hint: See Exercise 1.)
10. $y'' + 9y = 18e^{3x}; \quad y(0) = -1, \quad y'(0) = 6$
11. $y'' - y' - 2y = 0; \quad y(0) = 0, \quad y'(0) = 3$ (Hint: See Exercise 2.)
12. $y'' - y' - 2y = x^2; \quad y(0) = \dfrac{11}{4}, \quad y'(0) = \dfrac{1}{2}$
13. $y'' - 2y' + y = 2\sin x; \quad y(0) = -2, \quad y'(0) = 0$
14. $y''' - y'' + 4y' - 4y = 0; \quad y(0) = 0, \quad y'(0) = 5, \quad y''(0) = 5$
    (Hint: See Exercise 5.)
15. Show that for any continuous function $f(x)$ and any constant $a \neq 0$, the solution of the initial value problem

$$y'' + a^2 y = f(x); \quad y(0) = 0, \quad y'(0) = 0$$

is

$$y(x) = \left\{\frac{1}{a}\sin ax\right\} * f(x) = \frac{1}{a}\int_0^x \sin a(x-\xi) f(\xi)\, d\xi$$

$$= \frac{1}{a}\left\{\sin ax \int_0^x f(\xi)\cos a\xi\, d\xi - \cos ax \int_0^x f(\xi)\sin a\xi\, d\xi\right\}.$$

16. Show that for any continuous function $f(x)$ and any constant $a \neq 0$, the solution of the initial value problem

$$y'' - a^2 y = f(x); \quad y(0) = 0, \quad y'(0) = 0$$

is

$$y(x) = \left\{ \frac{1}{2a} e^{ax} \right\} * f(x) - \left\{ \frac{1}{2a} e^{-ax} \right\} * f(x)$$

$$= \frac{1}{2a} \left\{ e^{ax} \int_0^x e^{-a\xi} f(\xi)\, d\xi - e^{-ax} \int_0^x e^{a\xi} f(\xi)\, d\xi \right\}.$$

17. Show that for any continuous function $f(x)$ and any constant $a \neq 0$, the solution of the initial value problem

$$y'' - 2ay' + a^2 y = f(x); \quad y(0) = 0, \quad y'(0) = 0$$

is

$$y(x) = \{x e^{ax}\} * f(x) = \int_0^x (x - \xi) e^{a(x-\xi)} f(\xi)\, d\xi$$

$$= e^{ax} \left\{ x \int_0^x e^{-a\xi} f(\xi)\, d\xi - \int_0^x \xi e^{-a\xi} f(\xi)\, d\xi \right\}.$$

5.4 The Unit Function and Time-Delay Function

The Laplace transform method is useful not only for solving nonhomogeneous linear differential equations and initial value problems with constant coefficients when the nonhomogeneity—which is also called the **forcing function**—is a solution of some homogeneous linear differential equation with constant coefficients but also when the forcing function is a discontinuous function or an impulse function which has a Laplace transform. Forcing functions which are discontinuous functions or impulse functions occur frequently in electrical and mechanical systems. In this section, we shall consider forcing functions which are discontinuous functions and in the next section we shall consider forcing functions which are impulses.

One of the simplest functions which has a jump discontinuity of 1 at $x = c \geq 0$ is the **unit step function** or the **Heaviside function,** $u(x - c)$, which is named in honor of Oliver Heaviside (1850-1925).

The **unit step function** or **Heaviside function** is the function

$$u(x - c) = \begin{cases} 0, & x < c \\ 1, & c \leq x \end{cases}$$

a. $y(x) = u(x - 1)$

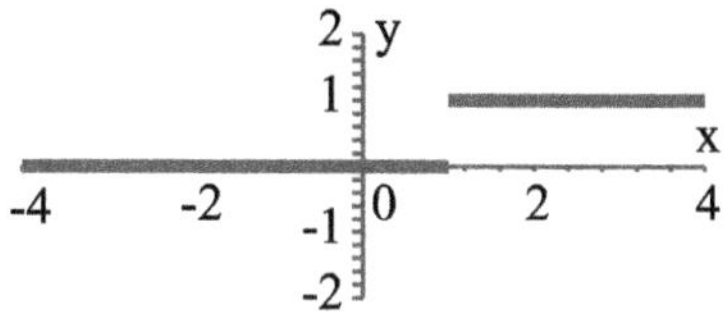

b. $y(x) = 2\,u(x - 1)$

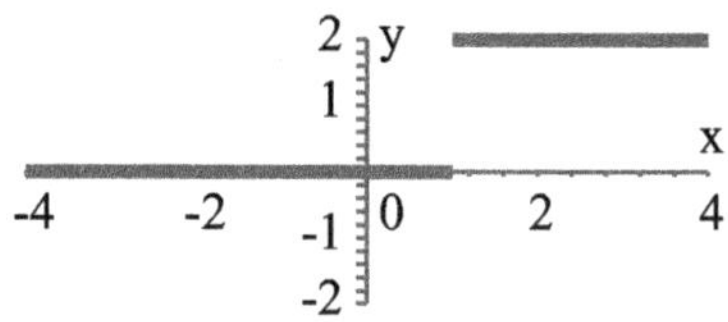

c. $y(x) = -\,u(x - 2)$

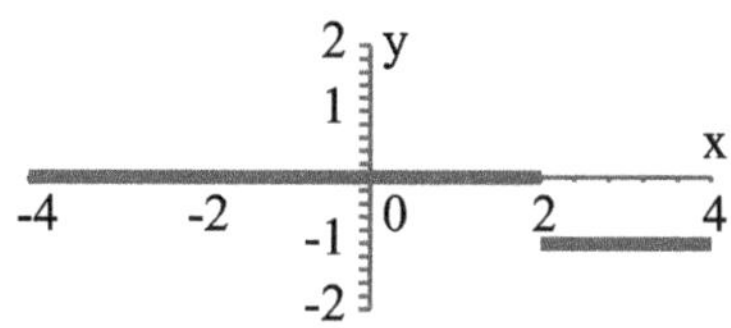

d. $y(x) = 2\,u(x - 1) - u(x - 2)$

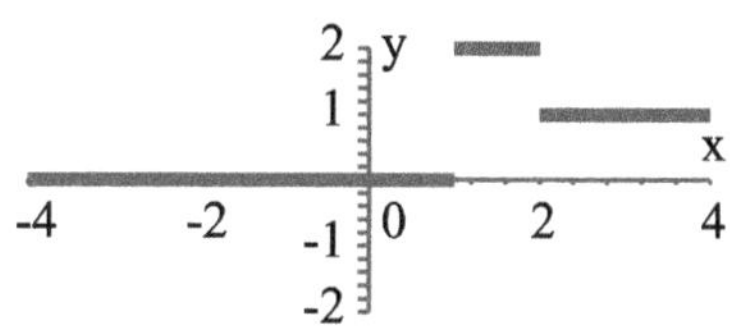

FIGURE 5.3
Graphs of Step Functions.

Graphs of $u(x-1)$, $2u(x-1)$, and $-u(x-2)$ are displayed in Figure 5.3. Other discontinuous step functions can be written as a linear combination of unit step functions. For instance,

$$(1) \qquad h(x) = \begin{cases} 0, & x < 1 \\ 2, & 1 \le x < 2 \\ 1, & 2 \le x \end{cases}$$

can be written as $h(x) = 2u(x-1) - u(x-2)$. Thus, the unit step function is the basic function for constructing other step functions. A graph of $h(x)$ is shown in Figure 5.3d.

We easily calculate the Laplace transform of the unit step function as follows:

$$\mathcal{L}[u(x-c)] = \int_0^\infty u(x-c)e^{-sx}\,dx = \int_c^\infty e^{-sx}\,dx = \frac{e^{-cx}}{s} \quad \text{for } s > 0.$$

Using the linearity property of the Laplace transforms, we will be able to calculate the Laplace transform of any particular step function once it is written as a linear combination of unit step functions. For example, the Laplace transform of $h(x) = 2u(x-1) - u(x-2)$ is calculated as follows:

$$\mathcal{L}[h(x)] = \mathcal{L}[2u(x-1) - u(x-2)] = 2\mathcal{L}[u(x-1)] - \mathcal{L}[u(x-2)] = \frac{2e^{-s}}{s} - \frac{e^{-2s}}{s}$$

for $s > 0$.

Jump discontinuities occur in physical problems such as electrical circuits which include "on/off" switches. Consider the function $g(x) = u(x-c)f(x)$. When $x < c$, $u(x-c) = 0$, and therefore $g(x) = 0$. When $x \geq c$, $u(x-c) = 1$, and consequently $g(x) = f(x)$. That is,

$$g(x) = u(x-c)f(x) = \begin{cases} 0, & x < c \\ f(x), & c \leq x \end{cases}$$

Hence, multiplication of the function $f(x)$ by the unit step function $u(x-c)$ "turns off" the function for $x < c$ and "turns on" the function for $x \geq c$. Graphs of $y(x) = \cos x$ and $y(x) = u(x-\pi)\cos x$ are displayed in Figure 5.4.

a. y(x) = cos x

b. y(x) = u(x - π) cos x

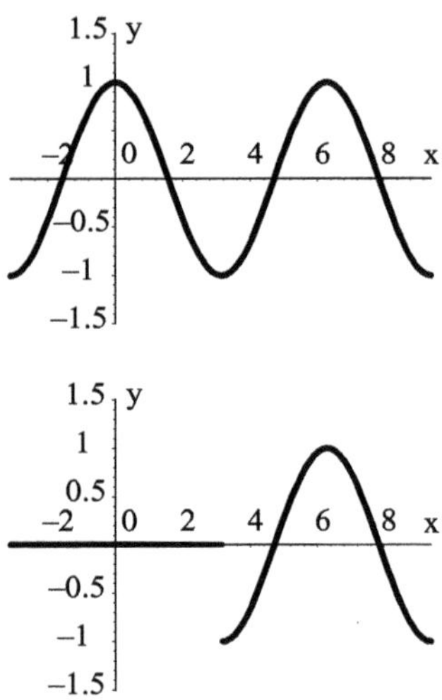

FIGURE 5.4
Graphs of $y(x) = \cos x$ and $y(x) = u(x-\pi)\cos x$.

Observe that the piecewise function

$$h(x) = \begin{cases} f(x), & x < c \\ g(x), & c \le x \end{cases}$$

can be written using unit step functions as

$$h(x) = f(x) - u(x-c)f(x) + u(x-c)g(x).$$

For $a \le x < b$ the "filter" function $[u(x-a) - u(x-b)]$ when multiplying the function $f(x)$ "turns off" the function $f(x)$ for $x < a$, "turns on" the function $f(x)$ for $a \le x < b$, and "turns off" the function $f(x)$ for $b \le x$. That is,

$$[u(x-a) - u(x-b)]f(x) = \begin{cases} 0, & x < a \\ f(x), & a \le x < b \\ 0, & b \le x \end{cases}$$

Graphs of $\cos x$, $[u(x-\pi) - u(x-2\pi)]$, and $[u(x-\pi) - u(x-2\pi)]\cos x$ are shown in Figure 5.5.

a. y(x) = cos x

b. y(x) = [u(x - π) - u(x - 2π)]

c. y(x) = [u(x - π) - u(x - 2π)] cos x

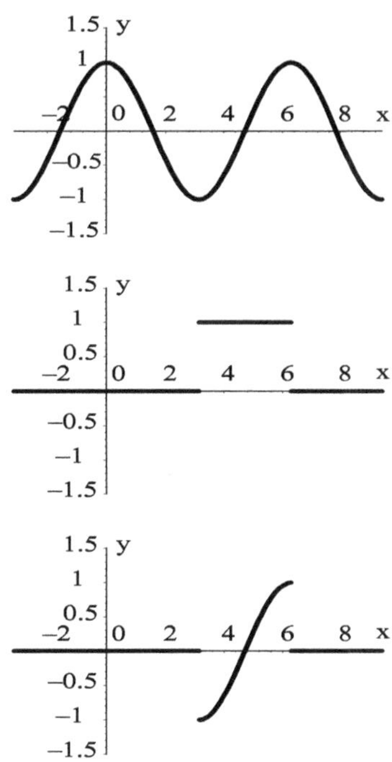

FIGURE 5.5
Graphs of $\cos x$, $[u(x-\pi) - u(x-2\pi)]$, and $[u(x-\pi) - u(x-2\pi)]\cos x$.

Now suppose that two identical sensing devices are placed at points A and B which are separated from one another by a relatively large distance. Further suppose that at time $x = 0$ a signal is sent from point A. Let the signal received by the sensing device at point A be the function $f(x)$ shown in Figure 5.6a. Assuming that no distortion or attenuation has occurred, the signal received by the sensing device at point B will be the function

$$g(x) = \begin{cases} 0, & 0 \leq x < c \\ f(x-c), & c \leq x, \end{cases}$$

where $c > 0$ is the time it takes the signal to travel from point A to point B. The function $g(x)$ is shown in Figure 5.6b. The function $g(x)$ is called the **c-time delay function of** $f(x)$. That is, the function $g(x)$ is the function $f(x)$ delayed by c units of time. One encounters time-delayed functions in many physical circumstances. In most physical situations the function $f(x)$ is defined only for $x \geq 0$. We have defined the c-time delay function $g(x)$ of $f(x)$ so that $g(x) = 0$ for $0 \leq x < c$, since no disturbance at point B is caused by the signal sent from point A during the interval of time $[0, c)$. Notice that $g(x)$ can be defined more concisely by $g(x) = u(x-c)f(x-c)$.

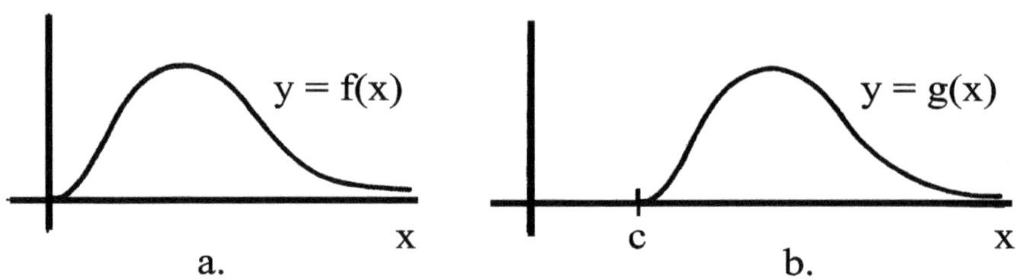

FIGURE 5.6
The Function $f(x)$ and Associated c-time Delay Function $g(x)$.

The following theorem is called the **Second Translation Property** and provides us with the relationship between the Laplace transform of a function $f(x)$ and the Laplace transform of the c-time delay function of $f(x)$, which is $u(x-c)f(x-c)$.

Theorem 5.3 If $\mathcal{L}[f(x)]$ exists for $s > a \geq 0$ and if $c > 0$, then

$$\mathcal{L}[u(x-c)f(x-c)] = e^{-cs}\mathcal{L}[f(x)] \quad \text{for} \quad s > a.$$

Proof: By definition,

$$\mathcal{L}[u(x-c)f(x-c)] = \int_0^\infty u(x-c)f(x-c)e^{-sx}\,dx = \int_c^\infty f(x-c)e^{-sx}\,dx.$$

Making the change of variable $\xi = x - c$, we find

$$\begin{aligned}\mathcal{L}[u(x-c)f(x-c)] &= \int_0^\infty f(\xi)e^{-s(\xi+c)}\,d\xi \\ &= e^{-cs}\int_0^\infty f(\xi)e^{-s\xi}\,d\xi = e^{-cs}\mathcal{L}[f(x)] \quad \text{for} \quad s > a.\end{aligned}$$

Example 1 Use the Second Translation Property to calculate the Laplace transform of

$$f(x) = \begin{cases} \cos x, & 0 \le x < \pi \\ 0, & \pi \le x \end{cases}$$

Solution

Using the trigonometric identity $\cos(x - \pi) = -\cos x$, we see that

$$f(x) = \cos x - u(x-\pi)\cos x = \cos x + u(x-\pi)\cos(x-\pi).$$

Then taking the Laplace transform, using the linearity property, and the second translation property, we find

$$\begin{aligned}\mathcal{L}[f(x)] &= \mathcal{L}[\cos x + u(x-\pi)\cos(x-\pi)] = \mathcal{L}[\cos x] + \mathcal{L}[u(x-\pi)\cos(x-\pi)] \\ &= \mathcal{L}[\cos x] + e^{-\pi s}\mathcal{L}[\cos x] = (1 + e^{-\pi s})\mathcal{L}[\cos x] = \frac{(1+e^{-\pi s})s}{s^2+1}. \ \blacksquare\end{aligned}$$

Example 2 Use the Second Translation Property to find a function f(x) whose Laplace transform is $\frac{e^{-\pi s}}{s^2+4}$. Solution

Since $\frac{2}{s^2+4}$ is the Laplace transform of $\sin 2x$ and since the factor $e^{-\pi s}$ indicates the function $\sin 2x$ should be delayed π units, we calculate the Laplace transform of $\frac{1}{2}u(x-\pi)\sin 2(x-\pi)$ and find

$$\mathcal{L}[\frac{1}{2}u(x-\pi)\sin 2(x-\pi)] = \frac{1}{2}e^{-\pi s}\mathcal{L}[\sin 2s] = \frac{e^{-\pi s}}{s^2+4}.$$

Consequently, $f(x) = \frac{1}{2}u(x-\pi)\sin 2(x-\pi)$. Noting that $\sin(2x - 2\pi) = \sin 2x$, we can write $f(x)$ more conventionally as

$$f(x) = \begin{cases} 0, & 0 \le x < \pi \\ \frac{1}{2}\sin 2x, & \pi \le x \end{cases} \qquad \blacksquare$$

Now let us consider finding a solution of the initial value problem

(2) $\quad y'' + y = 2u(x-1) - u(x-2) = h(x); \quad y(0) = 0, \quad y'(0) = 1.$

In this instance, the forcing function is the step function $h(x)$ of equation (1). This function is discontinuous at $x = 1$ and $x = 2$ and, therefore, is not the solution of any linear homogeneous differential equation with constant coefficients. So here we have our first example of an initial value problem which we cannot solve easily using the method of undetermined coefficients (See Section 4.4.). If we were to use the method of undetermined coefficients to solve this initial value problem, we would have to consider the differential equation of (2) to be three distinct differential equations defined on the three intervals $(-\infty, 1]$, $[1, 2]$, $[2, \infty)$—see the definition of $h(x)$ given in equation (2). Then we would have to require that (i) the solution $y_1(x)$ on the interval $(-\infty, 1]$ satisfy the given initial conditions, (ii) the solution $y_2(x)$ on the interval $[1, 2]$ satisfy $y_2(1) = y_1(1)$ and $y_2'(1) = y_1'(1)$, and (iii) the solution $y_3(x)$ on the interval $[2, \infty]$ satisfy $y_3(2) = y_2(2)$ and $y_3'(2) = y_2'(2)$. In this manner, we could obtain a solution,

$$y(x) = \begin{cases} y_1(x), & -\infty < x \leq 1 \\ y_2(x), & 1 \leq x \leq 2 \\ y_3(x), & 2 \leq x < \infty \end{cases}$$

of the given initial value problem which would be valid on $(-\infty, \infty)$. That is, $y(x)$ and $y'(x)$ would be continuous on $(-\infty, \infty)$—in particular at $x = 1$ and $x = 2$—and $y(x)$ would satisfy the differential equation and initial conditions of equation (2). Hence, in order to solve the initial value problem (2) using the method of undetermined coefficients, we would in effect have to solve the following three initial value problems in succession:

(3a) $\quad y_1' + y_1 = 0; \quad y_1(0) = 0, \quad y_1'(0) = 0, \quad$ for $\; -\infty < x \leq 1$

(3b) $\quad y_2' + y_2 = 2; \quad y_2(1) = y_1(1), \quad y_2'(1) = y_1'(1), \quad$ for $\; 1 \leq x \leq 2$

(3c) $\quad y_3' + y_3 = 1; \quad y_3(2) = y_2(2), \quad y_3'(2) = y_2'(2), \quad$ for $\; 2 \leq x < \infty$.

The advantage of the Laplace transform method is that the solution of the initial value problem (2) can be obtained with one application of the method—not three separate applications. In addition, it should be noted that the solution will also simultaneously satisfy equations (3*a*), (3*b*), and (3*c*). And, therefore, these equations can serve to check the validity of the Laplace transform method solution. The following example illustrates how to use the Laplace transform method to solve the initial value problem (2).

Example 3 Use the Laplace transform method to solve the initial value problem

(2) $\quad \mathbf{y'' + y = 2u(x-1) - u(x-2); \quad y(0) = 0, \quad y'(0) = 1.}$

Solution

Taking the Laplace transform of the differential equation in (2) and using the linearity property of the Laplace transform, we get

$$\mathcal{L}[y''(x)] + \mathcal{L}[y(x)] = 2\mathcal{L}[u(x-1)] - \mathcal{L}[u(x-2)]$$

and

$$-y'(0) - sy(0) + s^2\mathcal{L}[y(x)] + \mathcal{L}[y(x)] = \frac{2e^{-s}}{s} - \frac{e^{-2s}}{s}.$$

Imposing the initial conditions of (2) and solving for $\mathcal{L}[y(x)]$, we find

$$(4) \qquad \mathcal{L}[y(x)] = \frac{\dfrac{2e^{-s}}{s} - \dfrac{e^{-2s}}{s} + 1}{s^2+1} = \frac{2e^{-s}}{s(s^2+1)} - \frac{e^{-2s}}{s(s^2+1)} + \frac{1}{s^2+1}.$$

By partial fraction expansion

$$(5) \qquad \frac{1}{s(s^2+1)} = \frac{1}{s} - \frac{s}{s^2+1}.$$

From Table 5.1, we see that

$$\mathcal{L}[1] = \frac{1}{s}, \quad \mathcal{L}[\cos x] = \frac{s}{s^2+1}, \quad \text{and} \quad \mathcal{L}[\sin x] = \frac{1}{s^2+1} \quad \text{for} \quad s > 0.$$

Consequently, from equation (5) the Laplace transform of the function $f(x) = 1 - \cos x$ is $\dfrac{1}{s(s^2+1)}$. Substituting $\mathcal{L}[1 - \cos x]$ for $\dfrac{1}{s(s^2+1)}$ and $\mathcal{L}[\sin x]$ for $\dfrac{1}{s^2+1}$ in equation (4), we obtain

$$(6) \qquad \mathcal{L}[y(x)] = 2e^{-s}\mathcal{L}[1 - \cos x] - e^{-2s}\mathcal{L}[1 - \cos x] + \mathcal{L}[\sin x].$$

Applying the second translation property to equation (6), we get

$$\begin{aligned}\mathcal{L}[y(x)] &= \mathcal{L}[2u(x-1)(1-\cos(x-1))] - \mathcal{L}[u(x-2)(1-\cos(x-2))] + \mathcal{L}[\sin x] \\ &= \mathcal{L}[2u(x-1)(1-\cos(x-1)) - u(x-2)(1-\cos(x-2)) + \sin x].\end{aligned}$$

Hence,

$$(7) \qquad y(x) = 2u(x-1)(1-\cos(x-1)) - u(x-2)(1-\cos(x-2)) + \sin x$$

is the solution of the initial value problem (2). Or, writing $y(x)$ in a more conventional way, we have

$$y(x) = \begin{cases} \sin x, & -\infty < x \le 1 \\ 2 - 2\cos(x-1) + \sin x, & 1 \le x \le 2 \\ 1 - 2\cos(x-1) + \cos(x-2) + \sin x, & 2 \le x < \infty \end{cases}$$

The reader should verify that this solution simultaneously satisfies equations (3*a*), (3*b*), and (3*c*).

5.4.1 Exercises

1. Express each of the following functions in terms of unit step functions.

a. $f_1(x) = \begin{cases} 2, & 0 \le x < 1 \\ 1, & 1 \le x \end{cases}$

b. $f_2(x) = \begin{cases} 1, & 2 \le x < 4 \\ 0, & \text{otherwise} \end{cases}$

c. $f_3(x) = \begin{cases} 0, & 0 \le x < 1 \\ (x-1)^2, & 1 \le x \end{cases}$

d. $f_4(x) = \begin{cases} 0, & 0 \le x < 1 \\ x^2 - 2x + 3, & 1 \le x \end{cases}$

e. $f_5(x) = \begin{cases} 0, & 0 \le x < \pi \\ \sin 3(x-\pi), & \pi \le x \end{cases}$

f. $f_6(x) = \begin{cases} x, & 0 \le x < 1 \\ 1, & 1 \le x \end{cases}$

g. $f_7(x) = \begin{cases} x, & 0 \le x < 1 \\ 0, & 1 \le x \end{cases}$

2. Find the Laplace transform of the functions in Exercise 1.

3. Use Theorem 5.3 and Table 5.1 to find an inverse Laplace transform of the following functions.

a. $\dfrac{e^{-s}}{s+2}$ b. $\dfrac{1-e^{-2s}}{s^2}$

c. $\dfrac{se^{-\pi s}}{s^2+9}$ d. $\dfrac{se^{-\pi s}}{s^2-9}$

e. $\dfrac{e^{-2s}}{s^2+2s+2}$ f. $\dfrac{se^{-3s}}{s^2+2s+2}$

g. $\dfrac{e^{-3s}}{s^2+2s-3}$ h. $\dfrac{e^{-s}}{s^2-2s+1}$

4. Solve the following initial value problems. The functions $f_i(x)$ are as defined in Exercise 1.

a. $y' + 2y = f_1(x); \quad y(0) = 1$

b. $y'' - y' - 2y = f_2(x); \quad y(0) = 0, \quad y'(0) = 1$

c. $y'' - 2y' = f_3(x); \quad y(0) = 1, \quad y'(0) = 0$

d. $y'' - 2y' + y = f_4(x); \quad y(0) = 0, \quad y'(0) = 1$

e. $y'' + 4y = f_5(x); \quad y(0) = 1, \quad y'(0) = 1$

f. $y'' - 4y = f_6(x); \quad y(0) = 0, \quad y'(0) = 0$

g. $y'' - 4y' + 5y = f_7(x); \quad y(0) = 1, \quad y'(0) = 0$

5.4.1.1 Comments on Computing

In Example 3, we solved the initial value problem $y'' + y = h(x); \quad y(0) = 0, \quad y'(0) = 1$, where $h(x)$ is the piecewise defined function

$$h(x) = \begin{cases} 0, & x < 1 \\ 2, & 1 \le x < 2 \\ 1, & 2 \le x \end{cases}$$

The following ten MAPLE statements also solve this initial value problem and output some important intermediate results.

```
with(inttrans):
alias(L(y(x)) =laplace(y(x), x, s)) :
y(0) := 0:
D(y)(0) := 1:
h :=piecewise(x < 1, 0, 1 <= x and x < 2, 2, 2 <= x, 1):
h :=convert(%, Heaviside);
DE:=diff(y(x), x$2) + y(x) = h;
laplace(DE, x, s);
L(y(x)) =solve(%, L(y(x)));
invlaplace(%, s, x);
```

The third and fourth statements specify the initial condition values. The fifth statement defines the piecewise function $h(x)$, while the sixth statement tells the computer to write $h(x)$ in terms of Heaviside functions (unit step functions). MAPLE uses the notation "Heaviside($x - c$)" instead of the notation $u(x - c)$ for the unit step function. The seventh statement defines the differential equation to be solved. The eighth statement instructs the computer to apply the Laplace transform to the differential equation and to substitute the initial values into the transformed equation. The ninth statement tells the computer to solve the output of the eighth statement algebraically for $L(y(x))$. The last statement causes the computer to calculate the inverse Laplace transform and to print the solution to the initial value problem. The solution printed is expressed in terms of Heaviside functions and is equivalent to equation (7) of Example 3.

5.5 Impulse Function

A force which is of relatively large magnitude and which acts on a system for a relative short period of time is called an **impulse force.** A golf club striking a golf ball, a hammer striking a mass suspended on a spring, and a voltage source connected to an electrical circuit for a short interval of time are all examples of impulse forces. A function $f(x)$ which represents an impulse force is naturally called an **impulse function.** Since impulse functions represent impulse forces, impulse functions are zero except for a short interval of time. Suppose that $f(x)$ is an impulse function and $f(x) = 0$ except for $x_1 < x < x_2$. Then the integral

$$I_f = \int_{-\infty}^{\infty} f(x)\,dx = \int_{x_1}^{x_2} f(x)\,dx$$

is called the total impulse of the function $f(x)$ and represents the total force imparted to the system.

Let us consider the set of impulse step functions $d_\epsilon(x)$ defined by

$$d_\epsilon(x) = \begin{cases} 1/\epsilon, & 0 < x < \epsilon \\ 0, & x \le 0 \quad \text{or} \quad x \ge \epsilon \end{cases}$$

where ϵ is a small positive constant. A graph of one function $d_\epsilon(x)$ is shown in Figure 5.7. Notice that for all $\epsilon > 0$, the total impulse of $d_\epsilon(x)$ is

$$I_{d_\epsilon} = \int_{-\infty}^{\infty} d_\epsilon(x)\,dx = \int_0^{\epsilon} \frac{1}{\epsilon}\,dx = 1.$$

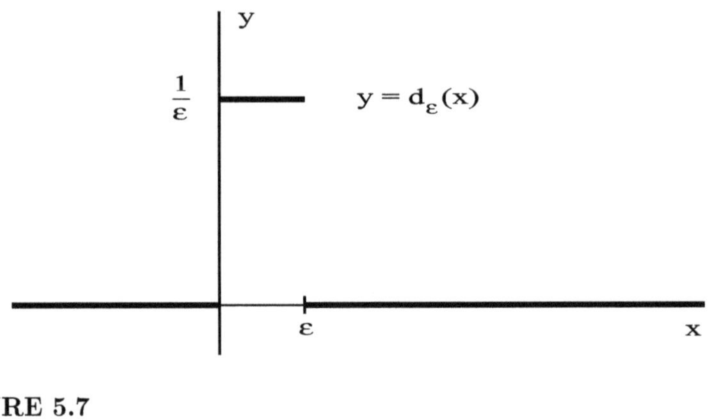

FIGURE 5.7
Graph of the Impulse Step Function $d_\epsilon(x)$.

We would like to define an idealized impulse function $\delta(x)$ to be the limit as $\epsilon \to 0^+$ of the impulse step functions $d_\epsilon(x)$ and we would like for the integral of the idealized step function $\delta(x)$ to be the limit as $\epsilon \to 0^+$ of I_{d_ϵ}. Thus, we define

$$\delta(x) = \lim_{\epsilon \to 0^+} d_\epsilon(x) = \begin{cases} \infty, & x = 0 \\ 0, & x \neq 0 \end{cases}$$

Notice that $\delta(x)$ is not a function in the usual sense but is a "generalized function." The function $\delta(x)$ is called the **Dirac delta function** in honor of the British physicist Paul Adrien Maurice Dirac (1902-1982), who won the Nobel prize in 1933. Laurent Schwartz developed a mathematical theory called *distribution theory* in the early 1950s. Within the framework of this theory, the Dirac delta function is an acceptable generalized function and is integrable with

$$(1) \qquad \int_{-\infty}^{\infty} \delta(x)\,dx = \lim_{\epsilon \to 0^+} \int_{-\infty}^{\infty} d_\epsilon(x)\,dx = 1.$$

The first integral of equation (1) is not the Riemann integral with which we are familiar but the generalized integral of distribution theory. Because of the property of the Dirac delta function exhibited in equation (1), the "function" $\delta(x)$ is often referred to as the **unit impulse function.** Following the notation developed earlier for translating functions, we have

$$\delta(x - c) = \begin{cases} \infty, & x = c \\ 0, & x \neq c \end{cases}$$

and

$$\int_{-\infty}^{\infty} \delta(x - c)\,dx = 1.$$

Let $f(x)$ be a function which is continuous on some interval about $x = c$. We shall define the generalized integral of the product of $\delta(x - c)$ and $f(x)$ as follows:

$$\begin{aligned} \int_{-\infty}^{\infty} \delta(x - c) f(x)\,dx &= \lim_{\epsilon \to 0^+} \int_{-\infty}^{\infty} d_\epsilon(x - c) f(x)\,dx \\ &= \lim_{\epsilon \to 0^+} \int_{c}^{c+\epsilon} \frac{1}{\epsilon} f(x)\,dx = \lim_{\epsilon \to 0^+} \frac{1}{\epsilon} f(\xi)(c + \epsilon - c) \end{aligned}$$

where $c < \xi < c + \epsilon$. The last equality was obtained by using the mean value theorem for integrals. Since $c < \xi < c+\epsilon$ and $f(x)$ is assumed to be continuous on some interval about $x = c$, we have $\lim_{\epsilon \to 0^+} f(\xi) = f(c)$. Consequently, in the context of distribution theory, we have for any function $f(x)$ which is continuous at $x = c$,

$$\int_{-\infty}^{\infty} \delta(x - c) f(x)\,dx = f(c).$$

Therefore, letting $f(x) = e^{-sx}$, we have some justification for defining the Laplace transform of the Dirac delta function $\delta(x-c)$, where $c > 0$, to be

$$\mathcal{L}[\delta(x-c)] = \int_0^\infty \delta(x-c)e^{-sx}\,dx = e^{-sc}.$$

In what follows, we will operate with the Dirac delta function, $\delta(x-c)$, as though it were an ordinary function, and we will use the properties discussed and developed in the preceding paragraphs even though our development of these properties was not mathematically rigorous. It is perhaps comforting to know that in the context of distribution theory all these operations and properties can and have been proven rigorously.

Example 1 Solve the Following Initial Value Problem which has an Impulse Forcing Function

$$\mathbf{y'' + 2y' + 5y = 3\delta(x-1); \quad y(0) = 0, \quad y'(0) = 0.}$$

Solution

This initial value problem could represent the damped motion of a mass on a spring which is initially at rest but which is set in motion at time $x = 1$ by an impulse force. Taking the Laplace transform of the given differential equation and imposing the initial conditions, we find successively

$$\mathcal{L}[y'' + 2y' + 5y] = \mathcal{L}[3\delta(x-1)]$$

$$\mathcal{L}[y''(x)] + 2\mathcal{L}[y'(x)] + 5\mathcal{L}[y(x)] = 3e^{-s}$$

$$-y'(0) - sy(0) + s^2\mathcal{L}[y(x)] - 2y(0) + 2s\mathcal{L}[y(x)] + 5\mathcal{L}[y(x)] = 3e^{-s}$$

$$(s^2 + 2s + 5)\mathcal{L}[y(x)] = 3e^{-s}.$$

Solving for $\mathcal{L}[y(x)]$ and using information found in Table 5.1, we see

$$\mathcal{L}[y(x)] = \frac{3e^{-s}}{s^2 + 2s + 5} = \frac{3e^{-s}}{2}\frac{2}{(s+1)^2 + 2^2} = \frac{3}{2}e^{-s}\mathcal{L}[e^{-x}\sin 2x].$$

Applying Theorem 5.3 yields

$$\mathcal{L}[y(x)] = \frac{3}{2}e^{-s}\mathcal{L}[e^{-x}\sin 2x] = \mathcal{L}[\frac{3}{2}u(x-1)e^{-(x-1)}\sin 2(x-1)].$$

Hence, the solution of the initial value problem is

$$y(x) = \frac{3}{2}u(x-1)e^{-(x-1)}\sin 2(x-1) = \begin{cases} 0, & 0 \le x \le 1 \\ \frac{3}{2}e^{-(x-1)}\sin 2(x-1), & 1 < x \end{cases}$$

A graph of this solution is displayed in Figure 5.8. For $0 \le x < 1$, the initial value problem is $y'' + 2y' + 5y = 0$; $y(0) = 0$, $y'(0) = 0$, and it obviously has the unique solution $y(x) \equiv 0$, which can easily be seen in Figure 5.8. The portion of the graph to the right of $x = 1$ is due to the impulse force which was applied to the system at the instant $x = 1$. The "solution" $y(x)$ is continuous for $x \ge 0$ but the first derivative has a jump discontinuity at $x = 1$, and the second derivative has an infinite discontinuity there.

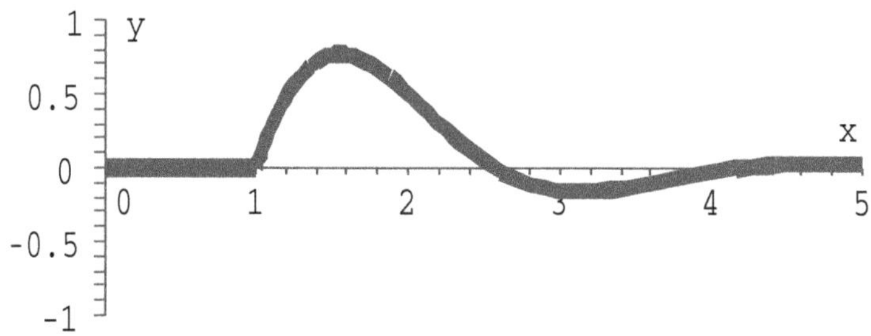

FIGURE 5.8
Graph of the Solution $y(x) = \dfrac{3}{2}u(x-1)e^{-(x-1)}\sin 2(x-1)$. ■

Example 2 Solve the Following Initial Value Problem which has an Impulse Forcing Function

$$\mathbf{y'' + 2y' + 5y = 5\delta(x - 2); \quad y(0) = 4, \quad y'(0) = 0.}$$

Solution

The solution process is the same as for Example 1. Taking the Laplace transform of the given differential equation and imposing the initial conditions, we find successively

$$\begin{aligned} \mathcal{L}[y'' + 2y' + 5y] &= \mathcal{L}[5\delta(x-2)] \\ \mathcal{L}[y''(x)] + 2\mathcal{L}[y'(x)] + 5\mathcal{L}[y(x)] &= 5e^{-2s} \\ -y'(0) - sy(0) + s^2\mathcal{L}[y(x)] - 2y(0) + 2s\mathcal{L}[y(x)] + 5\mathcal{L}[y(x)] &= 5e^{-2s} \\ -4s - 8 + (s^2 + 2s + 5)\mathcal{L}[y(x)] &= 5e^{-2s}. \end{aligned}$$

Solving for $\mathcal{L}[y(x)]$ and using partial fraction expansion and information found in Table 5.1, we see

$$\begin{aligned} \mathcal{L}[y(x)] &= \frac{5e^{-2s} + 4s + 8}{s^2 + 2s + 5} \\ &= \frac{5e^{-2s}}{2}\frac{2}{(s+1)^2 + 2^2} + 4\frac{s+1}{(s+1)^2 + 2^2} + 2\frac{2}{(s+1)^2 + 2^2} \\ &= \frac{5}{2}e^{-2s}\mathcal{L}[e^{-x}\sin 2x] + 4\mathcal{L}[e^{-x}\cos 2x] + 2\mathcal{L}[e^{-x}\sin 2x]. \end{aligned}$$

Applying the second translation property yields

$$\mathcal{L}[y(x)] = \mathcal{L}[\frac{5}{2}u(x-2)e^{-(x-2)}\sin 2(x-2) + 4e^{-x}\cos 2x + 2e^{-x}\sin 2x].$$

Hence, the solution of the initial value problem is

$$y(x) = \frac{5}{2}u(x-2)e^{-(x-2)}\sin 2(x-2) + 4e^{-x}\cos 2x + 2e^{-x}\sin 2x$$

$$= \begin{cases} 4e^{-x}\cos 2x + 2e^{-x}\sin 2x, & 0 \le x \le 2 \\ \frac{5}{2}e^{-(x-2)}\sin 2(x-2) + 4e^{-x}\cos 2x + 2e^{-x}\sin 2x, & 2 < x \end{cases}$$

A graph of this solution is displayed in Figure 5.9. The portion of the graph to the right of $x = 2$ is due to the impulse force which was applied to the system at the instant $x = 2$. The "solution" $y(x)$ is continuous for $x \ge 0$ but the first derivative has a jump discontinuity at $x = 2$ and the second derivative has an infinite discontinuity there. The dotted curve appearing in Figure 5.9 for $x > 2$ shows the solution, if no impulse force were applied to the system at the instant $x = 2$.

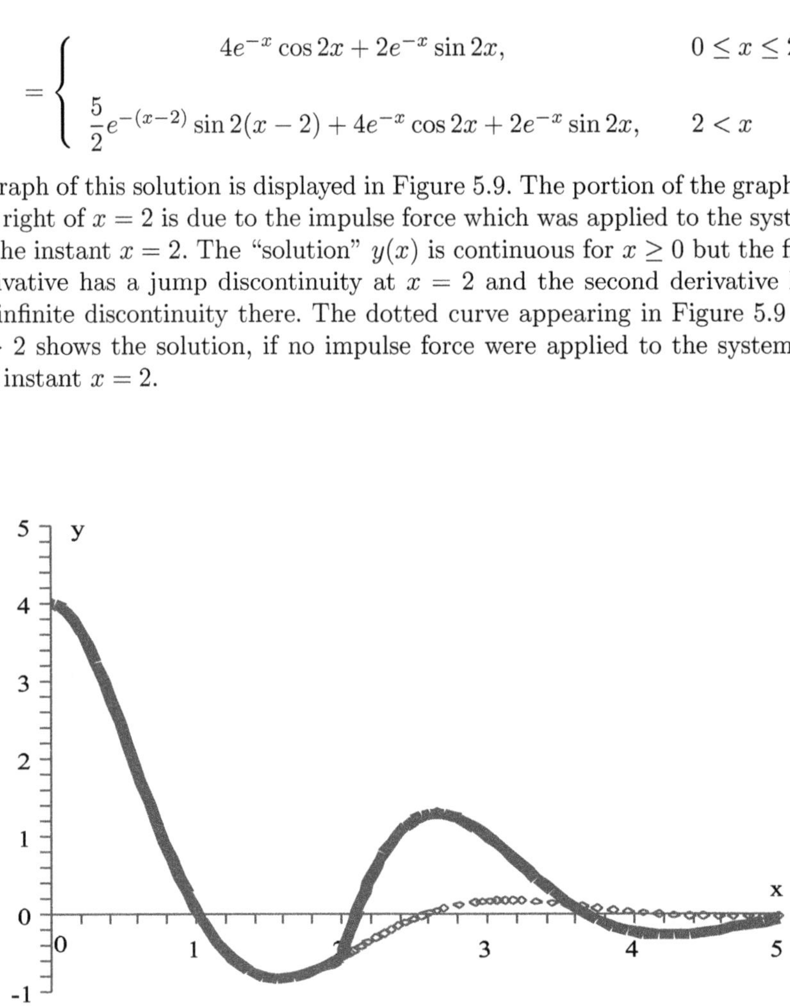

FIGURE 5.9
Graph of the Solution. y(x) $= \frac{5}{2}u(x-2)e^{-(x-2)}\sin 2(x-2) + 4e^{-x}\cos 2x + 2e^{-x}\sin 2x$ ■

5.5.1 Exercises

Find the solution of the following initial value problems using the Laplace transform method.

1. $y' + 3y = \delta(x-2); \quad y(0) = 0$
2. $y' - 3y = \delta(x-1) + 2u(x-2); \quad y(0) = 0$
3. $y'' + 9y = \delta(x-\pi) + \delta(x-3\pi); \quad y(0) = 0, \quad y'(0) = 0$
4. $y'' - 2y' + y = 2\delta(x-1); \quad y(0) = 0, \quad y'(0) = 1$
5. $y'' - 2y' + 5y = \cos x + \delta(x-\pi); \quad y(0) = 1, \quad y'(0) = 0$
6. $y'' + 4y = \delta(x-\pi)\cos x; \quad y(0) = 0, \quad y'(0) = 1$
7. $y'' + a^2 y = \delta(x-\pi)f(x); \quad y(0) = 0, \quad y'(0) = 0$
   where a is a real constant and $f(x)$ is a function that is continuous on some interval about the point $x = \pi$.

5.5.1.1 Comments on Computing

The solution of the initial value problem

$$y'' + 2y' + 5y = 5\delta(x-2); \quad y(0) = 4, \quad y'(0) = 0$$

of Example 2 can be computed using the following two MAPLE statements.

```
DE:=diff(y(x), x$2) + 2*diff(y(x), x) + 5 * y(x) = 5*Dirac(x - 2);
dsolve({DE, y(0) = 4, D(y)(0) = 0}, y(x), method=laplace);
```

The first statement specifies the differential equation to be solved. Observe that MAPLE uses the notation "Dirac($x-2$)" to represent the Dirac delta function $\delta(x-2)$. The second statement instructs the computer to use the Laplace transform method to solve the differential equation using the specified initial condition values and to print the solution. The printed solution contains the Heaviside function and is equivalent to the solution of Example 2.

Table 5.2, which appears on the following page, contains a few more functions and "generalized functions," $f(x)$, in the left column and the corresponding entry in the right column contains their Laplace transform, $F(s)$. Below Table 5.2 is a list of four properties of Laplace transforms and their inverse Laplace transforms.

TABLE 5.2
Inverse Laplace Transforms and Laplace Transforms

$f(x) = \mathcal{L}^{-1}(F(s))$	$F(s) = \mathcal{L}[f(x)]$
$u(x-c), \quad c \geq 0$	$\dfrac{e^{-cs}}{s}$ for $\quad s > 0$
$\delta(x-c), \quad c \geq 0$	e^{-cs}
$f(x)$	$F(s) = \mathcal{L}[f(x)]$
$f(cx), \quad c > 0$	$\frac{1}{c}F(\frac{s}{c})$
f'	$sF(s) - f(0)$
$f''(x)$	$s^2F(s) - sf(0) - f'(0)$
$f^{(n)}(x), \quad n$ a positive integer	$s^nF(s) - s^{n-1}f(0) - \cdots - f^{(n-1)}(0)$
$\int_0^x f(\xi)\, d\xi$	$\dfrac{F(s)}{s}$

Properties of Laplace Transforms

Linearity Property

$af(x) + bg(x), \quad a, b$ constants	$aF(s) + bG(s)$

Translation Property

$e^{ax}f(x)$	$F(s-a)$

Second Translation Property

$u(x-c)f(x-c), \quad c \geq 0$	$e^{-cs}F(s)$

Convolution Theorem

$\int_0^x f(x-\xi)g(\xi)\, d\xi$	$F(s)G(s)$

6

Applications of Linear Differential Equations with Constant Coefficients

In this chapter, we will examine a few linear differential equations with constant coefficients which arise in the study of various physical and electrical systems. Usually the independent variable, t, will represent time, and the dependent variable or solution, $y(t)$, will represent a parameter which describes the state of the system.

6.1 Second-Order Differential Equations

Second-order linear differential equations provide mathematical models for various physical phenomena. As a matter of fact, several physical phenomena often give rise to the same mathematical model. Therefore, it is sometimes possible to simulate the behavior of an expensive physical system such as an airplane, automobile, or bridge, etc., by another inexpensive physical system such as an electrical circuit. In this section, we present the mathematical models which arise from a simple pendulum, a mass on a spring, and an RLC electrical circuit, and discover that they are all the same model. This leads us to examine in detail the second-order linear differential equation with constant coefficients.

A Simple Pendulum A simple pendulum consists of a rigid, straight rod of negligible mass and length ℓ with a bob of mass m attached to one end. The other end of the rod is attached to a fixed support so that the pendulum is free to move in a vertical plane. Let y denote the angle, expressed in radians, which the rod makes with the vertical—the equilibrium position of the system. We arbitrarily choose y to be positive if the rod is to the right of vertical and negative if the rod is to the left of vertical as shown in Figure 6.1.

We will assume the only forces acting on the pendulum are the force of gravity and a force due to air resistance which is proportional to the velocity of the bob. Under these assumptions, it can be shown by applying Newton's second law of motion that the position of the pendulum satisfies the initial value problem

(1) $$my'' + cy' + k\sin y = 0; \quad y(0) = c_0, \quad y'(0) = c_1$$

DOI: 10.1201/9781003678014-6

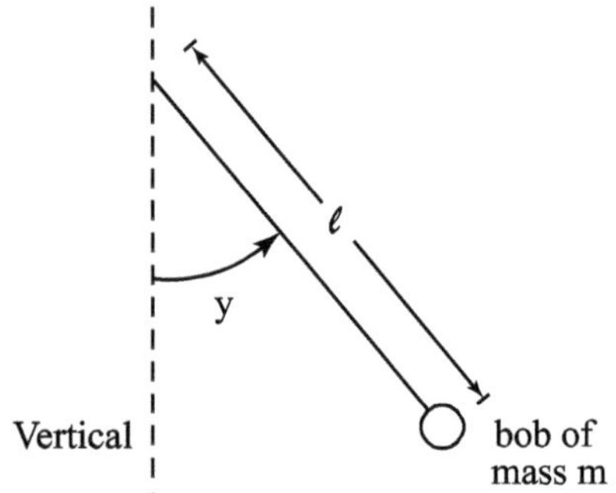

FIGURE 6.1
Simple Pendulum.

where $c \geq 0$ is a constant of proportionality, $k = mg/\ell$, g is the constant of acceleration due to gravity, c_0 is the initial displacement of the pendulum, and c_1 is the initial velocity of the pendulum.

The differential equation appearing in (1) is nonlinear because of the factor $\sin y$. An approximation which is made in order to linearize the differential equation is to replace $\sin y$ by y. This approximation is valid for small angles, say $|y| < .1$ radians $\approx 5.73°$. So for y small the following linear initial value problem approximately describes the motion of a simple pendulum

$$(2) \qquad my'' + cy' + ky = 0; \quad y(0) = c_0, \quad y'(0) = c_1.$$

If there is some external force $f(t)$ acting on the simple pendulum, such as a weight or main spring in a clock which is mechanically connected to and driving the rod of the pendulum, then the linearized homogeneous initial value problem (2) must be replaced by the nonhomogeneous initial value problem

$$(3) \qquad my'' + cy' + ky = f(t); \quad y(0) = c_0, \quad y'(0) = c_1.$$

Often $f(t)$ is periodic and representable as $f(t) = E \sin \omega t$ where E and ω are constants.

A Mass on a Spring A spring of natural length L is suspended by one end from a fixed support. A body of mass m, where m is small compared to the mass of the spring in order to avoid exceeding the elastic limit of the spring, is attached to the other end of the spring, and the resulting system is allowed to come to rest at its equilibrium position. Suppose in the equilibrium position the length of the elongated spring is $L + \ell$, where $\ell > 0$ and ℓ is small compared to L. See Figure 6.2. According to Hooke's law, the elongation ($\ell > 0$) and compression ($\ell < 0$) of a spring are directly proportional to the force that produces the elongation or compression. (The Englishman Robert Hooke (1635-1703), a friend of Isaac Newton, first published this result in 1676 as an anagram and then in detail in 1678. Thus, Hooke discovered Hooke's

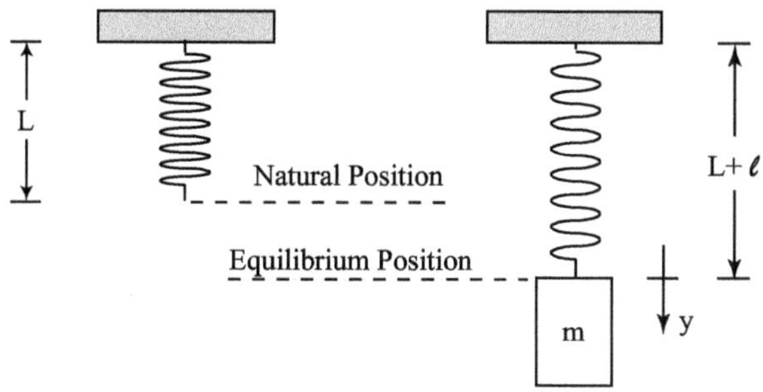

FIGURE 6.2
Mass on a Spring System.

law at about the same time calculus was invented.) In the equilibrium position the only force acting on the system is the force of gravity, so by **Hooke's law**

$$mg = k\ell$$

where $k > 0$ is the constant of proportionality for the particular spring. For a given spring, the spring constant k can easily be calculated by attaching a body of known mass m to the spring and accurately measuring the resulting elongation ℓ.

We arbitrarily choose the equilibrium position to be the origin and choose the positive direction to be measured vertically downward from the origin. See Figure 6.2. It has been shown experimentally that as long as the speed at which the mass is traveling is not too large, the damping force is proportional to the speed. The damping force is usually due to the resistance caused by the medium (air, perhaps) in which the system operates or by the resistance caused by adding some additional component to the system, such as a dashpot. Applying Newton's second law of motion to the mass on a spring system with damping, it can be shown that the position of the mass satisfies the initial value problem

$$(4) \qquad my'' + cy' + ky = 0; \quad y(0) = c_0, \quad y'(0) = c_1$$

where $c \geq 0$ is a constant (the damping constant), $k > 0$ is the spring constant, c_0 is the initial location of the mass, and c_1 is the initial velocity of the mass.

If there is an external force $f(t)$ acting on the spring-mass system, such as a motor which vibrates the support or a magnetic field which acts upon a suspended iron mass, then the homogeneous initial value problem (4) must be replaced by the nonhomogeneous initial value problem

$$(5) \qquad my'' + cy' + ky = f(t); \quad y(0) = c_0, \quad y'(0) = c_1.$$

TABLE 6.1
Electric Circuit Components and Quantities

Circuit component or quantity	Symbol: Alphabetic	Symbol: Graphic	Unit
Capacitor	C	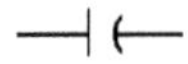	Farad (F)
Electric charge	q		Coulomb (C)
Electric current	i		Ampere (A)
Electromotive force			
Battery	E		Volt (V)
Generator	$E \sin \omega t$	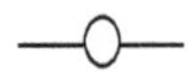	Volt (V)
Inductor	L		Henry (H)
Resistor	R		Ohm (Ω)
Time	t		Seconds (s)

The spring-mass mechanical systems (4) and (5) may be set in motion (i) by pulling the mass downward from its equilibrium position ($y(0) = c_0 > 0$) and releasing it without imparting any velocity ($y'(0) = c_1 = 0$), (ii) by lifting the mass upward from its equilibrium position ($y(0) = c_0 < 0$) and releasing it without imparting any velocity ($y'(0) = c_1 = 0$), (iii) by applying an instantaneous external force to the mass (say, by hitting the mass from below with a hammer) and thereby imparting a velocity to the mass ($y'(0) = c_1 \neq 0$) and dislodging the mass from the equilibrium position ($y(0) = c_0 = 0$), or (iv) by pulling the mass downward or lifting the mass upward ($y(0) = c_0 \neq 0$) and releasing the mass and imparting some velocity ($y'(0) = c_1 \neq 0$).

Electrical Circuits Now, let us consider the flow of electric current in some simple circuits. Table 6.1 contains a list of some common electric circuit components and quantities, the alphabetic symbols usually used to denote the numeric values of these components and quantities, the graphic symbols used to represent components in schematic drawings, and the units associated with each component or quantity. The units of the components were named in honor of the following physicists: André Marie Ampère (1775-1836, French), Charles Augustin De Coulomb (1736-1806, French), Michael Faraday (1791-1867, English), Joseph Henry (1797-1878, American), Georg Simon Ohm (1789-1854, German), and Allessandro Volta (1745-1827, Italian).

The electric charge, q, and the current, $i = dq/dt$, are functions of time. The positive quantities of capacitance, C, inductance, L, and resistance, R, are also functions of time. However, in many instances, these quantities are nearly constant. So we shall assume C, L, and R are positive constants.

The German physicist Gustav Kirchhoff (1824-1887) stated the following two laws regarding the behavior of electrical systems:

Kirchhoff's First Law (Current Law) At any junction in a network, the sum of the current flowing into the junction is equal to the sum of the current flowing out of the junction.

Kirchhoff's Second Law (Voltage Law) The algebraic sum of the voltage drops around any loop of a network is equal to the algebraic sum of the impressed electromotive forces around the loop.

And according to the fundamental laws of electricity, the voltage drop across a resistor is Ri, the voltage drop across a capacitor is q/C, and the voltage drop across an inductor is Ldi/dt.

The RLC Circuit Consider the simple RLC circuit consisting of a resistor with resistance R, an inductor with inductance L, a capacitor with capacitance C, an electromotive force $E(t)$, and a switch s connected in series as shown in Figure 6.3. The arrow in the figure provides an orientation for the current flow in the loop. The current i is positive if it is flowing in the direction of the arrow and negative if it is flowing in the opposite direction.

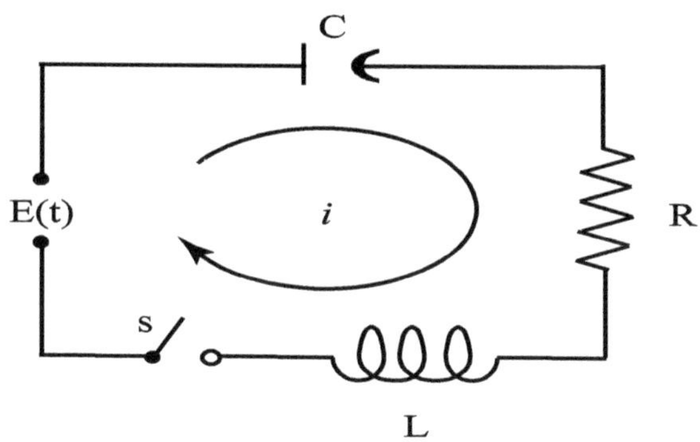

FIGURE 6.3
An RLC Series Circuit.

Since there are no junctions in this circuit, we need only apply Kirchhoff's second law regarding voltage drops to find

$$Li' + Ri + \frac{q}{C} = E(t). \tag{6}$$

Substituting $i = dq/dt = q'$ into equation (6), we find that the charge on the capacitor, q, satisfies the differential equation

$$Lq'' + Rq' + \frac{q}{C} = E(t). \tag{7}$$

Appropriate initial conditions are

$$(8) \qquad q(0) = q_0 \quad \text{and} \quad q'(0) = i(0) = i_0.$$

That is, appropriate initial conditions at some time $t = 0$ are the initial charge on the capacitor, q_0, and the initial current flowing in the system, i_0.

Differentiating equation (6) with respect to t and substituting $i = dq/dt = q'$, we see the current, i, satisfies the second-order linear differential equation

$$(9) \qquad Li'' + Ri' + \frac{i}{C} = E'(t).$$

Appropriate initial conditions for this problem are to specify the initial current, i_0, and the initial rate of change of the current, i_0', at some time $t = 0$. That is, appropriate initial conditions are

$$(10) \qquad i(0) = i_0 \quad \text{and} \quad i'(0) = i_0'.$$

The initial current, i_0, flowing in the circuit and the initial charge, q_0, on the capacitor are physically measurable quantities. But i_0' is not a physically measurable quantity. However, if at $t = 0$ we measure i_0, q_0, and $E_0 = E(0)$, then we can calculate i_0' from equation (6). Substituting $t = 0$ into equation (6) and solving for i_0', we find

$$i'(0) = i_0' = \frac{E_0 - Ri_0 - \dfrac{q_0}{C}}{L}.$$

Comparing equation (7) with equations (3) and (5), we discover the correspondence between electrical systems and mechanical systems (the simple pendulum and spring-mass system) displayed in Table 6.2. This correspondence allows us to simulate (model) mechanical systems, such as an airplane wings and suspension bridges—systems which would be expensive to actually construct—by electrical systems—systems which are relatively inexpensive to construct. By measuring quantities such as capacitance, inductance, resistance, current, and voltage, we can determine the response of the approximating electrical system and can thereby infer the response of the hypothetical mechanical system.

6.1.1 Free Motion

Since the simple pendulum, mass on a spring, and the *RLC* series circuit all lead to the initial value problem

$$(11) \qquad ay'' + by' + dy = f(t); \quad y(0) = c_0, \quad y'(0) = c_1$$

where $a > 0$, $b \geq 0$, and $d > 0$ are constants, we should examine this initial value problem in detail.

TABLE 6.2
Correspondence Between Electrical and Mechanical Systems

Electrical systems		**Mechanical systems**	
L	inductance	m	mass
q	charge	y	displacement
$dq/dt = i$	current	$dy/dt = v$	velocity
R	resistance	c	damping
$1/C$	reciprocal of capacitance	k	spring or pendulum constant
$E(t)$	electromotive force	$f(t)$	driving force

When there is no external force driving the system—that is, when $f(t) = 0$ for all t the system is said to execute **free motion**. In this case, the differential equation of (11) reduces to the homogeneous differential equation

(12) $$ay'' + by' + dy = 0$$

and the associated auxiliary equation is $ar^2 + br + d = 0$. The two roots of the auxiliary equation are

(13) $$r_1 = \frac{-b + \sqrt{b^2 - 4ad}}{2a} \quad \text{and} \quad r_2 = \frac{-b - \sqrt{b^2 - 4ad}}{2a}.$$

(Note: These roots can easily be calculated using a root finding routine instead of using the quadratic formula (13).)

6.1.1.1 Free Undamped Motion

Furthermore, when there is no damping force acting on the system—that is, when $b = 0$, the system executes **free undamped motion** which is also called **simple harmonic motion**. This type of motion occurs if the pendulum or spring-mass system operates in a vacuum or if all resistance, R, is removed from the RLC series circuit, making the circuit an LC series circuit. Later you may encounter simple harmonic oscillators (motion) in the study of quantum mechanics or in the study of the vibration of strings, membranes, and beams. When $f(t) = 0$ and $b = 0$, $r_1 = \sqrt{d/a}i = \omega i$ and $r_2 = -\sqrt{d/a}i = -\omega i$, so the general solution of $ay'' + dy = 0$ is

(14) $$y = A_1 \sin \omega t + A_2 \cos \omega t$$

where A_1 and A_2 are arbitrary constants. We would like to show that equation (14) may also be written in the equivalent form

$$y = A\sin(\omega t + \phi) \tag{15}$$

where A and ϕ are arbitrary constants. Applying the trigonometric formula for the sine of the sum of two angles, ωt and ϕ, to equation (15), we see

$$y = A\sin(\omega t + \phi) = A\cos\phi\sin\omega t + A\sin\phi\cos\omega t. \tag{16}$$

Equating coefficients of $\sin\omega t$ and $\cos\omega t$ in equations (14) and (16), we obtain the following two relationships between the constants A_1 and A_2 of (14) and A and ϕ of (16)

$$A_1 = A\cos\phi \tag{17a}$$

$$A_2 = A\sin\phi. \tag{17b}$$

Squaring both equations (17a) and (17b), adding the results, and solving for A, we find

$$A = \sqrt{A_1^2 + A_2^2}. \tag{18}$$

Substituting (18) into (17a) and (17b) and solving for $\cos\phi$ and $\sin\phi$, we see that ϕ must simultaneously satisfy

$$\cos\phi = \frac{A_1}{\sqrt{A_1^2 + A_2^2}} \quad \text{and} \quad \sin\phi = \frac{A_2}{\sqrt{A_1^2 + A_2^2}}.$$

The reason for wanting to write the general solution (14) in the form (15) is because it is easier to understand the physical significance of the arbitrary constants A and ϕ of (15). The constant A is the **amplitude of oscillation** and $-A \le y(t) \le A$ for all t. The constant ϕ is the **phase angle**. The phase angle is usually chosen with the restriction $-\pi < \phi \le \pi$. With this restriction when $0 < \phi \le \pi$, ϕ is the angle by which (15) $y = A\sin(\omega t + \phi)$ leads the function $y = A\sin\omega t$ and ϕ/ω is the time lead. See Figure 6.4. When $-\pi < \phi < 0$, ϕ is the angle by which $y = A\sin(\omega t + \phi)$ lags the function $y = A\sin\omega t$ and ϕ/ω is the time lag. Since the period of the function $y = A\sin(\omega t + \phi)$ is $2\pi/\omega$, the **period of oscillation** of a free undamped system is $P = 2\pi/\omega$. The period is the time interval between successive maxima (minima) of the system. So for a simple pendulum, the period is the time interval from one time the bob is farthest to the right (left) until the next time the bob is farthest to the right (left). And for a mass-spring system, the period is the time interval from one time the mass is at its lowest (highest) point until the next time the mass is at its lowest (highest) point. The reciprocal of the period is called the **frequency**. The frequency $F = 1/P = \omega/(2\pi)$ is the number of oscillations per unit of time. A pendulum clock which is keeping perfect time has a frequency of one cycle per second (1c/s).

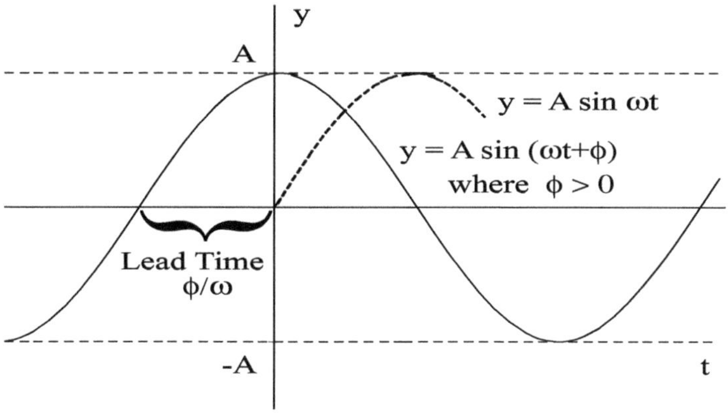

FIGURE 6.4
Simple Harmonic Motion.

6.1.1.2 Free Damped Motion

When there is a damping force acting on the system ($b \neq 0$) but no external force driving the system ($f(t) = 0$ for all t), the form of the general solution of equation (12) depends on the sign of the discriminant of the roots and, consequently, the types of roots of the auxiliary equation.

Case 1. When $b^2 - 4ad < 0$, the system executes **damped oscillatory motion**. The roots of the auxiliary equation are complex conjugates— $r_1 = -\alpha + \omega i$ and $r_2 = -\alpha - \omega i$ where $\alpha = b/(2a) > 0$ and $\omega = \sqrt{4ad - b^2}/(2a)$. So the general solution of (12) is

$$(19) \qquad y(t) = e^{-\alpha t}(A_1 \sin \omega t + A_2 \cos \omega t)$$

where A_1 and A_2 are arbitrary constants. As in the case of free undamped motion, the solution (19) may be rewritten in the equivalent form

$$(20) \qquad y(t) = Ae^{-\alpha t} \sin(\omega t + \phi)$$

where A and ϕ are arbitrary constants. The factor $Ae^{-\alpha t}$ is called the **damping factor**. The factor $\sin(\omega t + \phi)$ represents periodic, oscillatory motion with amplitude 1. Since $\alpha > 0$ and $|\sin(\omega t + \phi)| \leq 1$, $y(t) \to 0$ as $t \to \infty$. So the product of the two factors $Ae^{-\alpha t}$ and $\sin(\omega t + \phi)$ represents oscillatory motion in which the amplitude of oscillation decreases with increasing time. The time interval between two successive maxima is still called the period, P, and as in the free undamped case $P = 2\pi/\omega$. See Figure 6.5.

Case 2. When $b^2 - 4ad = 0$, the system is said to be **critically damped**. In this case, the roots of the auxiliary equation are real and equal, $r_1 = r_2 = -b/(2a)$. So the general solution of (12) is

$$(21) \qquad y = (A + Bt)e^{-\alpha t}$$

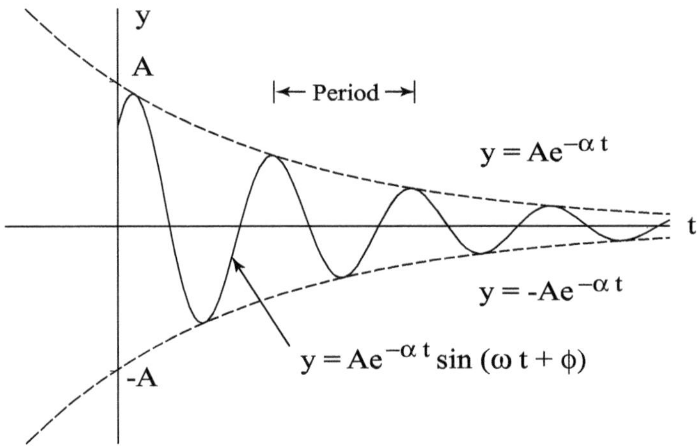

FIGURE 6.5
Damped, Oscillatory Motion.

where $\alpha = b/(2a) > 0$ and A and B are arbitrary constants. Due to the damping factor $e^{-\alpha t}$ and the fact that $\alpha > 0$, the solution $y(t) \to 0$ as $t \to \infty$. Since the factor $A + Bt$ is linear, the motion is not oscillatory and, in fact, the solution can cross the t-axis (the time axis) at most once. The graph of (21) depends on the constants A and B which, of course, are determined by the initial conditions. Three typical graphs of equation (21)—critically damped motion—are sketched in Figure 6.6.

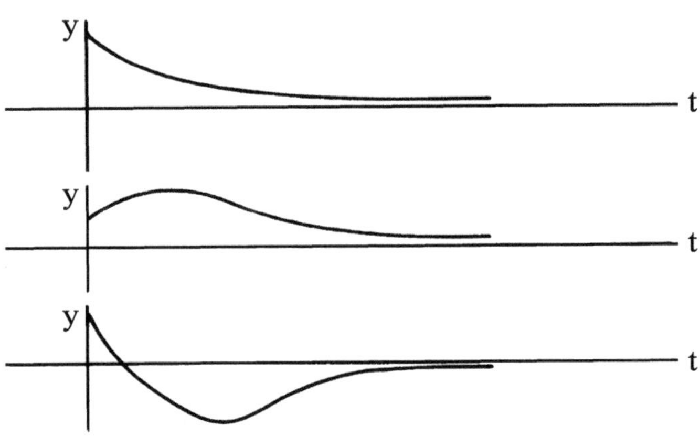

FIGURE 6.6
Typical Critically Damped and Overdamped Motion.

Case 3. When $b^2 - 4ad > 0$, the system is said to be **overdamped**. The roots of the auxiliary equation are real, distinct, and negative. See equation (13). So the general solution of (12) is

$$y(t) = Ae^{r_1 t} + Be^{r_2 t}. \tag{22}$$

Since $r_1 < 0$ and $r_2 < 0$, both $e^{r_1 t}$ and $e^{r_2 t}$ are monotone decreasing functions. Therefore, $y(t) \to 0$ as $t \to \infty$ and no oscillation occurs. The three graphs shown in Figure 6.6 are also typical of the graph of equation (22)—overdamped motion.

Example 1 Equation of Motion of an Undamped Pendulum An undamped pendulum of length .9 meters (m) with a bob of mass .2 kilograms (kg) is moved to the left of vertical so that the pendulum makes an angle of −.6 radians (rads) with the vertical and is released imparting a velocity of .3 radians/second (rads/s) to the pendulum.

a. Write the equation of motion for the pendulum. (The expression "write the equation of motion for the pendulum" is another way of saying "write the general solution of the initial value problem for the pendulum.")

b. What is the amplitude, period, and frequency of oscillation?

c. What is the phase angle?

Solution

a. On the earth's surface $g = 9.8$ meters/second^2 (m/s^2) is the mean gravitational constant. So the pendulum constant is

$$k = mg/\ell = (.2 \text{ kg})(9.8 \text{ m/s}^2)/.9 \text{ m} = 2.178 \text{ kg/s}^2.$$

Since the pendulum is undamped and there is no forcing function, the initial value problem which we need to solve is

$$my'' + ky = 0; \quad y(0) = -.6 \text{ rads}, \quad y'(0) = .3 \text{ rads/s}. \tag{23}$$

The roots of the auxiliary equation, $mr^2 + k = 0$, are $r_1 = \sqrt{k/m}i = \omega i$ and $r_2 = -\sqrt{k/m}i = -\omega i$. So the general solution of the differential equation in (23) is either

$$y(t) = A_1 \sin \omega t + A_2 \cos \omega t$$

where A_1 and A_2 are constants which are to be chosen to satisfy the initial conditions or equivalently

$$y(t) = A \sin(\omega t + \phi)$$

where A and ϕ are constants which are to be chosen to satisfy the initial conditions.

Differentiating the second form of the general solution of the differential equation, we see that

$$y'(t) = \omega A \cos(\omega t + \phi).$$

In order to satisfy the initial conditions given in (23), the constants A and ϕ must be chosen to simultaneously satisfy

$$y(0) = A \sin \phi = -.6 \text{ rads} \quad \text{and} \quad y'(0) = \omega A \cos \phi = .3 \text{ rads/s}.$$

Dividing the first equation by the second, we find ϕ must satisfy

$$\frac{\sin \phi}{\omega \cos \phi} = \frac{-.6}{.3} = -2 \text{ s}.$$

Multiplying by $\omega = \sqrt{k/m} = \sqrt{2.178/.2} = 3.3$ rads/s, we see $\tan \phi = -6.6$. So $\phi = -1.42$ radians. Next, solving the equation $A \sin \phi = -.6$ for A, yields

$$A = \frac{-.6}{\sin \phi} = \frac{-.6}{\sin(-1.42)} = \frac{-.6}{-.98865} = .607 \text{ rads}.$$

Thus, the equation of motion for this pendulum is

$$y(t) = A \sin(\omega t + \phi) = .607 \sin(3.3t - 1.42).$$

b. The amplitude is $A = .607$ radians.

The period is $P = 2\pi/\omega = 2\pi/3.3 = 1.904$ seconds.

The frequency is $F = 1/P = .5252$ cycles/second.

c. The phase angle is $\phi = -1.42$ radians. ■

Example 2 Equation of Motion of a Damped Mass on a Spring
A .1 kg mass is attached to one end of a spring, the other end is attached to a fixed support, and the system is allowed to come to rest. In the equilibrium position, the spring is stretched .25 m. The mass is pulled down an additional .5 m and released without imparting any velocity. If the damping constant of the system is $c = 1$ kg/s, write the equation of motion of the system.

Solution

Since the spring is stretched .25 m $= \ell$ by a mass of .1 kg $= m$, by Hooke's law the spring constant k satisfies $mg = k\ell$. Solving for k, we have

$$k = mg/\ell = (.1 \text{ kg})(9.8 \text{ m/s}^2)/(.25 \text{ m}) = 3.92 \text{ kg-m}^2/\text{s}^2.$$

So the equation of motion of this spring-mass system satisfies the initial value problem

$$my'' + cy' + ky = 0; \quad y(0) = .5 \text{ m}, \quad y'(0) = 0 \text{ m/s}$$

or

$$.1y'' + y' + 3.92y = 0; \quad y(0) = .5 \text{ m}, \quad y'(0) = 0 \text{ m/s}.$$

Using the computing software, we find the roots of the auxiliary equation $.1r^2 + r + 3.92 = 0$ are $r_1 = -5 + 3.7683i$ and $r_2 = -5 - 3.7683i$. Since the roots are complex conjugate roots, the system is executing damped oscillatory motion, and the equation of motion is

$$y(t) = Ae^{-5t} \sin(3.7683t + \phi)$$

where A and ϕ are constants which are to be determined to satisfy the initial conditions. Differentiating the equation of motion, we see

$$y'(t) = Ae^{-5t}[3.7683 \cos(3.7683t + \phi) - 5 \sin(3.7683t + \phi)].$$

To satisfy the given initial conditions A and ϕ must simultaneously satisfy

$$y(0) = A \sin \phi = .5 \text{ m} \quad \text{and} \quad y'(0) = A[3.7683 \cos \phi - 5 \sin \phi] = 0 \text{ m/s}.$$

Dividing the second equation by the first, we see ϕ must satisfy

$$3.7683 \cot \phi - 5 = 0 \quad \text{or} \quad \cot \phi = 5/3.7683 = 1.3269.$$

So $\phi = .6458$ rads. Solving the equation $A \sin \phi = .5$ m for A, we find $A = .5$ m$/ \sin(.6458) = .831$ m. Therefore, the equation of motion is

$$y(t) = .831e^{-5t} \sin(3.7683t + .6458). \blacksquare$$

Example 3 Finding the Equation for the Charge on a Capacitor Find the equation for the charge, q, on the capacitor and the current flowing in the RLC series circuit of Figure 6.3, if $E = 0\,V$ (volts), $L = .04\,H$ (Henry), $R = 20\Omega$ (Ohms), $C = 4 \times 10^{-4} F$ (Farad), if the initial charge on the capacitor is $q(0) = 2 \times 10^{-3} F$, and if the initial current flowing in the circuit is $i(0) = q'(0) = 0\,A$ (amps).

Solution

To find the equation for the charge on the capacitor, we must solve the following initial value problem which results from equations (7)-(8) by replacing E, L, R, C, q_0, and i_0 by the values given in the problem.

$$.04q'' + 20q' + 2500q = 0; \quad q(0) = 2 \times 10^{-3} F, \quad q'(0) = 0\,A.$$

Using computing software, we find the roots of the auxiliary equation $.04r^2 + 20r + 2500 = 0$ are $r_1 = r_2 = -250$. Since the roots are real and equal, this system provides an example of critically damped motion and the equation for the charge on the capacitor (the solution of the differential equation) is

$$q(t) = (A + Bt)e^{-250t}$$

where A and B must be chosen to satisfy the given initial conditions. Differentiating, we find

$$q'(t) = i(t) = (-250A + B - 250Bt)e^{-250t}.$$

To satisfy the given initial conditions, A and B must be chosen to simultaneously satisfy

$$q(0) = 2 \times 10^{-3} = A \quad \text{and} \quad q'(0) = i(0) = 0 = -250A + B.$$

So $A = 2 \times 10^{-3}$ F, $B = 250\,A = .5$ F/s, and the charge on the capacitor as a function of time after the switch is closed is

$$q(t) = (.002 + .5t)e^{-250t}\text{coulombs}.$$

And the current flowing in the circuit is

$$i(t) = q'(t) = -125e^{-250t}\text{amps}.$$

The negative sign in the equation for the current indicates that the current is flowing in the direction opposite of the arrow in Figure 6.3. ■

6.1.2 Exercises

Exercises 1–9 pertain to simple pendulums. In these exercises assume the angular displacement, y, is small enough so that the approximation *sin* y ∼ y is valid, and use the linearized initial value problem (2) to describe the motion of the pendulum. On the earth, g = 9.8 meters/second^2 (m/s^2) is the mean surface gravitational constant, and on the moon, g = 1.62 m/s^2.

1. A child whose mass is 40 kilograms (kg) is swinging in a park on a swing of length 5 meters (m). Assuming air resistance is negligible, calculate the period and frequency of oscillation.

2. If an undamped pendulum on the earth has a period of 1 second, how long is the pendulum in meters? How long must a pendulum be in order to have a period of 2 seconds? How long would an undamped pendulum need to be on the moon to have a period of 1 second?

3. Does an undamped pendulum of length ℓ with a .2 kg bob oscillate faster or slower than an undamped pendulum of length ℓ with a .1 kg bob? If both pendulums are subjected to the same damping, which will oscillate faster?

4. Two identical undamped pendulums, one on the earth and the other on the moon, are started with the same initial displacement and velocity. Which pendulum will oscillate faster? Which pendulum will have the larger amplitude?

5. An undamped pendulum of length .5 m with a bob of mass .3 kg is moved to the right of vertical so that the pendulum makes an angle of 1 radian with the vertical and is then released without imparting any velocity to the pendulum.

   a. Write the equation of motion for the pendulum.
   b. What is the amplitude and period of oscillation?
   c. What is the phase angle?
   d. What is the velocity and acceleration of the pendulum when the bob is at the vertical and headed toward the left? right?

6. An undamped pendulum of length 1/6 m with a bob of mass .2 kg is started with an angular displacement of .5 radians and released with an angular velocity of 1 radian/second.

   a. Write the equation of motion for the pendulum.
   b. What is the amplitude and period of oscillation?
   c. What is the phase angle?

7. A damped pendulum of length .6 m with a bob of mass .5 kg is started with initial position $y(0) = 0$ and initial velocity $y'(0) = .2$ radians/second in a medium with damping constant $c = .5$ kg/s.

   a. Write the equation of motion. (Hint: Use your computing software to find the roots of the auxiliary equation.)
   b. What is the maximum angular displacement from the vertical?

8. A pendulum of length .2 m with a bob of mass .5 kg oscillates in a viscous medium. Determine if the pendulum will execute damped oscillatory motion, critically damped motion, or overdamped motion for the following damping constants

   a. $c = 10$ kg/s b. $c = 7$ kg/s c. $c = 5$ kg/s

   (Hint: Use computing software to find the roots of the auxiliary equation and based on the type of roots determine the type of motion.)

9. In an experiment a pendulum of length .7 m with a bob of mass .3 kg oscillating in a viscous medium was observed to execute damped oscillatory motion. Two successive maxima angular displacements were measured to be $y(t_1) = 1/5$ radian and $y(t_2) = 1/6$ radian. Find the damping constant, c, of the medium. (Hint: The time interval between successive maxima, $t_2 - t_1$, is $2\pi/\omega$. Write equation (20) for t_1 and t_2, divide $y(t_1)$ by $y(t_2)$, and solve for c.)

Exercises 10–13 pertain to spring-mass systems. Assume the mass attached to the spring is executing simple harmonic motion—that is, assume there is no damping so c = 0.

10. a. What is the velocity of the mass at the instant the displacement from the equilibrium position is a maximum?

    b. What is the position of the mass when the velocity is a maximum?

11. A mass m is attached to a spring whose spring constant is 3.2 kg/s^2. If the period of oscillation is 2 seconds, determine the mass m.

12. The top of a spring is attached to a fixed support. A 2 kg mass is attached to the bottom of the spring. After coming to rest at the equilibrium position, the mass is pulled down X meters below the equilibrium position and released with an initial downward velocity of .3 m/s. If the amplitude of the resulting harmonic motion is .1 m and the period is 1 s, calculate the spring constant k and the distance X.

13. When a 1 kg mass is attached to a certain spring and the system set in motion, the period of oscillation is 2 seconds. Later the 1 kg mass is replaced with an unknown mass m. The new system is set in motion, and the new period is found to be 4 seconds. What is the mass m?

14. A .1 kg mass is attached to a spring, the other end is attached to a fixed support, and the system is allowed to come to rest. In the equilibrium position, the spring is stretched .05 m. Determine if the spring-mass system will execute damped oscillatory motion, critically damped motion, or overdamped motion for the following values of the damping constant

    a. $c = 2.8$ kg/s b. $c = 1.6$ kg/s c. $c = 3.5$ kg/s

    (Hint: Use computing software to find the roots of the auxiliary equation, and based on the type of roots, determine the type of motion.)

15. A .8 kg mass is attached to one end of a spring. The other end is attached to the ceiling. The system is allowed to come to rest. In the equilibrium position, the spring is stretched .3 m. The system is to be set in motion by pulling the mass downward and releasing it. What value of the damping constant c will result in damped oscillatory motion? Critically damped motion? overdamped motion?

16. A .6 kg mass is attached to the lower end of a spring. The upper end is attached to a fixed support. When the system is set in motion, and there is no damping, the period of oscillation is 2 seconds. When the damping constant is c kg/s, and the system is set in motion, the period of oscillation is 4 seconds. Determine the damping constant c and the spring constant k.

17. For the following RLC series circuits, the electromotive force is zero, $E = 0$. In each case, determine if the equation for the charge on the capacitor, q,

represents simple harmonic motion, damped oscillatory motion, critically damped motion, overdamped motion, or none of these.

	R (Ohms)	L (Henry)	C (Coulomb)
a.	0	0.5	2.0×10^{-5}
b.	10	0.0	3.0×10^{-4}
c.	20	0.1	1.0×10^{-3}
d.	20	0.1	0.5×10^{-3}
e.	30	0.2	1.0×10^{-2}

18. If $E = 0$ in an RLC series circuit, what relationship between the parameters R, L, and C results in an equation for the charge on the capacitor which represents simple harmonic motion? Damped oscillatory motion? Critically damped motion? Overdamped motion?

6.1.3 Forced Motion

Suppose some mechanical or electrical system whose state is represented by the parameter $y(t)$ is mathematically modeled by the differential equation

$$ay'' + by' + dy = f(t) \tag{24}$$

where $a > 0$, $b \geq 0$, and $d > 0$ are constants and $f(t) \neq 0$. Since $f(t) \neq 0$, the system is said to execute **forced motion**. We will assume the forcing function $f(t)$ is periodic. More specifically, we will assume $f(t) = E \sin \omega^* t$, where E and ω^* are constants. Other periodic functions such as $E \cos \omega^* t$ or $E \sin(\omega^* t + \theta)$ will serve as the forcing function as well as the function which we have chosen. For a pendulum system, the forcing function might represent a weight or main spring in a clock which is connected to and driving the pendulum. For a spring-mass system, the forcing function might represent a motor which vibrates the support or a magnetic field which acts upon a suspended iron mass. For an electrical circuit or network, the forcing function represents the electromotive force applied to the system by a battery or a generator.

6.1.3.1 Undamped Forced Motion

When there is no damping force ($b = 0$) and the forcing function $f(t) = E \sin \omega^* t$, equation (24) becomes

$$ay'' + dy = E \sin \omega^* t. \tag{25}$$

The general solution of the associated homogeneous equation $ay'' + dy = 0$,

$$y_c(t) = A \sin \omega t + B \cos \omega t = C \sin(\omega t + \phi)$$

where $\omega = \sqrt{d/a}$ and A, B, C, and ϕ are arbitrary constants, is the complementary solution of equation (25).

Case 1. If $\omega^* \neq \omega$, then a particular solution of (25) will have the form

$$y_p(t) = F \sin \omega^* t + G \cos \omega^* t.$$

Differentiating twice, we find

$$y_p'(t) = \omega^* F \cos \omega^* t - \omega^* G \sin \omega^* t$$

and

$$y_p''(t) = -(\omega^*)^2 F \sin \omega^* t - (\omega^*)^2 G \cos \omega^* t.$$

Substituting y_p and y_p'' into equation (25) and rearranging, we see that the constants F and G must satisfy

$$F[d - a(\omega^*)^2] \sin \omega^* t + G[d - a(\omega^*)^2] \cos \omega^* t = E \sin \omega^* t.$$

Since the set $\{\sin \omega^* t, \cos \omega^* t\}$ is a linearly independent set, we equate coefficients and find

$$F = \frac{E}{d - a(\omega^*)^2} \quad \text{and} \quad G = 0.$$

So when $\omega^* \neq \omega$, a particular solution of equation (25) is

$$y_p(t) = \frac{E}{d - a(\omega^*)^2} \sin \omega^* t,$$

and the general solution of equation (25) is

$$y(t) = y_c(t) + y_p(t) = A \sin \omega t + B \cos \omega t + \frac{E}{d - a(\omega^*)^2} \sin \omega^* t$$

where the constants A and B depend on the initial conditions. Notice that for $\omega^* \neq \omega$, the solution, $y(t)$, of equation (25) remains bounded for all time, t.

Case 2. If $\omega^* = \omega$—that is, if the frequency of the forcing function is identical to the natural frequency of the system, then a phenomenon known as **resonance** occurs, and the particular solution of equation (25) will have the form

$$y_p(t) = Mt \sin \omega^* t + Nt \cos \omega^* t.$$

Differentiating this equation twice, substituting y_p and y_p'' into equation (25), and solving, we get $M = 0$ and $N = -E/(2a\omega^*)$. So when $\omega^* = \omega$ the general solution of equation (25) is

$$y(t) = A \sin \omega t + B \cos \omega t - \frac{E}{2a\omega^*} t \cos \omega^* t.$$

Due to the factor $t \cos \omega^* t$, the solution $y(t)$ oscillates with unbounded amplitude as $t \to \infty$ regardless of the initial conditions which merely determine the constants A and B.

6.1.3.2 Damped Forced Motion

When the periodic external force $f(t) = E\sin\omega^* t$ and when there is a damping force ($b \neq 0$),which is the case for all realizable systems, equation (24) becomes

$$ay'' + by' + dy = E\sin\omega^* t \tag{26}$$

where a, b, and d are positive constants and E and ω^* are constants. In the previous section, we found that the solution of the associated homogeneous equation, the complementary solution y_c of equation (26), depends on the sign of $b^2 - 4ad$.

If $b^2 - 4ad < 0$, then

$$y_c(t) = e^{-\alpha t}(A\sin\omega t + B\cos\omega t) = Ce^{-\alpha t}\sin(\omega t + \phi)$$

where $\alpha = b/(2a)$, $\omega = \sqrt{4ad - b^2}/(2a)$, and A, B, C, and ϕ are arbitrary constants.

If $b^2 - 4ad = 0$, then

$$y_c(t) = (A + Bt)e^{-\alpha t}$$

where $\alpha = b/(2a)$ and A and B are arbitrary constants.

If $b^2 - 4ad > 0$, then

$$y_c(t) = Ae^{r_1 t} + Be^{r_2 t}$$

where $r_1 = (-b + \sqrt{b^2 - 4ad})/(2a) < 0$, $r_2 = (-b - \sqrt{b^2 - 4ad})/(2a) < 0$, and A and B are arbitrary constants.

Notice that as $t \to \infty$, $y_c(t) \to 0$ regardless of the value of the quantity $b^2 - 4ad$ and regardless of the form of the complementary solution $y_c(t)$. Because of this property the term $y_c(t)$ of the general solution $y(t) = y_c(t) + y_p(t)$ is called the **transient solution**—the function $y_c(t)$ and its effects die out with increasing time.

There is a particular solution of equation (26) of the form

$$y_p(t) = F\sin\omega^* t + G\cos\omega^* t.$$

Differentiating twice; substituting y_p, y_p', and y_p'' into equation (26); equating coefficients of $\sin\omega^* t$ and $\cos\omega^* t$; and solving for F and G it can be shown that

$$F = \frac{(d - a(\omega^*)^2)E}{H(\omega^*)} \quad \text{and} \quad G = \frac{-b\omega^* E}{H(\omega^*)}$$

where $H(\omega^*) = [d - a(\omega^*)^2]^2 + b^2(\omega^*)^2$. So a particular solution of equation (26) is

$$y_p(t) = E[(d - a(\omega^*)^2)\sin\omega^* t - b\omega^*\cos\omega^* t]/H(\omega^*)$$

or equivalently

$$y_p(t) = [E\sin(\omega^* t + \phi)]/\sqrt{H(\omega^*)} \tag{27}$$

where ϕ simultaneously satisfies

$$\cos\phi = (d - a(\omega^*)^2)/\sqrt{H(\omega^*)} \quad \text{and} \quad \sin\phi = -b\omega^*/\sqrt{H(\omega^*)}.$$

The general solution of equation (26) is $y(t) = y_c(t) + y_p(t)$. The transient solution (complementary solution, y_c) contains arbitrary constants A and B which depend upon the initial conditions under which the system was started. Since the initial conditions affect only the transient solution and since the transient solution approaches zero after a sufficiently long period of time, the initial conditions influence the solution of equation (26) only for a "short" period of time. For all practical purposes after a sufficiently long period of time the general solution becomes the particular solution. For this reason, the particular solution is called the **steady state solution**. Consequently, after a sufficiently long period of time, the solution depends only on the external force driving the system and not upon the conditions under which the system was started—that is, not upon the initial conditions.

We see from equation (27) that the steady state solution is oscillatory with frequency ω^*—the same frequency as the forcing function $f(t)$—and with amplitude $A(\omega^*) = E/\sqrt{H(\omega^*)}$. For $0 < \omega^* < \infty$ as $\omega^* \to 0$, $A(\omega^*) \to E/d$, and as $\omega^* \to \infty$, $A(\omega^*) \to 0$. Also since $b > 0$, for all positive ω^*, $A(\omega^*)$ is finite. Consequently, when there is damping present in a system ($b > 0$), the amplitude of oscillation remains finite; whereas, when there is no damping in the system ($b = 0$) and when resonance occurs ($\omega^* = \omega$) the amplitude of oscillation increases without bound until the system is destroyed. The amplitude of oscillation $A(\omega^*)$ will be a maximum when $H(\omega^*)$ is a minimum. Differentiating $H(\omega^*)$ with respect to ω^*, setting the result equal to zero, and solving for ω^*, we find

$$\omega^* = \frac{\sqrt{ad - b^2/2}}{a}.$$

When ω^* has this value the forcing function is said to be in **resonance** with the system. Resonance can occur only when $ad - b^2/2 > 0$, which implies $b^2 < 2ad < 4ad$, which in turn implies $b^2 - 4ad < 0$. Thus, resonance can occur only if the corresponding free system ($f(t) = 0$) executes damped oscillatory motion. Resonance will not occur if the free system executes critically damped motion or overdamped motion. For damped oscillatory motion the resonance frequency is

$$F_R = \frac{\omega^*}{2\pi} = \frac{\sqrt{ad - b^2/2}}{2\pi a}.$$

This frequency is less than the frequency of the corresponding free system which is

$$F = \frac{\omega}{2\pi} = \frac{\sqrt{ad - b^2/4}}{2\pi a}.$$

6.1.4 Exercises

1. A .6 kg mass is attached to one end of a spring. The other end is attached to a movable support. The support is held fixed and the system is permitted to come to rest. In the equilibrium position the spring is stretched .3 meters. The mass is pulled down .2 meters below the equilibrium position and released without imparting any velocity, and at the same instant, a motor starts to drive the support with a force $f(t) = 10 \sin \omega^* t$ kg-m/s^2.

   a. What value of ω^* causes resonance?

   b. If $\omega^* = 10$ cycles/second, what is the general solution?

   c. Suppose damping is added to the system, and the damping constant is computed to be $c = .4$ kg/s. What is the resonance frequency, F_R?

2. Suppose the equation of motion of some system satisfies the differential equation
$$y'' + cy' + 15y = 4 \sin \omega^* t.$$

   a. What value of c produces resonance?

   b. If $c = 4$, what value of ω^* produces resonance?

3. In the RLC series circuit suppose the electromotive force has a constant value of E.

   a. Find the steady state solution for the charge on the capacitor, $q_p(t)$.

   b. What is $\lim_{t\to\infty} q_p(t)$?

   c. What is $\lim_{t\to\infty} i(t)$?

4. In the RLC series circuit suppose the electromotive force is $f(t) = E \sin \omega^* t$.

   a. For $R = 0$ (an LC series circuit) and $\omega^* \neq 1/\sqrt{LC}$, find the steady state solution for the charge on the capacitor, $q_p(t)$. What is $\lim_{t\to\infty} q_p(t)$?

   b. For $R = 0$ and $\omega^* = 1/\sqrt{LC}$, find the steady state solution for the charge on the capacitor, $q_p(t)$. What is $\lim_{t\to\infty} q_p(t)$?

   c. For $L = 0$ (an RC series circuit) find the steady state solution for the charge on the capacitor. What is $\lim_{t\to\infty} q_p(t)$?

5. Find the steady state current, $i_p(t)$, flowing in an RL series circuit ($C = 0$) if

   a. $f(t) = E$, a constant. What is $\lim_{t\to\infty} i_p(t)$?

   b. $f(t) = E \cos \omega^* t$.

6. Find the steady state current, $i_p(t)$, for an RLC series circuit if $E(t) = E \cos \omega^* t$. For what value of ω^* will the amplitude of the steady state current be a maximum?

6.2 Higher-Order Differential Equations

In this section we will present some applications which require the solution of linear differential equations with constant coefficients of order greater than two. And we will show how to solve a linear system of differential equations with constant coefficients by writing the system as a single higher-order linear differential equation with constant coefficients.

A Coupled Spring-Mass System Suppose a mass m_1 is attached to one end of a spring with spring constant k_1. The other end of this spring is attached to a fixed support. A second mass m_2 is attached to one end of a second spring with spring constant k_2. The other end of the second spring is attached to the bottom of mass m_1, and the resulting system is permitted to come to rest in the equilibrium position as shown in Figure 6.7.

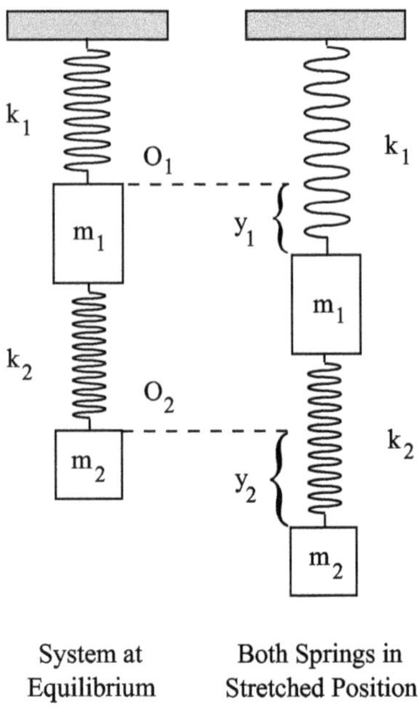

FIGURE 6.7
A Coupled Spring-Mass System.

For $i = 1, 2$, let y_i represent the vertical displacement of mass m_i from its equilibrium position. As before we will assign downward displacement from equilibrium to be positive and upward displacement from equilibrium to be negative. Applying Newton's second law of motion and assuming no damping is present, it can be shown that the equations of motion for this coupled spring-mass system satisfy the linear system of differential equations

(28a) $$m_1 y_1'' = -k_1 y_1 + k_2(y_2 - y_1)$$

(28b) $$m_2 y_2'' = -k_2(y_2 - y_1).$$

Solving equation (28a) for y_2, we get

(29) $$y_2 = \frac{m_1 y_1''}{k_2} + (1 + \frac{k_1}{k_2})y_1.$$

Differentiating twice, we find

(30) $$y_2'' = \frac{m_1 y_1^{(4)}}{k_2} + (1 + \frac{k_1}{k_2})y_1''.$$

Substituting these last two expressions for y_2 and y_2'' into equation (28b), we obtain the following single fourth-order differential equation for y_1

$$\frac{m_2 m_1 y_1^{(4)}}{k_2} + m_2(1 + \frac{k_1}{k_2})y_1'' = -m_1 y_1'' - (k_1 + k_2)y_1 + k_2 y_1$$

or multiplying by k_2 and rearranging

(31) $$m_1 m_2 y_1^{(4)} + [m_2(k_1 + k_2) + m_1 k_2]y_1'' + k_1 k_2 y_1 = 0.$$

Exercise 5.

a. For $m_1 = .2$ kg, $m_2 = .7$ kg, $k_1 = 5$ kg-m^2/s^2, and $k_2 = 11$ kg-m^2/s^2 find the general solution of equation (31). (Hint: Use computing software to find the roots of the auxiliary equation associated with equation (31).)

b. Use equation (29) and the answer to part a. to find $y_2(t)$.

c. Find the solution to the initial value problem consisting of the system of two first-order differential equations (28) and the initial conditions:

$$y_1(0) = .1 \text{ m}, \quad y_1'(0) = .3 \text{ m/s}, \quad y_2(0) = -.15 \text{ m}, \quad \text{and} \quad y_2'(0) = .4 \text{ m/s}.$$

Another Coupled Spring-Mass System A second coupled spring-mass system which consists of two masses, m_1 and m_2, connected to two fixed supports by three springs which have spring constants k_1, k_2, and k_3 is shown in Figure 6.8. Neglecting the effects of damping, the system of differential equations which describes the displacements y_1 and y_2 of masses m_1 and m_2, respectively, from their equilibrium positions is

(32a) $$m_1 y_1'' = -k_1 y_1 + k_2(y_2 - y_1)$$

(32b) $$m_2 y_2'' = -k_2(y_2 - y_1) - k_3 y_2.$$

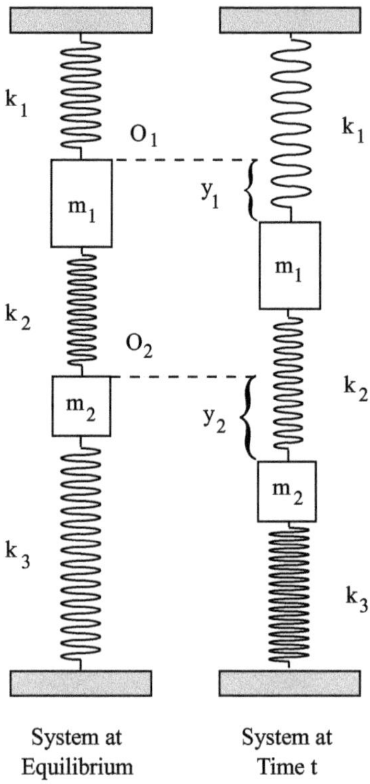

FIGURE 6.8
A Coupled Spring-Mass System.

Exercise 6. For $m_1 = .4$ kg, $m_2 = .25$ kg, $k_1 = 7$ kg-m^2/s^2, $k_2 = 6$ kg-m^2/s^2, and $k_3 = 9$ kg-m^2/s^2 find the solution to the initial value problem consisting of the system of differential equations (32) and the initial conditions $y_1(0) = -.6$ m, $y_1'(0) = .45$ m/s, $y_2(0) = .3$ m, and $y_2'(0) = -.37$ m/s. (Hint: Equation (32a) is the same as equation (28a). Eliminate y_2 and y_2'' from equation (32b) and obtain a fourth-order linear differential equation in y_1. Use computing software to find the roots of the associated auxiliary equation. Write the general solution y_1, then find y_2 and satisfy the initial conditions.)

A Double Pendulum A double pendulum consists of a bob of mass m_1 attached to a fixed support by a rod of length ℓ_1 and a second bob of mass m_2 attached to the first bob by a rod of length ℓ_2 as shown in Figure 6.9. Let y_1 and y_2 denote the displacement from the vertical of the rods of length ℓ_1 and ℓ_2, respectively. Assuming the double pendulum oscillates in a vertical plane and neglecting the mass of the rods and any damping forces, it can be shown

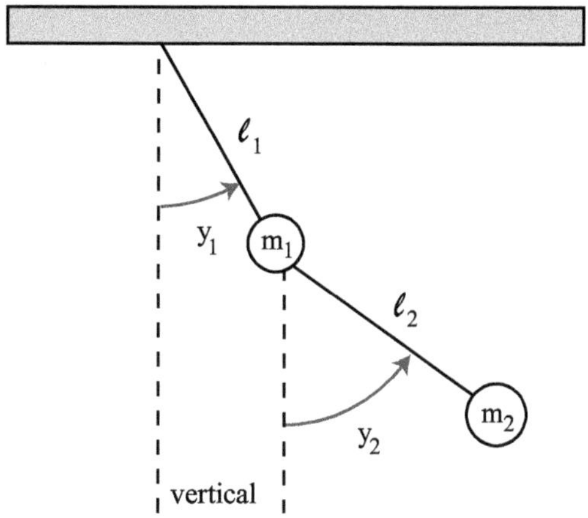

FIGURE 6.9
A Double Pendulum.

that the displacements, y_1 and y_2, satisfy the following system of differential equations

$$(m_1 + m_2)\ell_1^2 y_1'' + m_2\ell_1\ell_2 y_2'' + (m_1 + m_2)\ell_1 g y_1 = 0 \tag{33a}$$

$$m_2\ell_1\ell_2 y_1'' + m_2\ell_2^2 y_2'' + m_2\ell_2 g y_2 = 0 \tag{33b}$$

where g is the constant of gravitational acceleration.

Exercise 7. For $m_1 = .3$ kg, $m_2 = .2$ kg, $\ell_1 = .5$ m, $\ell_2 = .25$ m, and $g = 9.8$ m/s^2 solve the system of differential equations (33) subject to the initial conditions $y_1(0) = .05$ rad, $y_1'(0) = .15$ rad/s, $y_2(0) = .1$ rad, and $y_2'(0) = -.2$ rad/s. (Hint: Multiply equation (33a) by ℓ_2 and multiply equation (33b) by ℓ_1. Subtract one of these new equations from the other to eliminate the term containing y_2'' as a factor. Solve the resulting equation for y_2. Differentiate twice to get y_2' and y_2''. Substitute the expressions for y_2 and y_2'' into equation (33b) and obtain a fourth-order differential equation in y_1. Use computing software to find the roots of the associated auxiliary equation. Write the general solution y_1, then find y_2, and finally satisfy the initial conditions.)

The Path of an Electron In 1897, J. J. Thomson demonstrated the existence of the electron by determining the ratio of the charge of an electron to its mass. Let the ratio be $R = q/m$ where q is the charge of an electron and m is its mass. The position (x, y) of an electron in the plane satisfies the system of differential equations

$$x'' = -HRy' + ER \tag{34a}$$

$$y'' = HRx' \tag{34b}$$

where H is the intensity of the magnetic field and E is the intensity of an electric field acting on the electron.

Exercise 8. Find the position of an electron as a function of time t, if $HR = 2$, if $ER = 3$, and if the electron is initially at rest at the origin. That is, solve the system (34) subject to the initial conditions: $x(0) = 0$, $x'(0) = 0$, $y(0) = 0$, and $y'(0) = 0$. (Hint: Differentiate (34b) and substitute (34a) into the resulting equation. Then solve the third-order differential equation in y, etc.)

Compartmental Analysis Many complex biological and physical processes can be subdivided into several distinct phases. Then the complex process can be studied by analyzing each phase individually and the interaction between the phases. Each phase or stage in the overall process is called a **compartment.** It is assumed that material which moves from one compartment to another does so in a negligible amount of time and that the material itself is immediately dispersed throughout the entire compartment. A **closed** compartmental system is one in which there is no input to or output from any compartment in the system. An **open** system is one in which there is an input to or output from at least one compartment in the system. Engineers sometimes refer to compartmental systems as **block diagrams**.

Let $Y_1, Y_2, \ldots, Y_n$ denote the n separate compartments in a compartmental system. Each compartment is assumed to have a constant volume v_i which may vary in size from compartment to compartment. At time t the concentration of a particular substance S in the compartment Y_i is $y_i(t)$. It is assumed that at all times the substance is uniformly distributed throughout each of the compartments. The rate of change of concentration of the substance in compartment i at time t, $y_i'(t)$, is equal to the sum over all inputs to the i-th compartment of the concentration of each input times the rate of flow per volume of the input minus the concentration of the substance in compartment i, $y_i(t)$, times the sum of the rates of flow per volume of output from the compartment.

As an example, let $y_1(t)$, $y_2(t)$, and $y_3(t)$ denote the concentration of a substance S in compartments Y_1, Y_2, and Y_3 at time t. And let a, b, c, d, e, and f be the rates per volume at which the fluids containing the substance S flow into and out of the compartments of the open compartmental system shown in Figure 6.10. Also let u denote the constant concentration of the substance S in the fluid flowing into compartment Y_1. Under the assumptions that the volume of each compartment remains constant, the time for material to flow from one compartment to another is negligible, and the fluid in each

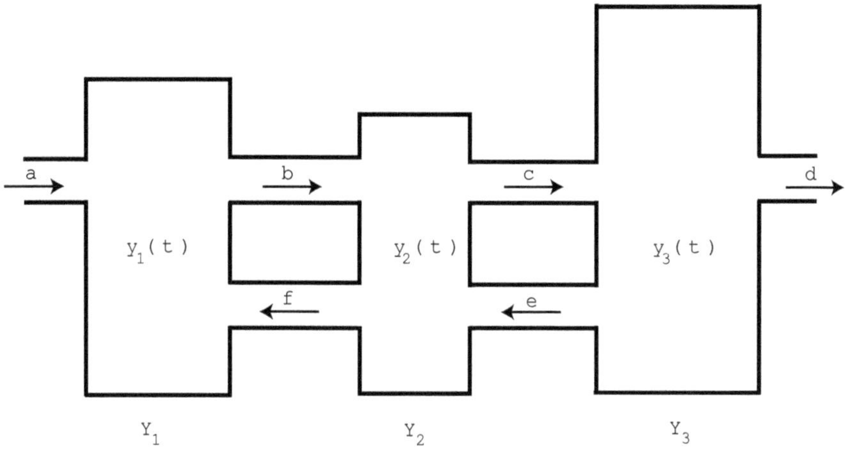

FIGURE 6.10
An Open Compartmental System.

compartment has uniform concentration, the concentrations y_i satisfy the following system of differential equations

(35a) $$y_1' = au + fy_2 - y_1 b$$

(35b) $$y_2' = by_1 + ey_3 - y_2(c+f)$$

(35c) $$y_3' = cy_2 - y_3(d+e).$$

Solving equation (35a) for y_2 and differentiating, we find

$$y_2 = (y_1' - au + by_1)/f \quad \text{and} \quad y_2' = (y_1'' + by_1')/f.$$

Substituting these expressions for y_2 and y_2' into equation (35b) and multiplying by f, we get

$$y_1'' + by_1' = bfy_1 + efy_3 - (c+f)(y_1' - au + by_1).$$

Solving this equation for y_3 and differentiating, we obtain

$$y_3 = [y_1'' + (b+c+f)y_1' + bcy_1 - (c+f)au]/(ef)$$

and

$$y_3' = [y_1^{(3)} + (b+c+f)y_1'' + bcy_1']/(ef).$$

Substituting from the above expressions for y_2, y_3, and y_3' in equation (35c), we find y_1 satisfies the single differential equation

$$[y_1^{(3)} + (b+c+f)y_1'' + bcy_1']/(ef) =$$
$$c(y_1' - au + by_1)/f - (d+e)[y_1'' + (b+c+f)y_1' + bcy_1 - (c+f)au]/(ef).$$

Multiplying by ef and rearranging, we see that y_1 satisfies the third-order, linear differential equation

$$(36)\quad y_1^{(3)} + (b+c+d+e+f)y_1'' + [b(c+d+e)+cd+f(d+e)]y_1' + bcdy_1 =$$

$$[cd+f(d+e)]au.$$

Since the volume of each compartment remains constant, the sum of the input rates per volume into a compartment must equal the sum of the output rates per volume from the compartment. So the rates per volume a, b, c, d, e, and f must also satisfy the equations

$$\begin{aligned} a+f &= b && \text{(for compartment } Y_1) \\ b+e &= c+f && \text{(for compartment } Y_2) \\ c &= d+e && \text{(for compartment } Y_3) \end{aligned}$$

Exercise 9. Find the concentration of a substance in each of the compartments of Figure 6.10 as a function of time, if $a = 1$ min^{-1}, $b = 4$ min^{-1}, $c = 7$ min^{-1}, $d = 1$ min^{-1}, $e = 6$ min^{-1}, $f = 3$ min^{-1}, $y_1(0) = .25$, $y_2(0) = .4$, and $y_3 = .7$. (Hint: Use computing software to find the roots of the auxiliary equation associated with equation (36). Write the general solution of equation (36). Then find y_2 and y_3. And finally, satisfy the initial conditions.)

Beams and Columns Beams and columns are common structural elements used in the construction of airplanes, bridges, buildings, and ships. Because of their importance in construction, beams and columns were studied extensively by ancient Greek and Roman architects. Galileo and Coulomb both made contributions to the early theory of the deflection of beams and bending of columns. However, our modern engineering theory regarding beams and columns has its origin during the 18th century in studies conducted by Euler and the Bernoullis.

An **ideal beam** is a long, slender, nonvertical rod which is supported at one or both ends and which is usually subject to external forces acting on it. The external forces may act at any point or points along the length of the beam and thereby cause a displacement of the beam from its unloaded position. The displacement of a beam from its unloaded position is described by a fourth-order ordinary differential equation.

Consider a horizontally placed beam which does not rest on an elastic foundation as shown in Figure 6.11. Let x denote the horizontal distance from the left end of the beam of length L, and let $y(x)$ denote the vertical downward deflection of the beam.

A fundamental equation of beam theory is

$$(37)\qquad \frac{y''}{[1+(y')^2]^{3/2}} = \frac{-M}{EI}$$

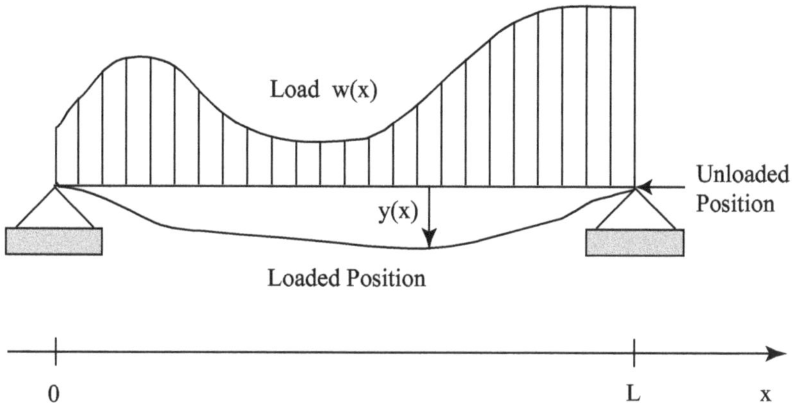

FIGURE 6.11
Deflection of a Beam.

where M is the bending moment, E is Young's modulus, and I is the moment of the area of the cross section about the neutral axis. (I is also sometimes called the moment of inertia.) Since a beam's cross section may vary, the moment of area, I, may not be constant. Two additional fundamental equations from beam theory are

$$(38) \qquad \frac{dM}{dx} = V \quad \text{and} \quad \frac{dV}{dx} = -w(x)$$

where V is the shearing force and $w(x)$ is the load on the beam. The fourth-order differential equation which results from differentiating equation (37) twice and substituting equations (38), depends upon the assumptions made regarding y' and I.

Case 1. If y' is not assumed to be small relative to 1 (that is, if the bending of the beam is not small), two differentiations of (37) yield the following nonlinear fourth-order differential equation for the deflection of the beam

(39)

$$y^{(4)}F^{-3/2} - 9y'y''y^{(3)}F^{-5/2} - 3(y'')^3F^{-5/2} + 15(y')^2(y'')^3F^{-7/2} = \frac{d^2}{dx^2}\left(\frac{-M}{EI}\right)$$

where $F = 1 + (y')^2$. The final form of the differential equation to be solved depends upon whether the moment of area of the cross section, I, is assumed to be constant or variable.

Case 2. If y' is assumed to be small compared to 1 (that is, if the bending of the beam is small), then $(y')^2$ is very small compared to 1 and may be neglected. So equation (37) becomes

$$(40) \qquad y'' = \frac{-M}{EI}.$$

Differentiating this equation twice, we obtain

$$y^{(4)} = \frac{d^2}{dx^2}\left(\frac{-M}{EI}\right).$$

a. If, in addition, I is assumed to be constant, then substituting from equations (38), we obtain the following simple linear fourth-order differential equation for the deflection of the beam

$$(41) \qquad y^{(4)} = \frac{-1}{EI}\left(\frac{d^2M}{dx^2}\right) = \frac{w(x)}{EI}.$$

Once the load, $w(x)$, and the initial or boundary conditions are specified, the deflection of the beam, $y(x)$, at any point can easily be calculated by finding the general solution of (41) and satisfying the initial or boundary conditions.

b. If I is assumed to be a variable, then one differentiation of (40) followed by substitution from (38) and (40) yields

$$(42) \qquad y^{(3)} = \frac{MI'}{EI^2} - \frac{M'}{EI} = -\frac{I'}{I}y'' - \frac{V}{EI}.$$

Differentiation of (42) followed by some algebraic rearrangement and substitution from (38) and (41), gives

$$y^{(4)} = -\frac{I'}{I}y^{(3)} + \left(\frac{I'}{I}\right)^2 y'' - \frac{I''}{I}y'' + \frac{VI'}{EI^2} - \frac{V'}{EI}$$

$$= -\frac{I'}{I}y^{(3)} - \frac{I''}{I}y'' + \frac{I'}{I}\left(\frac{I'}{I}y'' + \frac{V}{EI}\right) - \frac{V'}{EI}$$

$$= -\frac{2I'}{I}y^{(3)} - \frac{I''}{I}y'' + \frac{w}{EI}.$$

So if we assume y' is small compared to 1 and I is a variable, we obtain the following linear fourth-order differential equation for the deflection of the beam

$$(43) \qquad y^{(4)} + \frac{2I'}{I}y^{(3)} + \frac{I''}{I}y'' = \frac{w}{EI}.$$

To solve this differential equation and find the deflection of the beam, we need to know the load, $w(x)$, the moment of area of the cross section, $I(x)$, and the initial or boundary conditions. Unless $I(x)$ is an exponential function, equation (43) will be a linear differential equation with variable coefficients. The solution of the linear differential equations with variable coefficients is discussed in Chapter 7. Let us suppose $I(x)$ is an exponential function, say,

$I(x) = be^{ax}$. Then $I' = abe^{ax}$, $I'' = a^2be^{ax}$ and upon substitution (43) reduces to

$$y^{(4)} + 2ay^{(3)} + a^2y^{(2)} = \frac{w(x)}{Ebe^{ax}}. \tag{44}$$

If a beam is **simply supported** at $x = 0$, then $y(0) = 0$ and $y''(0) = 0$. The condition $y(0) = 0$ means the end of the beam is fixed—that is, it cannot move vertically. The condition $y''(0) = 0$ means the beam can rotate about $x = 0$ in the plane of deformation. A simply supported end of a beam is sometimes called a **pinned end**—see Figure 6.12a. Of course, if a beam is simply supported at $x = L$, then $y(L) = 0$ and $y''(L) = 0$.

If a beam is **clamped** at $x = 0$, then $y(0) = 0$ and $y'(0) = 0$. The condition $y'(0) = 0$ means at $x = 0$ the slope of the beam is zero—that is, at $x = 0$ the beam is horizontal. A clamped beam is often a built-in beam as shown in Figure 6.12b. If a beam is built-in at $x = L$, then $y(L) = 0$ and $y'(L) = 0$.

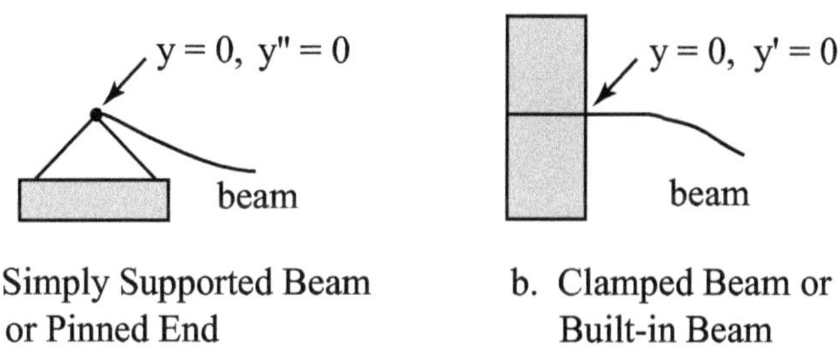

FIGURE 6.12
Types of Beam Support at an End.

Exercise 10. Find the equation for the deflection of a horizontal beam of length L, assuming y' is small, and I is constant (that is, solve equation (41)) under the following conditions:

a. Both ends are simply supported and the load is

    i. uniform, $w(x) = w_0$, where w_0 is constant
    ii. $w(x) = w_0 \sin(\pi x/L)$
    iii. $w(x) = w_0 \cos(\pi x/L)$

b. The end at $x = 0$ is simply supported, the end at $x = L$ is clamped, and the load is

    i. $w(x) = w_0$ ii. $w(x) = w_0 \sin(\pi x/L)$ iii. $w(x) = w_0 \cos(\pi x/L)$

c. Both ends are clamped and the load is

    i. $w(x) = w_0$ ii. $w(x) = w_0 \sin(\pi x/L)$ iii. $w(x) = w_0 \cos(\pi x/L)$

Exercise 11. Find the equation for the deflection of a horizontal beam of length $L = 120$ inches, assuming y' is small, $I(x) = 400e^{-x/L}$ in^4, $E = 15 \times 10^6$ lbs/in^2, and the load is uniform, $w(x) = w_0$ where w_0 is a constant (that is, solve equation (44)). (Hint: Use computing software to find the complementary solution of equation (44).)

A **cantilevered beam** is one which is clamped at one end and completely free at the other end. Suppose a cantilevered beam is clamped at $x = 0$ and free at $x = L$. Then two conditions which must be satisfied by the equation for the deflection of the beam are $y(0) = 0$ and $y'(0) = 0$. Suppose further that y' is assumed to be small relative to 1. Then from equations (40) and (42)—regardless of whether I is constant or variable, we obtain the following two additional initial conditions

$$y''(0) = \frac{-M(0)}{EI(0)} \tag{45}$$

and

$$y^{(3)}(0) = -\frac{I'(0)}{I(0)}y''(0) - \frac{M'(0)}{EI(0)}. \tag{46}$$

The bending moments $M(x)$ for various loads $w(x)$ on a cantilevered beam of length L are given in Figure 6.13.

Exercise 12. For each of the loads A, B, C, D, and E shown in Figure 6.13, find the equation for the deflection for a horizontal cantilevered beam of length $L = 240$ inches, assuming y' is small, $I(x) = 500e^{-x/L}$ in^4, $P = 500$ lbs, $d = 50$ in, $w_0 = 60$ lbs/in, and $E = 25 \times 10^6$ lbs/in^2. (Hint: Solve equation (44) subject to the initial conditions $y(0) = 0$, $y'(0) = 0$, and equations (45) and (46).)

Exercise 13. If y' is small and the moment of area of the cross section, I, is constant, then the **deflection of a horizontal beam resting on an elastic foundation** satisfies the differential equation

$$EIy^{(4)} + k^2 y = w(x) \tag{47}$$

where E is Young's modulus, k is the spring constant of the elastic foundation, and $w(x)$ is the load on the beam. A simple example of such a beam is a single rail of a railroad track. Find the deflection of a horizontal beam resting on an elastic foundation if the beam is pinned at $x = -L$ and $x = L$, if $k^2/EI = .09$, and if the load is $w(x) = w_0 \cos(\pi x/L)$ where w_0 is a constant. (Hint: Use your computing software to find the complementary solution of equation (47).)

An **ideal column** is a long, slender elastic rod which is held vertically in place by a support at the base and often by an additional support at the top. Suppose a weight is placed on the top of the column, thus inducing a force P on both ends of the column. If P is sufficiently small, then the column will

	Load Diagram	**Load, w(x)**	**Bending Moment, M(x)**
A.	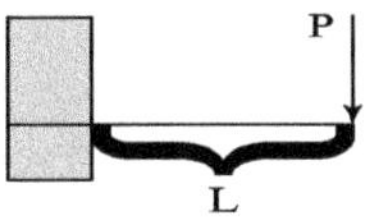	$\begin{cases} 0, & x \neq L \\ P, & x = L \end{cases}$	$-P(L-x)$
B.	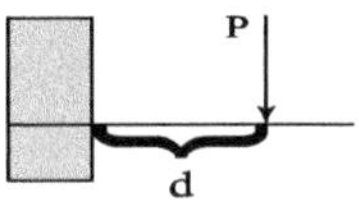	$\begin{cases} 0, & x \neq d \\ P, & x = d \end{cases}$	$\begin{cases} -P(d-x), & 0 \leq x \leq d \\ 0, & d < x \leq L \end{cases}$
C.	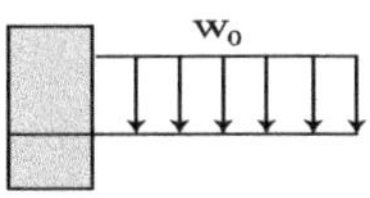	w_0, a constant	$-w_0[L(L-x)+\frac{x^2}{2}]$
D.	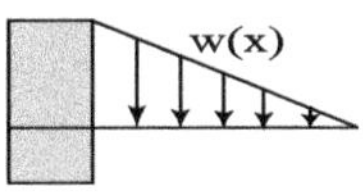	$w_0(1-\frac{x}{L})$	$-w_0(\frac{L^2}{6}-\frac{Lx}{2}+\frac{x^2}{2}-\frac{x^3}{6L})$
E.	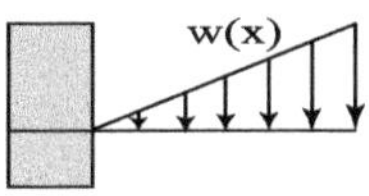	$\frac{w_0 x}{L}$	$-w_0(\frac{L^2}{3}-\frac{Lx}{2}+\frac{x^3}{6L})$

FIGURE 6.13
Bending Moments for Various Loads on a Cantilevered Beam of Length L.

support the weight and the column will deflect only slightly from the vertical. However, if P is sufficiently large, then the column cannot support the weight and will buckle—that is, the column will suddenly bow out from the vertical with large amplitude.

Consider the column of length L shown in Figure 6.14. Assume the column is constrained to move in the plane. Let the origin be located at the base of the column with the positive x-axis directed vertically upward and the positive y-axis directed to the right. So $y(x)$ represents the lateral displacement of the column. In the 18th century, Euler showed that a column carrying a load P satisfies the fourth-order linear differential equation

$$y^{(4)} + \frac{Py^{(2)}}{EI} = 0 \tag{48}$$

where E is Young's modulus and I is the moment of the area of the column's cross section. (Here I is assumed to be a positive constant.) Notice that regardless of the boundary conditions, $y(x) = 0$ is a solution of equation (48). The smallest value for the load P which will cause buckling is called the **critical load** or the **Euler load**.

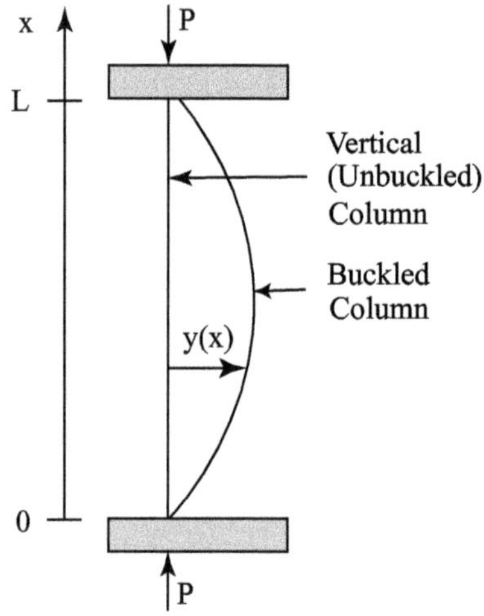

FIGURE 6.14
Buckling of a Column.

Example 1 Find the critical load (Euler load) for a column which is clamped at the base and pinned at the top.

Solution

Let $\lambda^2 = P/EI$. Then the auxiliary equation associated with equation (48) is $r^4 + \lambda^2 r^2 = 0$. The roots of this equation are $r = 0,\ 0,\ \lambda i$, and $-\lambda i$. So the general solution of (48) is

$$y(x) = A + Bx + C\sin\lambda x + D\cos\lambda x.$$

Differentiating twice, we find

$$y'(x) = B + \lambda C\cos\lambda x - \lambda D\sin\lambda x \quad \text{and} \quad y''(x) = -\lambda^2 C\sin\lambda x - \lambda^2 D\cos\lambda x.$$

Since the column is clamped at the base, two boundary conditions are $y(0) = 0$ and $y'(0) = 0$. And since the column is pinned at the top, two additional boundary conditions are $y(L) = 0$ and $y''(L) = 0$. In order to satisfy the boundary conditions, the constants A, B, C, and D must be chosen to simultaneously satisfy the following four equations.

$$\begin{aligned}
A + D &= 0 &\quad& \text{(from the condition } y(0) = 0)\\
B + \lambda C &= 0 && \text{(from the condition } y'(0) = 0)\\
A + BL + C\sin\lambda L + D\cos\lambda L &= 0 && \text{(from the condition } y(L) = 0)\\
-\lambda^2 C\sin\lambda L - \lambda^2 D\cos\lambda L &= 0 && \text{(from the condition } y''(L) = 0)
\end{aligned}$$

Written in matrix-vector notation, this system becomes

$$\begin{pmatrix} 1 & 0 & 0 & 1 \\ 0 & 1 & \lambda & 0 \\ 1 & L & \sin\lambda L & \cos\lambda L \\ 0 & 0 & -\lambda^2\sin\lambda L & -\lambda^2\cos\lambda L \end{pmatrix} \begin{pmatrix} A \\ B \\ C \\ D \end{pmatrix} = \begin{pmatrix} 0 \\ 0 \\ 0 \\ 0 \end{pmatrix}.$$

(If you are unfamiliar with matrix-vector notation, see Section 8.1.) Or symbolically, $M\mathbf{u} = \mathbf{0}$, where M denotes the 4×4 matrix; $\mathbf{u}$ denotes the column vector with entries A, B, C, D, and $\mathbf{0}$ is the vector of zeroes. Notice that $\mathbf{u} = \mathbf{0}$ is always a solution of $M\mathbf{u} = \mathbf{0}$. The solution $\mathbf{u} = \mathbf{0}$ is called the **trivial solution** and means $A = B = C = D = 0$. So corresponding to the trivial solution, $\mathbf{u} = \mathbf{0}$, of $M\mathbf{u} = \mathbf{0}$ is the solution $y(x) = 0$ of equation (48). The equation $M\mathbf{u} = \mathbf{0}$ has a **nontrivial solution** (a nonzero solution) if and only if the determinant of the matrix M (*det* M) is zero. For this example, the condition *det* $M = 0$ is the condition

$$-\lambda^2(-\lambda L\cos\lambda L + \sin\lambda L) = 0.$$

Since $\lambda^2 = P/EI \neq 0$, for *det* M to be zero λ must satisfy

$$-\lambda L\cos\lambda L + \sin\lambda L = 0.$$

Letting $z = \lambda L$ and dividing by $\cos\lambda L = \cos z$, we see z must satisfy

$$f(z) = -z + \tan z = 0.$$

There are an infinite number of positive solutions of the equation $f(z) = 0$. To verify this fact, graph $w = z$ and $w = \tan z$ and notice that the graphs intersect exactly once in each of the intervals $((2n-1)\pi/2, (2n+1)\pi/2)$ for $n = 1, 2, 3, \ldots$. So the smallest positive root of $f(z)$ lies in the interval $(\pi/2, 3\pi/2)$. Using Newton's method, we find the smallest positive root, accurate to six decimal places, to be $z = \lambda L = 4.493409$. Since $\lambda^2 = P/EI$, the critical load (Euler load) for a column which is clamped at the base and pinned at the top is

$$P_{cr} = EI\lambda^2 = EI(4.493409/L)^2 = 20.190724EI/L^2.$$

Observe that a shorter column can support a larger load without buckling. ■

Exercise 14. Find the critical load (Euler load) for a column of length L supported in the following ways.

a. Base and top pinned.

b. Base and top clamped.

c. Base clamped and top free to move vertically. (The boundary conditions at the top are $y(L) = \delta$, where $\delta \neq 0$ and $y''(L) = 0$.)

7

Systems of First-Order Differential Equations

In this chapter we shall attempt to answer the following questions:

"What is a system of first-order differential equations?"

"What is a linear system of first-order differential equations?"

"What is a solution of a system of first-order differential equations?"

"What is an initial value problem for a system of first-order differential equations?"

"Under what conditions does a solution to a system initial value problem exist and under what conditions is the solution unique?"

"Where—that is, on what interval or what region—does the solution to a system initial value problem exist and where is the solution unique?"

"How can an n-th order differential equation be rewritten as an equivalent system of first-order differential equations?"

7.1 Properties of Systems of Differential Equations

A **system of n first-order differential equations** has the form

$$
\begin{aligned}
y_1' &= f_1(x, y_1, y_2, \ldots, y_n) \\
y_2' &= f_2(x, y_1, y_2, \ldots, y_n) \\
\vdots\ & \quad \vdots \qquad\qquad \vdots \\
y_n' &= f_n(x, y_1, y_2, \ldots, y_n)
\end{aligned} \tag{1}
$$

where each dependent variable y_i is a real-valued function of the independent variable x and each f_i is a real-valued function of x, y_1, y_2, $\ldots$, y_n.

Systems of differential equations of this type arise naturally when there is one independent variable, such as time, and several dependent variables, such as position and velocity in multidimensional space.

In Chapter 3, we derived a system of two, first-order differential equations

DOI: 10.1201/9781003678014-7

for the quantities, $q_1(t)$ and $q_2(t)$, of dye in two tanks as a function of time t. The system was

$$\text{(2)} \qquad \begin{aligned} \frac{dq_1}{dt} &= .5 - \frac{q_1}{30} \\ \frac{dq_2}{dt} &= \frac{q_1}{30} - \frac{q_2}{15}. \end{aligned}$$

In this system, q_1 and q_2 are the dependent variables and t is the independent variable. Recall we were able to solve this system by solving the first equation explicitly for q_1, substituting this result into the second equation and then solving the resulting differential equation for q_2.

In Chapter 6, we solved two coupled spring-mass systems, a double pendulum system, a system for the path of an electron, and systems resulting from compartmental analysis by rewriting each system as a single higher order differential equation. The first coupled spring-mass system was the following system of two, second-order differential equations

$$\text{(3)} \qquad \begin{aligned} m_1 y_1'' &= -\,k_1 y_1 + k_2(y_2 - y_1) \\ m_2 y_2'' &= -\,k_2(y_2 - y_1). \end{aligned}$$

Here m_1 and m_2 are the masses attached to the springs with spring constants k_1 and k_2 respectively and y_1 and y_2 are the displacements of the masses from equilibrium. (See Figure 6.7.) We can rewrite this system as a system of first-order differential equations in the following manner. Let $u_1 = y_1$, $u_2 = y_1'$, $u_3 = y_2$, and $u_4 = y_2'$. So u_1 is the position of the first mass and u_2 is its velocity. While u_3 is the position of the second mass and u_4 is its velocity. Differentiating $u_1 = y_1$, $u_2 = y_1'$, $u_3 = y_2$, and $u_4 = y_2'$ and then substituting for y_1, y_1', y_2, and y_2' in terms of u_1, u_2, u_3, and u_4, we find

$$\begin{aligned} u_1' &= y_1' = u_2 \\ u_2' &= y_1'' = \frac{-k_1 y_1 + k_2(y_2 - y_1)}{m_1} = \frac{-k_1 u_1 + k_2(u_3 - u_1)}{m_1} \\ u_3' &= y_2' = u_4 \\ u_4' &= y_2'' = \frac{-k_2(y_2 - y_1)}{m_2} = \frac{-k_2(u_3 - u_1)}{m_2}. \end{aligned}$$

Thus, the system of two, second-order differential equations (3) is equivalent to the following system of four, first-order differential equations

$$\text{(4)} \qquad \begin{aligned} u_1' &= u_2 \\ u_2' &= \frac{-k_1 u_1 + k_2(u_3 - u_1)}{m_1} \\ u_3' &= u_4 \\ u_4' &= \frac{-k_2(u_3 - u_1)}{m_2}. \end{aligned}$$

In this instance, the independent variable is time, t, the dependent variables are u_1, u_2, u_3, and u_4 and for $i = 1, 2, 3, 4$ the functions $f_i(t, u_1, u_2, u_3, u_4)$ are

$$\begin{aligned} f_1(t, u_1, u_2, u_3, u_4) &= u_2 \\ f_2(t, u_1, u_2, u_3, u_4) &= \frac{-k_1 u_1 + k_2(u_3 - u_1)}{m_1} \\ f_3(t, u_1, u_2, u_3, u_4) &= u_4 \\ f_4(t, u_1, u_2, u_3, u_4) &= \frac{-k_2(u_3 - u_1)}{m_2}. \end{aligned}$$

In Chapter 10, we will show how to solve system (4) numerically, given the initial positions and velocities of the two masses.

The first-order systems (2) and (4) are examples of linear systems. Specifically a linear system of first-order equations is one in which each function f_i is linear in the dependent variables. Stated more precisely:

A **linear system of n first-order differential equations** has the general form

(5)

$$\begin{aligned} y_1' &= a_{11}(x)y_1 + a_{12}(x)y_2 + \cdots + a_{1n}(x)y_n + b_1(x) &= f_1(x, y_1, y_2, \ldots, y_n) \\ y_2' &= a_{21}(x)y_1 + a_{22}(x)y_2 + \cdots + a_{2n}(x)y_n + b_2(x) &= f_2(x, y_1, y_2, \ldots, y_n) \\ \vdots \quad &\vdots \qquad\qquad \vdots \qquad\qquad \vdots & \vdots \\ y_n' &= a_{n1}(x)y_1 + a_{n2}(x)y_2 + \cdots + a_{nn}(x)y_n + b_n(x) &= f_n(x, y_1, y_2, \ldots, y_n) \end{aligned}$$

where $a_{ij}(x)$ and $b_i(x)$ are all known, real-valued functions of the independent variable x.

In system (2), $a_{11}(x) = -1/30$, $a_{12}(x) = 0$, $b_1(x) = .5$, $a_{21}(x) = 1/30$, $a_{22}(x) = -1/15$, and $b_2(x) = 0$. In system (4), $a_{11}(x) = a_{13}(x) = a_{14}(x) = b_1(x) = 0$ and $a_{12}(x) = 1$; $a_{21}(x) = -(k_1 + k_2)/m_1$, $a_{22}(x) = a_{24}(x) = b_2(x) = 0$, and $a_{23}(x) = k_2/m_1$; $a_{31}(x) = a_{32}(x) = a_{33}(x) = b_3(x) = 0$ and $a_{34}(x) = 1$; and $a_{41}(x) = k_2/m_2$, $a_{42}(x) = a_{44}(x) = b_4(x) = 0$, and $a_{43}(x) = -k_2/m_2$. Since all $a_{ij}(x)$ and all $b_i(x)$ in both systems (2) and (4) are constant, as opposed to variable, these systems are referred to as linear systems of differential equations with constant coefficients. The following system is a simple example of a linear system with variable coefficients (a linear system in which not all of the coefficient functions are constant).

(6)
$$\begin{aligned} y_1' &= x^2 y_1 - e^x y_2 + 25 \\ y_2' &= -3y_1 + (x + 1/x)y_2 + \sin x. \end{aligned}$$

Here $a_{11}(x) = x^2$, $a_{12}(x) = -e^x$, $b_1(x) = 25$, $a_{21}(x) = -3$, $a_{22}(x) = x + 1/x$, and $b_2(x) = \sin x$.

In the linear system of n first-order differential equations (5):

If all of the functions $b_i(x)$ are identically zero, then the linear system (5) is said to be **homogeneous**.

If any $b_i(x)$ is not identically equal to zero, then the linear system (5) is said to be **nonhomogeneous**.

All systems of first-order differential equations not having the form of system (5) are called **nonlinear systems**.

As an example of a nonlinear system of differential equations, we introduce the Volterra prey-predator model, one of the fundamental models of mathematical ecology. Let $y_1(x)$ represent the prey population at time x, and let $y_2(x)$ represent the predator population at time x. In the absence of predators, the population y_1 is assumed to grow according to the Malthusian population model: $dy_1/dx = ay_1$ where $a > 0$ is the growth rate. The loss of population due to predation is assumed to be proportional to the number of encounters between prey and predator—that is, proportional to the product y_1y_2. Thus, the rate of change of the prey population becomes $dy_1/dx = ay_1 - by_1y_2$ where $b > 0$ is a constant which represents the proportion of encounters between the prey and predators which result in death to the prey. Without the prey to feed upon, it is assumed that the predators would die off according to the Malthusian population model: $dy_2/dx = -cy_2$ where $c > 0$ is the death rate. Due to predation, the predator population is assumed to grow at a rate which is proportional to the number of encounters between prey and predator. Hence, the rate of change in predator population becomes $dy_2/dx = -cy_2 + dy_1y_2$ where $d > 0$ is a constant which represents the proportion of encounters between prey and predator which is beneficial to the predator. Therefore, the Volterra prey-predator model is

$$\text{(7)} \qquad \begin{aligned} \frac{dy_1}{dx} &= \quad ay_1 - by_1y_2 \\ \frac{dy_2}{dx} &= -\,cy_2 + dy_1y_2 \end{aligned}$$

where $a, b, c,$ and d are positive constants. The system (7) is nonlinear due to the terms $-by_1y_2$ and dy_1y_2.

A system of n first-order differential equations

$$\text{(1)} \qquad \begin{aligned} y_1' &= f_1(x, y_1, y_2, \ldots, y_n) \\ y_2' &= f_2(x, y_1, y_2, \ldots, y_n) \\ \vdots\ &\ \ \vdots \qquad\qquad \vdots \\ y_n' &= f_n(x, y_1, y_2, \ldots, y_n) \end{aligned}$$

has a **solution** on the interval I, if there exists a set of n functions $\{y_1(x), y_2(x), \ldots, y_n(x)\}$ which all have continuous first derivatives on the interval

I and which satisfy (1) on I. The set of functions $\{y_1(x), y_2(x), \ldots, y_n(x)\}$ is called a **solution of system (1) on the interval I**.

For example, the set of two functions $\{y_1(x) = \sin x,\ \ y_2(x) = \cos x\}$ is a solution of the system of two, first-order, linear differential equations

$$\text{(8)} \qquad \begin{aligned} y_1' &= \quad y_2 = f_1(x, y_1, y_2) \\ y_2' &= -\,y_1 = f_2(x, y_1, y_2) \end{aligned}$$

on the interval $(-\infty, \infty)$. (You should verify this fact by showing that the derivatives of $y_1(x) = \sin x$ and $y_2(x) = \cos x$ are both continuous on the interval $I = (-\infty, \infty)$ and that $y_1'(x) = y_2(x)$ and $y_2'(x) = -y_1(x)$ for all $x \in I$.) The set of two functions $\{z_1(x) = A \sin x + B \cos x,\ \ z_2(x) = A \cos x - B \sin x\}$, where A and B are arbitrary constants, is also a solution to the system (8) on the interval $(-\infty, \infty)$. Verify this fact. The solution $\{z_1(x), z_2(x)\}$ of the system (8) is the general solution of (8). In Section 8.3, we will show how to find the general solution of linear systems of first-order equations with constant coefficients. (System (8) is an example of a system of this type.) And in Chapter 9, we will solve several applications which involve linear systems with constant coefficients.

An **initial value problem** for a system of first-order differential equations consists of solving a system of equations of the form (1) subject to a set of constraints, called **initial conditions** (IC), of the form $y_1(c) = d_1$, $y_2(c) = d_2, \ldots,\ y_n(c) = d_n$.

For example, the problem of finding a solution to the system

$$\text{(8)} \qquad \begin{aligned} y_1' &= \quad y_2 = f_1(x, y_1, y_2) \\ y_2' &= -\,y_1 = f_2(x, y_1, y_2) \end{aligned}$$

subject to the initial conditions

$$\text{(9)} \qquad y_1(0) = 2, \quad y_2(0) = 3$$

is an initial value problem. The solution of the initial value problem (8)-(9) on the interval $(-\infty, \infty)$ is

$$\{y_1(x) = 3 \sin x + 2 \cos x,\quad y_2(x) = 3 \cos x - 2 \sin x\}.$$

Verify this fact by showing that $y_1(x)$ and $y_2(x)$ satisfy the system (8) on the interval $(-\infty, \infty)$ and the initial conditions (9).

Notice that in an initial value problem, all n conditions to be satisfied are specified at a single value of the independent variable—the value we have called c. The problem of solving the system (8) subject to the constraints

$$\text{(10)} \qquad y_1(0) = 2, \quad y_2(\pi) = 3$$

is an example of a boundary value problem. Observe that the constraints for y_1 and y_2 are specified at two different values of the independent variable—namely, 0 and π. Verify that the set

$$\{y_1(x) = -3\sin x + 2\cos x, \quad y_2(x) = -3\cos x - 2\sin x\}$$

is a solution of the boundary value problem (8)-(10) on the interval $(-\infty, \infty)$.

The following **Fundamental Existence and Uniqueness Theorem for System Initial Value Problems** is analogous to the fundamental existence and uniqueness theorem for the scalar initial value problem: $y' = f(x, y)$; $y(c) = d$. In this theorem, R denotes a generalized rectangle in $xy_1y_2 \cdots y_n$-space.

Theorem 7.1 Let $R = \{(x, y_1, y_2, \ldots, y_n) \mid \alpha < x < \beta \text{ and } \gamma_i < y_i < \delta_i\}$ where α, β, γ_i and δ_i are all finite real constants. If each of the n functions $f_i(x, y_1, y_2, \ldots, y_n)$, $i = 1, 2, \ldots, n$ is a continuous function of x, y_1, y_2, $\ldots$, and y_n in R, if each of the n^2 partial derivatives $\partial f_i/\partial y_j$, $i, j = 1, 2, \ldots, n$ is a continuous function of x, y_1, $y_2, \ldots$, and y_n in R, and if $(c, d_1, d_2, \ldots, d_n) \in R$, then there exists a unique solution to the system initial value problem

$$\begin{aligned} y_1' &= f_1(x, y_1, y_2, \ldots, y_n) \\ y_2' &= f_2(x, y_1, y_2, \ldots, y_n) \\ \vdots \;\; & \;\; \vdots \qquad\qquad \vdots \\ y_n' &= f_n(x, y_1, y_2, \ldots, y_n) \end{aligned} \tag{11a}$$

$$y_1(c) = d_1, \; y_2(c) = d_2, \ldots, \; y_n(c) = d_n \tag{11b}$$

on some interval $I = (c - h, c + h)$ where I is a subinterval of (α, β).

The hypotheses of the fundamental existence and uniqueness theorem are sufficient conditions and guarantee the existence of a unique solution to the initial value problem on some interval of length $2h$. An expression for calculating a value for h is not specified by the theorem and so the interval of existence and uniqueness of the solution may be very small or very large. Furthermore, the conditions stated in the hypotheses are not necessary conditions. Therefore, if some condition stated in the hypotheses is not fulfilled (perhaps $\partial f_i/\partial y_j$ is not continuous for a single value of i and j and in only one particular variable), we **cannot** conclude a solution to the initial value problem does not exist or is not unique. It might exist and be unique. It might exist and not be unique. Or, it might not exist. We essentially have no information with respect to solving the initial value problem, if the hypotheses of the fundamental theorem are not satisfied, and little information concerning the interval of existence and uniqueness, if the hypotheses are satisfied. Finally, observe that the hypotheses of the theorem do not include any requirements on the functions $\partial f_1/\partial x, \partial f_2/\partial x, \ldots, \partial f_n/\partial x$.

Example 1 Analyze the following initial value problem.

(12a)
$$y_1' = \frac{y_1 y_2}{2}$$
$$y_2' = \frac{8}{y_1}$$

(12b)
$$y_1(1) = -2, \quad y_2(1) = 4.$$

Solution

In this example, $f_1(x, y_1, y_2) = y_1 y_2/2$ and $f_2(x, y_1, y_2) = 8/y_1$. Hence, $\partial f_1/\partial y_1 = y_2/2$, $\partial f_1/\partial y_2 = y_1/2$, $\partial f_2/\partial y_1 = -8/y_1^2$, and $\partial f_2/\partial y_2 = 0$. The functions f_1, f_2, $\partial f_1/\partial y_1$, $\partial f_1/\partial y_2$, $\partial f_2/\partial y_1$, and $\partial f_2/\partial y_2$ are all defined and continuous in any region of xy_1y_2-space which does not include the plane $y_1 = 0$—that is, which does not include any point of the form $(x, 0, y_2)$. Since the initial condition for y_1 is $y_1(1) = -2 < 0$, we let

$$R_1 = \{(x, y_1, y_2) \mid -A < x < B, \ -C < y_1 < -\epsilon < 0, \text{ and } -D < y_2 < E\}$$

where A, B, C, D, and E are positive constants which are as large as we choose and ϵ is a positive constant which is as small as we choose. Since f_1, f_2, $\partial f_1/\partial y_1$, $\partial f_1/\partial y_2$, $\partial f_2/\partial y_1$, and $\partial f_2/\partial y_2$ are all defined and continuous on R_1 and since $(1, -2, 4) \in R_1$, there exists a unique solution to the initial value problem (12) on some interval $(1 - h, 1 + h)$. ■

The fundamental existence and uniqueness theorem is a "local" theorem because the solution of the initial value problem is guaranteed to exist and be unique only on a "small" interval. The following theorem, which we state without proof, is called a **Continuation Theorem for System Initial Value Problems** and tells us how far we can extend (continue) the unique solution of an initial value problem, assuming the hypotheses of the fundamental theorem are satisfied.

Theorem 7.2 Under the hypotheses of the fundamental existence and uniqueness theorem, the solution of the initial value problem (11) can be extended until the boundary of R is reached.

The generalized rectangle R mentioned in the fundamental theorem can be enlarged in all directions until one of the $n+n^2$ functions f_i, $i = 1, 2, \ldots, n$ or $\partial f_i/\partial y_j$, $i, j = 1, 2, \ldots, n$ is not defined or not continuous on a bounding "side" of R or until the bounding "side" of R approaches infinity. Thus, the continuation theorem guarantees the existence of a unique solution to the initial value problem (11) which extends from one bounding "side" of R through the initial point $(c, d_1, d_2, \ldots, d_n)$, which is in R, to another bounding "side" of R. The two bounding "sides" may be the same "side."

According to the continuation theorem, the solution to the IVP (12) through $(1, -2, 4)$ can be extended uniquely until it reaches a boundary of R_1. Thus, the solution can be extended until at least two of the following six things occur: $x \to -A$, $x \to B$, $y_1 \to -C$, $y_1 \to -\epsilon < 0$, $y_2 \to -D$, or $y_2 \to E$. Enlarging R_1 by letting A, B, C, D, and E approach $+\infty$ and letting $\epsilon \to 0^+$ (the hypotheses of the fundamental theorem and continuation theorem regarding f_i and $\partial f_i/\partial y_j$ are still valid on the enlarged generalized rectangle), we find the conditions become $x \to -\infty$, $x \to \infty$, $y_1 \to -\infty$, $y_1 \to 0^-$, $y_2 \to -\infty$, or $y_2 \to \infty$. The solution of the IVP (12) is

$$\{y_1(x) = -2x^2, \quad y_2(x) = \frac{4}{x}\}. \tag{13}$$

(Verify this fact.) The functions $y_1(x)$ and $y_2(x)$ and their derivatives are simultaneously defined and continuous on the intervals $(-\infty, 0)$ and $(0, \infty)$. Since $c = 1 \in (0, \infty)$, the set $\{y_1(x), y_2(x)\}$ is a solution to (12) on $(0, \infty)$. From the solution, we see that the conditions which limit the interval of existence and uniqueness in this case are $x \to \infty$, $y_1 \to 0^-$, and $y_2 \to \infty$—the latter two occur simultaneously as $x \to 0^+$.

Linear system initial value problems—system initial value problems in which each function $f_i(x, y_1, y_2, \ldots, y_n)$, $i = 1, 2, \ldots, n$ in the system is a linear function of $y_1, y_2, \ldots, y_n$—are an important special class of initial value problems. We state a separate existence and uniqueness theorem for linear system initial value problems due to the special results that can be obtained for these problems regarding the interval of existence and uniqueness of the solution.

A **linear system initial value problem** consists of solving the system of n first-order equations

(14a)

$$\begin{aligned}
y_1' &= a_{11}(x)y_1 + a_{12}(x)y_2 + \cdots + a_{1n}(x)y_n + b_1(x) &&= f_1(x, y_1, y_2, \ldots, y_n) \\
y_2' &= a_{21}(x)y_1 + a_{22}(x)y_2 + \cdots + a_{2n}(x)y_n + b_2(x) &&= f_2(x, y_1, y_2, \ldots, y_n) \\
\vdots\ & \ \vdots && \vdots \\
y_n' &= a_{n1}(x)y_1 + a_{n2}(x)y_2 + \cdots + a_{nn}(x)y_n + b_n(x) &&= f_n(x, y_1, y_2, \ldots, y_n)
\end{aligned}$$

subject to the n initial conditions

$$y_1(c) = d_1, \; y_2(c) = d_2, \ldots, \; y_n(c) = d_n. \tag{14b}$$

Calculating the partial derivatives of f_i with respect to y_j in the linear system (14a), we find $\partial f_i/\partial y_j = a_{ij}(x)$. If the $n^2 + n$ functions $a_{ij}(x)$, $i, j = 1, 2, \ldots, n$ and $b_i(x)$, $i = 1, 2, \ldots, n$ are all defined and continuous on some interval $I = (\alpha, \beta)$ which contains c, then all of the functions f_i and $\partial f_i/\partial y_j$

will be defined and continuous functions of $x, y_1, y_2, \ldots, y_n$ on the generalized rectangle

$$R = \{(x, y_1, y_2, \ldots, y_n) \mid \alpha < x < \beta \text{ and } -\infty < y_i < \infty \text{ for } i = 1, 2, \ldots, n\}.$$

Therefore, by the fundamental existence and uniqueness theorem and by the continuation theorem for system initial value problems, there exists a unique solution to the initial value problem (14) on the interval $I = (\alpha, \beta)$. Hence, we have the following **Fundamental Existence and Uniqueness Theorem for Linear System Initial Value Problems.**

If the functions $a_{ij}(x)$, $i, j = 1, 2, \ldots, n$ and $b_i(x)$, $i = 1, 2, \ldots, n$ are all defined and continuous on the interval $I = (\alpha, \beta)$ and if $c \in I$, then there exists a unique solution to the linear system initial value problem (14) on the interval I.

Notice that this theorem guarantees the existence of a unique solution on the entire interval I on which the functions $a_{ij}(x)$ and $b_i(x)$ are all defined and continuous. This is the major difference between the results which one can obtain for linear system initial value problems versus nonlinear system initial value problems. For linear system initial value problems, we can explicitly determine the interval of existence and uniqueness of the solution from the problem itself; whereas, for the nonlinear system initial value problem, we can only conclude that there exists a unique solution on some "small" undetermined interval centered at $x = c$. The following example illustrates the type of result we can obtain for linear system initial value problems.

Example 2 Analyze the linear initial value problem:

(15a)
$$\begin{aligned} y_1' &= xy_1 + \sqrt{x}\, y_2 + \frac{2}{x-1} &&= f_1(x, y_1, y_2) \\ y_2' &= (\tan x) y_1 - x^2 y_2 + \frac{3}{x^2+1} &&= f_2(x, y_1, y_2) \end{aligned}$$

(15b)
$$y_1(1.5) = -1, \qquad y_2(1.5) = 2.$$

Solution

The system (15a) is linear with $a_{11}(x) = x$, $a_{12}(x) = \sqrt{x}$, $b_1(x) = 2/(x-1)$, $a_{21}(x) = \tan x$, $a_{22}(x) = -x^2$, and $b_2(x) = 3/(x^2+1)$. The functions $a_{11}(x)$, $a_{22}(x)$, and $b_2(x)$ are defined and continuous for all real x—that is, on the interval $J_1 = (-\infty, \infty)$. The function $a_{12}(x) = \sqrt{x}$ is defined and continuous for $x \geq 0$—that is, on $J_2 = [0, \infty)$. So the interval of existence and uniqueness I must be a subinterval of $J_1 \cap J_2 = [0, \infty)$. The function $b_1(x)$ is defined and continuous for $x \neq 1$. So $b_1(x)$ is defined and continuous on $J_3 = (-\infty, 1)$ and $J_4 = (1, \infty)$. Since $J_1 \cap J_2 \cap J_3 = [0, 1)$ and $J_1 \cap J_2 \cap J_4 = (1, \infty)$, the interval I will be a subinterval of $[0, 1)$ or $(1, \infty)$. Since the initial condition is specified at $c = 1.5 \in (1, \infty)$, I must be a subinterval of $(1, \infty)$. The function $a_{21}(x) = \tan x$ is defined and continuous for $x \neq (2n+1)\pi/2$ where

n is any integer. So $a_{21}(x)$ is defined and continuous on the intervals $K_n = ((2n-1)\pi/2,\ (2n+1)\pi/2)$. Since the initial condition is specified at $c = 1.5$ and since $1.5 \in (-\pi/2,\ \pi/2)$, the linear IVP (15) has a unique solution on the interval $(-\pi/2, \pi/2) \cap (1, \infty) = (1, \pi/2)$. The interval $(1, \pi/2)$ is the largest interval containing $c = 1.5$ on which the functions $a_{ij}(x),\ i, j = 1, 2$ and $b_i(x),\ i = 1, 2$ are simultaneously defined and continuous.

If the initial conditions were specified at $c = 2$, then the interval of existence and uniqueness would be $(\pi/2, 3\pi/2)$, since this is the largest interval containing $c = 2$ on which all the functions $a_{ij}(x)$ and $b_i(x)$ are simultaneously defined and continuous. If the initial conditions were specified at $c = .5$, then the interval of existence and uniqueness would be $(0, 1)$. If the initial condition were specified at any $c < 0$, then there would be no solution because the system (15a) would be undefined at c, since $a_{12}(x) = \sqrt{x}$ is undefined for $x < 0$. ■

7.1.1 Exercises

1. Verify that $\{y_1(x) = 3e^x,\ y_2(x) = e^x\}$ is a solution on the interval $(-\infty, \infty)$ of the system of differential equations

$$\begin{aligned} y_1' &= 2y_1 - 3y_2 \\ y_2' &= y_1 - 2y_2 \end{aligned}$$

2. Verify that $\{y_1(x) = e^{-x},\ y_2(x) = e^{-x}\}$ is also a solution on the interval $(-\infty, \infty)$ of the system in Exercise 1.

3. Verify that $\{y_1(x) = -e^{2x}(\cos x + \sin x),\ y_2(x) = e^{2x}\cos x\}$ is a solution on $(-\infty, \infty)$ of the system of differential equations

$$\begin{aligned} y_1' &= y_1 - 2y_2 \\ y_2' &= y_1 + 3y_2 \end{aligned}$$

4. Verify that $\{y_1(x) = 3x - 2,\ y_2(x) = -2x + 3\}$ is a solution on $(-\infty, \infty)$ of the system initial value problem

$$\begin{aligned} y_1' &= y_1 + 2y_2 + x - 1 \\ y_2' &= 3y_1 + 2y_2 - 5x - 2 \end{aligned}$$

$$y_1(0) = -2, \quad y_2(0) = 3$$

5. Consider the system initial value problem

(16a)
$$\begin{aligned} y_1' &= \frac{2y_1}{x} - \frac{y_2}{x^2} - 3 + \frac{1}{x} - \frac{1}{x^2} \\ y_2' &= 2y_1 + 1 - 6x \end{aligned}$$

(16b)
$$y_1(1) = -2, \quad y_2(1) = -5.$$

a. Is the system of differential equations (16a) linear or nonlinear?

b. Apply the appropriate theorem from this section to determine the interval on which a unique solution to the IVP (16) exists.

c. Verify that

$$\{y_1(x) = -2x, \;\; y_2(x) = -5x^2 + x - 1\} \tag{17}$$

is the solution to the initial value problem (16). What is the largest interval on which the functions $y_1(x)$ and $y_2(x)$ of (17) and their derivatives are defined and continuous? Why is this interval not the same interval as the one in which the theorem guarantees existence and uniqueness of the solution? Is the set $\{y_1(x), y_2(x)\}$ a solution of the IVP (16) on $(-\infty, \infty)$? Why or why not?

6. Consider the system initial value problem

$$\begin{aligned} y_1' &= \frac{5y_1}{x} + \frac{4y_2}{x} - 2x \\ y_2' &= \frac{-6y_1}{x} - \frac{5y_2}{x} + 5x \end{aligned} \tag{18a}$$

$$y_1(-1) = 3, \quad y_2(-1) = -3. \tag{18b}$$

a. Is the system of differential equations (18a) linear or nonlinear?

b. Apply the appropriate theorem from this section to determine the interval on which a unique solution to the IVP (18) exists.

c. Verify that

$$\{y_1(x) = 2x^2 + x - \frac{2}{x}, \;\; y_2(x) = -x^2 - x + \frac{3}{x}\} \tag{19}$$

is the solution to the initial value problem (18). On what intervals are the functions $y_1(x)$ and $y_2(x)$ of (19) and their derivatives simultaneously defined and continuous? How do these intervals compare with the interval that the appropriate theorem guarantees the existence of a unique solution?

7. Consider the system initial value problem

$$\begin{aligned} y_1' &= y_1 - 2y_1y_2 + \frac{1}{x+2} \\ y_2' &= y_1 + y_2 + y_2^2 - \tan x \end{aligned} \tag{20a}$$

$$y_1(0) = 1, \quad y_2(0) = 2. \tag{20b}$$

a. Is the system (20a) linear or nonlinear?

b. Applying the appropriate theorem from this section, what can be said about the interval on which a unique solution to this problem exists?

8. Consider the system initial value problem

(21a)
$$\begin{aligned} y_1' &= e^{-x} y_2 \\ y_2' &= e^{y_1} \end{aligned}$$

(21b)
$$y_1(0) = 0 \quad y_2(0) = 1.$$

a. Is the system (21a) linear or nonlinear?

b. Applying the appropriate theorem from this section, what can be said about the interval on which a unique solution to this problem exists?

c. Show that $\{y_1(x) = x,\ y_2(x) = e^x\}$ is the solution to the IVP (21). On what interval is this the solution to the initial value problem?

9. Consider the system initial value problem

(22a)
$$\begin{aligned} y_1' &= \frac{y_1}{2 - y_2} + \frac{y_2}{x + 3} \\ y_2' &= \frac{y_2}{2 + y_1} - \frac{y_1}{x - 4} \end{aligned}$$

(22b)
$$y_1(0) = 1, \quad y_2(0) = 1.$$

a. Is system (22a) linear or nonlinear?

b. What can be said about the interval on which a unique solution to this problem exists?

c. Analyze this initial value problem and complete the following statement. The interval of existence and uniqueness will terminate at the point $x = a$ if any of the following occurs as x approaches a, $x \to$ ____, $x \to$ ____, $y_1(x) \to$ ____, $y_1(x) \to$ ____, $y_2(x) \to$ ____, $y_2(x) \to$ ____.

10. Consider the system initial value problem

(23a)
$$\begin{aligned} y_1' &= 2\sqrt{x+4}\, y_1 y_2 - (\sin x) y_2 + 3e^x \\ y_2' &= -y_1^2 + 4\sqrt{5 - x}\, y_1 - 5 \ln 7x \end{aligned}$$

(23b)
$$y_1(2) = 3, \quad y_2(2) = -6.$$

a. Is the system (23a) linear or nonlinear?

b. What can be said about the interval on which a unique solution to this problem exists?

c. Analyze this initial value problem and complete the following statement. The interval of existence and uniqueness will terminate at the point $x = a$ if any of the following occurs as x approaches a, $x \to$ ____, $x \to$ ____, $y_1(x) \to$ ____, $y_1(x) \to$ ____, $y_2(x) \to$ ____, $y_2(x) \to$ ____.

7.2 Writing Systems as Equivalent First-Order Systems

At this point the only question remaining to be answered from the collection of questions at the beginning of this chapter is, "How can an n-th order differential equation be rewritten as an equivalent system of first-order differential equations?" Perhaps, we should first answer the ultimate question: "Why is there a need to write an n-th order differential equation as a system of n first-order differential equations?" The simple reason is because most computer programs are written to solve the general system initial value problem (11) and not an n-th order differential equation. The general n-th order differential equation has the form

$$y^{(n)} = g(x, y, y^{(1)}, \ldots, y^{(n-1)}). \tag{24}$$

Letting $u_1 = y$, $u_2 = y^{(1)}, \ldots,$ $u_n = y^{(n-1)}$, differentiating each of these equations, and substituting for y, $y^{(1)}$, $\ldots, y^{(n-1)}$ in terms of u_1, u_2, $\ldots, u_n$, we see that equation (24) may be rewritten as the equivalent system

$$\begin{array}{lllll} u_1' & = & u_2 & = & f_1(x, u_1, u_2, \ldots, u_n) \\ u_2' & = & u_3 & = & f_2(x, u_1, u_2, \ldots, u_n) \\ \vdots & \vdots & \vdots & \vdots & \vdots \\ u_{n-1}' & = & u_n & = & f_{n-1}(x, u_1, u_2, \ldots, u_n) \\ u_n' & = & g(x, u_1, u_2, \ldots, u_n) & = & f_n(x, u_1, u_2, \ldots, u_n). \end{array} \tag{25}$$

Observe that this system is a special case of the system (11a). The initial conditions corresponding to (11b) are $u_1(c) = d_1$, $u_2(c) = d_2, \ldots,$ and $u_n(c) = d_n$. When we apply the inverse transformation, these conditions in terms of y and its derivatives become $y(c) = d_1$, $y^{(1)}(c) = d_2$, $y^{(2)}(c) = d_3, \ldots,$ and $y^{(n-1)}(c) = d_n$. Hence, the general n-th order initial value problem

$$y^{(n)} = g(x, y, y^{(1)}, \ldots, y^{(n-1)}) \tag{26a}$$

$$y(c) = d_1,\ y^{(1)}(c) = d_2, \ldots,\ y^{(n-1)}(c) = d_n \tag{26b}$$

is equivalent to the system initial value problem

$$\begin{array}{lll} u_1' & = & u_2 \\ u_2' & = & u_3 \\ \vdots & \vdots & \vdots \\ u_{n-1}' & = & u_n \\ u_n' & = & g(x, u_1, u_2, \ldots, u_n) \end{array} \tag{27a}$$

$$u_1(c) = d_1,\ u_2(c) = d_2, \ldots,\ u_n(c) = d_n. \tag{27b}$$

Example 1 Write the n-th order initial value problem

$$y^{(4)} = 7x^2y + y^{(1)}y^{(3)} - e^x(y^{(2)})^3$$

$$y(0) = 1, \quad y^{(1)}(0) = -1, \quad y^{(2)}(0) = -2, \quad y^{(3)}(0) = 4$$

as an equivalent system initial value problem.

Solution

In this instance, $g(x, y, y^{(1)}, y^{(2)}, y^{(3)}) = 7x^2y + y^{(1)}y^{(3)} - e^x(y^{(2)})^3$. Letting $u_1 = y$, $u_2 = y^{(1)}$, $u_3 = y^{(2)}$, and $u_4 = y^{(3)}$, we obtain the desired equivalent system initial value problem

$$\begin{aligned} u_1' &= u_2 \\ u_2' &= u_3 \\ u_3' &= u_4 \\ u_4' &= 7x^2u_1 + u_2u_4 - e^xu_3^3 \end{aligned}$$

$$u_1(0) = 1, \quad u_2(0) = -1, \quad u_3(0) = -2, \quad u_4(0) = 4. \blacksquare$$

Notice in system (25) that $\partial f_i/\partial y_j = 0$ for $i = 1, 2, \ldots, n-1$ and $j = 1, 2, \ldots, n$ but $j \neq i+1$. And also $\partial f_i/\partial y_{i+1} = 1$ for $i = 1, 2, \ldots, n-1$. Thus, all of the $n(n-1)$ partial derivatives $\partial f_i/\partial y_j$, $i = 1, 2, \ldots, n-1$, and $j = 1, 2, \ldots, n$, are defined and continuous functions of x, u_1, $u_2, \ldots, u_n$ in all of $xu_1u_2\ldots u_n$-space. Applying the fundamental existence and uniqueness theorem to system (25), we see that it will have a unique solution on a small interval about c provided in some generalized rectangle R in $xu_1u_2\ldots u_n$-space, the function $f_n(x, u_1, u_2, \ldots, u_n) = g(x, u_1, \ldots, u_n)$ and the n partial derivatives $\partial f_n/\partial u_i = \partial g/\partial u_i$, $i = 1, 2, \ldots, n$ are all continuous functions. Substituting for u_1, $u_2, \ldots, u_n$ in terms of y, $y^{(1)}, \ldots, y^{(n-1)}$, we note that the initial value problem (26) will have a unique solution on some small interval centered about $x = c$ provided the function $g(x, y, y^{(1)}, \ldots, y^{(n-1)})$ and the partial derivatives $\partial g/\partial y$, $\partial g/\partial y^{(1)}, \ldots, \partial g/\partial y^{(n-1)}$ are all continuous functions of x, y, $y^{(1)}, \ldots, y^{(n)}$ on some generalized rectangle R in $xyy^{(1)}\ldots y^{(n-1)}$-space. Thus, we have the following **Existence and Uniqueness Theorem for the General n-th Order Initial Value Problem (26)**.

Theorem 7.3 Let R be the generalized rectangle

$$\{(x, y, y^{(1)}, \ldots, y^{(n-1)}) \mid \alpha < x < \beta \text{ and } \gamma_i < y^{(i-1)} < \delta_i, i = 1, 2, \ldots, n\}$$

where α, β, γ_i, and δ_i are all finite real constants. If $g(x, y, y^{(1)}, \ldots, y^{(n-1)})$ is a continuous function of x, y, $y^{(1)}, \ldots, y^{(n-1)}$ in R, if $\partial g/\partial y$, $\partial g/\partial y^{(1)}, \ldots,$ $\partial g/\partial y^{(n-1)}$ are all continuous functions of x, y, $y^{(1)}, \ldots, y^{(n-1)}$ in R, and if $(c, d_1, d_2, \ldots, d_n) \in R$, then there exists a unique solution to the initial value problem

(26a) $$y^{(n)} = g(x, y, y^{(1)}, \ldots, y^{(n-1)})$$

(26b) $$y(c) = d_1,\ y^{(1)}(c) = d_2, \ldots,\ y^{(n-1)}(c) = d_n$$

on some interval $I = (c - h, c + h)$ where I is a subinterval of (α, β) and the solution can be continued in a unique manner until the boundary of R is reached.

Recall that the n-th order differential equation $y^{(n)} = g(x, y, y^{(1)}, \ldots, y^{(n-1)})$ is linear if and only if g has the form $g(x, y, y^{(1)}, \ldots, y^{(n-1)}) = a_1(x)y + a_2(x)y^{(1)} + \cdots + a_n(x)y^{(n-1)} + b(x)$. So, the linear n-th order initial value problem

(28a) $$y^{(n)} = a_1(x)y + a_2(x)y^{(1)} + \cdots + a_n(x)y^{(n-1)} + b(x)$$

(28b) $$y(c) = d_1, \ y^{(1)}(c) = d_2, \ldots, y^{(n-1)}(c) = d_n$$

is equivalent to the linear system initial value problem

(29a) $$\begin{aligned} u_1' &= u_2 \\ u_2' &= u_3 \\ \vdots \quad & \quad \vdots \quad \vdots \\ u_{n-1}' &= u_n \\ u_n' &= a_1(x)u_1 + a_2(x)u_2 + \cdots + a_n(x)u_n + b(x) \end{aligned}$$

(29b) $$u_1(c) = d_1, \ u_2(c) = d_2, \ldots, \ u_n(c) = d_n.$$

Applying the fundamental existence and uniqueness theorem for linear system initial value problems to system (29), we see that there exists a unique solution to (29)—equivalently to (28)—on any interval which contains the point c and on which the functions $a_1(x), a_2(x), \ldots, a_n(x)$ and $b(x)$ are simultaneously defined and continuous. Hence, we have the following **Existence and Uniqueness Theorem for Linear n-th Order Initial Value Problems.**

Theorem 7.4 If the functions $a_1(x), a_2(x), \ldots, a_n(x)$ and $b(x)$ are all defined and continuous on some interval I which contains the point c, then there exists a unique solution on the entire interval I to the linear n-th order initial value problem

(28a) $$y^{(n)} = a_1(x)y + a_2(x)y^{(1)} + \cdots + a_n(x)y^{(n-1)} + b(x)$$

(28b) $$y(c) = d_1, \ y^{(1)}(c) = d_2, \ldots, y^{(n-1)}(c) = d_n.$$

Example 2 Write the third-order, linear initial value problem

(30a) $$y^{(3)} = (\ln(x^2 - 4))y + 3e^{-x}y^{(1)} - \frac{2y^{(2)}}{\sin x} + x^2$$

(30b) $$y(2.5) = -3, \quad y^{(1)}(2.5) = 0, \quad y^{(2)}(2.5) = 1.2$$

as an equivalent first-order system initial value problem and determine the largest interval on which there exists a unique solution.

Solution

Letting $u_1 = y$, $u_2 = y^{(1)}$, and $u_3 = y^{(2)}$, we obtain the desired equivalent linear first-order system initial value problem

$$\begin{aligned} u_1' &= u_2 \\ u_2' &= u_3 \\ u_3' &= (\ln(x^2 - 4))u_1 + 3e^{-x}u_2 - \frac{2u_3}{\sin x} + x^2 \end{aligned}$$

$$u_1(2.5) = -3, \quad u_2(2.5) = 0, \quad u_3(2.5) = 1.2.$$

The function $a_1(x) = \ln(x^2 - 4)$ is defined and continuous on $(-\infty, -2)$ and $(2, \infty)$. The function $a_2(x) = 3e^{-x}$ is defined and continuous on $(-\infty, \infty)$. The function $a_3(x) = -2/(\sin x)$ is defined and continuous for $x \neq n\pi$ where n is an integer. And the function $b(x) = x^2$ is defined and continuous on $(-\infty, \infty)$. Since $(2, \pi)$ is the largest interval containing $c = 2.5$ on which the functions $a_1(x)$, $a_2(x)$, $a_3(x)$, and $b(x)$ are simultaneously defined and continuous, $(2, \pi)$ is the largest interval on which a unique solution to the IVP (30) and the linear system above exists. ■

Notice that a linear n-th order differential equation is equivalent to a linear system of first-order differential equations; likewise, a nonlinear n-th order differential equation is equivalent to a nonlinear system of first-order differential equations. Higher order system initial value problems may also be rewritten as equivalent first-order system initial value problems as we illustrated at the beginning of this chapter for the coupled spring-mass system (3). The following example further demonstrates this technique.

Example 3 In Chapter 6, we saw that the position (x, y) of an electron which was initially at rest at the origin and subject to a magnetic field of intensity H and an electric field of intensity E satisfied the second-order system initial value problem

$$\text{(31a)} \qquad \begin{aligned} x'' &= -HRy' + ER \\ y'' &= HRx' \end{aligned}$$

$$\text{(31b)} \qquad x(0) = 0, \quad x'(0) = 0, \quad y(0) = 0, \quad y'(0) = 0.$$

Write the second-order system initial value problem (31) as an equivalent first-order system initial value problem.

Solution

Let $u_1 = x$, $u_2 = x'$, $u_3 = y$, and $u_4 = y'$. So u_1 represents the x-coordinate of the position of the electron and u_2 represents the velocity of the electron in the x-direction. While u_3 represents the y-coordinate of the electron and

u_4 represents its velocity in the y-direction. Differentiating u_1, u_2, u_3, and u_4, substituting into (31a), and then substituting for x, x', y, and y' in terms of u_1, u_2, u_3, and u_4, we obtain the following first-order system of four equations which is equivalent to the system (31a).

$$\text{(32a)} \qquad \begin{aligned} u_1' &= x' &&= u_2 && \\ u_2' &= x'' &&= -HRy' + ER &&= -HRu_4 + ER \\ u_3' &= y' &&= u_4 && \\ u_4' &= y'' &&= HRx' &&= HRu_2 \end{aligned}$$

Substituting for x, x', y, and y' in terms of u_1, u_2, u_3, and u_4, we see that the initial conditions which are equivalent to (32b) are

$$\text{(32b)} \qquad u_1(0) = 0, \quad u_2(0) = 0, \quad u_3(0) = 0, \quad u_4(0) = 0.$$

Hence, the required equivalent first-order system initial value problem consists of the system (32a) together with the initial conditions (32b). ■

7.2.1 Exercises

In Exercises 1–7 rewrite each of the following initial value problems as an equivalent first-order system initial value problem.

1. $y^{(4)} = -3xy^2 + (y^{(1)})^3 - e^x y^{(2)} y^{(3)} + x^2 - 1$

   $y(1) = -1, \quad y^{(1)}(1) = 2, \quad y^{(2)}(1) = -3, \quad y^{(3)}(1) = 0$

2. $my'' + cy' + k \sin y = 0$, where m, c, and k are positive constants

   $y(0) = 1, \quad y'(0) = -2$

3. $xy'' - 3x^3 y' + (\ln x) y = \sin x$

   $y(1) = -1, \quad y'(1) = 0$

4. $(\cos(x - y)) y^{(2)} - e^x y^{(3)} + xy^{(1)} - 4 = 0$

   $y(-3) = 0, \quad y^{(1)}(-3) = 2, \quad y^{(2)}(-3) = 1$

5. $y'' = 2y - 3z'$

   $z'' = 3y' - 2z$

   $y(0) = 1, \quad y'(0) = -3, \quad z(0) = -1, \quad z'(0) = 2$

6. $m_1 y_1'' = -k_1 y_1 + k_2(y_2 - y_1)$

   $m_2 y_2'' = -k_2(y_2 - y_1) - k_3 y_2$

   where m_1, m_2, k_1, k_2, and k_3 are positive constants.

   $y_1(0) = 0, \quad y_1'(0) = -2, \quad y_2(0) = 1, \quad y_2'(0) = 0$

7. $y' = xy + z$

   $z'' = -x^2y + z' - 3e^x$

   $y(1) = -2, \quad z(1) = 3, \quad z'(1) = 0$

8. Specify which initial value problems in Exercises 1-7 are linear and which are nonlinear.

In Exercises 9–15 determine the largest interval on which there exists a unique solution to the given initial value problem.

9. $y_1' = 3y_1 - 2y_2$

   $y_2' = -y_1 + y_2$

   $y_1(0) = 1, \quad y_2(0) = -1$

10. $y_1' = (\sin x)y_1 + \sqrt{x}\, y_2 + \ln x$

    $y_2' = (\tan x)y_1 - e^x y_2 + 1$

    Initial conditions:

    (i) $y_1(1) = 1, \quad y_2(1) = -1$ (ii) $y_1(2) = 1, \quad y_2(2) = -1$

11. $y_1' = e^{-x}y_1 - \sqrt{x+1}\, y_2 + x^2$

    $y_2' = \dfrac{y_1}{(x-2)^2}$

    Initial conditions:

    (i) $y_1(0) = 0, \quad y_2(0) = 1$ (ii) $y_1(3) = 1, \quad y_2(3) = 0$

12. $xy'' - 3x^3y' + (\ln x)y = \sin x$

    $y(1) = -1, \quad y'(1) = 0$

13. $y'' = 2y - 3z'$

    $z'' = 3y' - 2z$

    $y(0) = 1, \quad y'(0) = -3, \quad z(0) = -1, \quad z'(0) = 2$

14. $m_1y_1'' = -k_1y_1 + k_2(y_2 - y_1)$

    $m_2y_2'' = -k_2(y_2 - y_1) - k_3y_2$

    where m_1, m_2, k_1, k_2, and k_3 are positive constants.

    $y_1(0) = 0, \quad y_1'(0) = -2, \quad y_2(0) = 1, \quad y_2'(0) = 0$

15. $y' = xy + z$

    $z'' = -x^2y + z' - 3e^x$

    $y(1) = -2, \quad z(1) = 3, \quad z'(1) = 0$

8

Linear Systems of First-Order Differential Equations

In this chapter we discuss linear systems of first-order differential equations. In the first section, Matrices and Vectors, we introduce matrix notation and terminology, we review some fundamental facts from matrix theory and linear algebra, and we discuss some computational techniques. In the second section, Eigenvalues and Eigenvectors, we define the concepts of eigenvalues and eigenvectors of a constant matrix, we show how to manually compute eigenvalues and eigenvectors, and we illustrate how to use computer software to calculate eigenvalues and eigenvectors. In the last section, Linear Systems with Constant Coefficients, we indicate how to write a system of linear first-order differential equations with constant coefficients using matrix-vector notation, we state existence and representation theorems regarding the general solution of both homogeneous and nonhomogeneous linear systems, and we show how to write the general solution in terms of eigenvalues and eigenvectors when the linear system has constant coefficients. In Chapter 9, we examine a few linear systems with constant coefficients which arise in various physical systems such as coupled spring-mass systems, pendulum systems, the path of an electron, and mixture problems.

8.1 Matrices and Vectors

In this section we shall review some facts and computational techniques from matrix theory and linear algebra. In subsequent sections, we will show how these facts and techniques relate to solving systems of first-order differential equations.

A **matrix** is a rectangular array. We will use a bold-faced capital letter such as $\mathbf{A}$, $\mathbf{B}$, $\mathbf{C}$, ... to denote a matrix.

If the matrix $\mathbf{A}$ has m rows and n columns, we will write $\mathbf{A}$ is an $m \times n$ matrix—where $m \times n$ is read "m by n." We also say $\mathbf{A}$ has **size** $m \times n$.

For our purposes, the **elements** or **entries** of a matrix will be real numbers, complex numbers, or functions. An element of an $m \times n$ matrix $\mathbf{A}$ which is in the ith row and jth column is denoted by a_{ij}. Hence, the matrix $\mathbf{A}$ may

DOI: 10.1201/9781003678014-8

be represented in any one of the following three equivalent ways.

$$\mathbf{A} = \begin{pmatrix} a_{11} & a_{12} & \cdots & a_{1n} \\ a_{21} & a_{22} & \cdots & a_{2n} \\ \vdots & \vdots & \ddots & \vdots \\ a_{m1} & a_{m2} & \cdots & a_{mn} \end{pmatrix} = (a_{ij})$$

The 2×3 matrix

$$\mathbf{B} = \begin{pmatrix} -1 & i & 0 \\ -5i & 3 & -4 \end{pmatrix}$$

is called a **constant matrix** because each entry is a constant.

A **square matrix** is a matrix with the same number of rows as columns ($m = n$). The square, 2×2 matrix

$$\mathbf{C} = \begin{pmatrix} 1 & 3e^x \\ 2x & \sin x \end{pmatrix}$$

in which each element is a function of x, is often written as $\mathbf{C}(x)$ to indicate that the entries of the matrix are functions. A **column vector** is an $m \times 1$ matrix and a **row vector** is a $1 \times n$ matrix. We will denote a column vector with a bold-faced, lowercase letter such as **a**, **b**, **c**, $\ldots$. The 3×1 matrix

$$\mathbf{c} = \begin{pmatrix} -1 \\ 0 \\ 3 \end{pmatrix}$$

is an example of a constant column vector. And the 2×1 matrix

$$\mathbf{d}(x) = \begin{pmatrix} 3x - 1 \\ \tan x \end{pmatrix}$$

is an example of a column vector whose entries are functions.

The essential algebraic properties for matrices are stated below.

Two matrices $\mathbf{A} = (a_{ij})$ and $\mathbf{B} = (b_{ij})$ are **equal** if and only if

(i) they are the same size

and

(ii) $a_{ij} = b_{ij}$ for all i and j.

In order to add or subtract two matrices they must necessarily be the same size.

If $\mathbf{A} = (a_{ij})$ and $\mathbf{B} = (b_{ij})$ are both $m \times n$ matrices, then

(i) the **sum** $\mathbf{S} = \mathbf{A} + \mathbf{B}$ is an $m \times n$ matrix with elements $s_{ij} = a_{ij} + b_{ij}$—that is,

$$\mathbf{A} + \mathbf{B} = (a_{ij}) + (b_{ij}) = (a_{ij} + b_{ij})$$

and

(ii) the **difference** $\mathbf{D} = \mathbf{A} - \mathbf{B}$ is an $m \times n$ matrix with elements $d_{ij} = a_{ij} - b_{ij}$—that is,

$$\mathbf{A} - \mathbf{B} = (a_{ij}) - (b_{ij}) = (a_{ij} - b_{ij}).$$

For example, if

$$\mathbf{A} = \begin{pmatrix} 2 & -1 \\ 0 & i \\ \pi & e \end{pmatrix} \quad \text{and} \quad \mathbf{B} = \begin{pmatrix} 3 & 4 \\ \sqrt{2} & -2 \\ 4 & 0 \end{pmatrix},$$

then

$$\mathbf{A} + \mathbf{B} = \begin{pmatrix} 2+3 & -1+4 \\ 0+\sqrt{2} & i-2 \\ \pi+4 & e+0 \end{pmatrix} = \begin{pmatrix} 5 & 3 \\ \sqrt{2} & i-2 \\ \pi+4 & e \end{pmatrix}$$

and

$$\mathbf{A} - \mathbf{B} = \begin{pmatrix} 2-3 & -1-4 \\ 0-\sqrt{2} & i+2 \\ \pi-4 & e-0 \end{pmatrix} = \begin{pmatrix} -1 & -5 \\ -\sqrt{2} & i+2 \\ \pi-4 & e \end{pmatrix}.$$

From the definition of addition, it is evident that matrix addition is commutative and associative, since addition of real numbers, complex numbers, and functions is commutative and associative. Thus, if **A**, **B**, and **C** are all $m \times n$ matrices, then

$$\mathbf{A} + \mathbf{B} = \mathbf{B} + \mathbf{A} \qquad \text{(matrix addition is commutative)}$$

$$\mathbf{A} + (\mathbf{B} + \mathbf{C}) = (\mathbf{A} + \mathbf{B}) + \mathbf{C} \qquad \text{(matrix addition is associative)}$$

The **zero matrix** is any matrix whose elements are all zero. The symbol **0** will be used to denote the zero matrix.
Thus, the 2×3 zero matrix is

$$\mathbf{0} = \begin{pmatrix} 0 & 0 & 0 \\ 0 & 0 & 0 \end{pmatrix}$$

and the 2×1 zero matrix (a zero column vector) is

$$\mathbf{0} = \begin{pmatrix} 0 \\ 0 \end{pmatrix}.$$

If **A** is any $m \times n$ matrix and **0** is the $m \times n$ zero matrix, then

$$\mathbf{A} + \mathbf{0} = \mathbf{0} + \mathbf{A} = \mathbf{A}.$$

If α is a real number, a complex number, or a scalar function and if **A** is an $m \times n$ matrix, then $\alpha\mathbf{A}$ is the $m \times n$ matrix with elements αa_{ij}—that is,

$$\alpha\mathbf{A} = \alpha(a_{ij}) = (\alpha a_{ij}).$$

For example,

$$3\begin{pmatrix} 4 \\ -x \end{pmatrix} = \begin{pmatrix} 12 \\ -3x \end{pmatrix} \qquad \text{and} \qquad e^x \begin{pmatrix} \sin x \\ i\cos x \\ 0 \end{pmatrix} = \begin{pmatrix} e^x \sin x \\ ie^x \cos x \\ 0 \end{pmatrix}.$$

For any matrix $\mathbf{A}$, $\alpha\mathbf{A} = \mathbf{A}\alpha$. Hence, the following two distributive laws are valid for scalar multiplication

$$\alpha(\mathbf{A} + \mathbf{B}) = \alpha\mathbf{A} + \alpha\mathbf{B} = \mathbf{A}\alpha + \mathbf{B}\alpha = (\mathbf{A} + \mathbf{B})\alpha$$

$$(\alpha + \beta)\mathbf{A} = \alpha\mathbf{A} + \beta\mathbf{A} = \mathbf{A}\alpha + \mathbf{A}\beta = \mathbf{A}(\alpha + \beta).$$

When the number of columns of the matrix $\mathbf{A}$ is equal to the number of rows of the matrix $\mathbf{B}$, the matrix product $\mathbf{AB}$ is defined. If $\mathbf{A}$ is an $m \times n$ matrix and $\mathbf{B}$ is an $n \times p$ matrix, then the matrix product $\mathbf{AB} = \mathbf{C} = (c_{ij})$ is an $m \times p$ matrix with entries

$$(1) \qquad c_{ij} = \sum_{k=1}^{n} a_{ik}b_{kj} = a_{i1}b_{1j} + a_{i2}b_{2j} + \cdots + a_{in}b_{nj}.$$

For example, let

$$\mathbf{A} = \begin{pmatrix} 1 & 2 \\ -3 & 5 \end{pmatrix} \qquad \text{and} \qquad \mathbf{B} = \begin{pmatrix} -3 & 1 & 2 \\ 2 & 0 & -3 \end{pmatrix}.$$

Observe that $\mathbf{A}$ is a 2×2 matrix and $\mathbf{B}$ is a 2×3 matrix, so the matrix product $\mathbf{AB}$ will be a 2×3 matrix. Using equation (1) to compute each entry of the product, we find

$$\mathbf{C} = \begin{pmatrix} 1(-3) + 2(2) & 1(1) + 2(0) & 1(2) + 2(-3) \\ -3(-3) + 5(2) & -3(1) + 5(0) & -3(2) + 5(-3) \end{pmatrix}$$
$$= \begin{pmatrix} 1 & 1 & -4 \\ 19 & -3 & -21 \end{pmatrix}.$$

Notice, in this example, that the matrix product $\mathbf{BA}$ is undefined, since the number of columns of $\mathbf{B}$—namely, 3—is not equal to the number of rows of $\mathbf{A}$—which is 2. Hence, this example illustrates the important fact that matrix multiplication is not commutative. That is, in general

$$\mathbf{AB} \neq \mathbf{BA}.$$

In order for both products $\mathbf{AB}$ and $\mathbf{BA}$ to exist and be the same size, it is necessary that $\mathbf{A}$ and $\mathbf{B}$ both be square matrices of the same size. Even then the matrix product $\mathbf{AB}$ may not equal the matrix product $\mathbf{BA}$ as the following simple example illustrates. Let

$$\mathbf{A} = \begin{pmatrix} 0 & 1 \\ 1 & 0 \end{pmatrix} \qquad \text{and} \qquad \mathbf{B} = \begin{pmatrix} 1 & 0 \\ 0 & 0 \end{pmatrix}.$$

Then

$$\mathbf{AB} = \begin{pmatrix} 0 & 1 \\ 1 & 0 \end{pmatrix} \begin{pmatrix} 1 & 0 \\ 0 & 0 \end{pmatrix} = \begin{pmatrix} 0(1)+1(0) & 0(0)+1(0) \\ 1(1)+0(0) & 1(0)+0(0) \end{pmatrix} = \begin{pmatrix} 0 & 0 \\ 1 & 0 \end{pmatrix}$$

and

$$\mathbf{BA} = \begin{pmatrix} 1 & 0 \\ 0 & 0 \end{pmatrix} \begin{pmatrix} 0 & 1 \\ 1 & 0 \end{pmatrix} = \begin{pmatrix} 1(0)+0(1) & 1(1)+0(0) \\ 0(0)+0(1) & 0(1)+0(0) \end{pmatrix} = \begin{pmatrix} 0 & 1 \\ 0 & 0 \end{pmatrix}.$$

So, even if $\mathbf{A}$ and $\mathbf{B}$ are both square matrices of the same size, $\mathbf{AB}$ need not equal $\mathbf{BA}$.

Although matrix multiplication is not commutative, it is associative and both left and right distributive. That is, if $\mathbf{A}$ is an $n \times m$ matrix, $\mathbf{B}$ is an $m \times p$ matrix, and $\mathbf{C}$ is a $p \times q$ matrix, then

$$\mathbf{A}(\mathbf{BC}) = (\mathbf{AB})\mathbf{C}.$$

If $\mathbf{A}$ is an $n \times m$ matrix, $\mathbf{B}$ is an $m \times p$ matrix, $\mathbf{C}$ is an $m \times p$ matrix, and $\mathbf{D}$ is a $p \times q$ matrix, then

$$\mathbf{A}(\mathbf{B}+\mathbf{C}) = \mathbf{AB}+\mathbf{AC} \quad \text{and} \quad (\mathbf{B}+\mathbf{C})\mathbf{D} = \mathbf{BD}+\mathbf{BD}.$$

The $n \times n$ **identity matrix**, $\mathbf{I}$, is the matrix

$$\mathbf{I} = \begin{pmatrix} 1 & 0 & 0 & \cdots & 0 \\ 0 & 1 & 0 & \cdots & 0 \\ 0 & 0 & 1 & \cdots & 0 \\ \vdots & \vdots & \vdots & \ddots & \vdots \\ 0 & 0 & 0 & \cdots & 1 \end{pmatrix}$$

If $\mathbf{A}$ is any $n \times n$ square matrix and $\mathbf{I}$ is the $n \times n$ identity, then by the definition of matrix multiplication

$$\mathbf{AI} = \mathbf{IA} = \mathbf{A}.$$

Let $\mathbf{A}$ be a square matrix of size n. The **determinant** of $\mathbf{A}$ is denoted by $|\mathbf{A}|$ or $det\,\mathbf{A}$.

For $n = 1$, we define

$$|\mathbf{A}| = det\,\mathbf{A} = det\,(a_{11}) = a_{11}.$$

For $n = 2$, we define

$$|\mathbf{A}| = det\,\mathbf{A} = det\begin{pmatrix} a_{11} & a_{12} \\ a_{21} & a_{22} \end{pmatrix} = a_{11}a_{22} - a_{21}a_{12}.$$

And for $n = 3$, we define

$$|\mathbf{A}| = det\,\mathbf{A} = det \begin{pmatrix} a_{11} & a_{12} & a_{13} \\ a_{21} & a_{22} & a_{23} \\ a_{31} & a_{32} & a_{33} \end{pmatrix}$$

$$= a_{11}a_{22}a_{33} + a_{12}a_{23}a_{31} + a_{13}a_{21}a_{32}$$

$$-a_{11}a_{23}a_{32} - a_{12}a_{21}a_{33} - a_{13}a_{22}a_{31}.$$

We could give a general definition for $det\,\mathbf{A}$ or a recursive definition for $det\,\mathbf{A}$; however, since we will only compute $det\,\mathbf{A}$ for $\mathbf{A}$ of size $n = 1$, 2, and 3, the definitions which we have given will suffice.

We now calculate a few determinants.

$$\begin{vmatrix} -1 & 2 \\ 4 & -3 \end{vmatrix} = det \begin{pmatrix} -1 & 2 \\ 4 & -3 \end{pmatrix} = (-1)(-3) - (4)(2) = 3 - 8 = -5,$$

$$\begin{vmatrix} x & x^2 \\ 1 & 2x \end{vmatrix} = det \begin{pmatrix} x & x^2 \\ 1 & 2x \end{pmatrix} = (x)(2x) - (1)(x^2) = 2x^2 - x^2 = x^2,$$

$$\begin{vmatrix} -1 & 0 & 2 \\ 3 & 1 & 0 \\ 4 & -2 & 3 \end{vmatrix} = det \begin{pmatrix} -1 & 0 & 2 \\ 3 & 1 & 0 \\ 4 & -2 & 3 \end{pmatrix}$$

$$= (-1)(1)(3) + (0)(0)(4) + (2)(3)(-2)$$
$$- (-1)(0)(-2) - (0)(3)(3) - (2)(1)(4)$$
$$= -3 + 0 - 12 - 0 - 0 - 8 = -23,$$

and

$$\begin{vmatrix} 2-\lambda & 3 & -1 \\ 4 & -1-\lambda & 0 \\ 1 & 2 & 3-\lambda \end{vmatrix} = det \begin{pmatrix} 2-\lambda & 3 & -1 \\ 4 & -1-\lambda & 0 \\ 1 & 2 & 3-\lambda \end{pmatrix}$$

$$= (2-\lambda)(-1-\lambda)(3-\lambda) + (3)(0)(1) + (-1)(4)(2)$$
$$- (2-\lambda)(0)(2) - (3)(4)(3-\lambda) - (-1)(-1-\lambda)(1)$$

$$= (2-\lambda)(-1-\lambda)(3-\lambda) + 0 - 8 - 0 - 12(3-\lambda) + (-1-\lambda)$$

$$= -\lambda^3 + 4\lambda^2 - \lambda - 6 - 8 - 36 + 12\lambda - 1 - \lambda$$

$$= -\lambda^3 + 4\lambda^2 + 10\lambda - 51.$$

The set of n simultaneous linear equations

$$
\begin{array}{lllllll}
a_{11}x_1 & + a_{12}x_2 + & \cdots & + a_{1n}x_n & = & b_1 \\
a_{21}x_1 & + a_{22}x_2 + & \cdots & + a_{2n}x_n & = & b_2 \\
\vdots & \vdots & & \vdots & \vdots & \vdots \\
a_{n1}x_1 & + a_{n2}x_2 + & \cdots & + a_{nn}x_n & = & b_n
\end{array}
\tag{2}
$$

in the n unknowns $x_1, x_2, \ldots, x_n$ may be rewritten in matrix notation as

$$
\mathbf{A}\mathbf{x} = \mathbf{b} \tag{2$'$}
$$

where $\mathbf{A}$ is the $n \times n$ matrix

$$
\mathbf{A} = \begin{pmatrix} a_{11} & a_{12} & \cdots & a_{1n} \\ a_{21} & a_{22} & \cdots & a_{2n} \\ \vdots & \vdots & \ddots & \vdots \\ a_{n1} & a_{n2} & \cdots & a_{nn} \end{pmatrix}
$$

and $\mathbf{x}$ and $\mathbf{b}$ are the $n \times 1$ column vectors

$$
\mathbf{x} = \begin{pmatrix} x_1 \\ x_2 \\ \vdots \\ x_n \end{pmatrix} \quad \text{and} \quad \mathbf{b} = \begin{pmatrix} b_1 \\ b_2 \\ \vdots \\ b_n \end{pmatrix}.
$$

If $\mathbf{b} = \mathbf{0}$, then systems (2) and (2$'$) are said to be **homogeneous**; otherwise, they are called **nonhomogeneous**.

The following facts are proven in linear algebra.

If the determinant of $\mathbf{A}$ is not zero (if $det\,\mathbf{A} \neq 0$), then there is a unique solution to the system (2$'$) $\mathbf{A}\mathbf{x} = \mathbf{b}$. In particular, if $det\,\mathbf{A} \neq 0$, the homogeneous system $\mathbf{A}\mathbf{x} = \mathbf{0}$ has only the trivial solution, $\mathbf{x} = \mathbf{0}$.

If the determinant of $\mathbf{A}$ is zero (if $det\,\mathbf{A} = 0$), then the system (2$'$) $\mathbf{A}\mathbf{x} = \mathbf{b}$ does not have a solution or it has infinitely many nonunique solutions. If $det\,\mathbf{A} = 0$, the homogeneous system $\mathbf{A}\mathbf{x} = \mathbf{0}$ has infinitely many nonzero solutions in addition to the trivial (zero) solution.

A set of m constant vectors $\{\mathbf{y}_1, \mathbf{y}_2, \ldots, \mathbf{y}_m\}$ which are all the same size, say $n \times 1$, is said to be **linearly dependent** if there exist constants $c_1, c_2, \ldots, c_m$ at least one of which is nonzero, such that

$$
c_1\mathbf{y}_1 + c_2\mathbf{y}_2 + \cdots + c_m\mathbf{y}_m = \mathbf{0}.
$$

Otherwise, the set $\{\mathbf{y}_1, \mathbf{y}_2, \ldots, \mathbf{y}_m\}$ is said to be linearly independent. Hence, the set of vectors $\{\mathbf{y}_1, \mathbf{y}_2, \ldots, \mathbf{y}_m\}$ is **linearly independent** if

$$
c_1\mathbf{y}_1 + c_2\mathbf{y}_2 + \cdots + c_m\mathbf{y}_m = \mathbf{0} \text{ implies } c_1 = c_2 = \cdots = c_m = 0.
$$

That is, the only way to express the zero vector, $\mathbf{0}$, as a linear combination of linearly independent vectors is for all of the coefficients, c_i, to be zero.

Now let us consider a set of n constant column vectors each having n components, $\{\mathbf{y}_1, \mathbf{y}_2, \ldots, \mathbf{y}_n\}$. We will let

$$\mathbf{y}_1 = \begin{pmatrix} y_{11} \\ y_{21} \\ \vdots \\ y_{n1} \end{pmatrix}, \quad \mathbf{y}_2 = \begin{pmatrix} y_{12} \\ y_{22} \\ \vdots \\ y_{n2} \end{pmatrix}, \ldots, \quad \mathbf{y}_n = \begin{pmatrix} y_{1n} \\ y_{2n} \\ \vdots \\ y_{nn} \end{pmatrix}.$$

Thus, y_{ij} denotes the ith component of the jth vector in the set. The set $\{\mathbf{y}_1, \mathbf{y}_2, \ldots, \mathbf{y}_n\}$ is linearly dependent if and only if there exist constants $c_1, c_2, \ldots, c_n$ not all zero such that

(3) $$c_1\mathbf{y}_1 + c_2\mathbf{y}_2 + \cdots + c_n\mathbf{y}_n = \mathbf{0}.$$

Hence, the set $\{\mathbf{y}_1, \mathbf{y}_2, \ldots, \mathbf{y}_n\}$ is linearly dependent if and only if the simultaneous homogeneous system of equations

$$\begin{array}{lllllll} c_1y_{11} & + c_2y_{12} + & \cdots & + c_ny_{1n} & = & 0 \\ c_1y_{21} & + c_2y_{22} + & \cdots & + c_ny_{2n} & = & 0 \\ \vdots & \quad \vdots & & \quad \vdots & \vdots & \vdots \\ c_1y_{n1} & + c_2y_{n2} + & \cdots & + c_ny_{nn} & = & 0 \end{array}$$

which is equivalent to (3), has a nontrivial solution—a solution in which not all c_i's are zero. This system may be rewritten in matrix notation as

$$\begin{pmatrix} y_{11} & y_{12} & \cdots & y_{1n} \\ y_{21} & y_{22} & \cdots & y_{2n} \\ \vdots & \vdots & & \vdots \\ y_{n1} & y_{n2} & \cdots & y_{nn} \end{pmatrix} \begin{pmatrix} c_1 \\ c_2 \\ \vdots \\ c_n \end{pmatrix} = \begin{pmatrix} 0 \\ 0 \\ \vdots \\ 0 \end{pmatrix}$$

or more compactly as

(4) $$\mathbf{Yc} = \mathbf{0}$$

where $\mathbf{Y}$ is the $n \times n$ matrix whose jth column is the vector $\mathbf{y}_j$. As we stated earlier, a homogeneous system, such as system (4), has a nontrivial solution if and only if $det\,\mathbf{Y} = 0$. Consequently, we have the following important results:

A set $\{\mathbf{y}_1, \mathbf{y}_2, \ldots, \mathbf{y}_n\}$ of n constant column vectors each of size $n \times 1$ is linearly dependent if and only if $det\,\mathbf{Y} = 0$ where $\mathbf{Y}$ is the $n \times n$ matrix whose jth column is $\mathbf{y}_j$.

Or equivalently, the set $\{\mathbf{y}_1, \mathbf{y}_2, \ldots, \mathbf{y}_n\}$ is linearly independent if and only if $det\,\mathbf{Y} \neq 0$.

Example 1 Determine whether the set of vectors

$$\mathbf{y}_1 = \begin{pmatrix} 1 \\ 0 \\ -1 \end{pmatrix}, \qquad \mathbf{y}_2 = \begin{pmatrix} 0 \\ 1 \\ 1 \end{pmatrix}, \qquad \mathbf{y}_3 = \begin{pmatrix} 2 \\ 0 \\ 1 \end{pmatrix}$$

is linearly dependent or linearly independent.

Solution

Forming the matrix $\mathbf{Y}$ whose jth column is $\mathbf{y}_j$ and computing $det\,\mathbf{Y}$, we find

$$det\,\mathbf{Y} = det \begin{pmatrix} 1 & 0 & 2 \\ 0 & 1 & 0 \\ -1 & 1 & 1 \end{pmatrix} = 1 + 0 + 0 - 0 - 0 + 2 = 3 \neq 0.$$

Therefore, the set of vectors $\{\mathbf{y}_1, \mathbf{y}_2, \mathbf{y}_3\}$ is linearly independent. ■

Let $\{\mathbf{y}_1(x), \mathbf{y}_2(x), \ldots, \mathbf{y}_m(x)\}$ be a set of vector functions which are all the same size—say, $n \times 1$. Thus, each component of $\mathbf{y}_j(x)$ is a function of x. Let

$$\mathbf{y}_j(x) = \begin{pmatrix} y_{1j}(x) \\ y_{2j}(x) \\ \vdots \\ y_{nj}(x) \end{pmatrix}.$$

Suppose each vector function $\mathbf{y}_j(x)$ is defined on the interval $[a, b]$—that is, suppose $y_{ij}(x)$ is defined on $[a, b]$ for all $i = 1, 2, \ldots, n$ and all $j = 1, 2, \ldots, m$. The concept of linear dependence and linear independence for vector functions is then defined as follows:

If there exists a set of scalar constants $c_1, c_2, \ldots, c_m$ not all zero such that

$$c_1\mathbf{y}_1(x) + c_2\mathbf{y}_2(x) + \cdots + c_m\mathbf{y}_m(x) = \mathbf{0} \qquad \text{for all } x \in [a, b],$$

then the set of vector functions $\{\mathbf{y}_1(x), \mathbf{y}_2(x), \ldots, \mathbf{y}_m(x)\}$ is **linearly dependent on the interval** $[a, b]$.

The set of vector functions $\{\mathbf{y}_1(x), \mathbf{y}_2(x), \ldots, \mathbf{y}_m(x)\}$ is **linearly independent on the interval** $[a, b]$, if

$$c_1\mathbf{y}_1(x) + c_2\mathbf{y}_2(x) + \cdots + c_m\mathbf{y}_m(x) = \mathbf{0} \qquad \text{for all } x \in [a, b]$$

implies $c_1 = c_2 = \cdots = c_m = 0$.

Determining linear independence or linear dependence of a set of vector functions on an interval is more complicated than determining linear independence or linear dependence for a set of constant vectors. If the set of vector functions $\{\mathbf{y}_1(x), \mathbf{y}_2(x), \ldots, \mathbf{y}_m(x)\}$ is linearly dependent on an interval, then it is linearly dependent at each point in the interval. However, if the set is linearly independent on an interval, it may or may not be linearly independent at each point in the interval. For example, the two vector functions

$$\mathbf{y}_1(x) = \begin{pmatrix} x \\ 0 \end{pmatrix} \quad \text{and} \quad \mathbf{y}_2(x) = \begin{pmatrix} x^2 \\ 0 \end{pmatrix}$$

are linearly independent on any interval $[a, b]$, but these two vectors are linearly dependent at every point $p \in [a, b]$. The vectors are linearly dependent at any nonzero p, since for $c_1 = p$ and $c_2 = -1$

$$c_1\mathbf{y}_1(p) + c_2\mathbf{y}_2(p) = c_1 \begin{pmatrix} p \\ 0 \end{pmatrix} + c_2 \begin{pmatrix} p^2 \\ 0 \end{pmatrix} = p \begin{pmatrix} p \\ 0 \end{pmatrix} + (-1) \begin{pmatrix} p^2 \\ 0 \end{pmatrix}$$

$$= \begin{pmatrix} p^2 \\ 0 \end{pmatrix} + \begin{pmatrix} -p^2 \\ 0 \end{pmatrix} = \begin{pmatrix} 0 \\ 0 \end{pmatrix}.$$

The vectors are linearly dependent at $p = 0$, since for any c_1 and c_2

$$c_1\mathbf{y}_1(0) + c_2\mathbf{y}_2(0) = c_1 \begin{pmatrix} 0 \\ 0 \end{pmatrix} + c_2 \begin{pmatrix} 0 \\ 0 \end{pmatrix} = \begin{pmatrix} 0 \\ 0 \end{pmatrix}.$$

Thus, the set $\{\mathbf{y}_1(x), \mathbf{y}_2(x)\}$ is linearly independent on the interval $(-\infty, \infty)$, yet it is linearly dependent at every point in $(-\infty, \infty)$!

The following theorem provides a sufficient condition for linear independence of vector functions.

Theorem 8.1 Let $\{\mathbf{y}_1(x), \mathbf{y}_2(x), \ldots, \mathbf{y}_n(x)\}$ be a set of n vector functions of size $n \times 1$ and let $\mathbf{Y}(x)$ be the matrix with jth column $\mathbf{y}_j(x)$. If $\mathit{det}\,\mathbf{Y}(x) \neq \mathbf{0}$ for any $x \in [a, b]$, then the set $\{\mathbf{y}_1(x), \mathbf{y}_2(x), \ldots, \mathbf{y}_n(x)\}$ is linearly independent on the interval $[a, b]$.

The converse of this theorem is false as the previous example illustrates, since the set of vector functions

$$\left\{ \begin{pmatrix} x \\ 0 \end{pmatrix}, \begin{pmatrix} x^2 \\ 0 \end{pmatrix} \right\}$$

is linearly independent on any interval $[a, b]$; yet, for this set

$$\mathit{det}\,\mathbf{Y} = \mathit{det} \begin{pmatrix} x & x^2 \\ 0 & 0 \end{pmatrix} = 0 \qquad \text{for all } x \in [a, b].$$

Example 2 Verify that the set of vector functions

$$\left\{ \begin{pmatrix} e^x \\ 0 \\ 2e^x \end{pmatrix}, \begin{pmatrix} e^{-x} \\ -e^{-x} \\ e^{-x} \end{pmatrix}, \begin{pmatrix} 2 \\ 1 \\ 3 \end{pmatrix} \right\}$$

is linearly independent on the interval $(-\infty, \infty)$.

Solution

Forming the matrix $\mathbf{Y}$ and computing $det\,\mathbf{Y}$, we find

$$det\,\mathbf{Y} = det\begin{pmatrix} e^x & e^{-x} & 2 \\ 0 & -e^{-x} & 1 \\ 2e^x & e^{-x} & 3 \end{pmatrix}$$

$$= e^x(-e^{-x})(3) + e^{-x}(1)(2e^x) + 2(0)(e^{-x})$$

$$- e^x(1)(e^{-x}) - e^{-x}(0)(3) - 2(-e^{-x})(2e^x)$$

$$= -3 + 2 + 0 - 1 - 0 + 4 = 2.$$

Since $det\,\mathbf{Y} \neq 0$ for any $x \in (-\infty, \infty)$, the given set of vector functions is linearly independent on $(-\infty, \infty)$. ■

8.1.1 Exercises

For Exercises 1–16, let

$$\mathbf{A} = \begin{pmatrix} -1 & 2 & 1 \\ 0 & 3 & -4 \end{pmatrix}, \quad \mathbf{B} = \begin{pmatrix} 2 & -1 \\ 0 & 3 \\ 1 & 0 \end{pmatrix}, \quad \mathbf{x} = \begin{pmatrix} 3 \\ -1 \\ 2 \end{pmatrix},$$

$$\mathbf{y} = \begin{pmatrix} 2 \\ -1 \end{pmatrix}, \text{ and } \quad \mathbf{z} = \begin{pmatrix} 1 & -2 \end{pmatrix}.$$

State whether it is possible to compute the given expression or not. When possible, compute the expression.

1. $\mathbf{AB}$
2. $\mathbf{BA}$
3. $\mathbf{Ax}$
4. $\mathbf{Ay}$
5. $\mathbf{Az}$
6. $\mathbf{Bx}$
7. $\mathbf{By}$
8. $\mathbf{Bz}$
9. $\mathbf{xy}$
10. $\mathbf{xz}$
11. $\mathbf{yz}$
12. $\mathbf{zy}$
13. $\mathbf{A} + \mathbf{B}$
14. $\mathbf{x} + \mathbf{y}$
15. $\mathbf{Ax} + \mathbf{y}$
16. $\mathbf{By} + \mathbf{x}$

In Exercises 17–28 compute the determinant of the given matrix.

17. $\mathbf{A} = (4)$
18. $\mathbf{B} = (-4)$
19. $\begin{pmatrix} 3 & 1 \\ 5 & -1 \end{pmatrix}$
20. $\begin{pmatrix} \sqrt{2} & 2 \\ \sqrt{3} & 3 \end{pmatrix}$
21. $\begin{pmatrix} 1-i & -i \\ i & 1+i \end{pmatrix}$
22. $\begin{pmatrix} 2-\lambda & 1 \\ 3 & 4-\lambda \end{pmatrix}$
23. $\begin{pmatrix} \sin x & \cos x \\ \cos x & -\sin x \end{pmatrix}$
24. $\begin{pmatrix} e^{3x} & e^{-x} \\ 3e^{3x} & -e^{-x} \end{pmatrix}$
25. $\begin{pmatrix} 0 & 1 & -3 \\ 2 & 4 & 0 \\ -1 & 0 & 2 \end{pmatrix}$

26. $\begin{pmatrix} 3 & 2 & 1 \\ 4 & -1 & 2 \\ 0 & -2 & 3 \end{pmatrix}$

27. $\begin{pmatrix} x & x^2 & x^3 \\ 1 & 2x & 3x^2 \\ 0 & 2 & 6x \end{pmatrix}$

28. $\begin{pmatrix} -\lambda & 4 & 2 \\ 3 & -2-\lambda & 5 \\ -1 & 0 & 1-\lambda \end{pmatrix}$

29. Let $\mathbf{A}$ be any square matrix of size $n \times n$ and let $\mathbf{I}$ be the same size identity matrix. Does $(\mathbf{A} + \lambda\mathbf{I})^2 = \mathbf{A}^2 + 2\lambda\mathbf{A} + \lambda^2\mathbf{I}$, where $\mathbf{A}^2 = \mathbf{AA}$ and λ is any scalar? (Hint: Consider the distributive law for $(\mathbf{A} + \lambda\mathbf{I})^2 = (\mathbf{A} + \lambda\mathbf{I})(\mathbf{A} + \lambda\mathbf{I})$ and then the commutative law for scalar multiplication and multiplication of a matrix by the identity matrix. If you think the result is not true, give an example which shows equality does not always hold.)

30. Let $\mathbf{A}$ and $\mathbf{B}$ be any two square matrices of the same size. Does $(\mathbf{A} + \mathbf{B})^2 = \mathbf{A}^2 + 2\mathbf{AB} + \mathbf{B}^2$? (Hint: Consider the distributive law for $(\mathbf{A} + \mathbf{B})^2 = (\mathbf{A} + \mathbf{B})(\mathbf{A} + \mathbf{B})$ and the commutativity of $\mathbf{AB}$.)

31. If $\mathbf{x}_1$ and $\mathbf{x}_2$ are both solutions of $\mathbf{Ax} = \mathbf{0}$ (that is, if $\mathbf{Ax}_1 = \mathbf{0}$ and $\mathbf{Ax}_2 = \mathbf{0}$), show that $\mathbf{y} = c_1\mathbf{x}_1 + c_2\mathbf{x}_2$ is a solution of $\mathbf{Ax} = \mathbf{0}$ for every choice of the scalars c_1 and c_2.

32. If $\mathbf{x}_1$ is a solution of $\mathbf{Ax} = \mathbf{0}$ and $\mathbf{x}_2$ is a solution of $\mathbf{Ax} = \mathbf{b}$ show that $\mathbf{y} = c\mathbf{x}_1 + \mathbf{x}_2$ is a solution of $\mathbf{Ax} = \mathbf{b}$ for every choice of the scalar c.

In Exercises 33-36 determine whether the given set of vectors is linearly independent or linearly dependent.

33. $\left\{ \begin{pmatrix} 2 \\ -1 \end{pmatrix}, \begin{pmatrix} -6 \\ 3 \end{pmatrix} \right\}$

34. $\left\{ \begin{pmatrix} 2 \\ 1 \end{pmatrix}, \begin{pmatrix} 1 \\ 0 \end{pmatrix} \right\}$

35. $\left\{ \begin{pmatrix} 1 \\ 0 \\ -1 \end{pmatrix}, \begin{pmatrix} 1 \\ 1 \\ 0 \end{pmatrix}, \begin{pmatrix} 0 \\ 1 \\ 1 \end{pmatrix} \right\}$

36. $\left\{ \begin{pmatrix} 1 \\ -1 \\ 1 \end{pmatrix}, \begin{pmatrix} -1 \\ 0 \\ 1 \end{pmatrix}, \begin{pmatrix} 1 \\ 1 \\ 1 \end{pmatrix} \right\}$

In Exercises 37–40 show that the given set of vector functions are linearly independent on the interval specified.

37. $\left\{ \begin{pmatrix} x \\ 3 \end{pmatrix}, \begin{pmatrix} x \\ x-1 \end{pmatrix} \right\}$ on $(-1, 1)$

38. $\left\{ \begin{pmatrix} 2e^x \\ e^x \end{pmatrix}, \begin{pmatrix} e^{3x} \\ -e^{-3x} \end{pmatrix} \right\}$ on $(-\infty, \infty)$

39. $\left\{ \begin{pmatrix} e^{2x} \\ -3e^{2x} \\ 2e^{2x} \end{pmatrix}, \begin{pmatrix} 2 \\ 3 \\ 1 \end{pmatrix}, \begin{pmatrix} 3x \\ -x \\ -2x \end{pmatrix} \right\}$ on $(0, \infty)$

40. $\left\{ \begin{pmatrix} 1 \\ 0 \\ 0 \end{pmatrix}, \begin{pmatrix} 0 \\ \sin x \\ \cos x \end{pmatrix}, \begin{pmatrix} 0 \\ -\cos x \\ \sin x \end{pmatrix} \right\}$ on $(-\infty, \infty)$

8.2 Eigenvalues and Eigenvectors

The scalar λ is an **eigenvalue** (or **characteristic value**) of the $n \times n$, constant matrix $\mathbf{A}$ and the nonzero $n \times 1$ vector $\mathbf{x}$ is an **eigenvector** (or **characteristic vector**) associated with λ if and only if $\mathbf{Ax} = \lambda \mathbf{x}$.

Thus, an eigenvector of the matrix $\mathbf{A}$ is a nonzero vector which when multiplied by $\mathbf{A}$ equals some constant, λ, times itself. A vector chosen at random will generally not have this property. However, if $\mathbf{x}$ is a vector such that $\mathbf{Ax} = \lambda \mathbf{x}$, then $\mathbf{y} = c\mathbf{x}$ also satisfies $\mathbf{Ay} = \lambda \mathbf{y}$ for any arbitrary constant c, since $\mathbf{Ay} = \mathbf{A}(c\mathbf{x}) = (\mathbf{A}c)\mathbf{x} = (c\mathbf{A})\mathbf{x} = c(\mathbf{Ax}) = c\lambda\mathbf{x} = \lambda c\mathbf{x} = \lambda\mathbf{y}$. That is, if $\mathbf{x}$ is an eigenvector of the matrix $\mathbf{A}$ associated with the eigenvalue λ, then so is $c\mathbf{x}$. Hence, eigenvectors are not uniquely determined but are determined only up to an arbitrary multiplicative constant. If the manner in which the arbitrary constant is to be chosen is specified, then the eigenvector is said to be **normalized**. For example, the eigenvectors $\mathbf{x}$ of an $n \times n$ matrix $\mathbf{A}$ could be normalized so that $(x_1^2 + x_2^2 + \cdots + x_n^2)^{1/2} = 1$ where $x_1, x_2, \ldots, x_n$ denote the components of $\mathbf{x}$.

There is a nonzero vector $\mathbf{x}$ which satisfies the equation

$$(1) \qquad \mathbf{Ax} = \lambda \mathbf{x}$$

or the equivalent equation $\mathbf{Ax} - \lambda\mathbf{x} = (\mathbf{A} - \lambda\mathbf{I})\mathbf{x} = \mathbf{0}$ if and only if

$$(2) \qquad \mathit{det}\,(\mathbf{A} - \lambda\mathbf{I}) = 0.$$

If $\mathbf{A}$ is an $n \times n$ constant matrix, then equation (2) is a polynomial of degree n in λ—called the **characteristic polynomial of A**. So each $n \times n$ matrix $\mathbf{A}$ has n eigenvalues $\lambda_1, \lambda_2, \ldots, \lambda_n$, some of which may be repeated. If λ is a root of equation (2) m times, then we say λ is an eigenvalue of $\mathbf{A}$ with **multiplicity** m. Every eigenvalue has at least one associated eigenvector. If λ is an eigenvalue of multiplicity $m > 1$, then there are k linearly independent eigenvectors associated with λ where $1 \leq k \leq m$. When all the eigenvalues $\lambda_1, \lambda_2, \ldots, \lambda_n$ of a matrix $\mathbf{A}$ have multiplicity one (that is, when the roots of equation (2) are distinct), then the associated eigenvectors $\mathbf{x}_1, \mathbf{x}_2, \ldots, \mathbf{x}_n$ are linearly independent. However, if some eigenvalue of $\mathbf{A}$ has multiplicity $m > 1$ but fewer than m linearly independent associated eigenvectors, then there will be fewer than n linearly independent eigenvectors associated with $\mathbf{A}$. As we shall see, this situation will lead to difficulties when trying to solve a system of differential equations which involves the matrix $\mathbf{A}$.

Example 1 Find the eigenvalues and associated eigenvectors of the matrix

$$\mathbf{A} = \begin{pmatrix} 2 & -3 \\ 1 & -2 \end{pmatrix}.$$

Solution

The characteristic equation of $\mathbf{A}$ is $det\,(\mathbf{A} - \lambda \mathbf{I}) = 0$. For the given matrix

$$det\,(\mathbf{A} - \lambda \mathbf{I}) = det\left(\begin{pmatrix} 2 & -3 \\ 1 & -2 \end{pmatrix} - \lambda \begin{pmatrix} 1 & 0 \\ 0 & 1 \end{pmatrix} \right)$$

$$= det \begin{pmatrix} 2-\lambda & -3 \\ 1 & -2-\lambda \end{pmatrix} = (2-\lambda)(-2-\lambda) - (1)(-3)$$

$$= -4 + \lambda^2 + 3 = \lambda^2 - 1 = 0.$$

Solving the characteristic equation $\lambda^2 - 1 = 0$, we see the eigenvalues of $\mathbf{A}$ are $\lambda_1 = 1$ and $\lambda_2 = -1$.

An eigenvector $\mathbf{x}_1$ of $\mathbf{A}$ associated with $\lambda_1 = 1$ must satisfy the equation $\mathbf{A}\mathbf{x}_1 = \lambda_1 \mathbf{x}_1$ or equivalently $\mathbf{A}\mathbf{x}_1 - \lambda_1 \mathbf{x}_1 = (\mathbf{A} - \lambda_1 \mathbf{I})\mathbf{x}_1 = \mathbf{0}$. Let $\mathbf{x}_1 = \begin{pmatrix} x_{11} \\ x_{21} \end{pmatrix}$. Then $\mathbf{x}_1$ must satisfy

$$(\mathbf{A} - \lambda_1 \mathbf{I})\mathbf{x}_1 = (\mathbf{A} - \mathbf{I})\mathbf{x}_1 = \begin{pmatrix} 2-1 & -3 \\ 1 & -2-1 \end{pmatrix} \begin{pmatrix} x_{11} \\ x_{21} \end{pmatrix}$$

$$= \begin{pmatrix} 1 & -3 \\ 1 & -3 \end{pmatrix} \begin{pmatrix} x_{11} \\ x_{21} \end{pmatrix} = \begin{pmatrix} 0 \\ 0 \end{pmatrix}.$$

Performing the matrix multiplication, we find that x_{11} and x_{21} must simultaneously satisfy the system of equations

$$\begin{aligned} x_{11} - 3x_{21} &= 0 \\ x_{11} - 3x_{21} &= 0. \end{aligned}$$

Since these equations are identical, we conclude $x_{11} = 3x_{21}$ and x_{21} is arbitrary. Choosing $x_{21} = 1$, we find an eigenvector associated with the eigenvalue $\lambda_1 = 1$ is

$$\mathbf{x}_1 = \begin{pmatrix} x_{11} \\ x_{21} \end{pmatrix} = \begin{pmatrix} 3 \\ 1 \end{pmatrix}.$$

As we noted earlier, the vector

$$c\mathbf{x}_1 = \begin{pmatrix} 3c \\ c \end{pmatrix}$$

where $c \neq 0$ is any arbitrary constant is also an eigenvector of $\mathbf{A}$ associated with the eigenvalue $\lambda_1 = 1$.

An eigenvector $\mathbf{x}_2$ of $\mathbf{A}$ associated with $\lambda_2 = -1$ must satisfy the equation $\mathbf{A}\mathbf{x}_2 = \lambda_2\mathbf{x}_2$ or $\mathbf{A}\mathbf{x}_2 - \lambda_2\mathbf{x}_2 = (\mathbf{A} - \lambda_2\mathbf{I})\mathbf{x}_2 = \mathbf{0}$. Letting $\mathbf{x}_2 = \begin{pmatrix} x_{12} \\ x_{22} \end{pmatrix}$, we see that $\mathbf{x}_2$ must satisfy

$$(\mathbf{A} - \lambda_2\mathbf{I})\mathbf{x}_2 = (\mathbf{A} + \mathbf{I})\mathbf{x}_2 = \begin{pmatrix} 2+1 & -3 \\ 1 & -2+1 \end{pmatrix} \begin{pmatrix} x_{12} \\ x_{22} \end{pmatrix}$$

$$= \begin{pmatrix} 3 & -3 \\ 1 & -1 \end{pmatrix} \begin{pmatrix} x_{12} \\ x_{22} \end{pmatrix} = \begin{pmatrix} 0 \\ 0 \end{pmatrix}.$$

Performing the required matrix multiplication, we see x_{12} and x_{22} must satisfy the system of equations

$$\begin{aligned} 3x_{12} - 3x_{22} &= 0 \\ x_{12} - x_{22} &= 0. \end{aligned}$$

Notice that the first equation of this system is three times the last equation. So actually there is only one equation—say, $x_{21} - x_{22} = 0$—to be satisfied by x_{12} and x_{22}. Hence, $x_{12} = x_{22}$ and x_{22} is arbitrary. Choosing $x_{22} = 1$, we find an eigenvector associated with the eigenvalue $\lambda_2 = -1$ is

$$\mathbf{x}_2 = \begin{pmatrix} x_{12} \\ x_{22} \end{pmatrix} = \begin{pmatrix} 1 \\ 1 \end{pmatrix}.$$

The vector $k\mathbf{x}_2$, where $k \neq 0$ is an arbitrary constant, is also an eigenvector associated with the eigenvalue $\lambda_2 = -1$.

Letting $\mathbf{X}$ be the 2×2 matrix whose columns are $\mathbf{x}_1$ and $\mathbf{x}_2$—that is, letting $\mathbf{X} = (\mathbf{x}_1 \ \mathbf{x}_2)$, we find

$$\det \mathbf{X} = \det(\mathbf{x}_1 \ \mathbf{x}_2) = \det \begin{pmatrix} 3 & 1 \\ 1 & 1 \end{pmatrix} = 3 - 1 = 2 \neq 0.$$

Consequently, the eigenvectors $\mathbf{x}_1$ and $\mathbf{x}_2$ are linearly independent. Earlier, we stated that eigenvectors associated with distinct eigenvalues are linearly independent. If we had remembered this fact, it would not have been necessary to show $\det \mathbf{X} \neq 0$. ■

Example 2 Find the eigenvalues and associated eigenvectors of the matrix

$$\mathbf{A} = \begin{pmatrix} 3 & 1 \\ -1 & 1 \end{pmatrix}.$$

Solution

The characteristic equation for $\mathbf{A}$ is

$$\det(\mathbf{A} - \lambda\mathbf{I}) = \det\left(\begin{pmatrix} 3 & 1 \\ -1 & 1 \end{pmatrix} - \lambda\begin{pmatrix} 1 & 0 \\ 0 & 1 \end{pmatrix}\right) = \det\begin{pmatrix} 3-\lambda & 1 \\ -1 & 1-\lambda \end{pmatrix}$$
$$= (3-\lambda)(1-\lambda) + 1 = \lambda^2 - 4\lambda + 4 = 0.$$

Solving the characteristic equation $\lambda^2 - 4\lambda + 4 = 0$, we find the eigenvalues of $\mathbf{A}$ are $\lambda_1 = \lambda_2 = 2$. Thus, $\lambda = 2$ is a root of multiplicity $m = 2$, and there will be one or two linearly independent eigenvectors associated with the eigenvalue $\lambda = 2$.

An eigenvector of $\mathbf{A}$ associated with $\lambda = 2$ must satisfy $\mathbf{Ax} = 2\mathbf{x}$ or $(\mathbf{A} - 2\mathbf{I})\mathbf{x} = \mathbf{0}$. Letting $\mathbf{x} = \begin{pmatrix} x_1 \\ x_2 \end{pmatrix}$ and substituting for $\mathbf{A}$ and $\mathbf{I}$, we see that x_1 and x_2 must satisfy

$$(\mathbf{A} - 2\mathbf{I})\mathbf{x} = \left(\begin{pmatrix} 3 & 1 \\ -1 & 1 \end{pmatrix} - 2\begin{pmatrix} 1 & 0 \\ 0 & 1 \end{pmatrix}\right)\begin{pmatrix} x_1 \\ x_2 \end{pmatrix}$$
$$= \begin{pmatrix} 3-2 & 1 \\ -1 & 1-2 \end{pmatrix}\begin{pmatrix} x_1 \\ x_2 \end{pmatrix} = \begin{pmatrix} 1 & 1 \\ -1 & -1 \end{pmatrix}\begin{pmatrix} x_1 \\ x_2 \end{pmatrix} = \begin{pmatrix} 0 \\ 0 \end{pmatrix}.$$

Multiplying, we find x_1 and x_2 must simultaneously satisfy

$$x_1 + x_2 = 0$$
$$-x_1 - x_2 = 0.$$

Since the second equation is -1 times the first, there is actually only one equation to be satisfied, say $x_1 + x_2 = 0$. Thus, $x_1 = -x_2$ and x_2 is arbitrary. Choosing $x_2 = 1$, we find

$$\mathbf{x} = \begin{pmatrix} x_1 \\ x_2 \end{pmatrix} = \begin{pmatrix} -1 \\ 1 \end{pmatrix}$$

is an eigenvector of $\mathbf{A}$ associated with $\lambda = 2$. Although for $k \neq 0$, the vector $k\mathbf{x}$ is an eigenvector associated with $\lambda = 2$, the vector $k\mathbf{x}$ is not linearly independent from $\mathbf{x}$, since $1(k\mathbf{x}) - k(\mathbf{x}) = \mathbf{0}$. That is, $\mathbf{x}$ and $k\mathbf{x}$ are linearly dependent. Hence, $\lambda = 2$ is an eigenvalue of $\mathbf{A}$ of multiplicity $m = 2$ which has only one associated eigenvector. ■

Example 3 Find the eigenvalues and associated eigenvectors of the matrix

$$\mathbf{A} = \begin{pmatrix} 1 & -1 & 1 \\ -1 & 1 & 1 \\ 1 & 1 & 1 \end{pmatrix}.$$

Solution

The characteristic equation for the given matrix $\mathbf{A}$ is

$$\det(\mathbf{A}-\lambda\mathbf{I}) = \det\begin{pmatrix} 1-\lambda & -1 & 1 \\ -1 & 1-\lambda & 1 \\ 1 & 1 & 1-\lambda \end{pmatrix}$$

$$\begin{aligned} &= (1-\lambda)^3 + (-1)(1)(1) + (1)(-1)(1) \\ &\quad - (1-\lambda)(1)(1) - (-1)(-1)(1-\lambda) - (1)(1-\lambda)(1) \end{aligned}$$

$$= 1 - 3\lambda + 3\lambda^2 - \lambda^3 - 1 - 1 - 1 + \lambda - 1 + \lambda - 1 + \lambda$$

$$= -4 + 3\lambda^2 - \lambda^3 = 0.$$

Since $\lambda^3 - 3\lambda^2 + 4 = (\lambda+1)(\lambda-2)^2$, the roots of the characteristic equation are $\lambda_1 = -1$ and $\lambda_2 = \lambda_3 = 2$. Thus, -1 is an eigenvalue of the matrix $\mathbf{A}$ of multiplicity $m = 1$ and 2 is an eigenvalue of the matrix $\mathbf{A}$ of multiplicity $m = 2$.

Let

$$\mathbf{x}_1 = \begin{pmatrix} x_{11} \\ x_{21} \\ x_{31} \end{pmatrix}$$

be the eigenvector of $\mathbf{A}$ associated with the eigenvalue $\lambda_1 = -1$. The vector $\mathbf{x}_1$ must satisfy $(\mathbf{A} - \lambda_1\mathbf{I})\mathbf{x}_1 = (\mathbf{A}+\mathbf{I})\mathbf{x}_1 = \mathbf{0}$ or

$$\begin{pmatrix} 1+1 & -1 & 1 \\ -1 & 1+1 & 1 \\ 1 & 1 & 1+1 \end{pmatrix}\begin{pmatrix} x_{11} \\ x_{21} \\ x_{31} \end{pmatrix} = \begin{pmatrix} 2 & -1 & 1 \\ -1 & 2 & 1 \\ 1 & 1 & 2 \end{pmatrix}\begin{pmatrix} x_{11} \\ x_{21} \\ x_{31} \end{pmatrix} = \begin{pmatrix} 0 \\ 0 \\ 0 \end{pmatrix}.$$

Multiplying, we see x_{11}, x_{21}, and x_{31} must simultaneously satisfy the system of equations

$$(3) \qquad \begin{aligned} 2x_{11} - x_{21} + x_{31} &= 0 \\ -x_{11} + 2x_{21} + x_{31} &= 0 \\ x_{11} + x_{21} + 2x_{31} &= 0. \end{aligned}$$

Replacing the first equation in this system by the sum of the first equation and two times the second equation and also replacing the third equation in this system by the sum of the second and third equations, we obtain the following equivalent system of simultaneous equations

$$\begin{aligned} 3x_{21} + 3x_{31} &= 0 \\ -x_{11} + 2x_{21} + x_{31} &= 0 \\ 3x_{21} + 3x_{31} &= 0. \end{aligned}$$

Since the first and third equations in this system are identical, we have a system which consists of only two independent equations in three variables. So the value of one of the variables—x_{11}, x_{21}, or x_{31}—may be selected arbitrarily and the values of the other two variables can be expressed in terms of that variable. Solving the first equation for x_{21} in terms of x_{31}, we find $x_{21} = -x_{31}$. Substituting $x_{21} = -x_{31}$ into the second equation and solving for x_{11}, we get $x_{11} = -x_{31}$. That is, system (3) is satisfied by any vector $\mathbf{x}_1$ in which the component x_{31} is selected arbitrarily, $x_{21} = -x_{31}$ and $x_{11} = -x_{31}$. Choosing $x_{31} = 1$, we get $x_{21} = -1$ and $x_{11} = -1$. So an eigenvector of the matrix $\mathbf{A}$ associated with the eigenvalue $\lambda_1 = -1$ is

$$\mathbf{x}_1 = \begin{pmatrix} x_{11} \\ x_{21} \\ x_{31} \end{pmatrix} = \begin{pmatrix} -1 \\ -1 \\ 1 \end{pmatrix}.$$

Let

$$\mathbf{z} = \begin{pmatrix} z_1 \\ z_2 \\ z_3 \end{pmatrix}$$

be an eigenvector of the matrix $\mathbf{A}$ associated with the eigenvalue $\lambda = 2$. The vector $\mathbf{z}$ must satisfy $(\mathbf{A} - 2\mathbf{I})\mathbf{z} = \mathbf{0}$ or

$$\begin{pmatrix} 1-2 & -1 & 1 \\ -1 & 1-2 & 1 \\ 1 & 1 & 1-2 \end{pmatrix} \begin{pmatrix} z_1 \\ z_2 \\ z_3 \end{pmatrix} = \begin{pmatrix} -1 & -1 & 1 \\ -1 & -1 & 1 \\ 1 & 1 & -1 \end{pmatrix} \begin{pmatrix} z_1 \\ z_2 \\ z_3 \end{pmatrix} = \begin{pmatrix} 0 \\ 0 \\ 0 \end{pmatrix}.$$

Multiplying, we see z_1, z_2, and z_3 must simultaneously satisfy the system of equations

$$\begin{aligned} -z_1 - z_2 + z_3 &= 0 \\ -z_1 - z_2 + z_3 &= 0 \\ z_1 + z_2 - z_3 &= 0. \end{aligned}$$

Observe that the first two equations are identical and the third equation is -1 times the first equation. Thus, this system reduces to the single equation

$$(4) \qquad -z_1 - z_2 + z_3 = 0.$$

Hence, the values of two of the three variables z_1, z_2, and z_3 may be chosen arbitrarily and the third is then determined by equation (4). Since two components of the vector $\mathbf{z}$ may be chosen arbitrarily, we make two different choices of two components in such a manner that the resulting two vectors are linearly independent. For example, choosing $z_2 = 1$, choosing $z_3 = 0$, substituting these values into equation (4) and solving for z_1, we get $z_1 = -1$. So one eigenvector of the matrix $\mathbf{A}$ associated with the eigenvalue $\lambda = 2$ is

$$\mathbf{x}_2 = \begin{pmatrix} z_1 \\ z_2 \\ z_3 \end{pmatrix} = \begin{pmatrix} -1 \\ 1 \\ 0 \end{pmatrix}.$$

Next, choosing $z_2 = 0$, choosing $z_3 = 1$, substituting these values into equation (4), and solving for z_1, we get $z_1 = 1$. Thus, a second eigenvector of the matrix $\mathbf{A}$ associated with the eigenvalue $\lambda = 2$ is

$$\mathbf{x}_3 = \begin{pmatrix} z_1 \\ z_2 \\ z_3 \end{pmatrix} = \begin{pmatrix} 1 \\ 0 \\ 1 \end{pmatrix}.$$

The following computation shows that the set of eigenvectors $\{\mathbf{x}_1, \mathbf{x}_2, \mathbf{x}_3\}$ is linearly independent. Let $\mathbf{X}$ be the 3×3 matrix $\mathbf{X} = (\mathbf{x}_1\ \mathbf{x}_2\ \mathbf{x}_3)$. Then

$$\det \mathbf{X} = \det \begin{pmatrix} -1 & -1 & 1 \\ -1 & 1 & 0 \\ 1 & 0 & 1 \end{pmatrix} = -3 \neq 0.$$

Thus, in this example, there are two linearly independent eigenvectors associated with the eigenvalue $\lambda = 2$ of multiplicity two. ■

Example 4 Find the eigenvalues and associated eigenvectors of the matrix

$$\mathbf{A} = \begin{pmatrix} 1 & -2 \\ 1 & 3 \end{pmatrix}.$$

Solution

The characteristic equation of the matrix $\mathbf{A}$ is

$$\det(\mathbf{A} - \lambda\mathbf{I}) = \det\left(\begin{pmatrix} 1 & -2 \\ 1 & 3 \end{pmatrix} - \lambda\begin{pmatrix} 1 & 0 \\ 0 & 1 \end{pmatrix}\right) = \det\begin{pmatrix} 1-\lambda & -2 \\ 1 & 3-\lambda \end{pmatrix}$$

$$= (1-\lambda)(3-\lambda) + 2 = 3 - 4\lambda + \lambda^2 + 2 = 5 - 4\lambda + \lambda^2 = 0.$$

Solving the quadratic equation $\lambda^2 - 4\lambda + 5 = 0$, we find the eigenvalues of the matrix $\mathbf{A}$ are $\lambda_1 = 2 + i$ and $\lambda_2 = 2 - i$.

An eigenvector

$$\mathbf{x}_1 = \begin{pmatrix} x_{11} \\ x_{21} \end{pmatrix}$$

of the matrix $\mathbf{A}$ corresponding to the eigenvalue $\lambda_1 = 2 + i$ must satisfy

$$(\mathbf{A} - \lambda_1\mathbf{I})\mathbf{x}_1 = \left(\begin{pmatrix} 1 & -2 \\ 1 & 3 \end{pmatrix} - (2+i)\begin{pmatrix} 1 & 0 \\ 0 & 1 \end{pmatrix}\right)\begin{pmatrix} x_{11} \\ x_{21} \end{pmatrix}$$

$$= \begin{pmatrix} 1-(2+i) & -2 \\ 1 & 3-(2+i) \end{pmatrix}\begin{pmatrix} x_{11} \\ x_{21} \end{pmatrix}$$

$$= \begin{pmatrix} -1-i & -2 \\ 1 & 1-i \end{pmatrix}\begin{pmatrix} x_{11} \\ x_{21} \end{pmatrix} = \begin{pmatrix} 0 \\ 0 \end{pmatrix}.$$

Multiplying, we see that x_{11} and x_{21} must simultaneously satisfy the system of equations

$$(5) \qquad \begin{aligned} (-1-i)x_{11} - 2x_{21} &= 0 \\ x_{11} + (1-i)x_{21} &= 0. \end{aligned}$$

Since *det* $(\mathbf{A} - \lambda_1 \mathbf{I}) = 0$, these two equations must be multiples of one another. (To check this fact, multiply the second equation by $(-1-i)$ and obtain the first equation.) Consequently, system (5) reduces to a single condition—either the first equation or the second equation of (5). Hence, one variable x_{11} or x_{21} is arbitrary, and the other is determined by one equation from (5). Choosing $x_{21} = 1$ and solving the second equation of (5) for x_{11}, we get $x_{11} = -1+i$. Hence, an eigenvector of the matrix $\mathbf{A}$ associated with the eigenvalue $\lambda_1 = 2+i$ is

$$\mathbf{x}_1 = \begin{pmatrix} x_{11} \\ x_{21} \end{pmatrix} = \begin{pmatrix} -1+i \\ 1 \end{pmatrix}.$$

Let

$$\mathbf{x}_2 = \begin{pmatrix} x_{12} \\ x_{22} \end{pmatrix}$$

be an eigenvector of the matrix $\mathbf{A}$ associated with the eigenvalue $\lambda_2 = 2-i$. The vector $\mathbf{x}_2$ must satisfy

$$(\mathbf{A} - \lambda_2 \mathbf{I})\mathbf{x}_2 = \left(\begin{pmatrix} 1 & -2 \\ 1 & 3 \end{pmatrix} - (2-i) \begin{pmatrix} 1 & 0 \\ 0 & 1 \end{pmatrix} \right) \begin{pmatrix} x_{12} \\ x_{22} \end{pmatrix}$$

$$= \begin{pmatrix} 1-(2-i) & -2 \\ 1 & 3-(2-i) \end{pmatrix} \begin{pmatrix} x_{12} \\ x_{22} \end{pmatrix}$$

$$= \begin{pmatrix} -1+i & -2 \\ 1 & 1+i \end{pmatrix} \begin{pmatrix} x_{12} \\ x_{22} \end{pmatrix} = \begin{pmatrix} 0 \\ 0 \end{pmatrix}.$$

Thus, x_{12} and x_{22} must simultaneously satisfy

$$(6) \qquad \begin{aligned} (-1+i)x_{12} - 2x_{22} &= 0 \\ x_{12} + (1+i)x_{22} &= 0. \end{aligned}$$

Since det $(\mathbf{A} - \lambda_2 \mathbf{I}) = 0$ these two equations must also be multiples of one another. Therefore, either x_{12} or x_{22} may be selected arbitrarily and the other variable then determined from either equation of (6). Choosing $x_{22} = 1$ and solving the second equation of (6) for x_{12}, we get $x_{12} = -1-i$. Consequently, an eigenvector of the matrix $\mathbf{A}$ associated with the eigenvalue $\lambda_2 = 2-i$ is

$$\mathbf{x}_2 = \begin{pmatrix} x_{12} \\ x_{22} \end{pmatrix} = \begin{pmatrix} -1-i \\ 1 \end{pmatrix}. \quad \blacksquare$$

8.2.1 Comments on Computing

Computer algebra systems (CAS) often include a routine to numerically compute all eigenvalues and eigenvectors of an $n \times n$ matrix with real entries (elements). The input for such programs is the size, n, of the matrix and the matrix itself. The output from the program is a set of n eigenvalues and associated eigenvectors. The typical CAS will easily compute the eigenvalues and eigenvectors of an $n \times n$ matrix with real entries where $2 \leq n \leq 6$. The next example shows the typical output for a generic CAS. When an eigenvalue has multiplicity $m > 1$, the CAS generates m associated vectors. When the m associated eigenvectors are linearly independent, the vectors produced by the program are also linearly independent, as they should be. When an eigenvalue of multiplicity $m > 1$ does not have m associated linearly independent eigenvectors, the associated m vectors produced by the CAS will not be linearly independent. In general, if there are $k < m$ linearly independent eigenvectors associated with a particular eigenvalue, k of the m vectors produced by the CAS will be linearly independent eigenvectors.

Example 5 Use your computing software to compute the eigenvalues and associated eigenvectors of the following matrices. Compare the results with the results obtained in Examples 1, 2, 3, and 4.

1. $\begin{pmatrix} 2 & -3 \\ 1 & -2 \end{pmatrix}$ 2. $\begin{pmatrix} 3 & 1 \\ -1 & 1 \end{pmatrix}$

3. $\begin{pmatrix} 1 & -1 & 1 \\ -1 & 1 & 1 \\ 1 & 1 & 1 \end{pmatrix}$ 4. $\begin{pmatrix} 1 & -2 \\ 1 & 3 \end{pmatrix}$

Solution

1. We used a CAS to calculate the eigenvalues and associated eigenvectors of the given matrix by setting the size of the matrix equal to two and then entering the values for the elements of the matrix. From Figure 8.1 we see that one eigenvalue is $\lambda_1 = 1 + 0i = 1$ and the associated eigenvector is

$$\mathbf{x}_1 = \begin{pmatrix} 1.66410 + 0i \\ .55470 + 0i \end{pmatrix} = \begin{pmatrix} 1.64410 \\ .55470 \end{pmatrix}.$$

We noted in Example 1 that any vector of the form

$$\mathbf{x} = \begin{pmatrix} 3c \\ c \end{pmatrix}$$

where $c \neq 0$ is an eigenvector associated with $\lambda_1 = 1$. Observe that the first component of $\mathbf{x}_1$ is 3 times the second component of $\mathbf{x}_1$. So $\mathbf{x}_1$ is clearly an eigenvector of the given matrix associated with the eigenvalue $\lambda_1 = 1$. When we computed an eigenvector associated with $\lambda_1 = 1$ by

```
THE MATRIX WHOSE EIGENVALUES AND EIGENVECTORS ARE TO BE CALCULATED IS

  2.0000E+00  -3.0000E+00
  1.0000E+00  -2.0000E+00

AN EIGENVALUE IS   1.00000E+00  +  0.00000E+00  I
  THE ASSOCIATED EIGENVECTOR IS
                   1.66410E+00  +  0.00000E+00  I
                   5.54700E-01  +  0.00000E+00  I

AN EIGENVALUE IS  -1.00000E+00  +  0.00000E+00  I
  THE ASSOCIATED EIGENVECTOR IS
                   1.80278E+00  +  0.00000E+00  I
                   1.80278E+00  +  0.00000E+00  I
```

FIGURE 8.1
Eigenvalues and Eigenvectors for Example 5.1.

hand, we selected $c = 1$ and got a different eigenvector than the eigenvector selected by the computer. But this is to be expected, since eigenvectors are not unique. They are unique only up to an arbitrary scalar multiple, as we proved earlier. From Figure 8.1 we see that the second eigenvalue is $\lambda_2 = -1$ and an associated eigenvector is

$$\mathbf{x}_2 = \begin{pmatrix} 1.80278 \\ 1.80278 \end{pmatrix}.$$

In Example 1, the associated eigenvector which we computed manually was $\mathbf{x} = \begin{pmatrix} 1 \\ 1 \end{pmatrix}$. Since the first and second components of $\mathbf{x}_2$ are identical, the vectors $\mathbf{x}_2$ and $\mathbf{x}$ are both eigenvectors associated with $\lambda_2 = -1$. Clearly, $\mathbf{x}_2 = 1.80278\mathbf{x}$.

2. We used a CAS to calculate the eigenvalues and associated eigenvectors of the given matrix by setting the size of the matrix equal to two and then entering the values for the elements of the matrix. The two eigenvalues and associated eigenvectors computed by our CAS are displayed in Figure 8.2. In this case, $\lambda_1 = \lambda_2 = 2$ as we found in Example 2. From Figure 8.2 we see that the computer has generated the following two associated vectors

$$\mathbf{x}_1 = \begin{pmatrix} .707107 \\ -.707107 \end{pmatrix} \quad \text{and} \quad \mathbf{x}_2 = \begin{pmatrix} .318453 \times 10^{16} \\ -.318453 \times 10^{16} \end{pmatrix}.$$

```
THE MATRIX WHOSE EIGENVALUES AND EIGENVECTORS ARE TO BE CALCULATED IS

  3.0000E+00    1.0000E+00
 -1.0000E+00    1.0000E+00

AN EIGENVALUE IS    2.00000E+00  +  0.00000E+00  I
  THE ASSOCIATED EIGENVECTOR IS
                    7.07107E-01  +  0.00000E+00  I
                   -7.07107E-01  +  0.00000E+00  I

AN EIGENVALUE IS    2.00000E+00  +  0.00000E+00  I
  THE ASSOCIATED EIGENVECTOR IS
                    3.18453E+15  +  0.00000E+00  I
                   -3.18453E+15  +  0.00000E+00  I
```

FIGURE 8.2
Eigenvalues and Eigenvectors for Example 5.2.

These vectors are linearly dependent, since $\mathbf{x}_2$ is a multiple of $\mathbf{x}_1$—that is, $\mathbf{x}_2 = k\mathbf{x}_1$. In Example 2, we found the matrix under consideration here has only one eigenvector associated with the eigenvalue $\lambda = 2$. The single eigenvector we calculated by hand in Example 2 was $\mathbf{x} = \begin{pmatrix} -1 \\ 1 \end{pmatrix}$. Notice that $\mathbf{x}_1$ and $\mathbf{x}_2$ are both scalar multiples of $\mathbf{x}$. This example is intended to show that we can sometimes determine directly from computer output when eigenvalues of multiplicity $m > 1$ have fewer than m linearly independent associated eigenvectors.

3. In Example 3 we manually calculated the eigenvalues of the given matrix to be $\lambda_1 = -1$, $\lambda_2 = 2$, and $\lambda_3 = 2$. And we found the following set of three linearly independent associated eigenvectors

$$\mathbf{x}_1 = \begin{pmatrix} -1 \\ -1 \\ 1 \end{pmatrix}, \qquad \mathbf{x}_2 = \begin{pmatrix} -1 \\ 1 \\ 0 \end{pmatrix}, \qquad \mathbf{x}_3 = \begin{pmatrix} 1 \\ 0 \\ 1 \end{pmatrix}.$$

Using our CAS, we found, as shown in Figure 8.3, that the eigenvalues of the given matrix are -1, 2, 2 and that the associated eigenvectors are respectively

$$\mathbf{z}_1 = \begin{pmatrix} .745356 \\ .745356 \\ -.745356 \end{pmatrix}, \qquad \mathbf{z}_2 = \begin{pmatrix} .894427 \\ -.447214 \\ .447214 \end{pmatrix}, \qquad \mathbf{z}_3 = \begin{pmatrix} .311803 \\ .344098 \\ .655902 \end{pmatrix}.$$

Notice that $\mathbf{z}_1 = -.745356\mathbf{x}_1$, but $\mathbf{z}_2$ is not a multiple of $\mathbf{x}_2$ or $\mathbf{x}_3$ and $\mathbf{z}_3$ is not a multiple of $\mathbf{x}_2$ or $\mathbf{x}_3$. The easiest way for us to show that $\mathbf{z}_2$ and

```
THE MATRIX WHOSE EIGENVALUES AND EIGENVECTORS ARE TO BE CALCULATED IS

  1.0000E+00  -1.0000E+00   1.0000E+00
 -1.0000E+00   1.0000E+00   1.0000E+00
  1.0000E+00   1.0000E+00   1.0000E+00

AN EIGENVALUE IS    2.00000E+00  +  0.00000E+00  I
  THE ASSOCIATED EIGENVECTOR IS
                    8.94427E-01  +  0.00000E+00  I
                   -4.47214E-01  +  0.00000E+00  I
                    4.47214E-01  +  0.00000E+00  I

AN EIGENVALUE IS   -1.00000E+00  +  0.00000E+00  I
  THE ASSOCIATED EIGENVECTOR IS
                    7.45356E-01  +  0.00000E+00  I
                    7.45356E-01  +  0.00000E+00  I
                   -7.45356E-01  +  0.00000E+00  I

AN EIGENVALUE IS    2.00000E+00  +  0.00000E+00  I
  THE ASSOCIATED EIGENVECTOR IS
                    3.11803E-01  +  0.00000E+00  I
                    3.44098E-01  +  0.00000E+00  I
                    6.55902E-01  +  0.00000E+00  I
```

FIGURE 8.3
Eigenvalues and Eigenvectors for Example 5.3.

$\mathbf{z}_3$ are linearly independent eigenvectors associated with the eigenvalue $\lambda = 2$ of multiplicity 2 is to show that the set $\{\mathbf{z}_1, \mathbf{z}_2, \mathbf{z}_3\}$ is linearly independent. We do so by computing $det\,(\mathbf{z}_1\ \mathbf{z}_2\ \mathbf{z}_3)$ and finding its value to be $-3(.745356)(.447214) = -1 \neq 0$.

4. In Example 4, we manually calculated the eigenvalues of the given matrix to be $\lambda_1 = 2 + i$ and $\lambda_2 = 2 - i$. And we found the associated eigenvectors to be

$$\mathbf{x}_1 = \begin{pmatrix} -1+i \\ 1 \end{pmatrix} \quad \text{and} \quad \mathbf{x}_2 = \begin{pmatrix} -1-i \\ 1 \end{pmatrix}.$$

When the entries of a matrix are all real, complex eigenvalues and their associated eigenvectors occur in complex conjugate pairs. Observe in this instance λ_1 and λ_2 are complex conjugate scalars and $\mathbf{x}_1$ and $\mathbf{x}_2$ are complex conjugate vectors. Results for the given matrix computed using our CAS are shown in Figure 8.4. The eigenvalues computed are identical with those computed by hand, but the associated eigenvectors are

$$\mathbf{z}_1 = \begin{pmatrix} -1-i \\ i \end{pmatrix} \quad \text{and} \quad \mathbf{z}_2 = \begin{pmatrix} -1+i \\ -i \end{pmatrix}.$$

Notice that $\mathbf{z}_1$ and $\mathbf{z}_2$ are complex conjugate vectors. Also notice $\mathbf{z}_1 = i\mathbf{x}_1$ and $\mathbf{z}_2 = -i\mathbf{x}_2$.

```
THE MATRIX WHOSE EIGENVALUES AND EIGENVECTORS ARE TO BE CALCULATED IS

  1.0000E+00  -2.0000E+00
  1.0000E+00   3.0000E+00

AN EIGENVALUE IS   2.00000E+00  +  1.00000E+00  I
  THE ASSOCIATED EIGENVECTOR IS
                  -1.00000E+00  + -1.00000E+00  I
                   0.00000E+00  +  1.00000E+00  I

AN EIGENVALUE IS   2.00000E+00  + -1.00000E+00  I
  THE ASSOCIATED EIGENVECTOR IS
                  -1.00000E+00  +  1.00000E+00  I
                   0.00000E+00  + -1.00000E+00  I
```

FIGURE 8.4
Complex Conjugate Eigenvalues and Eigenvectors for Example 5.4. ■

8.2.2 Exercises

Use your computing software to compute the eigenvalues and associated eigenvectors of the following matrices.

1. $\begin{pmatrix} -2 & -4 \\ 1 & 3 \end{pmatrix}$

2. $\begin{pmatrix} -3 & -1 \\ 2 & -1 \end{pmatrix}$

3. $\begin{pmatrix} 1 & 0 & 1 \\ 0 & 1 & -1 \\ -2 & 0 & -1 \end{pmatrix}$

4. $\begin{pmatrix} 3 & 1 & -1 \\ 1 & 3 & -1 \\ 3 & 3 & -1 \end{pmatrix}$

5. $\begin{pmatrix} 7 & -1 & 6 \\ -10 & 4 & -12 \\ -2 & 1 & -1 \end{pmatrix}$

6. $\begin{pmatrix} 1 & 1 & 1 & 1 \\ 1 & 1 & 1 & 1 \\ 1 & 1 & 1 & 1 \\ 1 & 1 & 1 & 1 \end{pmatrix}$

7. $\begin{pmatrix} 1 & 3 & 5 & 7 \\ 2 & 6 & 10 & 14 \\ 3 & 9 & 15 & 21 \\ 6 & 18 & 30 & 42 \end{pmatrix}$ 8. $\begin{pmatrix} 1 & 3 & 5 & 2 & 4 \\ 5 & 2 & 4 & 1 & 3 \\ 4 & 1 & 3 & 5 & 2 \\ 3 & 5 & 2 & 4 & 1 \\ 2 & 4 & 1 & 3 & 5 \end{pmatrix}$

8.3 Linear Systems with Constant Coefficients

Numerous physical phenomena can be modeled by systems of first-order linear differential equations with constant coefficients. We will show how to model coupled pendulums by such a system.

Coupled Pendulums A pair of identical pendulums with bobs of mass m and rods of length ℓ are coupled by a spring with spring constant k as shown in Figure 8.5.

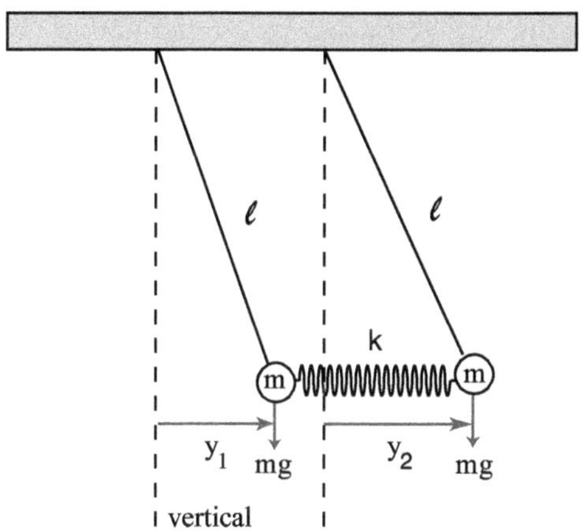

FIGURE 8.5
Coupled Pendulums.

Let y_1 and y_2 denote the displacement of each pendulum from the vertical (positive displacement is to the right). When the pendulums are in their equilibrium positions ($y_1 = y_2 = 0$), the spring is horizontal and not stretched nor compressed. Suppose at some time t, the positive horizontal displacement of the bobs are y_1 and y_2, so that the spring is stretched by the amount $y_2 - y_1$ and the spring exerts a force of $k(y_1 - y_2)$. Assuming the motion is undamped and the displacement of the pendulum bobs is small so that the restoring force due to the weight mg is mgy/ℓ, then the second-order system of equations satisfied by the displacements of the coupled pendulums is

$$\begin{aligned} my_1'' &= \frac{-mgy_1}{\ell} - k(y_1 - y_2) \\ my_2'' &= \frac{-mgy_2}{\ell} - k(y_2 - y_1). \end{aligned}$$

Letting $u_1 = y_1$ (the position of pendulum 1), $u_2 = y_1'$ (the velocity of pendulum 1), $u_3 = y_2$ (the position of pendulum 2), and $u_4 = y_2'$ (the velocity of pendulum 2) and differentiating each of these equations, we find

$$\begin{aligned} u_1' &= y_1' \\ u_2' &= y_1'' = -\frac{g}{\ell}y_1 - \frac{k}{m}(y_1 - y_2) = -\left(\frac{g}{\ell} + \frac{k}{m}\right)y_1 + \frac{k}{m}y_2 \\ u_3' &= y_2' \\ u_4' &= y_2'' = -\frac{g}{\ell}y_2 - \frac{k}{m}(y_2 - y_1) = \frac{k}{m}y_1 - \left(\frac{g}{\ell} + \frac{k}{m}\right)y_2. \end{aligned}$$

Next, replacing y_1 by u_1, y_1' by u_2, y_2 by u_3, and y_2' by u_4 on the right-hand side of the equation above, we obtain the following first-order system of differential equations

$$\begin{aligned} u_1' &= u_2 \\ u_2' &= -\left(\frac{g}{\ell} + \frac{k}{m}\right)u_1 + \frac{k}{m}u_3 \\ u_3' &= u_4 \\ u_4' &= \frac{k}{m}u_1 - \left(\frac{g}{\ell} + \frac{k}{m}\right)u_3. \end{aligned}$$

Written in matrix notation, this system of equations is

$$\begin{pmatrix} u_1' \\ u_2' \\ u_3' \\ u_4' \end{pmatrix} = \begin{pmatrix} 0 & 1 & 0 & 0 \\ -\left(\frac{g}{\ell} + \frac{k}{m}\right) & 0 & \frac{k}{m} & 0 \\ 0 & 0 & 0 & 1 \\ \frac{k}{m} & 0 & -\left(\frac{g}{\ell} + \frac{k}{m}\right) & 0 \end{pmatrix} \begin{pmatrix} u_1 \\ u_2 \\ u_3 \\ u_4 \end{pmatrix}.$$

Observe that the 4×4 matrix appearing in this system is a constant matrix.

If $\mathbf{y}$ is an $n \times 1$ column vector whose entries are functions of a single independent variable x, then the derivative of $\mathbf{y}$, denoted by $\mathbf{y}'$ or $d\mathbf{y}/dx$, is the $n \times 1$ column vector whose entries are the derivatives of the corresponding entries of $\mathbf{y}$. That is, if

$$\mathbf{y}(x) = \begin{pmatrix} y_1(x) \\ y_2(x) \\ \vdots \\ y_n(x) \end{pmatrix}, \quad \text{then} \quad \mathbf{y}'(x) = \begin{pmatrix} y_1'(x) \\ y_2'(x) \\ \vdots \\ y_n'(x) \end{pmatrix}.$$

Recall from Chapter 7 that an initial value problem for a system of n first-order differential equations is the problem of solving the system of n differential equations

$$\text{(1a)} \qquad \begin{aligned} y_1' &= f_1(x, y_1, y_2, \ldots, y_n) \\ y_2' &= f_2(x, y_1, y_2, \ldots, y_n) \\ \vdots \; &\;\; \vdots \qquad\qquad \vdots \\ y_n' &= f_n(x, y_1, y_2, \ldots, y_n) \end{aligned}$$

subject to the n constraints

$$\text{(1b)} \qquad y_1(c) = d_1, \; y_2(c) = d_2, \ldots, \; y_n(c) = d_n.$$

If we let

$$\mathbf{y} = \begin{pmatrix} y_1 \\ y_2 \\ \vdots \\ y_n \end{pmatrix}, \quad \mathbf{f}(x, \mathbf{y}) = \begin{pmatrix} f_1(x, y_1, y_2, \ldots, y_n) \\ f_2(x, y_1, y_2, \ldots, y_n) \\ \vdots \\ f_n(x, y_1, y_2, \ldots, y_n) \end{pmatrix}, \quad \text{and} \quad \mathbf{d} = \begin{pmatrix} d_1 \\ d_2 \\ \vdots \\ d_n \end{pmatrix},$$

then using vector notation, we can write the initial value problem (1) more concisely as

$$\text{(2)} \qquad \mathbf{y}' = \mathbf{f}(x, \mathbf{y}); \quad \mathbf{y}(c) = \mathbf{d}.$$

Notice that the vector initial value problem (2) is very similar in appearance to the scalar initial value problem

$$\text{(3)} \qquad y' = f(x, y); \quad y(c) = d,$$

which we studied in Chapters 2 and 3. In fact, when $n = 1$, the vector initial value problem (2) is exactly the scalar initial value problem (3).

Also recall from Chapter 7 that a linear system initial value problem is the problem of solving the system of n linear first-order differential equations

$$\text{(4a)} \qquad \begin{array}{ccccccccc} y_1' & = & a_{11}(x)y_1 & + a_{12}(x)y_2 + & \cdots & + a_{1n}(x)y_n + & b_1(x) \\ y_2' & = & a_{21}(x)y_1 & + a_{22}(x)y_2 + & \cdots & + a_{2n}(x)y_n + & b_2(x) \\ \vdots & \vdots & \vdots & \vdots & \vdots & \vdots & \vdots \\ y_n' & = & a_{n1}(x)y_1 & + a_{n2}(x)y_2 + & \cdots & + a_{nn}(x)y_n + & b_n(x) \end{array}$$

subject to the n constraints

(4b) $$y_1(c) = d_1,\; y_2(c) = d_2, \ldots,\; y_n(c) = d_n.$$

If we let $\mathbf{y}(x)$, $\mathbf{b}(x)$, and $\mathbf{d}$ be the $n \times 1$ column vectors

$$\mathbf{y}(x) = \begin{pmatrix} y_1(x) \\ y_2(x) \\ \vdots \\ y_n(x) \end{pmatrix}, \quad \mathbf{b}(x) = \begin{pmatrix} b_1(x) \\ b_2(x) \\ \vdots \\ b_n(x) \end{pmatrix}, \quad \mathbf{d} = \begin{pmatrix} d_1 \\ d_2 \\ \vdots \\ d_n \end{pmatrix}$$

and if we let $\mathbf{A}(x)$ be the $n \times n$ matrix

$$\mathbf{A}(x) = \begin{pmatrix} a_{11}(x) & a_{12}(x) & \cdots & a_{1n}(x) \\ a_{21}(x) & a_{22}(x) & \cdots & a_{2n}(x) \\ \vdots & \vdots & \ddots & \vdots \\ a_{n1}(x) & a_{n2}(x) & \cdots & a_{nn}(x) \end{pmatrix}$$

then using matrix notation, we can write the linear first-order system initial value problem (4) more concisely as

(5) $$\mathbf{y}' = \mathbf{A}(x)\mathbf{y} + \mathbf{b}(x); \quad \mathbf{y}(c) = \mathbf{d}.$$

For example, using matrix-vector notation, the system of equations

$$\begin{aligned} y_1' &= 3y_1 - 4y_2 + x \\ y_2' &= -2y_1 + y_2 - \sin x \end{aligned}$$

can be written as

$$\begin{pmatrix} y_1' \\ y_2' \end{pmatrix} = \begin{pmatrix} 3 & -4 \\ -2 & 1 \end{pmatrix} \begin{pmatrix} y_1 \\ y_2 \end{pmatrix} + \begin{pmatrix} x \\ -\sin x \end{pmatrix}.$$

The system of linear first-order differential equations

(6) $$\mathbf{y}' = \mathbf{A}(x)\mathbf{y} + \mathbf{b}(x)$$

is said to be **homogeneous** provided $\mathbf{b}(x) \equiv \mathbf{0}$ and **nonhomogeneous** provided $\mathbf{b}(x) \neq \mathbf{0}$. Thus, a homogeneous linear system is one of the form

(7) $$\mathbf{y}' = \mathbf{A}(x)\mathbf{y}.$$

The development of the theory for systems of linear first-order differential equations closely parallels that for n-th order linear differential equations. We state the following **Superposition Theorem for Homogeneous Linear Systems** without proof.

Theorem 8.2 If $\mathbf{y}_1$ and $\mathbf{y}_2$ are any two solutions of the homogeneous linear system (7) $\mathbf{y}' = \mathbf{A}(x)\mathbf{y}$, then $\mathbf{y}_3 = c_1\mathbf{y}_1 + c_2\mathbf{y}_2$ where c_1 and c_2 are arbitrary scalar constants is also a solution of (7).

The superposition theorem can easily be generalized to show that if $\mathbf{y}_1$, $\mathbf{y}_2, \ldots, \mathbf{y}_m$ are any m solutions of (7) $\mathbf{y}' = \mathbf{A}(x)\mathbf{y}$, then $\mathbf{y} = c_1\mathbf{y}_1 + c_2\mathbf{y}_2 + \cdots + c_m\mathbf{y}_m$ where $c_1, c_2, \ldots, c_m$ are arbitrary scalar constants is also a solution of the homogeneous system (7).

We state the following **Existence Theorem for Homogeneous Linear Systems** without proof.

Theorem 8.3 If $\mathbf{A}(x)$ is continuous on some interval (α, β)—that is, if $a_{ij}(x)$ is a continuous function on (α, β) for all $i, j = 1, 2, \ldots, n$, then there exist n linearly independent solutions of the homogeneous linear system (7) $\mathbf{y}' = \mathbf{A}(x)\mathbf{y}$ on the interval (α, β).

The existence theorem just stated tells us there are n linearly independent solutions $\mathbf{y}_1, \mathbf{y}_2, \ldots, \mathbf{y}_n$ of $\mathbf{y}' = \mathbf{A}(x)\mathbf{y}$ on the interval (α, β), provided $\mathbf{A}(x)$ is continuous on (α, β).

The following **Representation Theorem for Homogeneous Linear Systems** tells us how to write every other solution in terms of $\mathbf{y}_1, \mathbf{y}_2, \ldots, \mathbf{y}_n$.

Theorem 8.4 If $\mathbf{A}(x)$ is continuous on the interval (α, β), if $\mathbf{y}_1, \mathbf{y}_2, \ldots, \mathbf{y}_n$ are linearly independent solutions of the homogeneous linear system (7) $\mathbf{y}' = \mathbf{A}(x)\mathbf{y}$ on (α, β), and if $\mathbf{y}$ is any other solution of (7) on (α, β), then there exist scalar constants $c_1, c_2, \ldots, c_n$ such that $\mathbf{y}(x) = c_1\mathbf{y}_1(x) + c_2\mathbf{y}_2(x) + \cdots + c_n\mathbf{y}_n(x)$ on (α, β).

When $\mathbf{y}_1, \mathbf{y}_2, \ldots, \mathbf{y}_n$ are linearly independent solutions of the homogeneous linear system (7) $\mathbf{y}' = \mathbf{A}(x)\mathbf{y}$ on the interval (α, β), the linear combination $\mathbf{y}(x) = c_1\mathbf{y}_1(x) + c_2\mathbf{y}_2(x) + \cdots + c_n\mathbf{y}_n(x)$ where $c_1, c_2, \ldots, c_n$ are arbitrary scalar constants is called the **general solution** of (7) on (α, β).

Summarizing, the existence theorem says there are at least n linearly independent solutions of $\mathbf{y}' = \mathbf{A}(x)\mathbf{y}$, and the representation theorem states that there are at most n linearly independent solutions. Thus, the existence theorem gives us license to seek n linearly independent solutions of $\mathbf{y}' = \mathbf{A}(x)\mathbf{y}$, and the representation theorem tells us how to write the general solution (all other solutions) in terms of these solutions. So our task in solving $\mathbf{y}' = \mathbf{A}(x)\mathbf{y}$ is reduced to one of finding n linearly independent solutions—that is, our task becomes one of finding a set of n solutions and testing the set for linear independence. The following **Theorem on Linear Independence of Solutions of Homogeneous Linear Systems** aids in the determination of linear independence of solutions.

Theorem 8.5 Let $\mathbf{y}_1(x), \mathbf{y}_2(x), \ldots, \mathbf{y}_n(x)$ be solutions of the homogeneous linear system (7) $\mathbf{y}' = \mathbf{A}(x)\mathbf{y}$ on some interval (α, β). The set of functions $\{\mathbf{y}_1, \mathbf{y}_2, \ldots, \mathbf{y}_n\}$ is linearly independent on (α, β) if and only if for some $x_0 \in (\alpha, \beta)$ the determinant $det\,(\mathbf{y}_1\ \mathbf{y}_2 \cdots \mathbf{y}_n) \neq 0$.

Consequently, to check a set of n solutions $\{\mathbf{y}_1(x),\ \mathbf{y}_2(x), \ldots,\ \mathbf{y}_n(x)\}$ of $\mathbf{y}' = \mathbf{A}(x)\mathbf{y}$ for linear independence on (α, β), we only need to calculate the determinant of the matrix whose columns are $\mathbf{y}_1, \mathbf{y}_2, \ldots, \mathbf{y}_n$ evaluated at some convenient point $x_0 \in (\alpha, \beta)$. If $det\,(\mathbf{y}_1(x_0)\ \mathbf{y}_2(x_0) \cdots \mathbf{y}_n(x_0)) \neq 0$, then the solutions are linearly independent; whereas, if the determinant is zero, then the solutions are linearly dependent.

Example 1 Verify that

$$\mathbf{y}_1(x) = \begin{pmatrix} 1 \\ 2x \end{pmatrix} \quad \text{and} \quad \mathbf{y}_2(x) = \begin{pmatrix} x \\ x^2 \end{pmatrix}$$

are linearly independent solutions of the homogeneous linear system

$$\mathbf{y}' = \begin{pmatrix} \dfrac{2}{x} & \dfrac{-1}{x^2} \\ 2 & 0 \end{pmatrix} \mathbf{y} = \mathbf{A}(x)\mathbf{y} \tag{8}$$

on the interval $(0, \infty)$ and write the general solution of (8) on $(0, \infty)$.

Solution

Notice that $\mathbf{A}(x)$ is not defined for $x = 0$, but $\mathbf{A}(x)$ is defined and continuous on $(-\infty, 0)$ and $(0, \infty)$. So by the existence theorem, there are two linearly independent solutions of (8) on $(0, \infty)$.

Differentiating $\mathbf{y}_1$, we find

$$\mathbf{y}_1'(x) = \begin{pmatrix} 0 \\ 2 \end{pmatrix}.$$

Multiplying $\mathbf{A}(x)$ by $\mathbf{y}_1(x)$, we get for $x \neq 0$

$$\mathbf{A}(x)\mathbf{y}_1(x) = \begin{pmatrix} \dfrac{2}{x} & \dfrac{-1}{x^2} \\ 2 & 0 \end{pmatrix} \begin{pmatrix} 1 \\ 2x \end{pmatrix} = \begin{pmatrix} \dfrac{2}{x}(1) + \dfrac{-1}{x^2}(2x) \\ 2(1) + 0(2x) \end{pmatrix}$$

$$= \begin{pmatrix} \dfrac{2}{x} - \dfrac{2}{x} \\ 2 \end{pmatrix} = \begin{pmatrix} 0 \\ 2 \end{pmatrix}.$$

Since $\mathbf{y}_1' = \mathbf{A}(x)\mathbf{y}_1$ for $x \neq 0$, the vector $\mathbf{y}_1$ is a solution of (8) on $(0, \infty)$.

Differentiating $\mathbf{y}_2(x)$, we find

$$\mathbf{y}_2'(x) = \begin{pmatrix} 1 \\ 2x \end{pmatrix}.$$

Computing $\mathbf{A}(x)\mathbf{y}_2(x)$, we obtain for $x \neq 0$

$$\mathbf{A}(x)\mathbf{y}_2(x) = \begin{pmatrix} \frac{2}{x} & \frac{-1}{x^2} \\ 2 & 0 \end{pmatrix} \begin{pmatrix} x \\ x^2 \end{pmatrix} = \begin{pmatrix} \frac{2}{x}(x) + \frac{-1}{x^2}(x^2) \\ 2(x) + 0(x^2) \end{pmatrix}$$

$$= \begin{pmatrix} 2-1 \\ 2x \end{pmatrix} = \begin{pmatrix} 1 \\ 2x \end{pmatrix}.$$

Since $\mathbf{y}_2' = \mathbf{A}(x)\mathbf{y}_2$ for $x \neq 0$, the vector $\mathbf{y}_2$ is a solution of (8) on $(0, \infty)$.

To determine whether the set $\{\mathbf{y}_1(x), \mathbf{y}_2(x)\}$ is linearly dependent or linearly independent on $(0, \infty)$, we compute *det* $(\mathbf{y}_1(x)\ \mathbf{y}_2(x))$ at some convenient point $x_0 \in (0, \infty)$. As a matter of convenience, we decided to choose $x_0 = 1$. Computing, we get

$$\det(\mathbf{y}_1(1)\ \mathbf{y}_2(1)) = \det \begin{pmatrix} 1 & 1 \\ 2 & 1 \end{pmatrix} = 1 - 2 = -1.$$

Since $\det(\mathbf{y}_1(1)\ \mathbf{y}_2(1)) = -1 \neq 0$, the vectors $\mathbf{y}_1$ and $\mathbf{y}_2$ are linearly independent on $(0, \infty)$. Therefore, by the representation theorem, the general solution of (8) on $(0, \infty)$ is

$$\mathbf{y}(x) = c_1\mathbf{y}_1(x) + c_2\mathbf{y}_2(x) = c_1 \begin{pmatrix} 1 \\ 2x \end{pmatrix} + c_2 \begin{pmatrix} x \\ x^2 \end{pmatrix} = \begin{pmatrix} c_1 + c_2 x \\ 2c_1 x + c_2 x^2 \end{pmatrix}$$

where c_1 and c_2 are arbitrary scalar constants. ■

Now let us consider the nonhomogeneous linear system of equations

$$\mathbf{y}' = \mathbf{A}(x)\mathbf{y} + \mathbf{b}(x) \tag{9}$$

where $\mathbf{y}$, $\mathbf{y}'$, and $\mathbf{b}(x)$ are $n \times 1$ column vectors; $\mathbf{A}(x)$ is an $n \times n$ matrix; and $\mathbf{b}(x) \neq \mathbf{0}$.

The system of differential equations

$$\mathbf{y}' = \mathbf{A}(x)\mathbf{y} \tag{10}$$

is called the **associated homogeneous system** for the nonhomogeneous system (9) $\mathbf{y}' = \mathbf{A}(x)\mathbf{y} + \mathbf{b}(x)$. Any solution $\mathbf{y}_p(x)$ of the nonhomogeneous system (9) which includes no arbitrary constant is called a **particular solution** of (9).

The following **Representation Theorem for Nonhomogeneous Linear Systems** tells us how to write the general solution of the nonhomogeneous linear system (9).

Theorem 8.6 If $\mathbf{A}(x)$ is continuous on the interval (α, β), if $\mathbf{y}_p(x)$ is any particular solution on (α, β) of the nonhomogeneous linear system (9) $\mathbf{y}' = \mathbf{A}(x)\mathbf{y} + \mathbf{b}(x)$, and if $\mathbf{y}_1, \mathbf{y}_2, \ldots, \mathbf{y}_n$ are n linearly independent solutions on the interval (α, β) of the associated homogeneous linear system (10) $\mathbf{y}' = \mathbf{A}(x)\mathbf{y}$, then every solution of the nonhomogeneous system (9) on (α, β) has the form

$$\text{(11)} \qquad \mathbf{y}(x) = c_1\mathbf{y}_1 + c_2\mathbf{y}_2 + \cdots + c_n\mathbf{y}_n + \mathbf{y}_p$$

where $c_1, c_2, \ldots, c_n$ are scalar constants.

Since every solution of (9) can be written in the form of equation (11), this equation is called the **general solution** of the nonhomogeneous system (9). The general solution of the associated homogeneous system, namely $\mathbf{y}_c = c_1\mathbf{y}_1 + c_2\mathbf{y}_2 + \cdots + c_n\mathbf{y}_n$, is called the **complementary solution**. Thus, the general solution of the nonhomogeneous system (9) is $\mathbf{y} = \mathbf{y}_c + \mathbf{y}_p$ where $\mathbf{y}_c$ is the complementary solution and $\mathbf{y}_p$ is any particular solution of the nonhomogeneous system. So to find the general solution of the nonhomogeneous system (9) $\mathbf{y}' = \mathbf{A}(x)\mathbf{y} + \mathbf{b}(x)$, we first find the general solution of the associated homogeneous system (10) $\mathbf{y}' = \mathbf{A}(x)\mathbf{y}$, next we find a particular solution of the nonhomogeneous system (9), and then we add the results.

Example 2 Verify that

$$\mathbf{y}_p = \begin{pmatrix} -x \\ x^2 \end{pmatrix}$$

is a particular solution on the interval $(0, \infty)$ **of the nonhomogeneous system**

$$\text{(12)} \qquad \mathbf{y}' = \begin{pmatrix} \dfrac{2}{x} & \dfrac{-1}{x^2} \\ 2 & 0 \end{pmatrix} \mathbf{y} + \begin{pmatrix} 2 \\ 4x \end{pmatrix} = \mathbf{A}(x)\mathbf{y} + \mathbf{b}(x)$$

and write the general solution of (12) on $(0, \infty)$.

Solution

Differentiating $\mathbf{y}_p$, we find

$$\mathbf{y}_p' = \begin{pmatrix} -1 \\ 2x \end{pmatrix}.$$

Computing $\mathbf{A}(x)\mathbf{y}_p + \mathbf{b}(x)$, yields

$$\begin{pmatrix} \dfrac{2}{x} & \dfrac{-1}{x^2} \\ 2 & 0 \end{pmatrix} \begin{pmatrix} -x \\ x^2 \end{pmatrix} + \begin{pmatrix} 2 \\ 4x \end{pmatrix} = \begin{pmatrix} \dfrac{2}{x}(-x) + \dfrac{-1}{x^2}(x^2) \\ 2(-x) + 0(x^2) \end{pmatrix} + \begin{pmatrix} 2 \\ 4x \end{pmatrix}$$

$$= \begin{pmatrix} -2 - 1 + 2 \\ -2x + 4x \end{pmatrix} = \begin{pmatrix} -1 \\ 2x \end{pmatrix}.$$

Since $\mathbf{y}_p' = \mathbf{A}(x)\mathbf{y}_p + \mathbf{b}(x)$, the vector $\mathbf{y}_p$ is a particular solution of (12). In the previous example, we found the general solution of the associated homogeneous equation $\mathbf{y}' = \mathbf{A}(x)\mathbf{y}$ to be

$$\mathbf{y}_c = c_1 \begin{pmatrix} 1 \\ 2x \end{pmatrix} + c_2 \begin{pmatrix} x \\ x^2 \end{pmatrix}.$$

Therefore, the general solution of the nonhomogeneous linear system (12) is

$$\mathbf{y}(x) = \mathbf{y}_c + \mathbf{y}_p = c_1 \begin{pmatrix} 1 \\ 2x \end{pmatrix} + c_2 \begin{pmatrix} x \\ x^2 \end{pmatrix} + \begin{pmatrix} -x \\ x^2 \end{pmatrix} = \begin{pmatrix} c_1 + c_2 x - x \\ 2c_1 x + c_2 x^2 + x^2 \end{pmatrix}$$

where c_1 and c_2 are arbitrary constants. ■

Now let us consider the general, homogeneous linear system with constant coefficients

$$\text{(13)} \qquad \mathbf{y}' = \mathbf{A}\mathbf{y}$$

where $\mathbf{y}$ and $\mathbf{y}'$ are $n \times 1$ column vectors, and $\mathbf{A}$ is an $n \times n$ matrix of real numbers. In order to solve an n-th order linear homogeneous differential equation, we assumed there were solutions of the form $y = e^{rx}$, where r is an unknown constant. By analogy, we seek a solution of (13) of the form

$$\text{(14)} \qquad \mathbf{y} = \mathbf{v}e^{rx}$$

where $\mathbf{v}$ is an unknown constant vector and r is an unknown scalar constant. Differentiating (14), we find $\mathbf{y}' = r\mathbf{v}e^{rx}$. Substituting into (13), we see $\mathbf{v}$ and r must be chosen to satisfy

$$\mathbf{y}' = \mathbf{A}\mathbf{y} \quad \text{or} \quad r\mathbf{v}e^{rx} = \mathbf{A}(\mathbf{v}e^{rx}) = \mathbf{A}\mathbf{v}e^{rx}.$$

Cancelling the nonzero scalar factor e^{rx}, we find the unknowns $\mathbf{v}$ and r must satisfy $\mathbf{A}\mathbf{v} = r\mathbf{v}$. That is, for (14) $\mathbf{y} = \mathbf{v}e^{rx}$ to be a solution of (13), r must be an eigenvalue of $\mathbf{A}$ and $\mathbf{v}$ must be an associated eigenvector. Hence, we immediately have the following theorem.

Theorem 8.7 If $r_1, r_2, \ldots, r_n$ are the eigenvalues (not necessarily distinct) of an $n \times n$ constant matrix $\mathbf{A}$ and if $\mathbf{v}_1$, $\mathbf{v}_2, \ldots, \mathbf{v}_n$ are associated linearly independent eigenvectors, then the general solution of the homogeneous linear system $\mathbf{y}' = \mathbf{A}\mathbf{y}$ is

$$\text{(15)} \qquad \mathbf{y}(x) = c_1\mathbf{v}_1 e^{r_1 x} + c_2\mathbf{v}_2 e^{r_2 x} + \cdots + c_n\mathbf{v}_n e^{r_n x}$$

where $c_1, c_2, \ldots, c_n$ are arbitrary constants.

If the eigenvalues of $\mathbf{A}$ are distinct, then there are n linearly independent eigenvectors—one eigenvector corresponding to each eigenvalue. Moreover, if each eigenvalue of multiplicity $m > 1$ has m associated linearly independent eigenvectors, then there are a total of n linearly independent eigenvectors. In either of these cases, we can use the computer software to find the eigenvalues and associated linearly independent eigenvectors and thereby write the general solution of (13) $\mathbf{y}' = \mathbf{A}\mathbf{y}$ in the form of equation (15). It is only when there is some eigenvalue of multiplicity $m > 1$ with fewer than m linearly independent associated eigenvectors that we will not be able to write the solution of (13) $\mathbf{y}' = \mathbf{A}\mathbf{y}$ in the form of equation (15). In such a case, other techniques must be used to find the general solution of (13). These techniques will not be discussed in this text.

Example 3 Find the general solution of the homogeneous linear system

$$\text{(16)} \qquad \mathbf{y}' = \begin{pmatrix} 2 & 0 & 1 \\ 0 & 1 & 0 \\ 1 & 0 & 2 \end{pmatrix} \mathbf{y}.$$

Solution

We ran EIGEN by setting the size of the matrix equal to 3 and entering the values for the elements of the given matrix. The output of the program is displayed in Figure 8.6. From this figure, we see that the eigenvalues are 3, 1, 1 and the associated eigenvectors are respectively

$$\mathbf{v}_1 = \begin{pmatrix} .707107 \\ 0.000000 \\ .707107 \end{pmatrix}, \quad \mathbf{v}_2 = \begin{pmatrix} -.707107 \\ 0.000000 \\ .707107 \end{pmatrix}, \text{ and } \quad \mathbf{v}_3 = \begin{pmatrix} 0 \\ 1 \\ 0 \end{pmatrix}.$$

We see that the eigenvalue 1 has multiplicity two. Thus, we need to verify that the associated eigenvectors $\mathbf{v}_2$ and $\mathbf{v}_3$ are linearly independent. (In general, suppose $\mathbf{u} \neq \mathbf{0}$ and $\mathbf{v} \neq \mathbf{0}$ are linearly dependent vectors. Then, by definition of linearly dependent, there exist constants c_1 and c_2 not both zero such that $c_1\mathbf{u} + c_2\mathbf{v} = \mathbf{0}$. If $c_1 = 0$, then we would have $c_2\mathbf{v} = \mathbf{0}$ which implies $c_2 = 0$, since $\mathbf{v} \neq \mathbf{0}$. Hence, $c_1 \neq 0$, and by a similar argument $c_2 \neq 0$. Solving $c_1\mathbf{u} + c_2\mathbf{v} = \mathbf{0}$ for $\mathbf{u}$, we find $\mathbf{u} = -c_2\mathbf{v}/c_1 = k\mathbf{v}$, where $k \neq 0$. Thus, the only

```
THE MATRIX WHOSE EIGENVALUES AND EIGENVECTORS ARE TO BE CALCULATED IS

  2.0000E+00   0.0000E+00   1.0000E+00
  0.0000E+00   1.0000E+00   0.0000E+00
  1.0000E+00   0.0000E+00   2.0000E+00

AN EIGENVALUE IS    3.00000E+00  +  0.00000E+00  I
  THE ASSOCIATED EIGENVECTOR IS
                    7.07107E-01  +  0.00000E+00  I
                    0.00000E+00  +  0.00000E+00  I
                    7.07107E-01  +  0.00000E+00  I

AN EIGENVALUE IS    1.00000E+00  +  0.00000E+00  I
  THE ASSOCIATED EIGENVECTOR IS
                   -7.07107E-01  +  0.00000E+00  I
                    0.00000E+00  +  0.00000E+00  I
                    7.07107E-01  +  0.00000E+00  I

AN EIGENVALUE IS    1.00000E+00  +  0.00000E+00  I
  THE ASSOCIATED EIGENVECTOR IS
                    0.00000E+00  +  0.00000E+00  I
                    1.00000E+00  +  0.00000E+00  I
                    0.00000E+00  +  0.00000E+00  I
```

FIGURE 8.6
Eigenvalues and Eigenvectors of the Matrix in Equation (16).

way in which two nonzero vectors can be linearly dependent is for one of them to be a nonzero scalar multiple of the other.) Since there is no scalar constant k such that $\mathbf{v}_2 = k\mathbf{v}_3$, we conclude $\mathbf{v}_2$ and $\mathbf{v}_3$ are linearly independent and, therefore, the general solution of the linear homogeneous system (16) is

$$\mathbf{y}(x) = c_1 e^{3x} \begin{pmatrix} .707107 \\ 0.000000 \\ .707107 \end{pmatrix} + c_2 e^{x} \begin{pmatrix} -.707107 \\ 0.000000 \\ .707107 \end{pmatrix} + c_3 e^{x} \begin{pmatrix} 0 \\ 1 \\ 0 \end{pmatrix}$$

where c_1, c_2, and c_3 are arbitrary constants.

You may find it more appealing to write the general solution in the form

$$\mathbf{y}(x) = k_1 e^{3x} \begin{pmatrix} 1 \\ 0 \\ 1 \end{pmatrix} + k_2 e^{x} \begin{pmatrix} -1 \\ 0 \\ 1 \end{pmatrix} + k_3 e^{x} \begin{pmatrix} 0 \\ 1 \\ 0 \end{pmatrix}$$

where k_1, k_2, and k_3 are arbitrary constants. This representation was accomplished by multiplying the eigenvectors $\mathbf{v}_1$ and $\mathbf{v}_2$ by the scalar constant

1/.707107 to obtain the eigenvectors

$$\mathbf{w}_1 = \begin{pmatrix} 1 \\ 0 \\ 1 \end{pmatrix} \quad \text{and} \quad \mathbf{w}_2 = \begin{pmatrix} -1 \\ 0 \\ 1 \end{pmatrix}. \quad \blacksquare$$

In part 4 of Example 5 of Section 8.2, we ran our CAS and found that the eigenvalues of the matrix

$$\mathbf{A} = \begin{pmatrix} 1 & -2 \\ 1 & 3 \end{pmatrix}$$

are $\lambda_1 = 2 + i$, $\lambda_2 = 2 - i$ and the associated eigenvectors are

$$\mathbf{z}_1 = \begin{pmatrix} -1 - i \\ i \end{pmatrix} \quad \text{and} \quad \mathbf{z}_2 = \begin{pmatrix} -1 + i \\ -i \end{pmatrix}.$$

Therefore, the general solution of the homogeneous linear system $\mathbf{y}' = \mathbf{A}\mathbf{y}$ may be written as

(17) $$\mathbf{y}(x) = c_1 e^{(2+i)x} \begin{pmatrix} -1 - i \\ i \end{pmatrix} + c_2 e^{(2-i)x} \begin{pmatrix} -1 + i \\ -i \end{pmatrix},$$

where c_1 and c_2 are arbitrary constants. The function $\mathbf{y}(x)$ of equation (17) is a complex-valued function of the real variable x. Since the original system of differential equations, $\mathbf{y}' = \mathbf{A}\mathbf{y}$, is a real system (that is, since all of the entries of the matrix $\mathbf{A}$ are all real numbers), it is desirable to write the general solution of the system as a real-valued function of x, if possible. The following theorem specifies conditions under which we can write two real-valued, linearly independent solutions for $\mathbf{y}' = \mathbf{A}\mathbf{y}$ when $\mathbf{A}$ has a pair of complex conjugate eigenvalues.

Theorem 8.8 If A is a constant, real matrix (that is, if A has entries that are all real and constant) and if $\lambda = \alpha + i\beta$ where $\beta \neq 0$ is an eigenvalue of A with associated eigenvector $\mathbf{x} = \mathbf{u} + i\mathbf{v}$, then the linear homogeneous system $\mathbf{y}' = \mathbf{A}\mathbf{y}$ has two real-valued, linearly independent solutions of the form

$$\mathbf{y}_1(x) = (e^{\alpha x} \cos \beta x)\mathbf{u} - (e^{\alpha x} \sin \beta x)\mathbf{v}$$

$$\mathbf{y}_2(x) = (e^{\alpha x} \sin \beta x)\mathbf{u} + (e^{\alpha x} \cos \beta x)\mathbf{v}.$$

Proof: Since $\lambda = \alpha + i\beta$ is an eigenvalue of $\mathbf{A}$ and $\mathbf{x} = \mathbf{u} + i\mathbf{v}$ is an associated eigenvector, a complex-valued solution of $\mathbf{y}' = \mathbf{A}\mathbf{y}$ is

(18) $$\mathbf{z}(x) = e^{(\alpha + i\beta)x}(\mathbf{u} + i\mathbf{v}).$$

Recall that Euler's formula is $e^{i\theta} = \cos\theta + i \sin\theta$. Therefore,

(19) $$e^{(\alpha+i\beta)x} = e^{\alpha x + i\beta x} = e^{\alpha x} e^{i\beta x} = e^{\alpha x}(\cos \beta x + i \sin \beta x).$$

Substituting (19) into (18) and multiplying, we find

$$\begin{aligned}\mathbf{z}(x) &= e^{\alpha x}(\cos\beta x + i\sin\beta x)(\mathbf{u} + i\mathbf{v})\\ &= e^{\alpha x}[(\cos\beta x)\mathbf{u} - (\sin\beta x)\mathbf{v}] + ie^{\alpha x}[(\sin\beta x)\mathbf{u} + (\cos\beta x)\mathbf{v}]\\ &= \mathbf{y}_1(x) + i\mathbf{y}_2(x).\end{aligned}$$

Since $\mathbf{z}(x)$ is a complex-valued solution of $\mathbf{y}' = \mathbf{A}\mathbf{y}$, we have $\mathbf{z}'(x) = \mathbf{A}\mathbf{z}$ or

$$\text{(20)} \qquad \mathbf{y}_1' + i\mathbf{y}_2' = \mathbf{A}(\mathbf{y}_1 + i\mathbf{y}_2) = \mathbf{A}\mathbf{y}_1 + i(\mathbf{A}\mathbf{y}_2).$$

Equating the real parts and the imaginary parts of equation (20), we see that $\mathbf{y}_1' = \mathbf{A}\mathbf{y}_1$ and $\mathbf{y}_2' = \mathbf{A}\mathbf{y}_2$. That is, $\mathbf{y}_1$ and $\mathbf{y}_2$ are both solutions of $\mathbf{y}' = \mathbf{A}\mathbf{y}$. Clearly, they are real solutions. ■

We must still show that the solutions $\mathbf{y}_1$ and $\mathbf{y}_2$ are linearly independent. By definition, the complex conjugate of $\lambda = \alpha + i\beta$ is $\bar{\lambda} = \alpha - i\beta$. Since $\mathbf{A}$ is assumed to have real, constant entries, the characteristic polynomial of $\mathbf{A}$ will have real coefficients. So by the **complex conjugate root theorem** if $\lambda = \alpha + i\beta$ is a root of the characteristic polynomial of $\mathbf{A}$, then so is its complex conjugate $\bar{\lambda} = \alpha - i\beta$. The following computations show that the complex conjugate of $\mathbf{x}$, the vector $\bar{\mathbf{x}} = \mathbf{u} - i\mathbf{v}$, is an eigenvector of $\mathbf{A}$ associated with the eigenvalue $\bar{\lambda}$. Since λ is an eigenvalue of $\mathbf{A}$ and $\mathbf{x}$ is an associated eigenvector, λ and $\mathbf{x}$ satisfy $\mathbf{A}\mathbf{x} = \lambda\mathbf{x}$. Taking the complex conjugate of this equation, we find $(\overline{\mathbf{A}\mathbf{x}}) = (\overline{\lambda\mathbf{x}})$. Since the conjugate of a product equals the product of the conjugates, $\bar{\mathbf{A}}\bar{\mathbf{x}} = \bar{\lambda}\bar{\mathbf{x}}$. Because $\mathbf{A}$ has real entries $\bar{\mathbf{A}} = \mathbf{A}$, and we see that $\bar{\lambda}$ and $\bar{\mathbf{x}}$ satisfy $\mathbf{A}\bar{\mathbf{x}} = \bar{\lambda}\bar{\mathbf{x}}$. That is, $\bar{\lambda} = \alpha - i\beta$ is an eigenvalue of $\mathbf{A}$ and $\bar{\mathbf{x}} = \mathbf{u} - i\mathbf{v}$ is an associated eigenvector. Since $\beta \neq 0$, the eigenvalues λ and $\bar{\lambda}$ are distinct and, therefore, the associated complex-valued solutions

$$\mathbf{z}(x) = e^{(\alpha + i\beta)x}(\mathbf{u} + i\mathbf{v}) = \mathbf{y}_1(x) + i\mathbf{y}_2(x)$$

and

$$\mathbf{w}(x) = e^{(\alpha - i\beta)x}(\mathbf{u} - i\mathbf{v}) = \mathbf{y}_1(x) - i\mathbf{y}_2(x)$$

are linearly independent. Any two linear combinations of the solutions $\mathbf{z}(x)$ and $\mathbf{w}(x)$, $c_1\mathbf{z}(x) + c_2\mathbf{w}(x)$ and $c_3\mathbf{z}(x) + c_4\mathbf{w}(x)$, where c_1, c_2, c_3, and c_4 are complex constants, will also be linearly independent solutions provided $c_1c_4 - c_3c_2 \neq 0$. Choosing $c_1 = c_2 = 1/2$, one linear combination is

$$\frac{1}{2}\mathbf{z}(x) + \frac{1}{2}\mathbf{w}(x) = \frac{1}{2}(\mathbf{y}_1 + i\mathbf{y}_2) + \frac{1}{2}(\mathbf{y}_1 - i\mathbf{y}_2) = \mathbf{y}_1(x).$$

And choosing $c_3 = -i/2$ and $c_4 = i/2$, a second linear combination is

$$\frac{-i}{2}\mathbf{z}(x) + \frac{i}{2}\mathbf{w}(x) = \frac{-i}{2}(\mathbf{y}_1 + i\mathbf{y}_2) + \frac{i}{2}(\mathbf{y}_1 - i\mathbf{y}_2) = \mathbf{y}_2(x).$$

Since $c_1c_4 - c_3c_2 = (1/2)(i/2) - (-i/2)(1/2) = i/2 \neq 0$, the eigenvectors $\mathbf{y}_1(x)$ and $\mathbf{y}_2(x)$ are linearly independent solutions of $\mathbf{y}' = \mathbf{A}\mathbf{y}$. ■

Now, we use the theorem stated and proved above to produce a real general solution of the homogeneous linear system

$$\mathbf{y}' = \mathbf{A}\mathbf{y} = \begin{pmatrix} 1 & -2 \\ 1 & 3 \end{pmatrix} \mathbf{y}.$$

Earlier, we had found that one eigenvalue of the matrix $\mathbf{A}$ is $\lambda = 2+i = \alpha+\beta i$ and that the associated eigenvector is

$$\mathbf{z}_1 = \begin{pmatrix} -1-i \\ i \end{pmatrix} = \begin{pmatrix} -1 \\ 0 \end{pmatrix} + i \begin{pmatrix} -1 \\ 1 \end{pmatrix} = \mathbf{u} + i\mathbf{v}.$$

By the previous theorem, two real, linearly independent solutions of $\mathbf{y}' = \mathbf{A}\mathbf{y}$ are

$$\mathbf{y}_1(x) = e^{2x} \cos x \begin{pmatrix} -1 \\ 0 \end{pmatrix} - e^{2x} \sin x \begin{pmatrix} -1 \\ 1 \end{pmatrix} = \begin{pmatrix} -e^{2x} \cos x + e^{2x} \sin x \\ -e^{2x} \sin x \end{pmatrix}$$

and

$$\mathbf{y}_2(x) = e^{2x} \sin x \begin{pmatrix} -1 \\ 0 \end{pmatrix} + e^{2x} \cos x \begin{pmatrix} -1 \\ 1 \end{pmatrix} = \begin{pmatrix} -e^{2x} \sin x - e^{2x} \cos x \\ e^{2x} \cos x \end{pmatrix}.$$

Furthermore, the real general solution of $\mathbf{y}' = \mathbf{A}\mathbf{y}$ is $\mathbf{y}(x) = k_1\mathbf{y}_1(x)+k_2\mathbf{y}_2(x)$ where k_1 and k_2 are arbitrary real constants.

Example 4 Find the real general solution of the homogeneous linear system

$$\mathbf{y}' = \begin{pmatrix} 1 & 4 & 3 \\ 0 & 1 & -1 \\ -1 & 1 & 2 \end{pmatrix} \mathbf{y}. \tag{21}$$

Solution

We ran our computing software, setting the size of the matrix to three and entering the values for the elements of the matrix which appears in equation (21). The output of the program is shown in Figure 8.7. The eigenvalues are $1 + 2i = \alpha + \beta i$, $1 - 2i$, and 2 and the associated eigenvectors are

$$\mathbf{x}_1 = \begin{pmatrix} -.871550 \\ .0115057 \\ -.837032 \end{pmatrix} + i \begin{pmatrix} 1.23254 \\ -0.418516 \\ -0.0230115 \end{pmatrix} = \mathbf{u} + i\mathbf{v},$$

$$\mathbf{x}_2 = \bar{\mathbf{x}}_1, \quad \text{and} \quad \mathbf{x}_3 = \begin{pmatrix} .538516 \\ .538516 \\ -.538516 \end{pmatrix}.$$

Hence, the real general solution of (21) is

$$\mathbf{y}(x) = c_1[(e^x \cos 2x)\mathbf{u}-(e^x \sin 2x)\mathbf{v}]+c_2[(e^x \sin 2x)\mathbf{u}+(e^x \cos 2x)\mathbf{v}]+c_3e^{2x}\mathbf{x}_3.$$

```
THE MATRIX WHOSE EIGENVALUES AND EIGENVECTORS ARE TO BE CALCULATED IS

  1.0000E+00   4.0000E+00   3.0000E+00
  0.0000E+00   1.0000E+00  -1.0000E+00
 -1.0000E+00   1.0000E+00   2.0000E+00

AN EIGENVALUE IS   1.00000E+00  +  2.00000E+00  I
  THE ASSOCIATED EIGENVECTOR IS
                  -8.71550E-01  +  1.23254E+00  I
                   1.15057E-02  + -4.18516E-01  I
                  -8.37032E-01  + -2.30115E-02  I

AN EIGENVALUE IS   1.00000E+00  + -2.00000E+00  I
  THE ASSOCIATED EIGENVECTOR IS
                  -8.71550E-01  + -1.23254E+00  I
                   1.15057E-02  +  4.18516E-01  I
                  -8.37032E-01  +  2.30115E-02  I

AN EIGENVALUE IS   2.00000E+00  +  0.00000E+00  I
  THE ASSOCIATED EIGENVECTOR IS
                   5.38516E-01  +  0.00000E+00  I
                   5.38516E-01  +  0.00000E+00  I
                  -5.38516E-01  +  0.00000E+00  I
```

FIGURE 8.7
Eigenvalues and Eigenvectors of the Matrix in Equation (21). ■

8.3.1 Exercises

In Exercises 1–4 rewrite each of the linear systems of first-order differential equations using matrix-vector notation.

1. $y_1' = 2y_1 - 3y_2 + 5e^x$
   $y_2' = y_1 + 4y_2 - 2e^{-x}$

2. $y_1' = y_2 - 2y_1 + \sin 2x$
   $y_2' = -3y_1 + y_2 - 2\cos 3x$

3. $y_1' = 2y_2$
   $y_2' = 3y_1$
   $y_3' = 2y_3 - y_1$

4. $y_1' = 2xy_1 - x^2y_2 + 4x$
   $y_2' = e^x y_1 + 3e^{-x}y_2 - \cos 3x$

5. Consider the homogeneous linear system

$$\mathbf{y}' = \begin{pmatrix} 2 & -3 \\ 1 & -2 \end{pmatrix} \mathbf{y}. \tag{22}$$

   a. Verify that

$$\mathbf{y}_1 = \begin{pmatrix} 3 \\ 1 \end{pmatrix} e^x \quad \text{and} \quad \mathbf{y}_2 = \begin{pmatrix} 1 \\ 1 \end{pmatrix} e^{-x}$$

   are linearly independent solutions of (22).

b. Write the general solution of (22).

c. Verify that

$$\mathbf{y}_p = \begin{pmatrix} x \\ 2x-1 \end{pmatrix}$$

is a particular solution of the nonhomogeneous linear system

$$\mathbf{y}' = \begin{pmatrix} 2 & -3 \\ 1 & -2 \end{pmatrix} \mathbf{y} + \begin{pmatrix} 4x-2 \\ 3x \end{pmatrix}. \tag{23}$$

d. Write the general solution of (23).

6. Consider the homogeneous linear system

$$\mathbf{y}' = \begin{pmatrix} \dfrac{5}{x} & \dfrac{4}{x} \\ \dfrac{-6}{x} & \dfrac{-5}{x} \end{pmatrix} \mathbf{y}. \tag{24}$$

(a) Verify that

$$\mathbf{y}_1 = \begin{pmatrix} 1 \\ -1 \end{pmatrix} x \quad \text{and} \quad \mathbf{y}_2 = \begin{pmatrix} -2 \\ 3 \end{pmatrix} x^{-1}$$

are linearly independent solutions on $(0, \infty)$ of (24).

(b) Write the general solution of (24).

(c) Verify that

$$\mathbf{y}_p = \begin{pmatrix} 2 \\ -1 \end{pmatrix} x^2$$

is a particular solution of the nonhomogeneous linear system

$$\mathbf{y}' = \begin{pmatrix} \dfrac{5}{x} & \dfrac{4}{x} \\ \dfrac{-6}{x} & \dfrac{-5}{x} \end{pmatrix} \mathbf{y} + \begin{pmatrix} -2 \\ 5 \end{pmatrix} x. \tag{25}$$

(d) Write the general solution of (25).

In Exercises 7–16 find the real general solution of the homogeneous linear system $\mathbf{y}' = A\mathbf{y}$ for the given matrix A.

7. $\begin{pmatrix} 2 & 1 & -2 \\ 0 & 3 & -2 \\ 3 & 1 & -3 \end{pmatrix}$

8. $\begin{pmatrix} 5 & -5 & -5 \\ -1 & 4 & 2 \\ 3 & -5 & -3 \end{pmatrix}$

9. $\begin{pmatrix} 4 & 6 & 6 \\ 1 & 3 & 2 \\ -1 & -4 & -3 \end{pmatrix}$

10. $\begin{pmatrix} 1 & 2 & -3 \\ -3 & 4 & -2 \\ 2 & 0 & 1 \end{pmatrix}$

11. $\begin{pmatrix} -2 & -1 & 1 \\ -1 & -2 & -1 \\ 1 & -1 & -2 \end{pmatrix}$

12. $\begin{pmatrix} 1 & 1 & 2 \\ 1 & 1 & 2 \\ 2 & 2 & 4 \end{pmatrix}$

13. $\begin{pmatrix} 2 & 1 & 0 & 0 \\ -1 & 2 & 0 & 0 \\ 0 & 0 & 3 & -4 \\ 0 & 0 & 4 & 3 \end{pmatrix}$

14. $\begin{pmatrix} 0 & 1 & 0 & 0 \\ -3 & 0 & 2 & 0 \\ 0 & 0 & 0 & 1 \\ 2 & 0 & -5 & 0 \end{pmatrix}$

15. $\begin{pmatrix} 3 & 2 & 0 & 0 \\ -2 & 3 & 0 & 0 \\ 0 & 0 & 1 & 0 \\ 0 & 0 & 0 & 2 \end{pmatrix}$

16. $\begin{pmatrix} 0 & 1 & 0 & 1 \\ 1 & 0 & -1 & 0 \\ 0 & 0 & 0 & 1 \\ 0 & 0 & 1 & 0 \end{pmatrix}$

9

Applications of Linear Systems with Constant Coefficients

In this chapter we present several applications which require the solution of a linear system of first-order differential equations with constant coefficients.

9.1 Coupled Spring-Mass Systems

System 1 Suppose a mass m_1 is attached to one end of a spring with spring constant k_1. The other end of this spring is attached to a fixed support. A second mass m_2 is attached to one end of a second spring with spring constant k_2. The other end of the second spring is attached to the bottom of mass m_1 and the resulting system is allowed to come to rest in the equilibrium position as shown in Figure 9.1.

For $i = 1, 2$ let y_i represent the vertical displacement of mass m_i from its equilibrium position. We will assign downward displacement from equilibrium to be positive and upward displacement from equilibrium to be negative. Applying Newton's second law of motion, and assuming no damping is present, it can be shown that the equations of motion for this coupled spring-mass system satisfy the following linear system of differential equations

$$\begin{aligned} m_1 y_1'' &= -k_1 y_1 + k_2(y_2 - y_1) \\ m_2 y_2'' &= -k_2(y_2 - y_1). \end{aligned} \tag{1}$$

Dividing the first equation in system (1) by m_1, dividing the second equation of system (1) by m_2, and letting $u_1 = y_1$, $u_2 = y_1'$, $u_3 = y_2$, and $u_4 = y_2'$, we

DOI: 10.1201/9781003678014-9

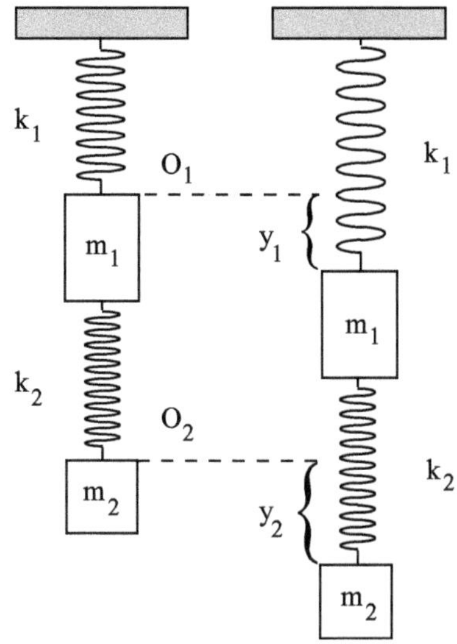

FIGURE 9.1
A Coupled Spring-Mass System.

see we may rewrite (1) as the following equivalent first-order system

$$
\begin{aligned}
u_1' &= u_2 \\
u_2' &= \frac{-k_1u_1 + k_2(u_3 - u_1)}{m_1} &&= \frac{-(k_1+k_2)u_1}{m_1} + \frac{k_2u_3}{m_1} \\
u_3' &= u_4 \\
u_4' &= \frac{-k_2(u_3 - u_1)}{m_2} &&= \frac{k_2u_1}{m_2} - \frac{k_2u_3}{m_2}.
\end{aligned} \tag{2}
$$

Next, using matrix-vector notation, we may rewrite (2) as

$$
\mathbf{u}' = \begin{pmatrix} 0 & 1 & 0 & 0 \\ \dfrac{-(k_1+k_2)}{m_1} & 0 & \dfrac{k_2}{m_1} & 0 \\ 0 & 0 & 0 & 1 \\ \dfrac{k_2}{m_2} & 0 & \dfrac{-k_2}{m_2} & 0 \end{pmatrix} \mathbf{u}. \tag{3}
$$

Example 1 Write the real general solution of system (3) for $\mathbf{m_1 = 3}$ g, $\mathbf{m_2 = 5}$ g, $\mathbf{k_1 = 18}$ g/s^2, and $\mathbf{k_2 = 3}$ g/s^2.

Solution

For the given values of m_1, m_2, k_1, and k_2 system (3) becomes

$$(4) \qquad \mathbf{u}' = \begin{pmatrix} 0 & 1 & 0 & 0 \\ -7 & 0 & 1 & 0 \\ 0 & 0 & 0 & 1 \\ .6 & 0 & -.6 & 0 \end{pmatrix} \mathbf{u}.$$

We used our computing software to find the eigenvalues and associated eigenvectors shown in Figure 9.2.

```
THE MATRIX WHOSE EIGENVALUES AND EIGENVECTORS ARE TO BE CALCULATED IS

  0.0000E+00   1.0000E+00   0.0000E+00   0.0000E+00
 -7.0000E+00   0.0000E+00   1.0000E+00   0.0000E+00
  0.0000E+00   0.0000E+00   0.0000E+00   1.0000E+00
  6.0000E-01   0.0000E+00  -6.0000E-01   0.0000E+00

AN EIGENVALUE IS    0.00000E+00  +  2.66316E+00  I
  THE ASSOCIATED EIGENVECTOR IS
                    3.75485E-01  +  0.00000E+00  I
                    0.00000E+00  +  9.99978E-01  I
                   -3.47007E-02  +  0.00000E+00  I
                    0.00000E+00  + -9.24135E-02  I

AN EIGENVALUE IS    0.00000E+00  + -2.66316E+00  I
  THE ASSOCIATED EIGENVECTOR IS
                    3.75485E-01  +  0.00000E+00  I
                    0.00000E+00  + -9.99978E-01  I
                   -3.47007E-02  +  0.00000E+00  I
                    0.00000E+00  +  9.24135E-02  I

AN EIGENVALUE IS    0.00000E+00  +  7.12450E-01  I
  THE ASSOCIATED EIGENVECTOR IS
                    2.13163E-01  +  0.00000E+00  I
                    0.00000E+00  +  1.51868E-01  I
                    1.38394E+00  +  0.00000E+00  I
                    0.00000E+00  +  9.85988E-01  I

AN EIGENVALUE IS    0.00000E+00  + -7.12450E-01  I
  THE ASSOCIATED EIGENVECTOR IS
                    2.13163E-01  +  0.00000E+00  I
                    0.00000E+00  + -1.51868E-01  I
                    1.38394E+00  +  0.00000E+00  I
                    0.00000E+00  + -9.85988E-01  I
```

FIGURE 9.2
Two Pair of Complex Conjugate Eigenvalues and Eigenvectors of the Constant 4×4 Matrix in Equation (4).

Using this information, we can write the real general solution of system (4) as

$$(5) \quad \mathbf{u}(x) = c_1[(\cos\beta x)\mathbf{v}_1 - (\sin\beta x)\mathbf{v}_2] + c_2[(\sin\beta x)\mathbf{v}_1 + (\cos\beta x)\mathbf{v}_2]$$

$$+c_3[(\cos\gamma x)\mathbf{v}_3 - (\sin\gamma x)\mathbf{v}_4] + c_4[(\sin\gamma x)\mathbf{v}_3 + (\cos\gamma x)\mathbf{v}_4]$$

where $\beta = 2.66316$, $\gamma = .712450$,

$$\mathbf{v}_1 = \begin{pmatrix} 0.375485 \\ 0.000000 \\ -0.0347007 \\ 0.000000 \end{pmatrix}, \quad \mathbf{v}_2 = \begin{pmatrix} 0.000000 \\ 0.999978 \\ 0.000000 \\ -0.0924135 \end{pmatrix},$$

$$\mathbf{v}_3 = \begin{pmatrix} 0.213163 \\ 0.000000 \\ 1.38394 \\ 0.000000 \end{pmatrix}, \text{ and } \quad \mathbf{v}_4 = \begin{pmatrix} 0.000000 \\ 0.151868 \\ 0.000000 \\ 0.985988 \end{pmatrix}. \ \blacksquare$$

Example 2 Find the solution to the initial value problem consisting of the system (4) and the initial conditions $\mathbf{u_1(0)} = 1$ cm, $\mathbf{u_2(0)} = 0$ cm/s, $\mathbf{u_3(0)} = -2$ cm, and $\mathbf{u_4(0)} = 0$ cm/s.

Solution

Written in vector notation, the given initial conditions are

$$\mathbf{u}(0) = \begin{pmatrix} 1 \\ 0 \\ -2 \\ 0 \end{pmatrix}.$$

Evaluating equation (5)—the general solution to system (4)—at $x = 0$, we see that the constants c_1, c_2, c_3, and c_4 must satisfy

$$\mathbf{u}(0) = c_1\mathbf{v}_1 + c_2\mathbf{v}_2 + c_3\mathbf{v}_3 + c_4\mathbf{v}_4 = \mathbf{0}$$

or equivalently

$$\begin{aligned} .375485c_1 + .213163c_3 &= 1 \\ .999978c_2 + .151868c_4 &= 0 \\ -.0347007c_1 + 1.38394c_3 &= -2 \\ -.0924135c_2 + .985988c_4 &= 0. \end{aligned}$$

Solving the first and third equations of this set for c_1 and c_3, we find $c_1 = 3.43474$ and $c_3 = -1.35903$. And solving the second and fourth equations of this set for c_2 and c_4, we find $c_2 = c_4 = 0$. Hence, the solution of the system (4) subject to the initial conditions given in this example is equation (5) with $c_1 = 3.43474$, $c_2 = 0$, $c_3 = -1.35903$, and $c_4 = 0$. Thus, the solution is

$$\mathbf{u}(x) = \begin{pmatrix} u_1(x) \\ u_2(x) \\ u_3(x) \\ u_4(x) \end{pmatrix} = \begin{pmatrix} 1.28069 \cos\beta x - .289695 \cos\gamma x \\ -3.43466 \sin\beta x + .206393 \sin\gamma x \\ -.119188 \cos\beta x - 1.88082 \cos\gamma x \\ .317416 \sin\beta x + 1.33999 \sin\gamma x \end{pmatrix}$$

where $\beta = 2.66316$ and $\gamma = .712450$. A graph of the position of the mass m_1, $u_1(x)$, and the position of the mass m_2, $u_3(x)$, on the interval $[0, 30]$ is displayed in Figure 9.3. Notice the interesting, oscillatory nature of both functions.

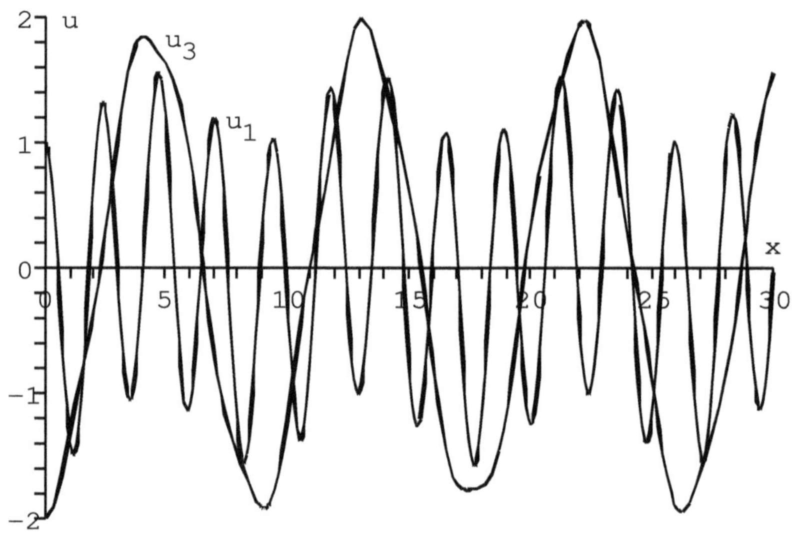

FIGURE 9.3
A Graph of the Positions of the Masses of a Coupled Spring-Mass System. ■

When damping is assumed to be present in the coupled spring-mass system shown in Figure 9.1, the equations of motion satisfy the following second-order linear system

$$\text{(6)} \qquad \begin{aligned} m_1 y_1'' &= -k_1 y_1 + k_2(y_2 - y_1) - d_1 y_1' \\ m_2 y_2'' &= -k_2(y_2 - y_1) - d_2 y_2' \end{aligned}$$

where, as before, y_1 and y_2 are the displacements of the masses from the equilibrium positions, m_1 and m_2 are the masses, and k_1 and k_2 are the respective spring constants. The positive constants d_1 and d_2 are due to the damping forces and are called the **damping constants**.

Exercise 15. Write the real general solution for the coupled spring-mass system (3) for

a. $m_1 = 5$ g, $m_2 = 10$ g, $k_1 = 10$ g/s, and $k_2 = 10$ g/s.

b. $m_1 = 5$ g, $m_2 = 10$ g, $k_1 = 5$ g/s, and $k_2 = 10$ g/s.

Exercise 16. Find the solution of the initial value problem consisting of the differential system of Exercise 1. a. subject to the initial conditions $u_1(0) = 3$ cm and $u_2(0) = u_3(0) = u_4(0) = 0$.

Exercise 17. Let $u_1 = y_1$, $u_2 = y_1'$, $u_3 = y_2$, and $u_4 = y_2'$.

a. Write system (6) as an equivalent system of four first-order differential equations.

b. Write the first-order system which is equivalent to (6) in matrix-vector notation.

c. Use your computing software to find the general solution to the homogeneous system of part b. for $m_1 = m_2 = 10$ g, $k_1 = 5$ g/s^2, $k_2 = 10$ g/s^2, and $d_1 = d_2 = 15$ g/s.

System 2 A second coupled spring-mass system which consists of two masses, m_1 and m_2, connected to two fixed supports by three springs which have spring constants k_1, k_2, and k_3 is shown in Figure 9.4.

Neglecting the effects of damping, the system of differential equations which describes the displacements y_1 and y_2 of masses m_1 and m_2, respectively, from their equilibrium positions is

$$(7)\qquad \begin{aligned} m_1 y_1'' &= -k_1 y_1 + k_2(y_2 - y_1) \\ m_2 y_2'' &= -k_2(y_2 - y_1) - k_3 y_2. \end{aligned}$$

When damping is assumed to be present in the coupled spring-mass system shown in Figure 9.4, the equations of motion satisfy the following second-order linear system

$$(8)\qquad \begin{aligned} m_1 y_1'' &= -k_1 y_1 + k_2(y_2 - y_1) - d_1 y_1' \\ m_2 y_2'' &= -k_2(y_2 - y_1) - k_3 y_2 - d_2 y_2' \end{aligned}$$

where d_1 and d_2 are the damping constants.

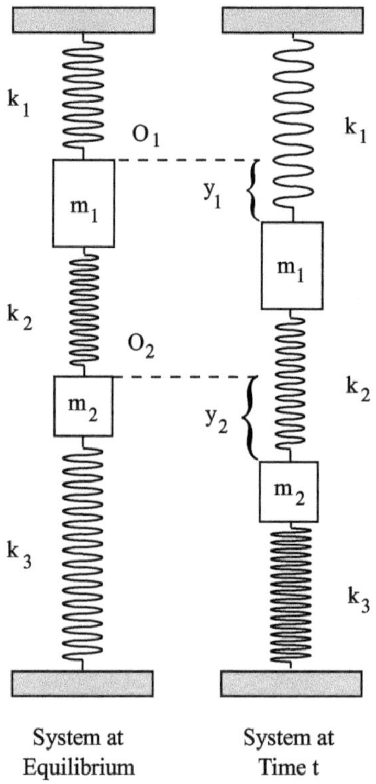

FIGURE 9.4
A Coupled Spring-Mass System.

Exercise 18. Let $u_1 = y_1$, $u_2 = y_1'$, $u_3 = y_2$, and $u_4 = y_2'$.

a. Write the system (7) as an equivalent system of four first-order differential equations.

b. Write the first-order system of equations which is equivalent to (7) in matrix-vector notation.

c. Use your computing software to find the general solution to the homogeneous system of part b. for $m_1 = m_2 = 2$ g, $k_1 = k_3 = 4$ g/s^2, and $k_2 = 6$ g/s^2.

Exercise 19.

a. Write system (8) as an equivalent system of first-order equations in matrix-vector notation.

b. Use your computing software to find the general solution of the resulting system for the following cases.

(i) $m_1 = m_2 = 2$ g, $k_1 = k_3 = 10$ g/s^2, $k_2 = 3$ g/s^2, and $d_1 = d_2 = 12$ g/s.

(ii) $m_1 = m_2 = 1$ g, $k_1 = k_3 = 3$ g/s^2, $k_2 = 1$ g/s^2, and $d_1 = d_2 = 4$ g/s.

(iii) $m_1 = m_2 = 1$ g, $k_1 = k_3 = 2$ g/s^2, $k_2 = 1$ g/s^2, and $d_1 = d_2 = 2$ g/s.

9.2 Pendulum Systems

Coupled Pendulums A coupled pendulum system consists of a pair of identical pendulums with bobs of mass m and rods of length ℓ coupled by a spring with spring constant k as shown in Figure 9.5. Let y_1 and y_2 denote the displacement of the pendulums from vertical (positive displacement is to the right). Neglecting damping and assuming the displacements y_1 and y_2 are small, we showed in Section 8.3 that the displacements satisfy the following system of second-order differential equations

$$
\begin{aligned}
my_1'' &= \frac{-mgy_1}{\ell} - k(y_1 - y_2) \\
my_2'' &= \frac{-mgy_2}{\ell} - k(y_2 - y_1).
\end{aligned}
$$

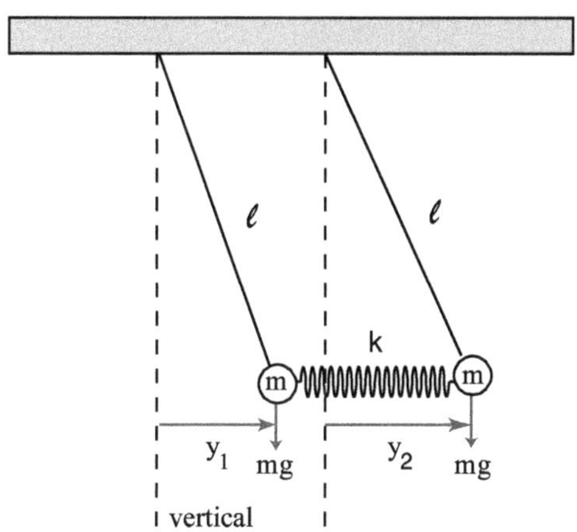

FIGURE 9.5
Coupled Pendulums.

Also in Section 8.3, we showed by letting $u_1 = y_1$, $u_2 = y_1'$, $u_3 = y_2$, and $u_4 = y_2'$ that the second-order system above can be rewritten in matrix notation as the following first-order system of differential equations.

$$(9) \qquad \begin{pmatrix} u_1' \\ u_2' \\ u_3' \\ u_4' \end{pmatrix} = \begin{pmatrix} 0 & 1 & 0 & 0 \\ -\left(\frac{g}{\ell} + \frac{k}{m}\right) & 0 & \frac{k}{m} & 0 \\ 0 & 0 & 0 & 1 \\ \frac{k}{m} & 0 & -\left(\frac{g}{\ell} + \frac{k}{m}\right) & 0 \end{pmatrix} \begin{pmatrix} u_1 \\ u_2 \\ u_3 \\ u_4 \end{pmatrix}.$$

Exercise 20. Use your computing software to find the general solution of system (9) for $m = 25$ g, $\ell = 50$ cm, $k = 400$ g/s^2, and $g = 980$ cm/s^2.

A Double Pendulum A double pendulum consists of a bob of mass m_1 attached to a fixed support by a rod of length ℓ_1 and a second bob of mass m_2 attached to the first bob by a rod of length ℓ_2 as shown in Figure 9.6. Let y_1 and y_2 denote the displacement from the vertical of the rods of length ℓ_1 and ℓ_2 respectively. Assuming the double pendulum oscillates in a vertical plane and neglecting the mass of the rods and any damping forces, it can be shown that the displacements, y_1 and y_2, satisfy the following system of differential

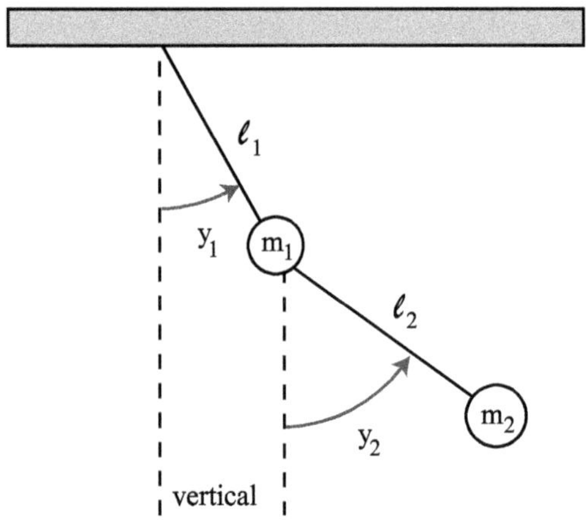

FIGURE 9.6
A Double Pendulum.

equations

$$(10) \qquad \begin{aligned} (m_1+m_2)\ell_1^2 y_1'' + m_2\ell_1\ell_2 y_2'' + (m_1+m_2)\ell_1 g y_1 &= 0 \\ m_2\ell_1\ell_2 y_1'' + m_2\ell_2^2 y_2'' + m_2\ell_2 g y_2 &= 0 \end{aligned}$$

where g is the constant of gravitational acceleration.

Exercise 21.

a. Write system (10) as an equivalent system of first-order differential equations in matrix notation.

b. Use your computing software to find the general solution of the resulting system for $m_1 = 30$ g, $m_2 = 20$ g, $\ell_1 = 50$ cm, $\ell_2 = 25$ cm, and $g = 980$ cm/s^2.

9.3 The Path of an Electron

In Chapter 6, we stated that the position (x, y) of an electron which was initially at rest at the origin and subject to a magnetic field of intensity H and an electric field of intensity E satisfied the second-order system initial value problem

$$(11a) \qquad \begin{aligned} x'' &= -HRy' + ER \\ y'' &= HRx' \end{aligned}$$

$$(11b) \qquad x(0) = 0, \qquad x'(0) = 0, \qquad y(0) = 0, \qquad y'(0) = 0$$

where $R = q/m$ is the ratio of the charge of an electron, q, to its mass, m. Here the independent variable is time, t, and the dependent variables are x and y. In Chapter 7 we showed how to write a system of first-order differential equations which is equivalent to system (9) by letting $u_1 = x$, $u_2 = x'$, $u_3 = y$, and $u_4 = y'$. Hence, in Chapter 7 we found that the following system of four first-order differential equations is equivalent to (11a)

$$\begin{aligned} u_1' &= u_2 \\ u_2' &= -HRu_4 + ER \\ u_3' &= u_4 \\ u_4' &= HRu_2. \end{aligned}$$

Using matrix-vector notation, this system may be rewritten as

$$\text{(12a)} \qquad \mathbf{u}' = \begin{pmatrix} 0 & 1 & 0 & 0 \\ 0 & 0 & 0 & -HR \\ 0 & 0 & 0 & 1 \\ 0 & HR & 0 & 0 \end{pmatrix} \mathbf{u} + \begin{pmatrix} 0 \\ ER \\ 0 \\ 0 \end{pmatrix}.$$

The initial conditions are $u_1(0) = 0$, $u_2(0) = 0$, $u_3(0) = 0$, and $u_4(0) = 0$. Or in vector notation the initial conditions are

$$\text{(12b)} \qquad \mathbf{u}(0) = \mathbf{0}.$$

Example 1 For HR = 2 and ER = 3, solve the nonhomogeneous initial value problem (12).

Solution

Replacing HR by 2 and ER by 3 in system (12), we see we must solve the initial value problem

$$\text{(13a)} \qquad \mathbf{u}' = \begin{pmatrix} 0 & 1 & 0 & 0 \\ 0 & 0 & 0 & -2 \\ 0 & 0 & 0 & 1 \\ 0 & 2 & 0 & 0 \end{pmatrix} \mathbf{u} + \begin{pmatrix} 0 \\ 3 \\ 0 \\ 0 \end{pmatrix} = \mathbf{A}\mathbf{u} + \mathbf{b}$$

$$\text{(13b)} \qquad \mathbf{u}(0) = \mathbf{0}.$$

In order to solve the nonhomogeneous initial value problem (13), we must first find the general solution, $\mathbf{u}_c$, of the associated homogeneous linear system $\mathbf{u}' = \mathbf{A}\mathbf{u}$. We will do this with the aid of our computer software. Next, we must find a particular solution, $\mathbf{u}_p$, of the nonhomogeneous system (13a) $\mathbf{u}' = \mathbf{A}\mathbf{u} + \mathbf{b}$. The general solution of (13a) is then $\mathbf{u} = \mathbf{u}_c + \mathbf{u}_p$. Finally, we determine values for the constants appearing in the general solution which will satisfy the initial conditions (13b).

We ran our computing software and found the eigenvalues of $\mathbf{A}$ to be 0, 0, and $\pm 2i$. We also found the associated linearly independent eigenvectors to be

$$\mathbf{v}_1 = \begin{pmatrix} 1 \\ 0 \\ 0 \\ 0 \end{pmatrix}, \quad \mathbf{v}_2 = \begin{pmatrix} 0 \\ 0 \\ 1 \\ 0 \end{pmatrix}, \quad \mathbf{v}_3 = \begin{pmatrix} 0.0 \\ -1.0 \\ 0.5 \\ 0.0 \end{pmatrix} + \begin{pmatrix} 0.5 \\ 0.0 \\ 0.0 \\ 1.0 \end{pmatrix} i = \mathbf{r} + \mathbf{s}i, \quad \mathbf{v}_4 = \mathbf{r} - \mathbf{s}i.$$

So the complementary solution—that is, the real general solution of the associated homogeneous equation $\mathbf{u}' = \mathbf{A}\mathbf{u}$—is

$$\text{(14)} \quad \mathbf{u}_c = c_1\mathbf{v}_1 + c_2\mathbf{v}_2 + c_3[(\cos 2t)\mathbf{r} - (\sin 2t)\mathbf{s}] + c_4[(\sin 2t)\mathbf{r} + (\cos 2t)\mathbf{s}].$$

We now use the method of undetermined coefficients to find a particular solution of the nonhomogeneous system of differential equations (13a). We used this method in Chapter 4 to find particular solutions of n-th order linear nonhomogeneous differential equations with constant coefficients. Recall that the method consisted of judiciously guessing the form of the particular solution with coefficients unknown and then determining specific values for the unknown coefficients in order to satisfy the differential equation. The method of undetermined coefficients works only when the matrix $\mathbf{A}$ is constant and the components of the nonhomogeneity $\mathbf{b}$ are polynomials, exponential or sinusoidal functions, or the sums and products of these functions.

Since the vector $\mathbf{b}$ of system (13a) is a constant, we attempt to find a particular solution, $\mathbf{u}_p$, which is also a constant. Hence, we assume

$$\mathbf{u}_p = \begin{pmatrix} \alpha \\ \beta \\ \gamma \\ \delta \end{pmatrix}$$

where α, β, γ, and δ are unknown constants. Differentiating, we find $\mathbf{u}_p' = \mathbf{0}$. Substituting for $\mathbf{u}_p$ and $\mathbf{u}_p'$ in (13a), we see that

$$\mathbf{u}_p' = \mathbf{A}\mathbf{u}_p + \mathbf{b} \quad \text{or} \quad \mathbf{0} = \mathbf{A}\mathbf{u}_p + \mathbf{b} \quad \text{or} \quad \mathbf{A}\mathbf{u}_p = -\mathbf{b}.$$

Computing $\mathbf{A}\mathbf{u}_p$ and setting the result equal to $-\mathbf{b}$, we find α, β, γ, and δ must simultaneously satisfy

$$\begin{aligned} \beta &= 0 \\ -2\delta &= -3 \\ \delta &= 0 \\ 2\beta &= 0 \end{aligned}$$

This is impossible, since we must have both $\delta = 3/2$ and $\delta = 0$. Thus, there is no particular solution of the form assumed. That is, no constant vector is a particular solution of (13a).

Having failed to find a constant particular solution of (13a), we next seek a particular solution in which each component is a linear function. Thus, we assume $\mathbf{u}_p$ has the form

$$\mathbf{u}_p = \begin{pmatrix} a + bt \\ c + dt \\ e + ft \\ g + ht \end{pmatrix}$$

where a, b, c, d, e, f, g, and h are unknown constants. Differentiating, we have

$$\mathbf{u}_p' = \begin{pmatrix} b \\ d \\ f \\ h \end{pmatrix}.$$

Substituting for $\mathbf{u}_p$ and $\mathbf{u}_p'$ in (13a), $\mathbf{u}_p' = \mathbf{A}\mathbf{u}_p + \mathbf{b}$, results in

$$\begin{pmatrix} b \\ d \\ f \\ h \end{pmatrix} = \begin{pmatrix} 0 & 1 & 0 & 0 \\ 0 & 0 & 0 & -2 \\ 0 & 0 & 0 & 1 \\ 0 & 2 & 0 & 0 \end{pmatrix} \begin{pmatrix} a+bt \\ c+dt \\ e+ft \\ g+ht \end{pmatrix} + \begin{pmatrix} 0 \\ 3 \\ 0 \\ 0 \end{pmatrix}$$

or equivalently

$$\begin{aligned} b &= c + dt \\ d &= -2(g+ht) + 3 &&= (-2g+3) - 2ht \\ f &= g + ht \\ h &= 2(c+dt) &&= 2c + 2dt. \end{aligned}$$

Equating coefficients in each of these four equations, we find the constants must simultaneously satisfy

$$\begin{array}{rclcrcl} & (\text{coefficients of } t^0) & & & & (\text{coefficients of } t) & \\ b &=& c & \text{and} & 0 &=& d \\ d &=& -2g+3 & \text{and} & 0 &=& -2h \\ f &=& g & \text{and} & 0 &=& h \\ h &=& 2c & \text{and} & 0 &=& 2d. \end{array}$$

Solving these eight equations, we find a and e are arbitrary, $b = c = d = h = 0$, and $f = g = 3/2$. Hence,

$$\mathbf{u}_p = \begin{pmatrix} a \\ 0 \\ e + 3t/2 \\ 3/2 \end{pmatrix}$$

is a particular solution of (13a) for any choice of a and e. Choosing $a = e = 0$, we obtain the simple particular solution

$$\mathbf{u}_p = \begin{pmatrix} 0 \\ 0 \\ 3t/2 \\ 3/2 \end{pmatrix}. \tag{15}$$

The general solution of (13a) is $\mathbf{u}(t) = \mathbf{u}_c(t) + \mathbf{u}_p(t)$, where $\mathbf{u}_c$ is given by equation (14) and where $\mathbf{u}_p$ is given by (15). Imposing the initial conditions (13b) $\mathbf{u}(0) = \mathbf{0}$ requires c_1, c_2, c_3, and c_4 to satisfy

$$\mathbf{u}(0) = \mathbf{u}_c(0) + \mathbf{u}_p(0) = c_1\mathbf{v}_1 + c_2\mathbf{v}_2 + c_3\mathbf{r} + c_4\mathbf{s} + \mathbf{u}_p(0) = \mathbf{0}.$$

Or

$$c_1\begin{pmatrix}1\\0\\0\\0\end{pmatrix}+c_2\begin{pmatrix}0\\0\\1\\0\end{pmatrix}+c_3\begin{pmatrix}0\\-1\\1/2\\0\end{pmatrix}+c_4\begin{pmatrix}1/2\\0\\0\\1\end{pmatrix}+\begin{pmatrix}0\\0\\0\\3/2\end{pmatrix}=\begin{pmatrix}0\\0\\0\\0\end{pmatrix}.$$

Thus, c_1, c_2, c_3 and c_4 must simultaneously satisfy

$$\begin{aligned}c_1+c_4/2&=0\\-c_3&=0\\c_2+c_3/2&=0\\c_4+3/2&=0.\end{aligned}$$

Solving this system of equations, we find $c_1=3/4$, $c_2=c_3=0$, and $c_4=-3/2$. Therefore, the solution of the initial value problem (13) is

$$\mathbf{u}(t)=\frac{3}{4}\mathbf{v}_1-\frac{3}{2}[(\sin 2t)\mathbf{r}+(\cos 2t)\mathbf{s}]+\mathbf{u}_p(t).$$

Consequently,

$$\mathbf{u}=\begin{pmatrix}u_1\\u_2\\u_3\\u_4\end{pmatrix}=\begin{pmatrix}x\\x'\\y\\y'\end{pmatrix}=\frac{3}{4}\begin{pmatrix}1\\0\\0\\0\end{pmatrix}-\frac{3}{2}\sin 2t\begin{pmatrix}0\\-1\\1/2\\0\end{pmatrix}-\frac{3}{2}\cos 2t\begin{pmatrix}1/2\\0\\0\\1\end{pmatrix}+\begin{pmatrix}0\\0\\3t/2\\3/2\end{pmatrix}.$$

Or, equivalently

$$\begin{pmatrix}x(t)\\x'(t)\\y(t)\\y'(t)\end{pmatrix}=\begin{pmatrix}\frac{3}{4}-\frac{3}{4}\cos 2t\\\frac{3}{2}\sin 2t\\\frac{-3}{4}\sin 2t+\frac{3t}{2}\\\frac{-3}{2}\cos 2t+\frac{3}{2}\end{pmatrix}.$$

Hence, the position of the electron in the xy-plane as a function of time, t, is

$$(x(t),\ y(t))=\left(\frac{3}{4}-\frac{3}{4}\cos 2t,\ \frac{-3}{4}\sin 2t+\frac{3t}{2}\right).$$

A graph of the path of the electron in the xy-plane is displayed in Figure 9.7. The electron is initially at the origin, travels along the path shown, and reaches the point $(0, 1.5\pi)$ in π units of time. ■

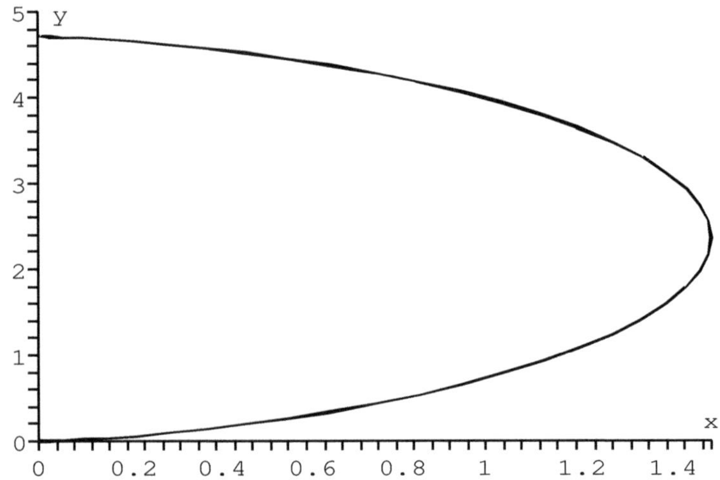

FIGURE 9.7
The Path of an Electron.

9.4 Mixture Problems

Many important problems in biology, chemistry, engineering, and physics can be formulated within the following general framework. Suppose at a given time $t = 0$ various containers (tanks, lakes, or rooms, for example) hold known amounts of a certain substance. Also suppose at time $t = 0$ any or all of the following three events occur: A solution (fluid) containing a specified concentration of the substance begins to flow into one or more containers in the system from outside the system. A solution begins to flow from one container in the system to another container. A solution begins to flow out of the system from some container. The mixture in each container is assumed to be kept at a uniform, but not necessarily constant, concentration throughout by a mixing device. The problem is to determine the amount of the substance in each container as a function of time. For each problem of the type just described, a system of differential equations to be solved is established using the following assumptions:

> The rate of change of the amount of substance in any container at time t is equal to the sum over all inputs to the container of the concentration of each input times the rate of input minus the concentration of the substance in the particular container times the sum of the rates of flow of outputs from the container.

That is, the rate of change of the amount of substance in container k at time t is

$$\frac{dq_k}{dt} = \sum_i c_i(t)r_i(t) - c_k(t)\sum_o r_o(t)$$

where q_k is the amount of substance in container k at time t, $c_i(t)$ is the concentration of the substance in input i and $r_i(t)$ is the corresponding rate of input, $c_k(t)$ is the concentration of the substance in container k at time t, and $r_o(t)$ is the rate of output to output o. Here the index of summation i is over all inputs and the index of summation o is over all outputs.

As an example, suppose at time $t = 0$ an amount A_1 of a particular substance is present in a solution that fills a container of volume V_1 and an amount A_2 of the same substance is present in a solution that fills a second container of volume V_2. Assume the two containers are connected by tubes of negligible length as shown in Figure 9.8. Further assume at time $t = 0$, (i) the solution in the first container which is kept at uniform concentration $c_1(t)$ is pumped into the second container at the constant rate r, and (ii) the solution in the second container which is kept at uniform concentration $c_2(t)$ is pumped back into the first container at the rate r. The problem is to determine the amount of substance $q_1(t)$ in the first container and the amount of substance $q_2(t)$ in the second container as a function of time.

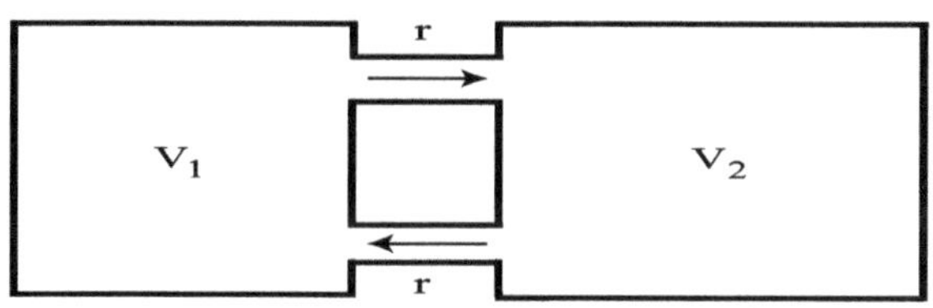

FIGURE 9.8
Mixture Problem for Two Interconnected Tanks.

Since the volume of both containers is constant, the concentration of the substance in the first container is the amount of substance in the container divided by the volume of the container—that is, $c_1(t) = q_1(t)/V_1$—and the concentration of the substance in the second container is $c_2(t) = q_2(t)/V_2$. Equating the rate of change of the amount of substance in each container to the concentration of the substance in the other container times the rate at which the solution enters the container from the other container minus the concentration of the substance in the container times the rate at which the solution leaves the container, we obtain the following system of differential equations:

$$\begin{aligned}\frac{dq_1}{dt} &= c_2(t)r - c_1(t)r = r\frac{q_2(t)}{V_2} - r\frac{q_1(t)}{V_1}\\ \frac{dq_2}{dt} &= c_1(t)r - c_2(t)r = r\frac{q_1(t)}{V_1} - r\frac{q_2(t)}{V_2}.\end{aligned}$$

Rewriting this system in matrix-vector notation, we see q_1 and q_2 satisfy

(16a)
$$\mathbf{q}' = \begin{pmatrix} q_1' \\ q_2' \end{pmatrix} = \begin{pmatrix} \frac{-r}{V_1} & \frac{r}{V_2} \\ \frac{r}{V_1} & \frac{-r}{V_2} \end{pmatrix} \begin{pmatrix} q_1 \\ q_2 \end{pmatrix}.$$

The initial conditions are $q_1(0) = A_1$ and $q_2(0) = A_2$ or

(16b)
$$\mathbf{q}(0) = \begin{pmatrix} q_1(0) \\ q_2(0) \end{pmatrix} = \begin{pmatrix} A_1 \\ A_2 \end{pmatrix}.$$

For our next example, suppose two containers are connected by tubes of negligible length as shown in Figure 9.9. Suppose at time $t = 0$ an amount A_1 of a substance is present in a solution that fills a container of constant volume V_1 and an amount A_2 of the same substance is present in a solution that fills a container of constant volume V_2.

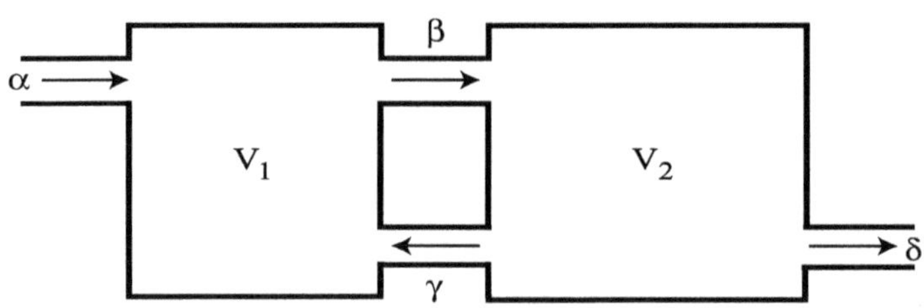

FIGURE 9.9
Mixture Problem for Two Interconnected Tanks.

Also assume at time $t = 0$ that (i) a solution containing a concentration c_α of the substance is allowed to enter the first container from an outside source at the rate α; (ii) the solution in the first container is kept at a uniform concentration $c_1(t)$ and is pumped into the second container at a constant rate β; and (iii) the solution in the second container, which is kept at a uniform concentration $c_2(t)$, is pumped back into the first container at the constant rate γ and is pumped out of the system at the constant rate δ. We assume both containers are always full. This means the sum of the input rates to a container must equal the sum of the output rates. Hence, for the system under consideration the following relationships between the rates of flow must hold:

(17)
$$\alpha + \gamma = \beta \qquad\qquad \beta = \gamma + \delta$$
$$\text{(for container 1)} \qquad \text{(for container 2).}$$

Let $q_1(t)$ be the amount of substance in the first container and $q_2(t)$ be the amount of substance in the second container at time t. The concentration of the substance in the first container is $c_1(t) = q_1(t)/V_1$ and the concentration of the substance in the second container is $c_2(t) = q_2(t)/V_2$. Equating the rate of change of the amount of substance in each container to the sum over all inputs to the container of the concentration of the input times the rate of input minus the concentration of the substance in the container times the sum of the rates of output from the container, we obtain the following system of differential equations for the mixture system under consideration:

$$\begin{aligned} \frac{dq_1}{dt} &= c_\alpha \alpha + c_2(t)\gamma - c_1(t)\beta &&= c_\alpha \alpha + \gamma \frac{q_2}{V_2} - \beta \frac{q_1}{V_1} \\ \frac{dq_2}{dt} &= c_1(t)\beta - c_2(t)(\gamma + \delta) &&= \beta \frac{q_1}{V_1} - (\gamma + \delta)\frac{q_2}{V_2}. \end{aligned}$$

Or, writing this system in matrix-vector notation

$$\text{(18a)} \qquad \mathbf{q}' = \begin{pmatrix} q_1' \\ q_2' \end{pmatrix} = \begin{pmatrix} \dfrac{-\beta}{V_1} & \dfrac{\gamma}{V_2} \\ \dfrac{\beta}{V_1} & \dfrac{-(\gamma+\delta)}{V_2} \end{pmatrix} \begin{pmatrix} q_1 \\ q_2 \end{pmatrix} + \begin{pmatrix} \alpha c_\alpha \\ 0 \end{pmatrix}.$$

The initial conditions for this system are

$$\text{(18b)} \qquad \mathbf{q}(0) = \begin{pmatrix} q_1(0) \\ q_2(0) \end{pmatrix} = \begin{pmatrix} A_1 \\ A_2 \end{pmatrix}.$$

Exercise 22.

a. Use your computing software to find the general solution to system (16a) for $V_1 = 100$ gal, $V_2 = 50$ gal, and $r = 10$ gal/min.

b. Solve the initial value problem (16) for V_1, V_2, and r as given in part a., $A_1 = 50$ lbs, and $A_2 = 0$ lbs. That is, determine values for the arbitrary constants in the general solution for part a. to satisfy the initial conditions (16b) when $A_1 = 50$ lbs and $A_2 = 0$ lbs.

c. What is $\lim_{t\to\infty} q_1(t)$? What is $\lim_{t\to\infty} q_2(t)$? Are these values the values you expect from strictly physical considerations?

Exercise 23.

a. Find the general solution to the homogeneous system associated with the nonhomogeneous system (18a) for $\alpha = 15$ gal/min, $\beta = 20$ gal/min, $\gamma = 5$ gal/min, $\delta = 15$ gal/min, $V_1 = 200$ gal, and $V_2 = 100$ gal.

b. Find the general solution of the nonhomogeneous system (18a) for α, β, γ, δ, V_1, and V_2 as given in part a. and $c_\alpha = 1$ lb/gal. (Hint: Assume there is a particular solution of (18a) in which each component is constant.)

c. Find the general solution of the initial value problem (18) for α, β, γ, δ, V_1, and V_2, as given in part a., for $c_\alpha = 1$ lb/min, and $A_1 = 10$ lbs and $A_2 = 0$ lbs.

d. What is $\lim_{t\to\infty} q_1(t)$? What is $\lim_{t\to\infty} q_2(t)$? How does $\lim_{t\to\infty} q_1(t)/V_1$ and $\lim_{t\to\infty} q_2(t)/V_2$ compare with c_α?

Exercise 24. Tank A is a 200 gallon tank and is initially filled with a brine solution which contains 50 pounds of salt. Tank B is a 200 gallon tank and is initially filled with pure water. Pure water flows into tank A from an outside source at 5 gal/min. Solution from tank A is pumped into tank B at 7 gal/min. Solution from tank B is pumped back into tank A at 2 gal/min and out of the system at 5 gal/min. Find the amount of salt in each tank as a function of time. What is the limiting value for the amount of salt in each tank?

Exercise 25. Tank A is a 100 gallon tank and is initially filled with a brine solution which contains 30 pounds of salt. Tank B is a 200 gallon tank and is initially filled with a brine solution which contains 15 pounds of salt. A brine solution containing 2 pounds of salt per gallon enters tank A from an outside source at a rate of 5 gal/min. Solution from tank A is pumped into tank B at a rate of 3 gal/min. Solution from tank B is pumped back into tank A at the same rate. Solution from tank A is pumped out of the system at the rate of 5 gal/min. Find the amount of salt in each tank as a function of time. What is the limiting value for the amount of salt in each tank?

Exercise 26. Three tanks are connected as shown in Figure 9.10. Each tank contains 100 gallons of brine. The rates of flow are $\alpha = 10$ gal/min, $\beta = 5$ gal/min, and $\gamma = 15$ gal/min. Find the amount of salt in each tank as a function of time, if initially tank 1 has 20 pounds, tank 2 has 5 pounds, and tank 3 has 10 pounds. What is the limiting amount of salt in each tank?

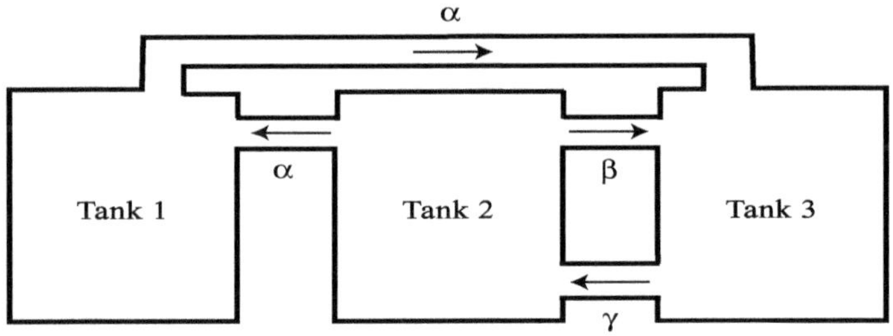

FIGURE 9.10
Mixture Problem for Three Interconnected Tanks.

Exercise 27. Three small ponds each containing 10,000 gallons of pure water are formed by a spring rain. Water containing .3 lbs/gal of salt enters pond A

at 10 gal/hr. Water evaporates from pond A at 3 gal/hr and flows into pond B at 7 gal/hr. Water evaporates from pond B at 2 gal/hr and flows into pond C at 5 gal/hr. Water evaporates from pond C at 2.5 gal/hr and flows out of the system at 2.5 gal/hr. Find the amount of salt in each pond as a function of time. What is the limiting amount of salt in each pond?

Exercise 28. Three 150-gallon tanks initially contain pure water. Brine with a concentration of 4 lbs/gal flows into tank A at a rate of 10 gal/min. Water is pumped from tank A into tank B at a rate of 6 gal/min and into tank C at a rate of 4 gal/min. Water is pumped from tank C into tank B at a rate of 4 gal/min. And water is pumped out of the system from tank B at the rate of 10 gal/min. Find the amount of salt in each tank as a function of time.

Pollution in the Great Lakes As we stated earlier, one of the major problems facing industrialized nations is water pollution. Rivers and lakes become polluted with various types of waste products such as DDT, mercury, and phosphorus which kill plant and marine life. Once pollution in a river is stopped, the river cleans itself fairly rapidly. However, as this example for the Great Lakes will illustrate, large lakes require much longer to clean themselves by the natural process of clean water flowing into the lake and polluted water flowing out. Figure 9.11 shows the Great Lakes, their volumes, and inflow and outflow rates.

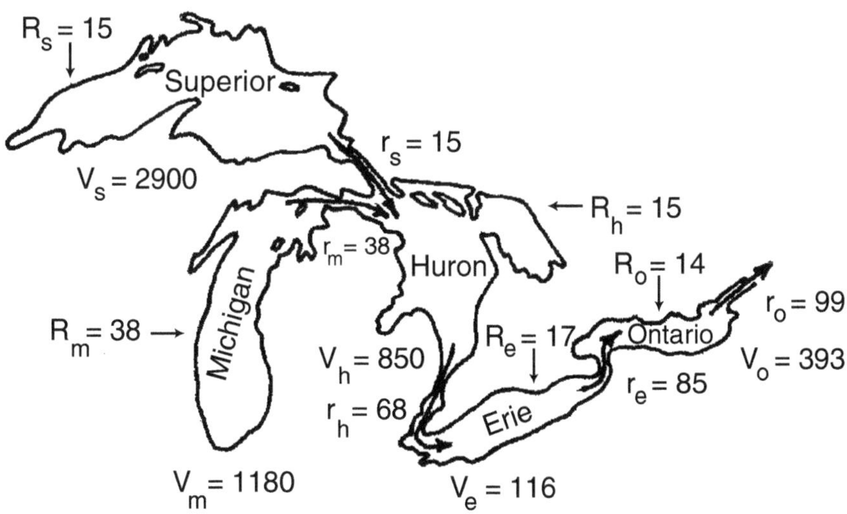

FIGURE 9.11
Volumes (mi^3) and Flow Rates (mi^3/yr) for the Great Lakes.

A simple mathematical model for pollution in the Great Lakes treats this system of lakes as a standard, perfect mixing problem. Thus, we make the following four assumptions:

1. The flow rates are constant. (Of course, these rates are not constant but variable, since they are affected seasonally by rainfall, snowfall, and evaporation.)

2. The volume of each lake is constant. (The volumes of the lakes are variable due to the variation in inflow and outflow rates and seasonal changes.)

3. Perfect mixing occurs in each lake so that the pollutants are uniformly distributed throughout the lake. (Pollutants in the lakes are not uniformly mixed. Incoming water tends to move from its source to the outlet of the lake in a pipeline fashion without mixing. So the clean up time for the main part of a lake will be less than predicted while the clean up time for isolated, slow moving portions of the lake will be longer than predicted.)

4. Pollutants are dissolved in water and enter and leave the lake by inflow and outflow. (DDT, for example, is ingested by higher predators and retained in their body fat. These animals are large and not apt to leave the lake with the outflow unless they choose to do so. When the predator dies most of its body fat is consumed by other organisms. So most DDT remains in the biosphere for an extended period of time. As a result DDT will remain in a lake in higher quantities than predicted by the model. Phosphorus, on the other hand, causes "algae bloom"—a sudden population explosion of algae. Later, the algae dies and settles to the bottom of the lake removing some phosphorus in the process. However, this removal is only temporary, since the decaying process returns the phosphorus to the lake water.)

We will use the following notation and information in deriving the system of differential equations for the amount of pollution in the Great Lakes.

LAKE	**SUBSCRIPT** (i)	**VOLUME** (V_i) mi^3	**OUTFLOW RATE** (r_i) mi^3/yr	**NON-LAKE INFLOW RATE** (R_i) mi^3/yr
Erie	e	116	85	17
Huron	h	850	68	15
Michigan	m	1180	38	38
Ontario	o	393	99	14
Superior	s	2900	15	15

Let

$q_i(t)$ be the amount of pollutant (DDT, phosphorus, or mercury) in lake i at time t;

$c_i(t)$ be the concentration of pollutant in lake i at time t; and

C_i be the concentration of pollutant in the inflow to lake i.

Since the amount of pollutant in each lake, $q_i(t)$, equals the concentration of pollutant, $c_i(t)$, times the volume, V_i, we have for each lake $q_i(t) = c_i(t)V_i$. Using Figure 9.11 to write an equation for the rate of change of the amount of pollutant in each lake, we find

for Lake Superior

$$\frac{dq_s}{dt} = \frac{d(c_sV_s)}{dt} = R_sC_s - r_sc_s$$

for Lake Michigan

$$\frac{dq_m}{dt} = \frac{d(c_mV_m)}{dt} = R_mC_m - r_mc_m$$

for Lake Huron

$$\frac{dq_h}{dt} = \frac{d(c_hV_h)}{dt} = R_hC_h + r_sc_s + r_mc_m - r_hc_h$$

for Lake Erie

$$\frac{dq_e}{dt} = \frac{d(c_eV_e)}{dt} = R_eC_e + r_hc_h - r_ec_e$$

for Lake Ontario

$$\frac{dq_o}{dt} = \frac{d(c_oV_o)}{dt} = R_oC_o + r_ec_e - r_oc_o.$$

Dividing each of these equations by the volume of the corresponding lake, we obtain the following system of differential equations for the concentration of pollution in the Great Lakes

$$\begin{aligned}
\frac{dc_s}{dt} &= \frac{R_sC_s - r_sc_s}{V_s} \\
\frac{dc_m}{dt} &= \frac{R_mC_m - r_mc_m}{V_m} \\
\frac{dc_h}{dt} &= \frac{R_hC_h + r_sc_s + r_mc_m - r_hc_h}{V_h} \qquad (19) \\
\frac{dc_e}{dt} &= \frac{R_eC_e + r_hc_h - r_ec_e}{V_e} \\
\frac{dc_o}{dt} &= \frac{R_oC_o + r_ec_e - r_oc_o}{V_o}.
\end{aligned}$$

Exercise 29.

a. Write system (19) using matrix-vector notation. (The answer will have the form $\mathbf{c}' = \mathbf{A}\mathbf{c} + \mathbf{b}$.)

b. Assume all pollution of the Great Lakes ceases—that is, assume $C_s = C_m = C_h = C_e = C_o = 0$. Use your computing software to find the general solution of $\mathbf{c}' = \mathbf{A}\mathbf{c}$ where the constant entries of $\mathbf{A}$ are calculated using the information given above for the Great Lakes.

c. Assume the initial concentration of pollution in each lake is .5%—that is, assume $c_i(0) = .005$. When will the concentration of pollution in each lake be reduced to .4%? .3%? (Hint: Solve the initial value problem: $\mathbf{c}' = \mathbf{A}\mathbf{c}$; $c_i(0) = .005$. Then graph the equation for the concentration of pollution in each lake and determine when the concentration drops below the specified levels.)

d. Assume the concentration of pollution entering the Great Lakes from outside the system is reduced from the current levels to .2%—that is, assume $C_s = C_m = C_h = C_e = C_o = .002$.

   (i) Find the general solution of $\mathbf{c}' = \mathbf{A}\mathbf{c} + \mathbf{b}$. (Hint: Assume there is a particular solution, $\mathbf{c}_p$, in which each component is a constant. The general solution is $\mathbf{c} = \mathbf{c}_c + \mathbf{c}_p$ where $\mathbf{c}_c$ is the answer to part b.)

   (ii) Now assume the initial concentration of pollution in each lake is $c_i(0) = 1\% = .01$. When will the concentration of pollution in each lake be reduced to .5%? .3%? (Hint: Solve the initial value problem: $\mathbf{c}' = \mathbf{A}\mathbf{c} + \mathbf{b}$; $c_i(0) = .01$. Then graph the equation for the concentration of pollution in each lake and determine when the concentration of pollution drops below the specified levels.)

10

Applications of Systems of Equations

In this chapter we initially present techniques for determining the behavior of solutions to systems of first-order differential equations without first finding the solutions. To this end, we define and discuss equilibrium points (critical points), various types of stability and instability, and phase-plane graphs. Next, we show how to use computer software to solve systems of first-order differential equations numerically, how to graph the solution components, and how to produce phase-plane graphs. We also state stability theorems for systems of first-order differential equations. Throughout this chapter we develop and discuss a wide variety of models and applications which can be written as vector initial value problems and then solved numerically.

10.1 Richardson's Arms Race Model

Throughout recorded history, there have been many discussions regarding the causes of war. In his account of the Peloponnesian war, written over two thousand years ago, Thucydides asserted that armaments cause war. He wrote:

> The real though unavowed cause I believe to have been the growth of Athenian power, which terrified the Lacedaemonians and forced them into war.

The mathematical model of an arms race which we will study in this section was developed by Lewis Fry Richardson. He was born at Newcastle-upon-Tyne, England, on October 11, 1881. Richardson attended Cambridge University where he was a pupil of Sir J. J. Thomson, the discoverer of the electron. In 1923, Richardson published his book, *Weather Prediction by Numerical Process.* This text is considered a classic work in the field of meteorology. In 1926, Richardson was awarded a Doctor of Science degree from the University of London for his contributions in the fields of meteorology and physics. The following year he was elected to the Fellowship of the Royal Society.

During World War I, Richardson served with an ambulance convoy in France. While serving in the war, he began to contemplate and write about the causes of war and how to avoid war. During the mid-1930s, Richardson developed his model for a two nation arms race. Later, he extended his model

DOI: 10.1201/9781003678014-10

to include n nations and applied this model to the contemporary situation in Europe. Hoping to avoid an impending war, Richardson submitted an article on the subject to an American journal and requested its immediate acceptance. The editors rejected the article. Shortly after the outbreak of World War II, Richardson retired. Early in his retirement he continued his research on the causes of war, but later he abandoned those efforts and returned to his studies in meteorology. He died on September 30, 1953. One idea underlying Richardson's two nation arms race model is the concept of "mutual fear." One nation begins to arm itself. The nation's stated reason for doing so is for the purpose of self-defense. A second nation, which fears the first nation, then begins to arm itself, also for the purpose of self-defense. Due to the "mutual fear" of each nation for the other, the armaments of both nations continue to increase indefinitely with time. Since no nation has infinite resources, this situation cannot actually occur. When expenditures for arms become too large a part of a nation's budget, the people of that nation force the government to limit the amount spent on arms and thereby "damp" and perhaps even "limit" the arms race. Richardson selected the amount of money spent on arms per year by a nation as the dependent variable in his arms race model. Stated verbally the assumptions of his model are

1. Arms expenditures increase because of mutual fear.

2. Societies resist ever-increasing expenditures for arms.

3. Considerations which are independent of the amounts being spent for arms contribute to the rate of change of the amount spent on arms.

If we let $x(t) \geq 0$ and $y(t) \geq 0$ represent the amounts spent on arms per year by two nations in some standard monetary unit, then according to Richardson's model, expenditures for arms per year must satisfy the following system of differential equations

$$
\begin{aligned}
\frac{dx}{dt} &= Ay - Cx + r \\
\frac{dy}{dt} &= Bx - Dy + s
\end{aligned} \tag{1}
$$

where A, B, C, and D are nonnegative real constants and r and s are real constants (negative, zero, or positive). The term Ay represents the rate of increase in yearly expenditures for arms by the first nation due to its fear of the second nation. That is, the rate of increase in yearly expenditures due to fear is modeled as the nonnegative constant A times the amount, y, currently being spent per year by the second nation for arms. The term $-Cx$ represents the rate of decrease in yearly expenditures for arms by the first nation due to the resistance of its people to increased spending for arms. Thus, the rate of decrease in expenditures due to societal resistance is modeled as the nonpositive constant $-C$ times the amount, x, currently being spent per year by the first nation for arms. The constant term r represents the underlying "grievance" which the first nation feels toward the second nation. If r is positive, then the first nation has a grievance against the second nation which causes it to increase arms expenditures. If r is negative, the first nation has a feeling of goodwill toward the second nation and, therefore, tends to decrease its expenditures for arms. The terms of the second equation of system (1) can be interpreted in an analogous manner. The system of differential equations (1) for Richardson's arms race model asserts that the rate of change in the amount spent on arms per year by one nation is increased in proportion to the amount the other nation is currently spending on arms per year, is decreased in proportion to the amount it is currently spending on arms per year, and is decreased or increased based on a feeling of goodwill or grievance against the other nation. Appropriate initial conditions for Richardson's arms race model (1) is the amount spent on arms per year by both nations at a particular time. Suppose at time $t = t_0$ the first nation spends $x(t_0) = x_0$ for arms per year and the second nation spends $y(t_0) = y_0$. Thus, the initial value problem to be solved is

$$(2) \qquad \begin{aligned} \frac{dx}{dt} &= -Cx + Ay + r; \qquad x(t_0) = x_0 \\ \frac{dy}{dt} &= Bx - Dy + s; \qquad y(t_0) = y_0. \end{aligned}$$

Or, written in matrix-vector notation

$$(2') \qquad \begin{pmatrix} x' \\ y' \end{pmatrix} = \begin{pmatrix} -C & A \\ B & -D \end{pmatrix} \begin{pmatrix} x \\ y \end{pmatrix} + \begin{pmatrix} r \\ s \end{pmatrix}; \quad \begin{pmatrix} x(t_0) \\ y(t_0) \end{pmatrix} = \begin{pmatrix} x_0 \\ y_0 \end{pmatrix}.$$

Using the techniques of Chapter 9, a general solution to this initial value problem can be found (i) by computing the eigenvalues and eigenvectors of the associated homogeneous system; (ii) by writing the complementary solution (the solution of the associated homogeneous system); (iii) by calculating a particular solution to (2′); (iv) by writing the general solution of system (2′), which is the sum of the complementary solution and the particular solution; and (v) by determining values for the arbitrary constants appearing in the general solution which satisfy the given initial conditions. Our general solution would then depend on the six parameters A, B, C, D, r, and s, and on the initial conditions x_0 and y_0.

However, instead of finding the general solution to (2′), we will use techniques from elementary calculus to answer important questions concerning the nature of the solutions without first finding the solutions. The first and most important question to ask is, "Is there an equilibrium point?" By an equilibrium point, we mean a point (x^*, y^*) which is a constant solution of (1). At an equilibrium point, the amount spent on arms per year by each nation remains constant for all time—the first nation always spends x^* and the second nation always spends y^*. If there is an equilibrium point, we then want to answer the question: "Is the equilibrium point stable or unstable?" That is, we want to know if an initial condition (x_0, y_0) is sufficiently close to the equilibrium point, does the associated solution $(x(t), y(t))$ remain close to (x^*, y^*) for all time—in which case, we say the equilibrium point is stable—or does the associated solution move away from the equilibrium point—in which case, we say the equilibrium point is unstable. For there to be an equilibrium point (constant solution) for Richardson's arms race model, the rate of change of the amount spent for arms per year by both nations must simultaneously be zero. Thus, we must have $dx/dt = dy/dt = 0$. Setting $dx/dt = 0$ and $dy/dt = 0$ in system (2), we see that an equilibrium point (x^*, y^*) must simultaneously satisfy

$$
\begin{aligned}
L_1 : \quad & -Cx + Ay + r &= 0 \\
L_2 : \quad & Bx - Dy + s &= 0.
\end{aligned} \tag{3}
$$

If $E = CD - AB \neq 0$, then there is a unique solution to (3)—namely,

$$
x^* = \frac{rD + sA}{E}, \qquad y^* = \frac{Cs + Br}{E}. \tag{4}
$$

In discussing Richardson's arms race model (2), you might think we should sketch graphs of x and y as functions of time and draw conclusions from these graphs. However, it is often more informative to look at a "phase-plane" diagram in which one dependent variable is graphed versus the other dependent variable. In this instance, we could plot y versus x or x versus y. The graph in the xy-plane of both L_1 and L_2 are lines—see equations (3).

Let us assume in the discussion which follows that the parameters A, B, C, and D are all nonzero. A typical sketch of the graph of L_1 for $r > 0$ is shown in Figure 10.1. The line L_1 divides the xy-plane into three sets—(i) the line L_1, itself, on which $dx/dt = -Cx + Ay + r = 0$, (ii) the half-plane where $dx/dt = -Cx + Ay + r > 0$, and (iii) the half-plane where $dx/dt = -Cx + Ay + r < 0$.

If we view Richardson's arms race model (2) as the equations of motion for a particle in the xy-plane, the first equation in (2), $dx/dt = -Cx + Ay + r$, tells us the horizontal component of the velocity, and the second equation in (2), $dy/dt = Bx - Dy + s$, tells us the vertical component of the velocity. The line L_1 divides the xy-plane into two half-planes. In one half-plane $dx/dt > 0$, so $x(t)$ is increasing in this region, and the particle will tend to move to the right. In the other half-plane, $dx/dt < 0$, so $x(t)$ is decreasing in this region and the particle will tend to move to the left.

Let (p, q) be any point below and to the right of the line L_1 and let (u, q) be the corresponding point on L_1 as shown in Figure 10.1. Since (u, q) is on L_1, $-Cu + Aq + r = 0$. At the point (p, q)

$$
\begin{aligned}
\frac{dx}{dt} = -Cp + Aq + r &= -Cp + Aq + r - 0 \\
&= -Cp + Aq + r - (-Cu + Aq + r) = -C(p - u).
\end{aligned}
$$

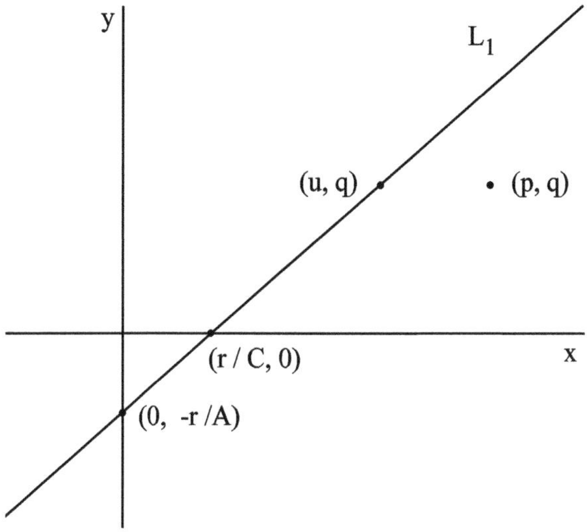

FIGURE 10.1
Graph of L_1: $-Cx + Ay + r = 0$ for $A > 0$, $C > 0$, and $r > 0$.

Since p is to the right of u, $p - u > 0$, and since we have assumed $C > 0$, $dx/dt = -C(p-u) < 0$ in the half-plane below and to the right of the line L_1. In this region the particle tends to move to the left toward the line L_1. In the half-plane above and to the left of L_1, $dx/dt > 0$ and in this region the particle tends to move to the right toward the line L_1. So the first nation in the arms race tries, at all times, to adjust its expenditures by moving them horizontally toward the line L_1. (The reader should note that if the point (u, q) is on the line L_1, then $dx/dt = 0$, but most likely $dy/dt \neq 0$ on L_1. If $dy/dt \neq 0$ at (u, q), the graph of y versus x will cross the line L_1 at (u, q).)

A similar argument shows that (i) L_2 divides the xy-plane into two half-planes, (ii) in the half-plane below and to the right of L_2, $dy/dt > 0$ so $y(t)$ is increasing in this region and the particle tends to move upward toward the line L_2, and (iii) in the half-plane above and to the left of L_2, $dy/dt < 0$, so $y(t)$ is decreasing in this region, and the particle tends to move downward toward the line L_2. That is, the second nation in the arms race tries, at all times, to adjust its expenditures by moving them vertically toward the line L_2.

We wish to examine the limiting behavior of the expenditures for arms in Richardson's model. We divide the limiting behavior into three categories and classify the types of arms races as follows:

1. If $x \to 0$ and $y \to 0$ as $t \to \infty$, then we say the arms race results in **mutual disarmament**.

2. If $x \to x^*$ and $y \to y^*$ as $t \to \infty$, then we say the arms race is a **stable arms race**.

3. If $x \to \infty$ and $y \to \infty$ as $t \to \infty$, then we say there is an **unstable arms race** or a **runaway arms race**.

We will now consider two cases under the assumptions that A, B, C, and D are nonzero. Solving the equations of L_1 and L_2 for y (see equations (3)), we find

$$y = \frac{Cx}{A} - \frac{r}{A} \qquad \text{(for } L_1\text{)}$$

$$y = \frac{Bx}{D} + \frac{s}{D} \qquad \text{(for } L_2\text{).}$$

Thus, the slope of line L_1 is C/A and the slope of L_2 is B/D.

Case 1. Suppose $r = s = 0$ and L_1 and L_2 are not parallel. Since $r = s = 0$, the equilibrium point is the origin, $(0, 0)$—see equations (4). Since L_1 and L_2 are not parallel, they intersect at the origin, and either (a.) the slope of L_1 is larger than the slope of L_2 or (b.) the slope of L_2 is larger than the slope of L_1. Representative graphs for case 1a. and case 1b. are shown in Figures 10.2a. and 10.2b., respectively. We numbered the regions in the first quadrant, I, II, and III, as shown in the figures.

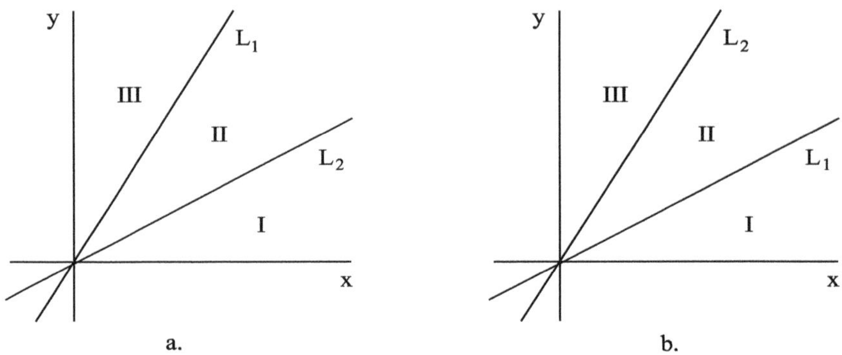

FIGURE 10.2
Graphs of Lines L_1 and L_2 for r = s= 0 and a. $C/A > B/D$ and b. $C/A < B/D$.

Case 1a. In this case, we are assuming $r = s = 0$ and the slope of L_1 is greater than the slope of L_2 ($C/A > B/D$). Regions I, II, and III refer to the regions shown in Figure 10.2 a. In region I, $dx/dt < 0$ and $dy/dt > 0$. So if a particle is in region I, it will tend to move to the left and up until it reaches the line L_2. On L_2, $dx/dt < 0$ and $dy/dt = 0$, so a particle from region I crosses L_2 into region II. If a particle is in region III, $dx/dt > 0$ and $dy/dt < 0$. So a particle in region III tends to move to the right and down until it reaches the line L_1. On L_1, $dx/dt = 0$ and $dy/dt < 0$, so a particle from region III crosses L_1 into region II. In region II, $dx/dt < 0$ and $dy/dt < 0$. So a particle in region II moves to the left and down. If at some instant in time the particle were to move from region II to the line L_1, then $dx/dt = 0$ and $dy/dt < 0$, and the particle would move downward into region II again. A similar argument holds for any particle in region II which approaches the line L_2. Such a particle would move to the left and return to region II. Thus, any particle in region II or any particle crossing into region II from either region I or region III remains in region II and approaches the origin as $t \to \infty$. Thus, for any initial conditions $x(t_0) = x_0 \geq 0$, $y(t_0) = y_0 \geq 0$, that is, for any initial point in the first quadrant, $x(t) \to 0$ and $y(t) \to 0$ as $t \to \infty$. Hence, Case 1a. results in mutual disarmament regardless of the initial conditions (the initial expenditures for arms).

Case 1b. This case is left as an exercise.

Case 2. Suppose $r < 0$, $s < 0$, and L_1 and L_2 are not parallel. Since $r < 0$ and $s < 0$ both nations have a permanent underlying feeling of "goodwill," toward the other nation. Since L_1 and L_2 are not parallel, they intersect at an equilibrium point (x^*, y^*), and either (a.) the slope of L_1, namely C/A, is greater than the slope of L_2, namely B/D, or (b.) the slope of L_1 is less than the slope of L_2 $(C/A < B/D)$.

Case 2a. This case is left as an exercise.

Case 2b. We now consider the case where $C/A < B/D$. Multiplying this inequality by AD which is positive, since A and D are assumed to be positive, we find $CD < AB$ or $E = CD - AB < 0$. Since $A > 0$, $B > 0$, $C > 0$, $D > 0$, $r < 0$, $s < 0$, and $E < 0$, we see from equations (4) that $x^* = (rD+sA)/E > 0$ and $y^* = (Cs + Br)/E > 0$. Thus, the equilibrium point (x^*, y^*) lies in the first quadrant. A sketch of the situation for this particular case and a table indicating the sign of dx/dt and dy/dt in each region is shown in Figure 10.3. The arrows in each region indicate the horizontal and vertical direction a particle in the region will take. Notice that all horizontal arrows point toward the line L_1 while all vertical arrows point toward the line L_2.

Let the initial values for arms expenditures be (x_0, y_0) and assume (x_0, y_0) lies in region I of Figure 10.3. In region I, $dx/dt > 0$ and $dy/dt > 0$. So a particle in region I will tend to move to the right and up. That is, the particle will tend to move farther out in region I. If the particle were to reach the line L_1, dx/dt would be zero, but dy/dt would be positive. Hence, at L_1 the particle would be forced back into region I and farther away from (x^*, y^*). The same holds true for a particle which reaches line L_2 from inside region I. At L_2, $dy/dt = 0$ but $dx/dt > 0$ so the particle moves back into region I and away from (x^*, y^*). Hence, a particle in region I remains in region I for all time and moves away from (x^*, y^*) as time increases. That is, $x \to \infty$ and $y \to \infty$ as $t \to \infty$. So if the initial arms expenditures are such that (x_0, y_0) falls in region I, a runaway arms race results. This example illustrates that Richardson's arms race model predicts an unstable arms race can result even when both nations have a feeling of "goodwill" toward the other nation ($r < 0$

and $s < 0$). A runaway arms race occurs if the initial arms expenditures by both nations are sufficiently large so that (x_0, y_0) lies in region I.

Region	dx/dt	dy/dt
I	+	+
II	+	−
III	−	−
IV	−	+

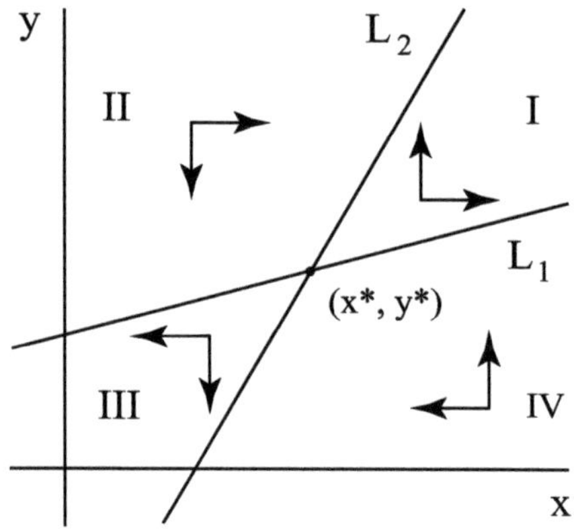

FIGURE 10.3
Graph of Lines L_1 and L_2 for $r < 0$, $s < 0$, and $C/A < B/D$.

In region III, $dx/dt < 0$ and $dy/dt < 0$. So a particle in region III tends to move to the left and down in this region, away from (x^*, y^*) and toward the origin $(0, 0)$. When the initial arms expenditures (x_0, y_0) lie in region III, it can be shown that mutual disarmament (total disarmament) always results—that is, as $t \to \infty$, $x \to 0$, and $y \to 0$.

Looking at the arrows in Figure 10.3, we see that a particle in region II will move to the right and down. Thus, a particle in region II, (i) will move toward the segment of line L_2 between regions I and II, (ii) will move toward the equilibrium point (x^*, y^*), or (iii) will move toward the segment of line L_1 between regions II and III. If a particle which starts in region II moves to the portion of the line L_2 between regions I and II, the particle will proceed into region I, and a runaway arms race will result, because on L_2, $dx/dt > 0$ and $dy/dt = 0$. If a particle in region II moves to the equilibrium point (x^*, y^*), it will remain there. Whereas, if a particle which starts in region II moves to the portion of the line L_1 between regions II and III, the particle will proceed into region III, and mutual disarmament will result, because on L_1, $dx/dt = 0$ and $dy/dt < 0$. So when the initial arms expenditures lie in region II, a runaway arms race may occur, a stable arms race may occur, or mutual disarmament may occur.

The results for initial arms expenditures which lie in region IV are similar to the results for expenditures which lie in region II. A runaway arms race, a stable arms race, or mutual disarmament may occur depending on where (x_0, y_0) lies within region IV.

10.1.1 Exercises

1. Perform an analysis for Case 1b. similar to the analysis performed in the text for Case 1a. What type of arms race results in this case?

2. Perform an analysis for Case 2a. That is, perform an analysis for Richardson's arms race model when $r < 0$, $s < 0$, and the slope of line L_1 is greater than the slope of L_2 $(C/A > B/D)$. In what quadrant does the equilibrium point (x^*, y^*) lie? How many distinct regions are there to consider in the first quadrant? Draw an appropriate figure and construct a table to aid you in performing your analysis. (Figure 10.3 includes an appropriate figure and table for Case 2b.) What kinds of arms races can occur in this instance?

3. Perform an analysis for Richardson's arms race model for $r > 0$, $s > 0$ and

   a. $C/A > B/D$ (the slope of line L_1 is greater than the slope of line L_2.)

   b. $C/A < B/D$ (the slope of line L_1 is less than the slope of line L_2.)

   In each case, answer the following questions:

   (i) Where is the equilibrium point?

   (ii) What kinds of arms races can occur?

4. Perform an analysis for Richardson's arms race model for $r < 0$, $s > 0$ and

   a. $C/A > B/D$ and b. $C/A < B/D$.

   Answer the following questions:

   (i) Where is the equilibrium point?

   (ii) What kinds of arms races can occur?

5. Using the results given in the text and the results of Exercises 1-4 complete the following table by entering the type of arms race that may occur in each case. (md = mutual disarmament, sar = stable arms race, rar = runaway arms race).

r	s	E = CD − AB	Possible Arms Race(s)
0	0	+	
0	0	−	
−	−	+	
−	−	−	
+	+	+	
+	+	−	
−	+	+	
−	+	−	

6. What kinds of arms races can develop if $r < 0$, $s < 0$, and lines L_1 and L_2 are parallel ($C/A = B/D$; $E = 0$)? (Hint: Consider two separate cases, where L_1 is above L_2 and vice versa.)

10.2 Phase-Plane Portraits

There are two fundamental subdivisions in the study of differential equations: (1) quantitative theory and (2) qualitative theory. The object of quantitative theory is (i) to find an explicit solution of a given differential equation or system of equations, (ii) to express the solution as a finite number of quadratures, or (iii) to compute an approximate solution. At an early stage in the development of the subject of differential equations, it appears to have been believed that elementary functions were sufficient for representing the solutions of differential equations which evolved from problems in geometry and mechanics. However, in 1725, Daniel Bernoulli published results concerning Riccati's equation which showed that even a first-order ordinary differential equation does not necessarily have a solution which is finitely expressible in terms of elementary functions. In the 1880s, Picard proved that the general linear differential equation of order n is not integrable by quadratures. At about the same time in a series of papers published between 1880 and 1886, Henri Poincaré (1854-1912) initiated the qualitative theory of differential equations. The object of this theory is to obtain information about an entire set of solutions without actually solving the differential equation or system of equations. For example, one tries to determine the behavior of a solution with respect to that of one of its neighbors. That is, one wants to know whether or not a solution $v(t)$ which is "near" another solution $w(t)$ at time $t = t_0$ remains "near" $w(t)$ for all $t \geq t_0$ for which both $v(t)$ and $w(t)$ are defined.

An **autonomous system of differential equations** has the form

$$\begin{aligned} \frac{dx}{dt} &= f(x,y) \\ \frac{dy}{dt} &= g(x,y). \end{aligned} \tag{1}$$

System (1) is called **autonomous** because the functions f and g depend explicitly only on the dependent variables x and y and not upon the independent variable t (which often represents time).

In performing a qualitative analysis of an autonomous system of the form (1), we are interested in answering the following questions:

> "Are there any equilibrium points?" That is, "Are there any constant solutions?" "Is a particular equilibrium point stable or unstable?"

In Chapter 7, we stated that if f, g, f_x, f_y, g_x, and g_y are continuous functions of x and y in some rectangle R in the xy-plane and if $(x_0, y_0) \in R$, then there is a unique solution to system (1) which satisfies the initial conditions $x(t_0) = x_0$, $y(t_0) = y_0$. Furthermore, the solution can be extended in a unique manner until the boundary of R is reached. In the discussions which follow, we will assume that $f(x, y)$ and $g(x, y)$ and their first partial derivatives are all continuous functions in some rectangle R.

The xy-plane is called the **phase-plane**. A solution $(x(t), y(t))$ of (1) traces out a curve in the phase-plane. This curve is called a **trajectory** or **orbit**. A **phase-plane portrait** is a sketch of a few trajectories of (1) together with arrows indicating the direction a particle will flow along the trajectory as t (time) increases. It follows from the uniqueness of solutions of (1) that at most one trajectory passes through any point in R. That is, trajectories do not intersect one another in the phase-plane.

A **critical point** or **equilibrium point** of system (1) is a point (x^*, y^*) where $f(x^*, y^*) = g(x^*, y^*) = 0$. Hence, by definition, the critical points of system (1) are determined by simultaneously solving

$$\begin{aligned} f(x, y) &= 0 \\ g(x, y) &= 0. \end{aligned} \tag{2}$$

A critical point (x^*, y^*) of system (1) is **stable** if every solution which is "near" (x^*, y^*) at time t_0 exists for all $t \geq t_0$ and remains "near" (x^*, y^*). A stable critical point (x^*, y^*) of system (1) is **asymptotically stable** if every solution which is "near" (x^*, y^*) at time t_0 exists for all $t \geq t_0$ and $\lim_{t\to\infty} x(t) = x^*$ and $\lim_{t\to\infty} y(t) = y^*$. A stable critical point which is not asymptotically stable is said to be **neutrally stable**. A critical point which is not stable is called **unstable**.

In order to better understand the concept of stability and to be able to sketch trajectories of an autonomous system near a critical point, we first need to discuss stability and phase-plane portraits for linear systems of the form

$$\begin{aligned} \frac{dX}{dt} &= AX + BY + r \\ \frac{dY}{dt} &= CX + DY + s \end{aligned} \tag{3}$$

where A, B, C, D, r, and s are real constants. In many, but not all cases, the type of stability at the critical point of a nonlinear system is the same as the type of stability of a corresponding linear system, and near the critical point the trajectories of the nonlinear system resemble the trajectories of the linear system.

Suppose (x^*, y^*) is a critical point of the linear system (3). Hence, (x^*, y^*) simultaneously satisfies

$$\begin{aligned} Ax^* + By^* + r = 0 \\ Cx^* + Dy^* + s = 0. \end{aligned}$$

In order to locate the origin of a new xy-coordinate system at (x^*, y^*) with x-axis parallel to the X-axis and y-axis parallel to the Y-axis, we make the changes of variables $x = X - x^*$ and $y = Y - y^*$. Differentiating and substituting into (3), we find

$$\begin{aligned}\frac{dx}{dt} = \frac{dX}{dt} &= A(x + x^*) + B(y + y^*) + r \\ &= Ax + By + (Ax^* + By^* + r) \\ &= Ax + By\end{aligned}$$

and

$$\begin{aligned}\frac{dy}{dt} = \frac{dY}{dt} &= C(x + x^*) + D(y + y^*) + s \\ &= Cx + Dy + (Cx^* + Dy^* + s) \\ &= Cx + Dy.\end{aligned}$$

Next, letting

$$\mathbf{z} = \begin{pmatrix} x \\ y \end{pmatrix} \quad \text{and} \quad \mathbf{A} = \begin{pmatrix} A & B \\ C & D \end{pmatrix}$$

we see that the stability of the nonhomogeneous linear system (3) at the critical point (x^*, y^*) and the behavior of the trajectories of (3) near (x^*, y^*) is the same as the stability and behavior of the linear homogeneous system

$$\text{(4)} \qquad \frac{d\mathbf{z}}{dt} = \mathbf{A}\mathbf{z}$$

at $(x, y) = (0, 0)$. Hence, we need to study only the various types of stability of system (4) at the origin and the associated behavior of the trajectories near the origin to understand the stability and behavior of systems of the form (3) at (x^*, y^*).

In Chapter 8, we saw how to write the general solution of (4) in terms of the eigenvalues and eigenvectors of the matrix $\mathbf{A}$. The equation $\mathbf{A}\mathbf{z} = \mathbf{0}$ has the unique solution $\mathbf{z} = \mathbf{0}$ if and only if $det\,\mathbf{A} \neq 0$. That is, the origin is the unique critical point of system (4) if and only if $det\,\mathbf{A} \neq 0$. If $det\,\mathbf{A} = 0$, there is a line of critical points which passes through the origin. Throughout, we will assume $det\,\mathbf{A} \neq 0$. Let us now consider several examples.

Example 1 Determine the type of stability of the origin and sketch a phase-plane portrait for the system

$$\text{(5)} \qquad \begin{aligned} x' &= -4x + y \\ y' &= 2x - 5y. \end{aligned}$$

Solution The eigenvalues and associated eigenvectors of the matrix

$$\mathbf{A} = \begin{pmatrix} -4 & 1 \\ 2 & -5 \end{pmatrix}$$

are $\lambda_1 = -3$, $\lambda_2 = -6$,

$$\mathbf{v}_1 = \begin{pmatrix} 1 \\ 1 \end{pmatrix} \quad \text{and} \quad \mathbf{v}_2 = \begin{pmatrix} -1 \\ 2 \end{pmatrix}.$$

(Check this by using your computing software to find the eigenvalues of **A** and their associated eigenvectors.) So the general solution of (5) is

$$(6) \quad \begin{pmatrix} x(t) \\ y(t) \end{pmatrix} = c_1 e^{\lambda_1 t}\mathbf{v}_1 + c_2 e^{\lambda_2 t}\mathbf{v}_2$$

$$= c_1 e^{-3t}\begin{pmatrix} 1 \\ 1 \end{pmatrix} + c_2 e^{-6t}\begin{pmatrix} -1 \\ 2 \end{pmatrix} = \begin{pmatrix} c_1 e^{-3t} - c_2 e^{-6t} \\ c_1 e^{-3t} + 2c_2 e^{-6t} \end{pmatrix}.$$

The vector $\mathbf{v}_1$ is a vector with its "tail" at the origin and its "head" at the point $(1, 1)$ in the xy-plane as shown in Figure 10.4. The graph of the vector equation $\mathbf{u} = k\mathbf{v}_1$, where k is a parameter, is the line m_1, which contains the vector $\mathbf{v}_1$. Likewise, $\mathbf{v}_2$ is the vector with its "tail" at the origin and its "head" at $(-1, 2)$ and the graph of the vector equation $\mathbf{w} = k\mathbf{v}_2$ is the line m_2, which contains the vector $\mathbf{v}_2$. If a particle starts at any point on the line m_1 in the first quadrant, then $c_2 = 0$ and $c_1 > 0$, since $x > 0$ in the first quadrant. Since $c_2 = 0$, the particle remains on the line m_1 (see the general solution of equation (5) – equation (6)). And since $\lambda_1 = -3 < 0$, the particle moves toward the origin along the line m_1 as $t \to \infty$. If a particle starts at a point on the line m_1 in the third quadrant, then $c_2 = 0$ and $c_1 < 0$, since $x < 0$ in the third quadrant. Because $c_2 = 0$, the particle remains on the line m_1 and since $\lambda_1 = -3 < 0$, the particle moves toward the origin as $t \to \infty$. By a similar argument, any particle which starts on the line m_2 remains on m_2, because $c_1 = 0$, and proceeds toward the origin as $t \to \infty$, because $\lambda_2 = -6 < 0$. Suppose a particle starts at some point not on line m_1 or line m_2—that is, suppose $c_1 \neq 0$ and $c_2 \neq 0$. Since $x(t) = c_1 e^{-3t} - c_2 e^{-6t} \to 0$ as $t \to \infty$ and since $y(t) = c_1 e^{-3t} + 2c_2 e^{-6t} \to 0$ as $t \to \infty$, the particle proceeds toward the origin. Since e^{-6t} approaches 0 faster than e^{-3t} as t approaches ∞, the trajectories of particles not on lines m_1 and m_2 approach $(0, 0)$ asymptotic to the line m_1 as shown in Figure 10.4. In this case, the origin is an **asymptotically stable** critical point, since $x(t) \to 0$ and $y(t) \to 0$ as $t \to \infty$. The phase-plane portrait of Figure 10.4 is typical of an **asymptotically stable node.** ■

Example 2 Determine the type of stability of the origin and sketch a phase-plane portrait of the system

$$(7) \quad \begin{aligned} x' &= 4x - y \\ y' &= -2x + 5y. \end{aligned}$$

Solution

The eigenvalues and associated eigenvectors of the matrix

$$\mathbf{A} = \begin{pmatrix} 4 & -1 \\ -2 & 5 \end{pmatrix}$$

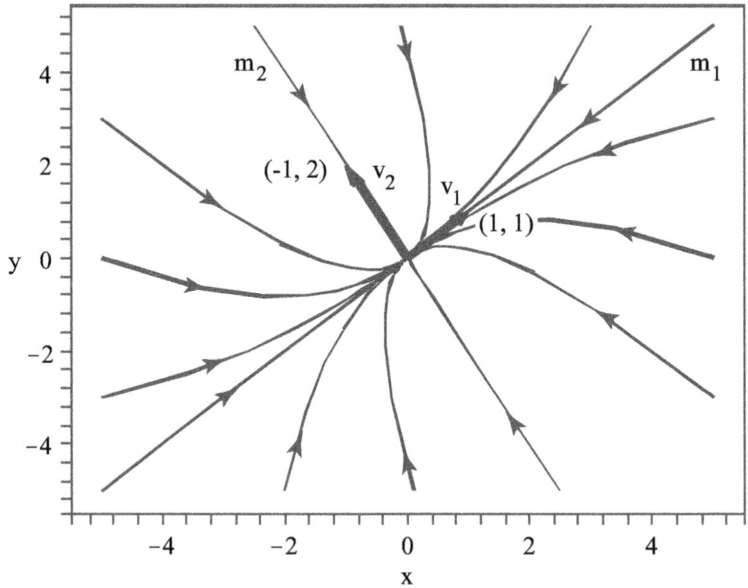

FIGURE 10.4
Asymptotically Stable Node.

are $\mu_1 = 3$, $\mu_2 = 6$,

$$\mathbf{v}_1 = \begin{pmatrix} 1 \\ 1 \end{pmatrix} \quad \text{and} \quad \mathbf{v}_2 = \begin{pmatrix} -1 \\ 2 \end{pmatrix}.$$

The eigenvectors of this example are identical to the eigenvectors of Example 1, but the eigenvalues have opposite sign. So the phase-plane portrait for this system looks exactly like the phase-plane portrait for Example 1 (Figure 10.4) except that the arrows on the trajectories, which indicate the direction that a particle will take, must be reversed. In this instance, the origin is an unstable critical point and as $t \to \infty$, the components $x(t) \to \pm\infty$ and $y(t) \to \pm\infty$. Figure 10.4 with the direction arrows reversed is a typical phase-plane portrait for an **unstable node.** ■

Example 3 Determine the type of stability of the origin and sketch a phase-plane portrait of the system

$$\begin{aligned} x' &= x + y \\ y' &= 4x - 2y. \end{aligned} \tag{8}$$

Solution

The eigenvalues and associated eigenvectors of the matrix

$$\mathbf{A} = \begin{pmatrix} 1 & 1 \\ 4 & -2 \end{pmatrix}$$

are $\lambda_1 = -3$, $\lambda_2 = 2$,

$$\mathbf{v}_1 = \begin{pmatrix} 1 \\ -4 \end{pmatrix} \quad \text{and} \quad \mathbf{v}_2 = \begin{pmatrix} 1 \\ 1 \end{pmatrix}.$$

Hence, the general solution of (8) is

$$\begin{aligned} \begin{pmatrix} x(t) \\ y(t) \end{pmatrix} &= c_1 e^{\lambda_1 t}\mathbf{v}_1 + c_2 e^{\lambda_2 t}\mathbf{v}_2 \\ &= c_1 e^{-3t}\begin{pmatrix} 1 \\ -4 \end{pmatrix} + c_2 e^{2t}\begin{pmatrix} 1 \\ 1 \end{pmatrix} \\ &= \begin{pmatrix} c_1 e^{-3t} + c_2 e^{2t} \\ -4c_1 e^{-3t} + c_2 e^{2t} \end{pmatrix}. \end{aligned}$$

The vector $\mathbf{v}_1$ is a vector with its "tail" at the origin and its "head" at $(1, -4)$. The vector $\mathbf{v}_1$ and the line m_1 containing $\mathbf{v}_1$ are drawn in Figure 10.5. The vector $\mathbf{v}_2$ and the line m_2 containing $\mathbf{v}_2$ are also drawn in Figure 10.5.

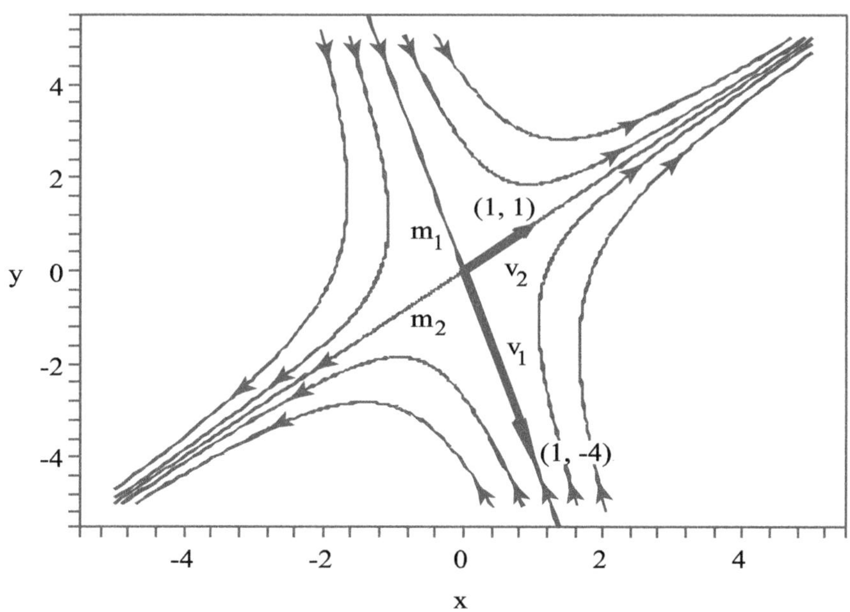

FIGURE 10.5
Saddle Point.

If a particle starts at a point on line m_1 in the fourth quadrant, then $c_2 = 0$ and $c_1 > 0$, since $x > 0$ in the fourth quadrant. Since $c_2 = 0$, the particle remains on the line m_1. And since $\lambda_1 = -3 < 0$, the particle moves toward the origin as $t \to \infty$. If a particle starts on line m_1 in the second quadrant, then $c_2 = 0$ and $c_1 < 0$, since $x < 0$ in the second quadrant. Since $c_2 = 0$, the particle remains on the line m_1 and since $\lambda_1 = -3 < 0$, the particle moves toward the origin as $t \to \infty$. Similarly a particle which starts at a point on line m_2 remains on m_2, but since $\lambda_2 = 2 > 0$, the particle moves away from the origin as $t \to \infty$. Next, suppose a particle starts at some point not on line m_1 or m_2—that is, suppose $c_1 \neq 0$ and $c_2 \neq 0$. As $t \to \infty$, $x(t) = c_1 e^{-3t} + c_2 e^{2t} \to c_2 e^{2t}$ and $y(t) = -4c_1 e^{-3t} + c_2 e^{2t} \to c_2 e^{2t}$. Thus, as $t \to \infty$,

$$\begin{pmatrix} x(t) \\ y(t) \end{pmatrix} \to \begin{pmatrix} c_2 e^{2t} \\ c_2 e^{2t} \end{pmatrix} = c_2 e^{2t} \begin{pmatrix} 1 \\ 1 \end{pmatrix}$$

asymptotically. That is, any particle which starts at a point not on line m_1 or m_2 approaches the line m_2 asymptotically as $t \to \infty$. Summarizing, we have found that a particle which starts on the line m_1 moves toward the origin as $t \to \infty$, while a particle which does not start on m_1 ultimately moves away from the origin and approaches the line m_2. In this example, the origin is an **unstable** critical point. The phase-plane portrait shown in Figure 10.5 is typical for autonomous systems for which the eigenvalues are real and of opposite sign. In such cases, the critical point (the origin) is called a **saddle point.** ■

Now suppose that $\lambda = \alpha + \beta i$ where $\beta \neq 0$ is an eigenvalue of the real matrix $\mathbf{A}$ and $\mathbf{w} = \mathbf{u} + i\mathbf{v}$ is an associated eigenvector. We saw in Chapter 8 that the general solution of the system (4) $d\mathbf{z}/dt = \mathbf{A}\mathbf{z}$ can be written as

$$\mathbf{z} = c_1\{(e^{\alpha t}\cos\beta t)\mathbf{u} - (e^{\alpha t}\sin\beta t)\mathbf{v}\} + c_2\{(e^{\alpha t}\sin\beta t)\mathbf{u} + (e^{\alpha t}\cos\beta t)\mathbf{v}\}$$

where c_1 and c_2 are arbitrary real constants. Recall that the first and second components of $\mathbf{z}$ are $x(t)$ and $y(t)$, respectively. Since $|\sin\beta t| \leq 1$, $|\cos\beta t| \leq 1$, and $\mathbf{u}$ and $\mathbf{v}$ are real constant vectors, if $\alpha < 0$, then as $t \to \infty$, $x(t) \to 0$ and $y(t) \to 0$ and the origin is an **asymptotically stable** critical point. The functions $\sin\beta t$ and $\cos\beta t$ cause the trajectories to rotate about the origin as they approach forming "spirals." If $\alpha > 0$, the trajectories spiral outward away from the origin and the origin is an unstable critical point. When $\alpha = 0$, the trajectories are elliptical in appearance with centers at the origin. In this case, the origin is a **neutrally stable** critical point. It can be shown that when the system (4) has complex conjugate eigenvalues $\lambda = \alpha \pm i\beta$, the trajectories are always spirals when $\alpha \neq 0$. The spirals may be elongated and skewed with respect to the coordinate axes and may spiral clockwise or counterclockwise about the origin.

Example 4 An Asymptotically Stable Critical Point at the Origin
The eigenvalues of the system

$$\begin{aligned} x' &= x - 6y \\ y' &= 3x - 5y \end{aligned}$$

are $\lambda = -2 \pm 3i$. Since $\alpha = -2 < 0$, the origin is an **asymptotically stable** critical point and the trajectories spiral inward toward the origin in a counterclockwise direction as shown in the phase-plane portrait of Figure 10.6.

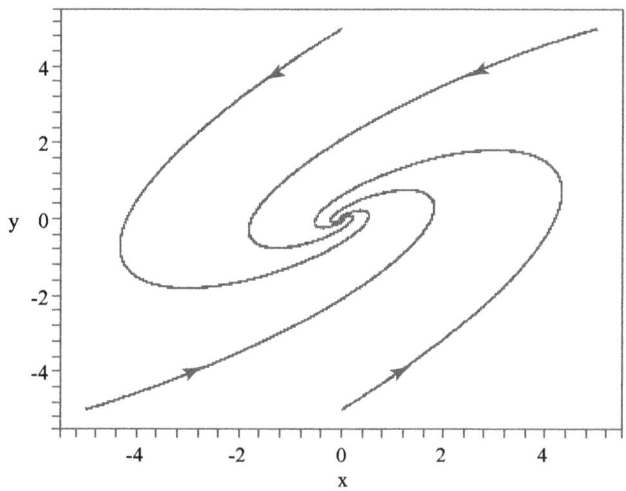

FIGURE 10.6
Asymptotically Stable Spiral Point. ■

Example 5 An Asymptotically Unstable Critical Point at the Origin
The eigenvalues of the system

$$\begin{aligned} x' &= -x + 6y \\ y' &= -3x + 5y \end{aligned}$$

are $\lambda = 2 \pm 3i$. Since $\alpha = 2 > 0$, the origin is an unstable critical point and the trajectories spiral outward away from the origin in a clockwise direction. The phase-plane portrait is the same as shown in Figure 10.6 except the direction arrows on the trajectories must be reversed. ■

Example 6 A Neutrally Stable Critical Point at the Origin The eigenvalues of the system

$$\begin{aligned} x' &= x + 2y \\ y' &= -5x - y \end{aligned}$$

are $\lambda = \pm 3i$. Since $\alpha = 0$, the origin is a **neutrally stable** critical point. In this case, the origin is called a **center**. The trajectories are "skewed ellipses" with centers at the origin. A phase-plane portrait for this system is displayed in Figure 10.7.

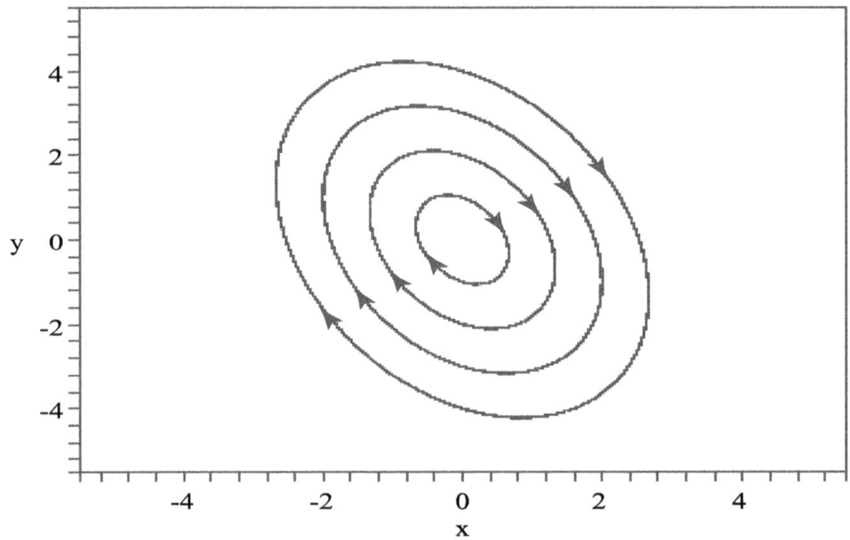

FIGURE 10.7
Neutrally Stable Center.

In examining the stability of the critical point at the origin for system (4), $d\mathbf{z}/dt = \mathbf{Az}$, we have assumed that $det\,\mathbf{A} \neq 0$. It follows from this assumption that $\lambda = 0$ cannot be an eigenvalue of the matrix $\mathbf{A}$. For if $\lambda = 0$ were an eigenvalue of $\mathbf{A}$, then $\lambda = 0$ would, by definition of an eigenvalue, be a root of the equation $det\,(\mathbf{A} - \lambda\mathbf{I}) = 0$. But substitution of $\lambda = 0$ into this equation yields the contradiction $det\,\mathbf{A} = 0$. Consequently, Table 10.1 summarizes the stability properties of the critical point at the origin for the system (4), $d\mathbf{z}/dt = \mathbf{Az}$, in terms of the eigenvalues of $\mathbf{A}$ under the assumption $det\,\mathbf{A} \neq 0$.

Now let us consider the autonomous system

$$(1) \qquad \begin{aligned} \frac{dx}{dt} &= f(x, y) \\ \frac{dy}{dt} &= g(x, y) \end{aligned}$$

TABLE 10.1
Stability Properties of the Linear System $\frac{d\mathbf{z}}{dt} = \mathbf{Az}$ when $det\ \mathbf{A} \neq 0$

Eigenvalues	**Stability**	**Type of critical point**
$\lambda_1, \lambda_2 < 0$	Asymptotically stable	Node
$\lambda_1, \lambda_2 > 0$	Unstable	Node
$\lambda_1 > 0, \quad \lambda_2 < 0$	Unstable	Saddle point
$\lambda = \alpha \pm i\beta \quad (\beta \neq 0)$		
$\alpha < 0$	Asymptotically stable	Spiral point
$\alpha > 0$	Unstable	Spiral point
$\lambda = \pm i\beta \quad (\beta \neq 0)$	Neutrally stable	Center

under the assumption that $f(x,y)$ or $g(x,y)$ is nonlinear. Recall that the point (x^*,y^*) is a critical point of (1), if $f(x^*,y^*) = g(x^*,y^*) = 0$. When f and g have continuous partial derivatives up to order two at (x^*,y^*), then we have from the two-dimensional version of Taylor series expansion for f and g about (x^*,y^*)

$$f(x,y) = f(x^*,y^*) + f_x(x^*,y^*)(x-x^*) + f_y(x^*,y^*)(y-y^*) + R_1(x,y)$$

$$g(x,y) = g(x^*,y^*) + g_x(x^*,y^*)(x-x^*) + g_y(x^*,y^*)(y-y^*) + R_2(x,y)$$

where $R_i(x,y)/\sqrt{(x-x^*)^2 + (y-y^*)^2} \to 0$ as $(x,y) \to (x^*,y^*)$ for $i = 1, 2$. Since $f(x^*,y^*) = g(x^*,y^*) = 0$, the nonlinear system (1) can be written as

$$\begin{aligned} \frac{dx}{dt} &= f_x(x^*,y^*)(x-x^*) + f_y(x^*,y^*)(y-y^*) + R_1(x,y) \\ \frac{dy}{dt} &= g_x(x^*,y^*)(x-x^*) + g_y(x^*,y^*)(y-y^*) + R_2(x,y). \end{aligned} \tag{9}$$

Since the functions $R_i(x,y)$ are "small" when (x,y) is near (x^*,y^*), we anticipate that the stability and type of critical point at (x^*,y^*) for the nonlinear system (9) will be similar to the stability and type of critical point at (x^*,y^*) for the linear system

$$\begin{aligned} \frac{dx}{dt} &= A(x-x^*) + B(y-y^*) \\ \frac{dy}{dt} &= C(x-x^*) + D(y-y^*) \end{aligned}$$

where $A = f_x(x^*, y^*)$, $B = f_y(x^*, y^*)$, $C = g_x(x^*, y^*)$, and $D = g_y(x^*, y^*)$. On letting $X = x - x^*$ and $Y = y - y^*$ and noting that $dX/dt = dx/dt$ and $dY/dt = dy/dt$, we anticipate that the stability characteristics of the nonlinear system (1) or (9) near (x^*, y^*) will be similar to the stability characteristics at $(0, 0)$ of the associated linear system

$$\begin{aligned} \frac{dX}{dt} &= AX + BY \\ \frac{dY}{dt} &= CX + DY. \end{aligned} \tag{10}$$

In fact, it turns out that the stability characteristics of the nonlinear system (9) are the same as the stability characteristics of the associated linear system (10) with the following two possible exceptions. (i) When the eigenvalues of the linear system (10) are purely imaginary ($\lambda = \pm i\beta$), the neutrally stable critical point of the associated linear system can become asymptotically stable, unstable, or remain neutrally stable in the nonlinear system and, correspondingly, the trajectories can become stable spirals, unstable spirals, or remain centers. So when the eigenvalues of the associated linear system are purely imaginary, the stability and behavior of the trajectories of the nonlinear system near (x^*, y^*) must be analyzed on a case-by-case basis. (ii) When the eigenvalues are equal, the stability properties of the nonlinear system and the associated linear system remain the same, but the critical point might change from a node in the linear system to a spiral point in the nonlinear system. ■

Example 7 Find all critical points of the nonlinear system

$$\begin{aligned} x' &= x - xy = f(x, y) \\ y' &= y - xy = g(x, y). \end{aligned} \tag{11}$$

For each critical point, determine the type of stability (asymptotically stable, neutrally stable, or unstable) and the type of critical point (node, saddle point, spiral point, or center).

Solution

Simultaneously solving

$$x - xy = x(1 - y) = 0 \tag{12a}$$

$$y - xy = y(1 - x) = 0, \tag{12b}$$

we find from (12a) that $x = 0$ or $y = 1$. Substituting $x = 0$ into (12b), we get $y = 0$. So $(0, 0)$ is a critical point of system (11). And substituting $y = 1$ into (12b) and solving for x, we get $x = 1$. So $(1, 1)$ is also a critical point of system (11). Since $f(x, y) = x - xy$, $f_x = 1 - y$, and $f_y = -x$. And since $g(x, y) = y - xy$, $g_x = -y$, and $g_y = 1 - x$.

At $(0, 0)$ the linear system associated with the nonlinear system (11) is

$$\begin{aligned} x' &= f_x(0,0)(x-0) + f_y(0,0)(y-0) = 1x + 0y = x \\ y' &= g_x(0,0)(x-0) + g_y(0,0)(y-0) = 0x + 1y = y. \end{aligned}$$

The eigenvalues of this system, which are the eigenvalues of the matrix

$$\mathbf{A} = \begin{pmatrix} 1 & 0 \\ 0 & 1 \end{pmatrix}$$

are $\lambda_1 = \lambda_2 = 1 > 0$. So, from Table 10.1, the origin is an unstable node of this linear system and an unstable critical point of the nonlinear system (11). At $(1, 1)$ the linear system associated with the nonlinear system (11) is

$$x' = f_x(1,1)(x-1) + f_y(1,1)(y-1) = 0(x-1) + (-1)(y-1) = (-1)(y-1)$$

$$y' = g_x(1,1)(x-1) + g_y(1,1)(y-1) = (-1)(x-1) + 0(y-1) = (-1)(x-1).$$

Translating the critical point $(1, 1)$ to the origin by letting $X = x - 1$ and $Y = y - 1$, we are lead to consider the stability characteristics at $(0, 0)$ of the linear system

$$\begin{aligned} X' &= -Y \\ Y' &= -X. \end{aligned}$$

The eigenvalues of this system, which are the eigenvalues of the matrix

$$\mathbf{A} = \begin{pmatrix} 0 & -1 \\ -1 & 0 \end{pmatrix}$$

are $\lambda_1 = 1 > 0$ and $\lambda_2 = -1 < 0$. From Table 10.1 the origin is an unstable saddle point of this linear system; and, therefore, the critical point $(1, 1)$ is an unstable saddle point of the nonlinear system (11).

We used a computer software program to produce the direction field shown in Figure 10.8a for

$$\frac{dy}{dx} = \frac{y'}{x'} = \frac{y - xy}{x - xy}$$

on the rectangle $R = \{(x, y) | -1.5 \le x \le 2.5 \text{ and } -1.5 \le y \le 2.5\}$. From this direction field and the stability properties of system (11) at the critical points $(0, 0)$ and $(1, 1)$, we were able to sketch the phase-plane portrait shown in Figure 10.8b for the nonlinear system (11).

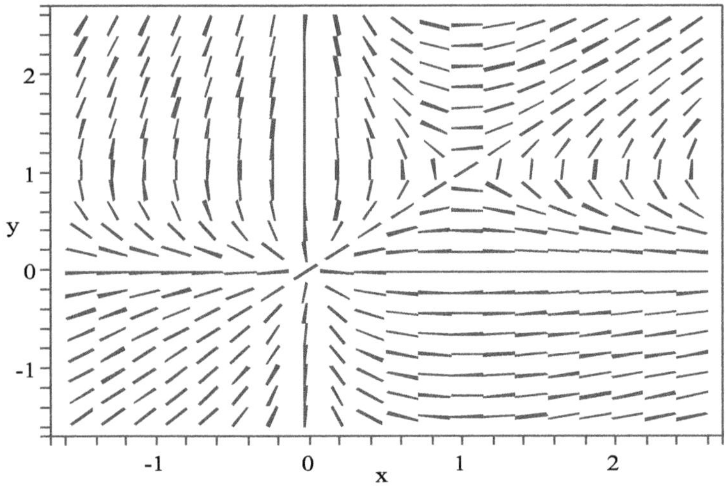

FIGURE 10.8a
Direction Field for System (11).

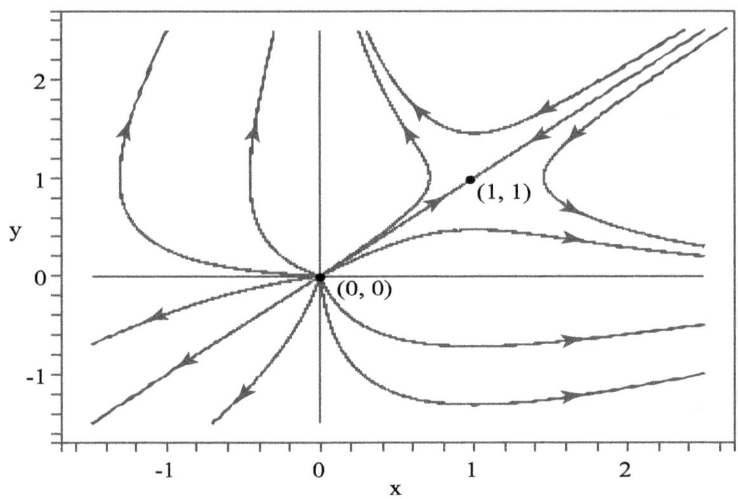

FIGURE 10.8b
Phase-Plane Portrait for System (11). ■

Example 8 Find all critical points of the nonlinear system

$$\begin{aligned} x' &= -y - x^3 &= f(x,y) \\ y' &= x - y^3 &= g(x,y). \end{aligned} \tag{13}$$

For each critical point determine the type of stability and type of critical point.

Solution

Solving $-y - x^3 = 0$ for y and substituting the result into $x - y^3 = 0$, we see that the x coordinate of the critical points must satisfy

$$x - (-x^3)^3 = x + x^9 = x(1 + x^8) = 0.$$

Since $x = 0$ is the only real solution of $x(1+x^8) = 0$, the only critical point of system (13) is $(0,0)$. Calculating first partial derivatives, we find $f_x = -3x^2$, $f_y = -1$, $g_x = 1$, and $g_y = -3y^2$. Evaluating these partial derivatives at $(0,0)$, we find the associated linear system is

$$\begin{aligned} x' &= f_x(0,0)x + f_y(0,0)y &&= 0x + (-1)y &&= -y \\ y' &= g_x(0,0)x + g_y(0,0)y &&= 1x + \quad 0y &&= x. \end{aligned}$$

The eigenvalues of this linear system are $\lambda = \pm i$, so $(0,0)$ is a neutrally stable center of this linear system. Since this is one of the exceptional cases, we have no information regarding the stability characteristics at the origin of the nonlinear system (13). To further analyze system (13), we let $r^2(t) = x^2(t) + y^2(t)$. Thus, $r(t)$ is the distance of a particle from the origin at time t. Differentiating with respect to t, we find $2rr' = 2xx' + 2yy'$. Dividing this equation by 2 and substituting for x' and y' from (13), we find

$$rr' = x(-y - x^3) + y(x - y^3) = -x^4 - y^4 = -(x^4 + y^4).$$

Since for $(x,y) \neq (0,0)$, $r = \sqrt{x^2 + y^2} > 0$, and $x^4 + y^4 > 0$, the derivative $r' = -(x^4 + y^4)/r < 0$ for all $(x,y) \neq (0,0)$. That is, for a particle which is not at the origin, $r' < 0$ for all t. So the particle always moves toward the origin as t increases. Hence, the origin is an asymptotically stable critical point of the nonlinear system (13) and consequently the origin is a node or spiral point. From the direction field shown in Figure 10.9 for

$$\frac{dy}{dx} = \frac{y'}{x'} = \frac{x - y^3}{-y - x^3}$$

it is fairly obvious that the origin is a spiral point and that the trajectories spiral counterclockwise and inward toward the origin.

10.2.1 Exercises

In Exercises 1–8 calculate the eigenvalues of the given linear system. Indicate whether the critical point is asymptotically stable, neutrally stable, or unstable. Specify if the critical point is a node, saddle point, spiral point, or center.

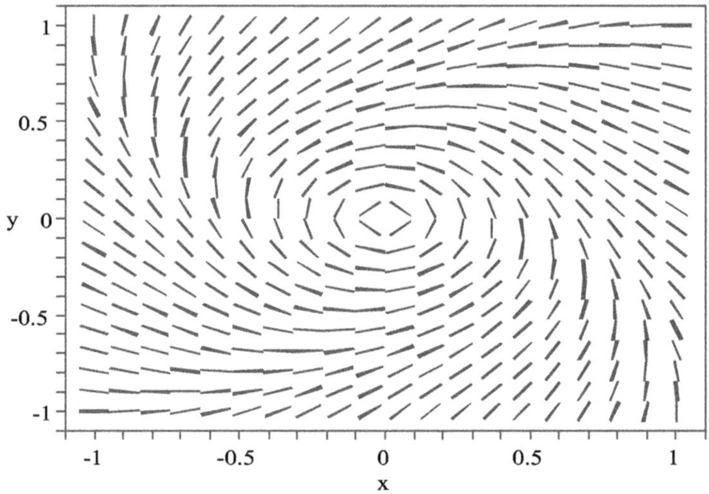

FIGURE 10.9
Direction Field for System (13).

1. $x' = -2x + 3y$
   $y' = -x + 2y$

2. $x' = -x + 2y$
   $y' = -2x + 3y$

3. $x' = x - 2y$
   $y' = 2x - 3y$

4. $x' = -x - 2y$
   $y' = 5x + y$

5. $x' = -x + 2y$
   $y' = -2x - y$

6. $x' = x - 2y$
   $y' = 2x + y$

7. $x' = -5x - y + 2$
   $y' = 3x - y - 3$

8. $x' = 3x - 2y - 6$
   $y' = 4x - y + 2$

In Exercises 9–15 find all real critical points. Compute the eigenvalues of the associated linear system at each critical point. Determine if the critical point is stable or unstable and, when possible, specify its type (node, saddle point, or spiral point).

9. $x' = -x + xy$
   $y' = -y + xy$

10. $x' = x + y$
    $y' = y + x^2$

11. $x' = -x + xy$
    $y' = y - 2xy$

12. $x' = x + 1$
    $y' = 1 - y^2$

13. $x' = y^2 - x^2$
    $y' = 1 - x$

14. $x' = y - xy$
    $y' = y^2 - x^2$

15. $x' = -x + xy$
    $y' = y - xy + y^2$

16. Consider the nonlinear autonomous system

$$
\begin{aligned}
x' &= y + ax^3 + axy^2 \\
y' &= -x + ay^3 + ayx^2.
\end{aligned}
\tag{14}
$$

a. Verify that the origin is a critical point of system (14).

b. Write the associated linear system at $(0,0)$, calculate the eigenvalues of the linear system, and verify that the origin is a neutrally stable center of the linear system.

c. Define $r^2(t) = x^2(t) + y^2(t)$.

(i) For $a < 0$, prove $dr/dt < 0$ provided $(x, y) \neq (0,0)$. Conclude, therefore, that the origin is an asymptotically stable critical point of the nonlinear system (14).

(ii) For $a > 0$, prove $dr/dt > 0$ provided $(x, y) \neq (0,0)$. Hence, conclude that the origin is an unstable critical point of the nonlinear system (14).

10.3 Modified Richardson's Arms Race Models

Let $x(t) \geq 0$ and $y(t) \geq 0$ represent the amounts spent per year on arms by two nations. According to Richardson's arms race model, the expenditures for arms per year satisfy the following linear autonomous system of differential equations

$$
\begin{aligned}
\frac{dx}{dt} &= Ay - Cx + r \\
\frac{dy}{dt} &= Bx - Dy + s
\end{aligned}
\tag{1}
$$

where A, B, C, and D are nonnegative, real constants, and r and s are real constants. The term Ay represents the rate of increase in yearly expenditures for arms by the first nation due to its fear of the second nation. The term $-Cx$ represents the rate of decrease in yearly expenditures for arms by the first nation due to the resistance of its people to increased spending for arms. The constant term r represents the underlying "grievance" which the first nation feels toward the second nation provided $r > 0$ or the underlying "feeling of goodwill" provided $r < 0$. The terms in the second equation in system (1) can be interpreted in an analogous manner.

We analyzed the linear arms race model (1) earlier. Suppose that the term representing the resistance of the first nation's people to increased spending for arms, $-Cx$, is changed to $-Cx^2$ and, likewise, the resistance of the second nation's people to increased spending for arms, $-Dy$, is changed to $-Dy^2$. Then the nonlinear arms race to be analyzed becomes

$$\begin{aligned} \frac{dx}{dt} &= Ay - Cx^2 + r \\ \frac{dy}{dt} &= Bx - Dy^2 + s. \end{aligned} \tag{2}$$

The critical points (x^*, y^*) of (2) simultaneously satisfy

$$Ay - Cx^2 + r = 0 \tag{3a}$$

and

$$Bx - Dy^2 + s = 0. \tag{3b}$$

For $A \neq 0$, the graph of equation (3a) is a parabola. The vertex of the parabola is at $(0, -r/A)$, the parabola's axis of symmetry is the y-axis, and the parabola opens upward. For $B \neq 0$, the graph of equation (3b) is also a parabola. The vertex of this parabola is at $(-s/B, 0)$, its axis of symmetry is the x-axis, and the parabola opens to the right. Visualizing these parabolas to be free to slide along the x-axis and y-axis, and free to open and close, we see that it is possible for (2) to have 0, 1, 2, 3, or 4 critical points, and we see that at most two critical points can lie in the first quadrant. Solving equation (3a) for y yields

$$y = \frac{Cx^2 - r}{A}. \tag{4}$$

Substituting this expression into equation (3b) and rearranging terms, we find x^* satisfies the fourth-degree equation

$$C^2Dx^4 - 2CDrx^2 - A^2Bx + Dr^2 - A^2s = 0. \tag{5}$$

The real, nonnegative critical points of system (2) can be found by using computer software to solve (5) and then substituting the real, nonnegative roots, x^*, into equation (4). Next, the stability characteristics of the nonlinear system (2) at a critical point (x^*, y^*) can often be determined from the stability characteristics of the associated linear system at (x^*, y^*). ■

Example 1 A Nonlinear Modified Richardson's Arms Race Model Consider the modified arms race model

$$\begin{aligned} \frac{dx}{dt} &= 3y - 2x^2 - 1 &= f(x, y) \\ \frac{dy}{dt} &= 8x - y^2 - 7 &= g(x, y). \end{aligned} \tag{6}$$

a. Find the critical points in the first quadrant.

b. Write the associated linear systems and determine the stability characteristics at each critical point. What are the stability characteristics of the nonlinear system (6) at each critical point?

Solution

a. System (6) is a special case of system (2) in which $A = 3$, $B = 8$, $C = 2$, $D = 1$, $r = -1$, and $s = -7$. Substituting these values into equation (5) and then into (4), we see that the x-coordinate of the critical points of system (6) must satisfy $4x^4 + 4x^2 - 72x + 64 = 0$ and the y-coordinates must satisfy $y = (2x^2 + 1)/3$. Using computing software to calculate the roots of the first equation, we find $x = 1, 2, -1.5 \pm 2.39792i$. To be in the first quadrant, x and y must both be real and nonnegative. Substituting $x = 1$ into the equation $y = (2x^2 + 1)/3$, we get $y = 1$. So $(1, 1)$ is a critical point in the first quadrant. Substituting $x = 2$ into the equation $y = (2x^2 + 1)/3$, we get $y = 3$. Thus, $(2, 3)$ is a second critical point in the first quadrant.

b. Calculating first partial derivatives, we see $f_x = -4x$, $f_y = 3$, $g_x = 8$, and $g_y = -2y$.

   At $(1, 1)$ the associated linear system is

$$\begin{aligned} x' &= f_x(1,1)(x-1) + f_y(1,1)(y-1) = -4(x-1) + 3(y-1) \\ y' &= g_x(1,1)(x-1) + g_y(1,1)(y-1) = \ 8(x-1) - 2(y-1). \end{aligned}$$

   The coefficient matrix of this linear system,

$$\mathbf{A} = \begin{pmatrix} f_x(1,1) & f_y(1,1) \\ g_x(1,1) & g_y(1,1) \end{pmatrix} = \begin{pmatrix} -4 & 3 \\ 8 & -2 \end{pmatrix}$$

   has eigenvalues 2 and -8. (Verify this fact by using your computing software to compute the eigenvalues of $\mathbf{A}$.) Since one eigenvalue is positive and the other is negative, the critical point $(1, 1)$ is a saddle point of both the associated linear system and the given nonlinear system (6). At $(2, 3)$ the associated linear system is

$$\begin{aligned} x' &= f_x(2,3)(x-2) + f_y(2,3)(y-3) = -8(x-2) + 3(y-3) \\ y' &= g_x(2,3)(x-2) + g_y(2,3)(y-3) = \ 8(x-2) - 6(y-3). \end{aligned}$$

   The coefficient matrix of this linear system

$$\mathbf{A} = \begin{pmatrix} -8 & 3 \\ 8 & -6 \end{pmatrix}$$

   has eigenvalues -2 and -12. Since both eigenvalues are negative, the critical point $(2, 3)$ is an asymptotically stable node of both the associated linear system and the nonlinear system (6). ∎

10.3.1 Comments on Computing

Various computer software packages include algorithms which numerically solve the system initial value problem

$$(7)\qquad \begin{aligned} y_1' &= f_1(t, y_1, y_2, \ldots, y_n); & y_1(c) &= d_1 \\ y_2' &= f_2(t, y_1, y_2, \ldots, y_n); & y_2(c) &= d_2 \\ \vdots\ & \ \vdots \qquad \vdots & \vdots\ & \ \vdots \quad \vdots \\ y_n' &= f_n(t, y_1, y_2, \ldots, y_n); & y_n(c) &= d_n \end{aligned}$$

on the interval $[a, b]$ for $c \in [a, b]$ and for $2 \leq n \leq N$ for some given maximum integer value N. After the numerical solution has been calculated by your computing software, you may elect (i) to print solution components, (ii) to graph any subset of the components in any order on a rectangle R where $a \leq t \leq b$ and YMIN $\leq y \leq$ YMAX and you select the values for YMIN and YMAX, or (iii) to produce a phase-plane portrait of y_i versus y_j for any two distinct components y_i and y_j on any rectangle in $y_j y_i$-space.

Example 2 Computer Solution of a Nonlinear Modified Richardson's Arms Race Model Solve the system initial value problem

$$(8)\qquad \begin{aligned} \frac{dx}{dt} &= 3y - 2x^2 - 1 &= f(x, y); \qquad x(0) = 2 \\ \frac{dy}{dt} &= 8x - \ y^2 - 7 &= g(x, y); \qquad y(0) = 0 \end{aligned}$$

on the interval $[0, 2.5]$. Display a graph of $x(t)$ and $y(t)$. Produce a phase-plane graph of $y(t)$ versus $x(t)$ on the rectangle

$$R = \{(x, y) \mid 0 \leq x \leq 5 \ \text{ and } \ 0 \leq y \leq 5\}.$$

Solution

Mathematical Analysis The functions $f_x = -4x$, $f_y = 3$, $g_x = 8$, and $g_y = -2y$ are all defined and continuous on the entire xy-plane. Hence, according to the existence and uniqueness theorem and the continuation theorem presented in Chapter 7, the system initial value problem (8) has a unique solution on some interval I centered about $c = 0$ and this solution can be extended uniquely until either $x(t) \to \pm\infty$ or $y(t) \to \pm\infty$. Since the functions $x(t)$ and $y(t)$ have physical significance only for $x(t) \geq 0$ and $y(t) \geq 0$, we will stop the numerical integration if $x(t)$ or $y(t)$ becomes negative.

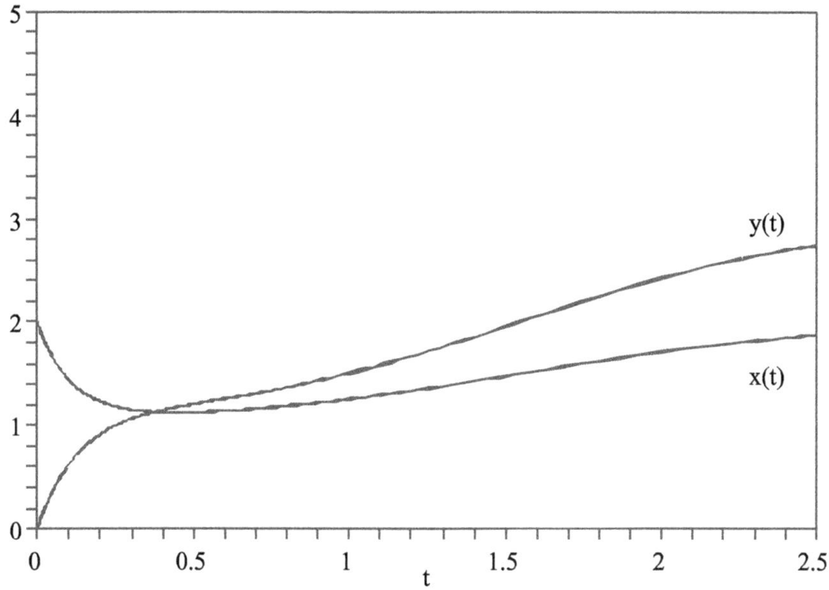

FIGURE 10.10
Graph of the Solution of System (8).

Computer Solution We input the two functions defining the system, $f(x,y)$ and $g(x,y)$; the interval of integration $[0, 2.5]$; and the initial conditions $x(0) = 2$ and $y(0) = 0$ into our computer software. After the integration was completed, we indicated we wanted to graph $x(t)$ and $y(t)$ on the rectangle $R = \{(t,x) \mid 0 \leq t \leq 2.5 \text{ and } 0 \leq x \leq 5\}$. The resulting graph is displayed in Figure 10.10. The graph of $x(t)$ starts at a height of 2, decreases to nearly 1, and then increases to 1.87154. The graph of $y(t)$ starts at a height of 0 and steadily increases to 2.743162.

Since we wanted to produce a phase-plane graph of y versus x, we indicated to our computer software that we wanted $x(t)$ assigned to the horizontal axis and $y(t)$ assigned to the vertical axis and that we wanted to display the graph on the rectangle R where $0 \leq x \leq 5$ and $0 \leq y \leq 5$. The resulting phase-plane graph of y versus x is shown in Figure 10.11. The arrow heads, point, and point coordinates which appear in this figure were added to the output of our computer algebra system using illustrating software.

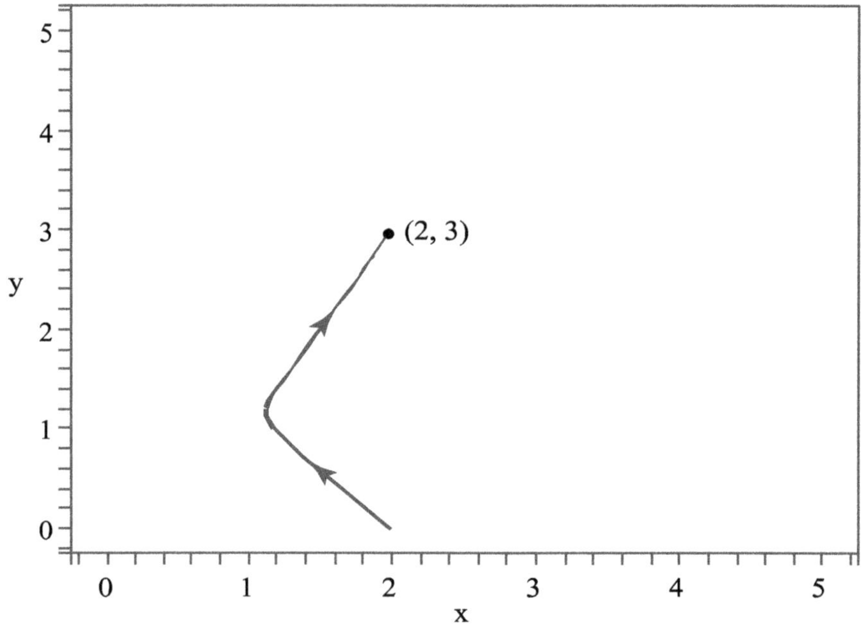

FIGURE 10.11
Phase-Plane Plot of y versus x for System (8). ■

We were able to produce the phase-plane portrait of y versus x shown in Figure 10.12 by using our software to solve the nine initial value problems consisting of system (8) and the following set of nine initial conditions on the interval $[0, 2.5]$ and plotting the results on the same graph. The initial conditions are

(i) $x(0) = 0$, $y(0) = 2.5$; (ii) $x(0) = 0$, $y(0) = 3$; (iii) $x(0) = 0$, $y(0) = 5$; (iv) $x(0) = 1.5$, $y(0) = 0$; (v) $x(0) = 2$, $y(0) = 0$; (vi) $x(0) = 5$, $y(0) = 0$; (vii) $x(0) = 5$, $y(0) = 3$; (viii) $x(0) = 1$, $y(0) = 5$; and (ix) $x(0) = 3$, $y(0) = 5$.

The arrow heads, points, and points coordinates which appear in Figure 10.12 were added to the output of our computer algebra system using illustrating software.

10.3.2 Comments on Computing

For $1 \leq i \leq 10$, a computer algebra system will attempt to solve numerically the i initial value problems consisting of the autonomous, two-component

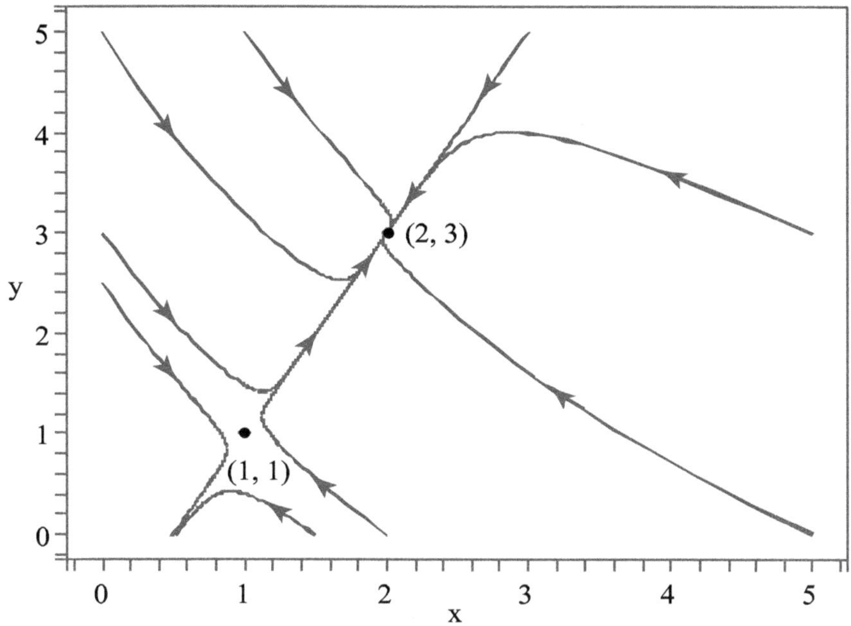

FIGURE 10.12
Phase-Plane Portrait for System (8).

system initial value problem

$$\begin{aligned} \frac{dy_1}{dx} &= f_1(y_1, y_2) \\ \frac{dy_2}{dx} &= f_2(y_1, y_2) \end{aligned}$$

and the initial conditions $y1(c_i) = d_{1i}$, $y2(c_i) = d_{2i}$ on the interval $[a_i, b_i]$ where $c_i \in [a_i, b_i]$. After the solution of an initial value problem has been calculated, you may elect for any initial value problem already solved (i) to print the solution components on the monitor, (ii) to graph any subset of the solution components in a rectangle R where $a_i \leq x \leq b_i$ and YMIN $\leq y \leq$ YMAX where you select the values for YMIN and YMAX, (iii) to produce a phase-plane portrait of any subset of IVPs already solved on any rectangle, (iv) to rerun the most recent initial value problem using a different interval of integration or initial conditions, or (v) input the initial conditions for the next initial value problem to be solved.

10.3.3 Exercises

1. Consider the nonlinear Richardson's arms race model

$$\begin{aligned}\frac{dx}{dt} &= 5y - x^2 + 5 \\ \frac{dy}{dt} &= 4x - y^2 - 4.\end{aligned} \tag{9}$$

   a. Find the critical point in the first quadrant.

   b. Write the associated linear system and determine the stability characteristics at the critical point. What can you conclude about the stability of the nonlinear system (9) at the critical point?

   c. Use your computing software to solve the system initial value problems consisting of system (9) and the following five initial conditions on the interval $[0, 2]$:
   (i) $x(0) = 0$, $y(0) = 7$; (ii) $x(0) = 2$, $y(0) = 1$; (iii) $x(0) = 5$, $y(0) = 0$; (iv) $x(0) = 7$, $y(0) = 0$; (v) $x(0) = 7$, $y(0) = 7$.
   Display a phase-plane portrait of y versus x on the rectangle with $0 \leq x \leq 10$ and $0 \leq y \leq 10$.

2. Consider the following nonlinear arms race model

$$\begin{aligned}\frac{dx}{dt} &= y - x^2 - 4 \\ \frac{dy}{dt} &= 4x - y^2 - 4.\end{aligned} \tag{10}$$

   a. Find the critical point in the first quadrant.

   b. Write the associated linear system and calculate the eigenvalues. What can you say about the stability of the associated linear system based on the eigenvalues? What can you conclude about the stability of the nonlinear system (10) at the critical point?

   c. Use your computing software to solve the system initial value problems consisting of system (10) and the following four initial conditions on the interval $[0, 4]$:

   (i) $x(0) = 1.8$, $y(0) = 1.8$; (ii) $x(0) = 1$, $y(0) = 4$; (iii) $x(0) = 3$, $y(0) = 1$; (iv) $x(0) = 3$, $y(0) = 4$.

   Graph the phase-plane portrait of y versus x on the rectangle with $0 \leq x \leq 5$ and $0 \leq y \leq 5$. What do you infer about the stability of the nonlinear system (10) from these results?

3. Consider the nonlinear arms race model

$$\begin{aligned}\frac{dx}{dt} &= 9y - x^2 - 9 \\ \frac{dy}{dt} &= 4x - y^2 - 8.\end{aligned} \tag{11}$$

a. Find the critical points in the first quadrant.

b. Write the associated linear systems and determine the stability characteristics of each critical point. What can you conclude about the stability of the nonlinear system (11) at each critical point?

c. Use your computing software to solve the system initial value problems consisting of (11) and the following seven initial conditions on the interval $[0, 2]$:
(i) $x(0) = 1$, $y(0) = 1$; (ii) $x(0) = 2$, $y(0) = 3$; (iii) $x(0) = 4$, $y(0) = 0$; (iv) $x(0) = 4$, $y(0) = 3$; (v) $x(0) = 4$, $y(0) = 5$; (vi) $x(0) = 6$, $y(0) = 3$; (vii) $x(0) = 6$, $y(0) = 6$.
Display the phase-plane graph of y versus x on the rectangle with $0 \le x \le 10$ and $0 \le y \le 10$.

4. Consider the following modification of Richardson's arms race model

$$\begin{aligned} \frac{dx}{dt} &= Ay - Cx + r \\ \frac{dy}{dt} &= Bx - Dy^2 + s \end{aligned} \tag{12}$$

where A, B, C, and D are positive real constants and r and s are real constants. Here each nation's fear of the other nation is modeled as being proportional to the amount the other nation spends annually for arms. The resistance of the first nation's people to increased spending for arms is modeled as being proportional to the amount spent yearly for arms, while the resistance of the second nation's people to increased spend- ing for arms is modeled as being proportional to the square of the amount spent yearly for arms.

a. What is the graph of $Ay - Cx + r = 0$? What is the graph of $Bx - Dy^2 + s = 0$? How many critical points can system (12) have? What is the maximum number of critical points system (12) can have in the first quadrant?

b. Write a single equation which the x-coordinate of a critical point must satisfy.

(i) Find the critical points in the first quadrant of the system

$$\begin{aligned} \frac{dx}{dt} &= 3y - x - 1 \\ \frac{dy}{dt} &= x - y^2 - 1. \end{aligned} \tag{13}$$

(ii) Write the associated linear system and determine the stability characteristics at each critical point. What can you conclude about the stability of the nonlinear system (13) at each critical point?

(iii) Use your computing software to solve system (13) and the following six initial conditions on the interval $[0, 2]$:

(i) $x(0) = 1$, $y(0) = 1$; (ii) $x(0) = 0$, $y(0) = 4$; (iii) $x(0) = 3$, $y(0) = 0$; (iv) $x(0) = 4$, $y(0) = 0$; (v) $x(0) = 7$, $y(0) = 0$; (vi) $x(0) = 3$, $y(0) = 5$.

Display the phase-plane graph of y versus x on the rectangle with $0 \leq x \leq 10$ and $0 \leq y \leq 5$.

5. Consider the following modification of Richardson's arms race model

$$\begin{aligned} \frac{dx}{dt} &= Ay^2 - Cx + r \\ \frac{dy}{dt} &= Bx^2 - Dy + s \end{aligned} \tag{14}$$

where A, B, C, and D are positive real constants, and r and s are real constants. Here the fear of each nation for the other is modeled as being proportional to the square of the other nation's yearly expenditures for arms.

a. What is the graph of $Ay^2 - Cx + r = 0$? What is the graph of $Bx^2 - Dy + s = 0$? How many critical points can system (14) have? What is the maximum number of critical points system (14) can have in the first quadrant?

b. Write a single equation which the x-coordinate of the critical point must satisfy.

c. (i) Find the critical point in the first quadrant of the system

$$\begin{aligned} \frac{dx}{dt} &= y^2 - 4x + 4 \\ \frac{dy}{dt} &= x^2 - 5y - 5. \end{aligned} \tag{15}$$

(ii) Write the associated linear system and determine the stability characteristics at the critical point. What can you conclude about the stability of the nonlinear system (15) at the critical point?

(iii) Use your computing software to solve system (15) and the following five initial conditions on the interval $[0, 2]$:
(i) $x(0) = 0$, $y(0) = 3$; (ii) $x(0) = 0$, $y(0) = 7$; (iii) $x(0) = 3$, $y(0) = 0$; (iv) $x(0) = 3$, $y(0) = 10$; (v) $x(0) = 7$, $y(0) = 10$.
Graph the phase-plane graph of y versus x on the rectangle with $0 \leq x \leq 10$ and $0 \leq y \leq 10$.

d. (i) Find the critical point in the first quadrant of the system

$$(16)\qquad \begin{aligned}\frac{dx}{dt} &= y^2 - 4x + 4\\ \frac{dy}{dt} &= x^2 - 4y + 4.\end{aligned}$$

(ii) Write the associated linear system and calculate the eigenvalues. What can you say about the stability of this linear system? What can you conclude about the stability of the nonlinear system (16) at the critical point?

(iii) Solve the system initial value problems consisting of system (16) and the following initial conditions on the interval $[0, 3]$: (i) $x(0) = 0$, $y(0) = 0$; (ii) $x(0) = 0$, $y(0) = 4$; (iii) $x(0) = 3$, $y(0) = 0$;

(iv) $x(0) = 3$, $y(0) = 4$. Display y versus x on the rectangle with $0 \leq x \leq 5$ and $0 \leq y \leq 5$. What do you infer about the stability of the nonlinear system (16) from these results?

Additional Information Needed to Solve Exercises 6 and 7

Recall that a homogeneous system of n linear first-order differential equations with constant coefficients can be written in matrix-vector notation as

$$(17)\qquad \mathbf{y}' = \mathbf{A}\mathbf{y}$$

where $\mathbf{y}$ is an $n \times 1$ vector and $\mathbf{A}$ is an $n \times n$ constant matrix. The origin, $\mathbf{y} = \mathbf{0}$, is a critical point of this system. We state the following theorem regarding the stability of the critical point at the origin.

Stability Theorem

(1) If the eigenvalues of $\mathbf{A}$ all have negative real part, then the origin is an asymptotically stable critical point of the linear system (17).

(2) If any eigenvalue of $\mathbf{A}$ has positive real part, then the origin is an unstable critical point of the linear system (17).

(3) Suppose the eigenvalues of $\mathbf{A}$ all have nonpositive real part and suppose those eigenvalues which have zero real part are purely imaginary eigenvalues. That is, suppose the eigenvalues which have zero real part are of the form $i\beta$ where $\beta \neq 0$. If every purely imaginary eigenvalue of multiplicity k has k linearly independent eigenvectors, then the origin is a stable critical point of system (17). Whereas, if any purely imaginary eigenvalue of multiplicity k has fewer than k linearly independent eigenvectors, then the origin is an unstable critical point of system (17).

The stability characteristics of the nonhomogeneous system of n linear first-order differential equations

$$\mathbf{z}' = \mathbf{A}\mathbf{z} + \mathbf{b} \tag{18}$$

at the critical point $\mathbf{z}^*$, which satisfies $\mathbf{A}\mathbf{z}^* + \mathbf{b} = \mathbf{0}$, is the same as the stability characteristics of the homogeneous linear system $\mathbf{y}' = \mathbf{A}\mathbf{y}$ at the origin, since the linear transformation $\mathbf{y} = \mathbf{z} - \mathbf{z}^*$ transforms the nonhomogeneous system (18) into the homogeneous system (17).

One possible extension of Richardson's arms race model from a system for two nations to a system for three nations is

$$\begin{aligned} \frac{dy_1}{dt} &= -c_1y_1 + a_{12}y_2 + a_{13}y_3 + r_1 \\ \frac{dy_2}{dt} &= a_{21}y_1 - c_2y_2 + a_{23}y_3 + r_2 \\ \frac{dy_3}{dt} &= a_{31}y_1 + a_{32}y_2 - c_3y_3 + r_3 \end{aligned} \tag{19}$$

where a_{ij} and c_i are nonnegative real constants and r_i are any real constants. In system (19) y_1, y_2, and y_3 are the yearly amounts spent by nations 1, 2, and 3 for arms; the terms $a_{ij}y_j$ represent the rate of increase in yearly expenditures for arms by nation i due to its fear of nation j; the terms $-c_iy_i$ represent the rate of decrease in yearly expenditures for arms by nation i due to the resistance of its people to increased spending for arms; and r_i represents the collective underlying "goodwill" or "grievance" which nation i feels toward the other two nations.

6. a. Suppose in system (19) $c_1 = c_2 = c_3 = 1$ and $a_{ij} = 2$ for all appropriate values of i and j. Is the critical point of system (19) stable or unstable? (Hint: Write system (19) in matrix-vector form, use your computer software to calculate the eigenvalues of the appropriate 3×3 constant matrix $\mathbf{A}$, and use the stability theorem to determine the answer to the question.)

   b. Suppose in system (19) $c_1 = c_2 = c_3 = 3$ and $a_{ij} = 1$. Is the critical point of system (19) stable or unstable?

7. Suppose in system (19) that nation 3 is a pacifist nation. This situation can be represented by setting $a_{31} = a_{32} = 0$.

   a. Let $c_1 = c_2 = c_3 = 1$ and $a_{12} = a_{13} = a_{21} = a_{23} = 2$.

      (i) Determine if the critical point of system (19) is stable or unstable.

      (ii) For $r_1 = r_2 = 2$ and $r_3 = -1$ solve the system initial value problem consisting of system (19) and the initial conditions $y_1(0) = 4$, $y_2(0) = 4$, $y_3(0) = 2$ on the interval $[0, 3]$. Display $y_1(t)$, $y_2(t)$, and $y_3(t)$ on the same graph. Produce phase-plane graphs of y_2 versus y_1, y_3 versus y_1, and y_3 versus y_2.

b. Do parts (i) and (ii) of part a. for $c_1 = c_2 = c_3 = 2$ and $a_{12} = a_{13} = a_{21} = a_{23} = 1$.

10.4 Lanchester's Combat Models

In 1916, during World War I, F. W. Lanchester authored the book *Aircraft in Warfare: The Dawn of the Fourth Arm.* In the text, Lanchester described some simple mathematical models for the then emerging art of air warfare. More recently, these models have been extended to general combat situations and are referred to as Lanchester's combat models.

In the elementary combat models, two forces are engaged in combat. Let $x(t)$ and $y(t)$ denote the number of combatants in the "x-force" and "y-force" respectively. The principle underlying Lanchester's combat models is that the rate of change of the number of combatants is equal to the **reinforcement rate** minus the **operational loss rate** minus the **combat loss rate**. The **reinforcement rate** is the rate at which new combatants enter or withdraw from the battle. The **operational loss rate** refers to noncombat losses due to such things as disease, accident, desertion, etc. Lanchester proposed that the operational loss rate be modeled as being proportional to the number of combatants. This assumption appears to be too simplistic. The **combat loss rate** is the rate at which combatants are killed in battle.

A "conventional force" is one which operates in the open and one whose members are all within the "kill range" of the enemy. As soon as the conventional force suffers a loss, the enemy concentrates its fire on the remaining conventional combatants. Thus, the combat loss rate of a conventional force is proportional to the number of the enemy. The constant of proportionality is called the "combat effectiveness coefficient."

The Lanchester model for two conventional forces engaged in battle is

$$\text{(1)} \qquad \begin{aligned} \frac{dx}{dt} &= f(t) - Ax - By \\ \frac{dy}{dt} &= g(t) - Cy - Dx \end{aligned}$$

where A, B, C, and D are nonnegative constants; where $f(t)$ is the reinforcement rate, Ax is the operational loss rate, and By is the combat loss rate of the x-force; and where $g(t)$ is the reinforcement rate, Cy is the operational loss rate, and Dx is the combat loss rate of the y-force. Here, B is the combat effectiveness coefficient for the y-force and D is the combat effectiveness coefficient of the x-force. First, let us consider the simplest case of system (1)—the case in which the battle takes place so rapidly that no reinforcements arrive, $f(t) = g(t) = 0$, and no operational losses occur, $A = C = 0$. Thus, we wish to consider the linear autonomous system

$$(2) \qquad \begin{aligned} \frac{dx}{dt} &= -By \\ \frac{dy}{dt} &= -Dx. \end{aligned}$$

Solving $-By = 0$ and $-Dx = 0$ simultaneously, we see that the origin is the only critical point of system (2). The coefficient matrix of the linear autonomous system (2)

$$\mathbf{A} = \begin{pmatrix} 0 & -B \\ -D & 0 \end{pmatrix}$$

has eigenvalues $\lambda = \pm\sqrt{BD}$. (Verify this fact.) Since one eigenvalue is positive and the other is negative, the origin is a saddle point of system (2).

The trajectories, $(x(t), y(t))$, of system (2) satisfy the first-order differential equation

$$(3) \qquad \frac{dy}{dx} = \frac{dy/dt}{dx/dt} = \frac{-Dx}{-By}.$$

Let $x_0 > 0$ and $y_0 > 0$ be the number of combatants of the two forces at the start $(t = 0)$ of the battle. Separating variables in equation (3) and integrating from the initial point (x_0, y_0) to $(x(t), y(t))$, we find

$$B \int_{y_0}^{y(t)} y\,dy = D \int_{x_0}^{x(t)} x\,dx$$

and, therefore,

$$B(y^2(t) - y_0^2) = D(x^2(t) - x_0^2).$$

So

$$(4) \qquad By^2(t) - Dx^2(t) = By_0^2 - Dx_0^2 \equiv K$$

where K is a constant. For $K \neq 0$ the graph of equation (4) is a hyperbola and for $K = 0$ the graph of equation (4) is two lines which intersect at the origin. The trajectories in the first quadrant defined by equation (4) are sketched in Figure 10.13. The direction $(x(t), y(t))$ moves along the hyperbolas (4) as t increases, which is indicated by the arrows in Figure 10.13, was determined by noting in system (2) that $dx/dt < 0$ and $dy/dt < 0$ for $x > 0$ and $y > 0$.

We will say that one force "wins" the battle if the other force vanishes first. We see from equation (4) that the y-force wins if $K > 0$; the x-force wins if $K < 0$; and there is a tie if $K = 0$. If $K > 0$, the x-force vanishes $(x = 0)$

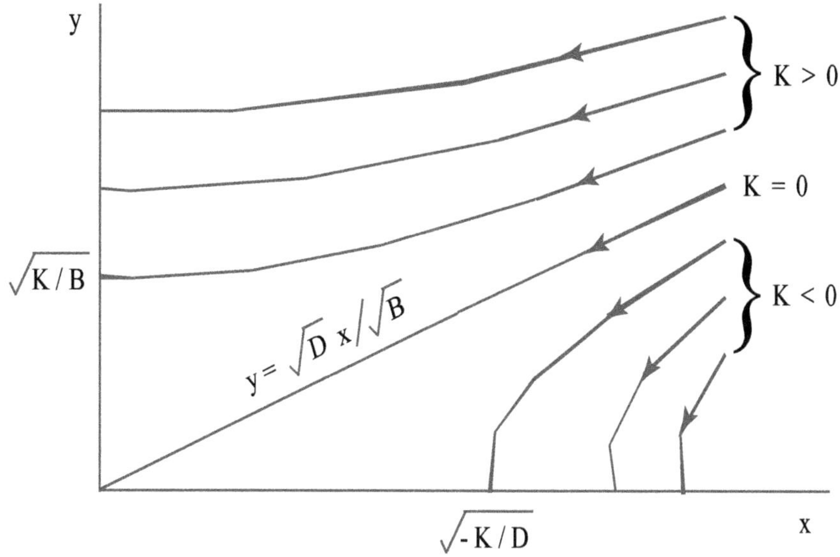

FIGURE 10.13
Phase-Plane Portrait of Lanchester's Combat Model for Two Conventional Forces with No Reinforcements and No Operational Losses.

when $y = \sqrt{K/B}$. Thus, for $K > 0$ the strength of the y-force at the end of the battle is $\sqrt{K/B}$. Likewise, if $K < 0$, the y-force vanishes (the x-force wins the battle) and at the end of the battle the strength of the x-force is $x = \sqrt{-K/D}$. In order to win the battle, the y-force seeks to establish conditions under which the inequality $K = By_0^2 - Dx_0^2 > 0$ or $By_0^2 > Dx_0^2$ holds. This can be accomplished by increasing its combat effectiveness coefficient, B—perhaps by using more powerful or more accurate weapons—or by increasing its initial number of combatants, y_0. Notice that doubling B causes By_0^2 to double while doubling y_0 causes By_0^2 to quadruple. Since the initial strengths of the opposing forces, x_0 and y_0, appear quadratically in equation (4) and, hence, effect the outcome of the battle in a quadratic manner, equation (4) is known as "Lanchester's square law."

A "guerrilla force" is one which is invisible to its enemy. When the enemy fires into a region containing the guerrilla force, the enemy does not know when a kill occurs, so the enemy is unable to concentrate its fire on the remaining guerrillas. It is reasonable to assume that the combat loss rate of a guerrilla force is jointly proportional to the number of guerrillas and the number of the enemy, since the probability that the enemy kills a guerrilla increases as the number of guerrillas in a given region increases, and it increases as the number of enemy firing into the region increases. Thus, the Lanchester model for a conventional force, x, engaged in battle with a guerrilla force, y, is

$$(5)\qquad \begin{aligned} \frac{dx}{dt} &= f(t) - Ax - By \\ \frac{dy}{dt} &= g(t) - Cy - Dxy \end{aligned}$$

where A, B, C, and D are nonnegative constants; where $f(t)$ is the reinforcement rate, Ax is the operational loss rate, and By is the combat loss rate of the conventional force, x; and where $g(t)$ is the reinforcement rate, Cy is the operational loss rate, and Dxy is the combat loss rate for the guerrilla force, y. Let us now consider a conventional force and a guerrilla force engaged in a battle in which no reinforcements occur, $f(t) = g(t) = 0$, and in which no operational losses occur, $A = C = 0$. That is, let us consider the nonlinear autonomous system

$$(6)\qquad \begin{aligned} \frac{dx}{dt} &= -By \\ \frac{dy}{dt} &= -Dxy. \end{aligned}$$

Dividing the second equation of (6) by the first, we see that

$$\frac{dy}{dx} = \frac{dy/dt}{dx/dt} = \frac{-Dxy}{-By} = \frac{Dx}{B}.$$

Multiplying by $B\,dx$ and integrating from (x_0, y_0) to $(x(t), y(t))$, we find

$$B\int_{y_0}^{y(t)} dy = D\int_{x_0}^{x(t)} x\,dx$$

or

$$B(y(t) - y_0) = \frac{D(x^2(t) - x_0^2)}{2}.$$

Hence,

$$(7)\qquad By(t) - \frac{Dx^2(t)}{2} = By_0 - \frac{Dx_0^2}{2} \equiv K$$

where K is a constant. The graph of equation (7) is a one-parameter (K is the parameter) family of parabolas with the y-axis as the axis of symmetry, with vertex at $(0, K/B)$, and which opens upward. The trajectories in the first quadrant defined by equation (7) are sketched in Figure 10.14. For $K > 0$, the guerrilla force, y, wins the battle and at the end of the battle the number of combatants in the y-force is K/B. For $K = 0$, there is a tie—that is, for some $t^* > 0$, $x(t^*) = y(t^*) = 0$. For $K < 0$, the conventional force, x, wins the battle and at the end of the battle the number of combatants in the x-force is $\sqrt{-2K/D}$.

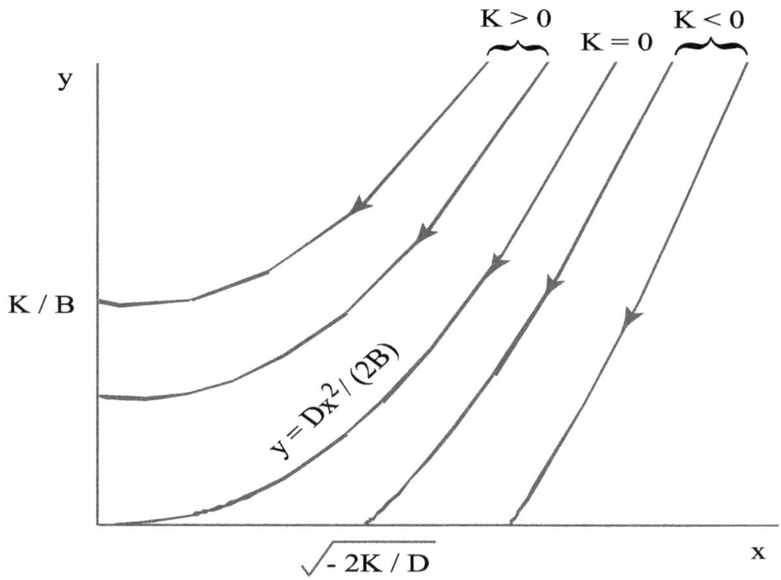

FIGURE 10.14
Phase-Plane Portrait of Lanchester's Combat Model for a Conventional Force Versus a Guerrilla Force with No Reinforcements and No Operational Losses.

10.4.1 Exercises

1. Suppose two conventional forces are engaged in a battle in which each force reinforces its combatants at a constant rate and no operational losses occur. The Lanchester model for this battle is

$$\text{(8)} \qquad \begin{aligned} \frac{dx}{dt} &= r - By \\ \frac{dy}{dt} &= s - Dx \end{aligned}$$

where the positive constants r and s are the reinforcement rates of the x-force and y-force and By and Dx are the combat loss rates for the x-force and y-force, respectively.

a. For $r = 3$, $s = 1$, $B = 2$, and $D = .75$ use your computing software to solve the following three initial value problems on the interval $[0, 5]$:

(i) $x(0) = 1.5, \quad y(0) = 1.754$ (ii) $x(0) = 1.5, \quad y(0) = 1.5$
(iii) $x(0) = 0, \quad y(0) = 0$

(Note: The numbers in this example have been scaled so that x, y, r, and s can represent hundreds, thousands, or tens of thousands, etc., combatants. Interpret the time as being in days.)

b. For each initial value problem in part a. determine (i) which force wins the battle, (ii) the time the battle is over, and (iii) the number of combatants in the winning force at the end of the battle.

2. Use your computing software to solve the Lanchester combat model

$$(9)\qquad \begin{aligned} \frac{dx}{dt} &= (3 - .2t) - .1x - 2y \\ \frac{dy}{dt} &= (1 + .3t) - .05y - .75x \end{aligned}$$

for two conventional forces engaged in battle for the three initial conditions:

(i) $x(0) = 1.5, \quad y(0) = 1.75$ (ii) $x(0) = 1.5, \quad y(0) = 1.5$

(iii) $x(0) = 0, \quad y(0) = 0$

In each case, determine (a) which force wins the battle, (b) the time the battle is over, and (c) the number of combatants in the winning force at the end of the battle. In this engagement, the x-force is being reinforced at the decreasing rate $f(t) = 3 - .2t$, their operational loss rate is $.1x$, and their combat loss rate is $2y$. The y-force is being reinforced at the increasing rate $g(t) = 1 + .3t$, their operational loss rate is $.05y$, and their combat loss rate is $.75x$.

3. Use your computing software to solve the Lanchester combat model

$$(10)\qquad \begin{aligned} \frac{dx}{dt} &= 2 - x - 2y \\ \frac{dy}{dt} &= 1 - .2y - .2xy \end{aligned}$$

for a conventional force, x, versus a guerrilla force, y, on the interval $[0, 5]$ for the initial conditions (i) $x(0) = 0$, $y(0) = 0$ and (ii) $x(0) = 4$, $y(0) = 1$. In each case, decide which force wins the battle, when victory occurs, and the number of combatants in the winning force at the time of victory.

4. The Lanchesterian model for two guerrilla forces engaged in battle is

$$(11)\qquad \begin{aligned} \frac{dx}{dt} &= f(t) - Ax - Bxy \\ \frac{dy}{dt} &= g(t) - Cy - Dxy \end{aligned}$$

where A, B, C, and D are nonnegative constants and where $f(t)$ and $g(t)$ are the reinforcement rates, Ax and Cy are the operational loss rates, and Bxy and Dxy are the combat loss rates.

a. Write the system of equations for two guerrilla forces engaged in combat when there are no reinforcements and no operational losses.

b. Divide one equation of your system by the other equation to obtain a differential equation in x and y.

c. Find the general solution to the differential equation of part b. That is, find the trajectories of the system.

d. Sketch a phase-plane portrait for the system in the first quadrant.

e. For each force determine conditions which ensure victory and determine the number of combatants in the winning force at the end of the battle. Specify conditions under which there is a tie.

5. Use your computing software to solve the Lanchester combat model

$$\begin{aligned} \frac{dx}{dt} &= 2 - .5x - .5xy \\ \frac{dy}{dt} &= 1 - .2y - .25xy \end{aligned} \tag{12}$$

for two guerrilla forces engaged in battle on the interval $[0, 5]$ for the initial conditions:

(i) $x(0) = 0, \quad y(0) = 0$ (ii) $x(0) = 0, \quad y(0) = 6$

(iii) $x(0) = 5, \quad y(0) = 0$ and (iv) $x(0) = 5, \quad y(0) = 6$

In each case, display a phase-plane graph of y versus x. Do you think either force will ever win?

10.5 Models for Interacting Species

We examined the Malthusian model and the logistic model (Verhulst–Pearl model) for population growth of a single species in Chapter 3. In this section, we will examine several models which attempt to represent population dynamics when two or more species interact in the same environment.

10.5.1 Lotka–Volterra Prey-Predator Model

The first model which we will study is named in honor of the American scientist and statistician Alfred Lotka and the Italian mathematician Vito Volterra. Both men studied this model at about the same time and arrived at similar conclusions. This particular model serves as the cornerstone for the study of population dynamics for interacting species.

On March 2, 1880, Alfred James Lotka (1880-1949) was born in Lemberg, Austria, to American parents. He received his elementary and secondary education in France. In 1901, he was granted a B.Sc. degree by Bingham University in England. During 1901-2, Lotka pursued graduate studies at the University of Leipzig in Germany. In 1902, he moved to the United States and worked as a chemist for the General Chemical Company until 1908. During 1908-9 he was an assistant in physics at Cornell University. In 1909, Cornell University granted him an M.A. degree. He worked briefly as an examiner for the U.S. Patent Office and from 1909-11 he was a physicist for the U.S. Bureau of Standards. In 1912, Lotka received his D.Sc. degree from Bingham University. He then returned to work for the General Chemical Company from 1914-19. In 1924, he joined the statistical bureau of the Metropolitan Life Insurance Company in New York City. He remained there until his death on December 5, 1949.

Alfred Lotka was the first person to systematically analyze the wide variety of relationships which occur between two interacting species and to formulate and study mathematical models to represent those interactions. Some of the models which we are about to study appeared in his 1925 book, *Elements of Physical Biology.*

The Italian mathematician Vito Volterra (1860-1940) independently constructed the same basic prey-predator population model as Lotka and arrived at many of the same conclusions. Volterra was born on May 3, 1860, in Ancona, Italy. He began studying geometry at age 11 and calculus at age 14. In 1882, he received his doctorate in physics from the University of Pisa. His first appointment was as professor of mechanics and mathematics at Pisa. Later, he was a faculty member of the University of Rome for a period of thirty years. Volterra's major contributions to mathematics are in the areas of functional analysis and integral equations. The following is an account of the events which led Volterra to his formulation of the prey-predator model.

From 1910 to 1923, Humberto D'Ancona collected data on the number of each species of fish sold in the markets of Trieste, Fiume, and Venice. D'Ancona assumed that the relative numbers of the various species available at the markets indicated the relative numbers of the species in the Adriatic Sea. He noticed that the percentage of predator fish (sharks, skates, rays, etc.) in the total fish population was higher during and immediately after World War I (1914-18). He concluded that the reduction in fishing due to the war caused the change in the ratio of predators to prey. He reasoned that during the war, the ratio was close to its natural state and that the decrease in this ratio before and after the war was due to fishing. D'Ancona hypothesized that the reduction in the level of fishing due to the war caused an increase in the number of prey fish which in turn caused a larger increase in the number of predator fish—thus, accounting for the increased percentage of predators. However, he could not give biological or ecological reasons why fishing should be more beneficial to prey than to their predators. D'Ancona requested his father-in-law, Vito Volterra, to construct some mathematical models to explain the situation. In a few months, Volterra formulated a set of models for the population dynamics for two or more interacting species.

The following sequence of assumptions and reasoning may be similar to those which led both Lotka and Volterra to their elementary prey-predator model. Let $x(t)$ be the population (number) of prey at time t and let $y(t)$ be the population of predators at time t.

Assumption 1 The rates of change of the populations depend only on x and y.

This assumption means the prey-predator population dynamics can be modeled by an autonomous system of differential equations of the form

$$\text{(1)} \qquad \begin{aligned} \frac{dx}{dt} &= f(x,y) \\ \frac{dy}{dt} &= g(x,y). \end{aligned}$$

Assumption 2 The functions f and g are quadratic in x and y.

This assumption means the functions f and g have the following forms

$$\text{(2)} \qquad f(x,y) = A + Bx + Cx^2 + Dy + Ey^2 + Hxy$$

and

$$\text{(3)} \qquad g(x,y) = K + Lx + Mx^2 + Ny + Py^2 + Qxy$$

where A, B, C, D, E, H, K, L, M, N, P, and Q are all real constants.

Assumption 3 If one species is not present, there is no change in its population.

Stated mathematically, this assumption is (i) if $x = 0$, then $f(0,y) = 0$ and (ii) if $y = 0$, $g(x,0) = 0$. Setting $x = 0$ and $f(0,y) = 0$ in equation (2) leads to the requirement $0 = A + Dy + Ey^2$ for all y. Thus, $A = D = E = 0$. Likewise, setting $y = 0$ and $g(x,0) = 0$ in equation (3), yields the requirement $0 = K + Lx + Mx^2$ for all x. Thus, $K = L = M = 0$.

Summarizing to this point, assumptions 1, 2, and 3 result in a system of the form

$$\text{(4)} \qquad \frac{dx}{dt} = Bx + Cx^2 + Hxy = f(x,y)$$

$$\text{(5)} \qquad \frac{dy}{dt} = Ny + Py^2 + Qxy = g(x,y).$$

Assumption 4 An increase in the prey population increases the growth rate of the predators.

Mathematically, this assumption is $g_x(x,y) > 0$ for all x, y. Since from equation (5) $g_x = Qy$, this assumption means Q is positive for y positive.

Assumption 5 An increase in the predator population decreases the growth rate of the prey.

This assumption is $f_y(x, y) < 0$ for all x, y. Since from equation (4) $f_y = Hx$, this assumption means H is negative for x positive.

Assumption 6 If there are no predators, then the prey population increases according to the Malthusian law.

Thus, if $y = 0$, then $f(x, 0) > 0$ and $f(x, 0) = rx$ where r is positive. From equation (4), this assumption leads to the requirement

$$f(x, 0) = Bx + Cx^2 = rx.$$

Hence, $B = r$, a positive constant, and $C = 0$.

Assumption 7 If there are no prey, then the predator population decreases according to the Malthusian law.

Mathematically, this assumption is: if $x = 0$, then $g(0, y) < 0$ and $g(0, y) = -sy$, where s is positive. From equation (5), this assumption leads to the requirement

$$g(0, y) = Ny + Py^2 = -sy.$$

Hence, $N = -s$, where s is positive, and $P = 0$.

These seven assumptions yield the **Lotka–Volterra prey-predator model**

$$\text{(6)} \qquad \begin{aligned} \frac{dx}{dt} &= rx - Hxy = f(x, y) \\ \frac{dy}{dt} &= -sy + Qxy = g(x, y) \end{aligned}$$

where r, s, H, and Q are all positive constants. Solving the two equations

$$\begin{aligned} rx - Hxy &= x(r - Hy) = 0 \\ -sy + Qxy &= y(-s + Qx) = 0 \end{aligned}$$

simultaneously, we see that system (6) has two critical points—namely, $(0, 0)$ and $(s/Q, r/H)$.

Let us now consider initial conditions (x_0, y_0) where $x_0 \geq 0$ and $y_0 \geq 0$. If $x_0 = 0$, then the solution of system (6) is $(x(t), y(t)) = (0, y_0 e^{-st})$. That is, if $x_0 = 0$, then the y-axis is a trajectory in the phase-plane and a solution which starts on the y-axis remains on the y-axis and moves toward the origin. Notice that this result is a consequence of assumption 7 which says: "In the absence of prey, the predator population decreases according to the Malthusian law." Likewise, if $y_0 = 0$, the solution to system (6) is $(x(t), y(t)) = (x_0 e^{rt}, 0)$. Thus, if $y_0 = 0$, the x-axis is a trajectory in the phase-plane and a solution which starts on the x-axis remains on the x-axis and moves away from the origin.

Observe that this result is a consequence of assumption 6. Since trajectories in the phase-plane cannot intersect one another and since the x-axis and the y-axis are both trajectories, any trajectory which begins in the first quadrant must remain in the first quadrant for all time.

Calculating first partial derivatives of f and g of system (6), we find $f_x = r - Hy$, $f_y = -Hx$, $g_x = Qy$, and $g_y = -s + Qx$. At the critical point $(s/Q, r/H)$ the associated linear system has coefficient matrix

$$\mathbf{A} = \begin{pmatrix} f_x(s/Q, r/H) & f_y(s/Q, r/H) \\ g_x(s/Q, r/H) & g_y(s/Q, r/H) \end{pmatrix} = \begin{pmatrix} 0 & -sH/Q \\ Qr/H & 0 \end{pmatrix}.$$

The eigenvalues of this matrix are $\lambda = \pm\sqrt{-rs} = \pm i\sqrt{rs}$, so $(s/Q, r/H)$ is a neutrally stable center of the associated linear system. Consequently, without any further analysis, we cannot tell if this critical point remains a stable center or becomes a stable or unstable spiral point of the nonlinear system (6).

A trajectory $(x(t), y(t))$ of an autonomous system is said to be **periodic** if for some positive T, $(x(t+T), y(t+T)) = (x(t), y(t))$ for all t.

Volterra summarized his findings for the prey-predator model (6) in three basic principles. First, based on the assumption that the coefficients of growth, r and s, and the coefficients of interaction, H and Q, remain constant, he was able to prove that "**the trajectories of the nonlinear system (6) are closed, periodic trajectories which enclose the critical point** $(\mathbf{s/Q, r/H})$." That is, the critical point $(s/Q, r/H)$ of the nonlinear system (6) remains a stable center and the prey and predator populations vary periodically. The period T depends only on the coefficients r, s, H, and Q and on the initial conditions $x_0 > 0$ and $y_0 > 0$. Let $(x(t), y(t))$ be a periodic solution of (6) with period $T > 0$. The average number of prey $\bar{x}$ and the average number of predators $\bar{y}$ over the period T is

$$\bar{x} = \frac{1}{T}\int_0^T x(t)\,dt \qquad \bar{y} = \frac{1}{T}\int_0^T y(t)\,dt.$$

Volterra's second principle, also based on the assumption that r, s, H, and Q remain constant, is that $\bar{x} = s/Q$ and $\bar{y} = r/H$. Thus, his second principle, the **law of conservation of averages**, says that "**the average values of** $x(t)$ **and** $y(t)$ **over the period** T **is equal to their critical point values.**" This law is fairly easy to verify as the following computations show. Dividing the first equation of system (6) by x and integrating from 0 to T, we obtain

$$\int_0^T \frac{x'(t)}{x(t)}\,dt = \ln x(t)\Big|_0^T = \ln x(T) - \ln x(0) = \int_0^T (r - Hy)\,dt.$$

Since x is periodic with period T, we have $x(T) = x(0)$ and $\ln x(T) = \ln x(0)$; therefore,

$$\int_0^T (r - Hy)\,dt = 0.$$

Consequently,

$$rT = H \int_0^T y\,dt \qquad \text{or} \qquad \bar{y} = \frac{1}{T}\int_0^T y\,dt = \frac{r}{H}.$$

Dividing the second equation of system (6) by y and proceeding as above, we also find $\bar{x} = s/Q$.

We are now ready to determine the effect of fishing on the prey-predator model (6). The simplest model which includes the effects of fishing (harvesting) assumes **indiscriminate, constant-effort harvesting** in which fishermen keep whatever fish they catch. Therefore, in indiscriminate, constant-effort harvesting it is assumed that the number of fish harvested (caught) of each species is proportional to the population of that species. The **prey-predator model with indiscriminate, constant-effort harvesting** is

$$\begin{aligned} \frac{dx}{dt} &= rx - Hxy - hx &&= (r-h)x - Hxy \\ \frac{dy}{dt} &= -sy + Qxy - hy &&= (-s-h)y + Qxy \end{aligned} \tag{7}$$

where h, the **harvesting coefficient**, is a positive constant. For $r - h > 0$ (i.e., for $h < r$) system (7) is the same as system (6) with r replaced by $r - h$ and $-s$ replaced by $-s - h$. So, the critical point in the first quadrant of system (7) is at $((s+h)/Q, (r-h)/H)$. Since $h > 0$, we see that $(s+h)/Q > s/Q$ and $(r-h)/H < r/H$. Thus, we arrive at Volterra's third principle which states: "**Indiscriminate, constant-effort harvesting increases the average prey population and decreases the average predator population.**" This third principle substantiates the conclusion reached by D'Ancona—namely, that the predator population, on the average, increases when fishing decreases and decreases when fishing increases and, conversely, the prey population, on the average, decreases when fishing decreases and increases when fishing increases.

Since its initial formulation, the Lotka–Volterra prey-predator model has been supported and challenged by many ecologists and biologists. Critics of the model cite the fact that most prey-predator systems found in nature tend to an equilibrium state. However, this model is a good model for the prey-predator fish system of the Adriatic Sea, since these fish do not compete within their own species for available resources. This model also adequately represents population dynamics for other prey-predator systems in which the individual populations do not compete within their own species for resources. Shortly, we will examine systems which include terms to reflect internal competition.

Example 1 Computer Solution of a Lotka–Volterra Prey-Predator System Solve the Lotka–Volterra prey-predator system

$$\begin{aligned} \frac{dx}{dt} &= x - .5xy \\ \frac{dy}{dt} &= -2y + .25xy \end{aligned}$$

on the interval $[0, 5]$ for the initial conditions $x(0) = 10$, $y(0) = 5$. Display $x(t)$ and $y(t)$ on the same graph over the interval $[0, 5]$. Produce a phase-plane graph of y versus x.

Solution

We input the two functions defining the system, $f(x, y) = x - .5xy$ and $g(x, y) = -2y + .25xy$, into our computer software. Then, we input the interval of integration $[0, 5]$ and the initial conditions $x(0) = 10$ and $y(0) = 5$. After the integration was completed, we indicated we wanted to graph $x(t)$ and $y(t)$ on the interval $[0, 5]$ with $0 \leq y \leq 20$. The resulting graph is displayed in Figure 10.15. In this graph, the solution $x(t)$ lies above the solution $y(t)$.

Since we wanted to produce a phase-plane graph of y versus x, we indicated to our software that we wanted $x(t)$ assigned to the horizontal axis and $y(t)$ assigned to the vertical axis and that we wanted to display the graph on the rectangle where $0 \leq x \leq 20$ and $0 \leq y \leq 10$. The resulting phase-plane graph of y versus x is shown in Figure 10.16.

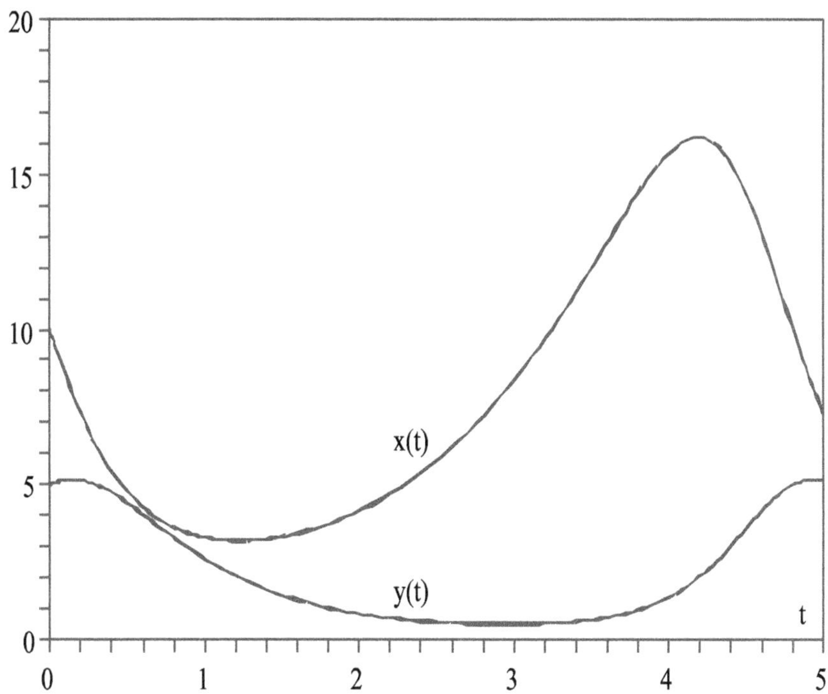

FIGURE 10.15
Graph of the Solution of a Lotka–Volterra Prey-Predator System.

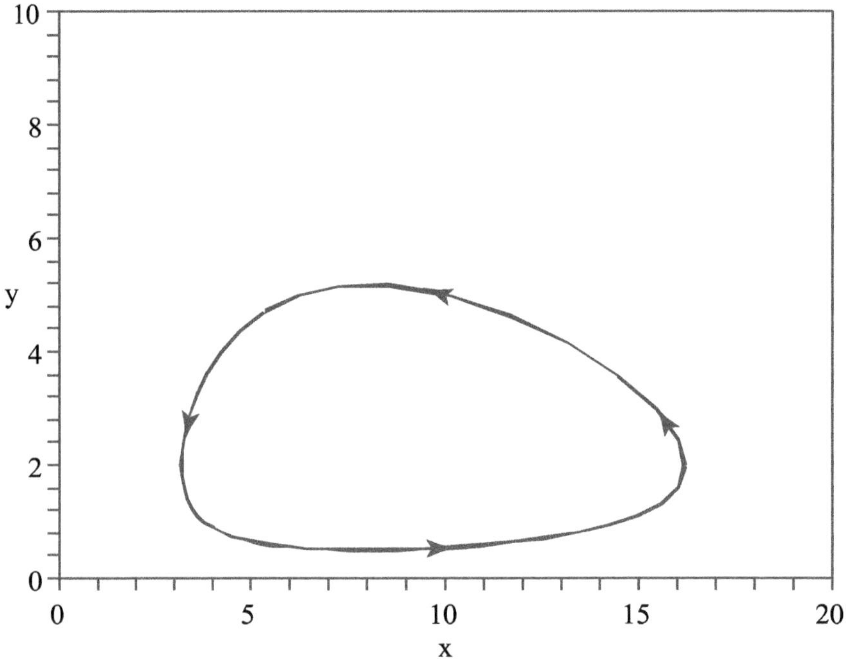

FIGURE 10.16
Phase-Plane Plot of Predator Population Versus Prey Population. ■

10.5.1.1 Exercise

1. a. Use your computing software to solve the prey-predator initial value problem

$$\frac{dx}{dt} = 2x - .5xy; \qquad x(0) = 2$$

$$\frac{dy}{dt} = -2y + xy; \qquad y(0) = 3.5$$

on the interval $[0, 4]$. Assume that the unit of time is measured in years and the unit of population is measured in thousands. What is the period, T? What is the minimum and maximum population of the prey, x, and the predators, y? What is the average prey and predator population?

b. Solve the indiscriminate, constant-effort harvesting, prey-predator initial value problem

$$\begin{aligned} \frac{dx}{dt} &= 2x - .5xy - .3x; \qquad x(0) = 2 \\ \frac{dy}{dt} &= -2y + xy - .3y; \qquad y(0) = 3.5 \end{aligned}$$

on the interval $[0, 4]$. What is the period, T? How does this period compare with the period for part a? What is the minimum and maximum population for the prey and predators? How do these values compare with the corresponding answers in part a? What is the average prey and predator population? How do these values compare with the corresponding answers in part a?

10.5.2 Modified Prey-Predator Models

In this section, we will consider several modifications to the Lotka–Volterra prey-predator model.

Internal Prey Competition Model First, suppose that in the absence of a predator, the prey population grows so rapidly that internal competition within the prey population for important resources such as food and living space becomes a factor. This internal competition can be modeled by changing the assumption of Malthusian population growth for the prey (assumption 6 of the previous section) to logistic population growth. The resulting **prey-predator model with internal prey competition** is

$$(1) \qquad \begin{aligned} \frac{dx}{dt} &= rx - Cx^2 - Hxy \\ \frac{dy}{dt} &= -sy + Qxy \end{aligned}$$

where r, H, C, s, and Q are positive constants.

Internal Prey and Internal Predator Competition with Harvesting Model Now, suppose that in the absence of the other population both the prey and predator populations obey the logistic law model and that each population is harvested. The system of differential equations to be investigated then becomes

$$(2) \qquad \begin{aligned} \frac{dx}{dt} &= rx - Cx^2 - Hxy - h_1x \\ \frac{dy}{dt} &= -sy - Py^2 + Qxy - h_2y \end{aligned}$$

where r, C, H, h_1, s, P, Q, and h_2 are all nonnegative constants. Notice that the Lotka–Volterra prey-predator system, the Lotka–Volterra prey-predator system with harvesting, and the prey-predator system with internal prey competition are all special cases of this system and may be obtained by setting various combinations of the constants equal to zero. Consequently, system (2) is the most general model for prey-predator population dynamics that we have encountered thus far.

Three Species Models Depending upon the assumptions made, there are several ways to formulate a system of differential equations to represent the population dynamics for three interacting species. First of all, suppose a species with population y_1 is prey for two other species with populations y_2 and y_3. Suppose the species with population y_2 is prey for the species with population y_3. And suppose no other predation occurs. Further, suppose the population growth for each species, in the absence of the other species, satisfies the Malthusian law. The system of differential equations to be studied under these assumptions is

$$\begin{aligned} \frac{dy_1}{dt} &= ay_1 - by_1y_2 - cy_1y_3 \\ \frac{dy_2}{dt} &= -dy_2 + ey_1y_2 - fy_2y_3 \\ \frac{dy_3}{dt} &= -gy_3 + hy_1y_3 + iy_2y_3 \end{aligned} \tag{3}$$

where a, b, c, d, e, f, g, h, and i are positive constants.

M. Braun discussed the following system of differential equations for representing the population dynamics for three interacting species which live on the island of Komodo in Malaysia. One species with population y_1 is a plant. A second species with population y_2 is a mammal. And the third species with population y_3 is a reptile. The plants are prey for the mammals and the mammals are prey for the reptiles. No other predation occurs. In the absence of the mammals, the plants are assumed to grow according to the logistic law, since they compete with one another for space in which to grow. In the absence of the plants, the mammals are assumed to die out according to the Malthusian law. And in the absence of the mammals, the reptiles are also assumed to die out according to the Malthusian law. Hence, the population dynamics for these three interacting species is represented by the system of differential equations

$$\begin{aligned} \frac{dy_1}{dt} &= ay_1 - by_1^2 - cy_1y_2 \\ \frac{dy_2}{dt} &= -dy_2 + ey_1y_2 - fy_2y_3 \\ \frac{dy_3}{dt} &= -gy_3 + hy_2y_3 \end{aligned} \tag{4}$$

where a, b, c, d, e, f, g, and h are positive constants.

Next, suppose that one species preys upon the adults of a second species but not upon the young of that species. This situation can occur when the young prey are protected in some manner—perhaps by their coloration, by their smaller size, by their living quarters, or simply by the physical intervention of the adults. The model for this prey-predator system with protected young prey leads to a system of differential equations with three components. Let y_1 denote the number of young prey, y_2 denote the number of adult prey, and y_3 denote the number of predators. Thus, the total prey population at time t is $y_1(t) + y_2(t)$. We assume the birth rate for the prey is proportional to the number of adult prey—that is, we assume a Malthusian type of growth for the prey. We assume the number of young prey maturing to adulthood is proportional to the number of young prey. And we assume the death rate for the young prey is proportional to the number of young prey. Hence, the rate of change for the number of young prey is

$$\text{(5a)} \qquad \frac{dy_1}{dt} = ay_2 - by_1 - cy_1$$

where a, b, and c are positive constants—constants of proportionality for the processes of birth, maturation, and death. Next, we assume the adult prey population increases due to the maturation of the young into adults, so dy_2/dt includes the term by_1 to reflect the maturation process. We assume the death rate of the adult population due to causes other than predation to be proportional to the number of adults, so dy_2/dt includes the term $-dy_2$ to model the death of adults. And we model the predation by including the term $-ey_2y_3$. Thus, the equation for the rate of change of the adult prey population is

$$\text{(5b)} \qquad \frac{dy_2}{dt} = by_1 - dy_2 - ey_2y_3$$

where b, d, and e are positive constants—constants of proportionality for the processes of maturation, death, and predation. We assume the death rate for the predator population is proportional to the number of predators, so dy_3/dt includes the term $-fy_3$. In other words, we assume the Malthusian law holds for the predators. We also assume the rate of increase in the number of predators due to predation can be modeled as gy_2y_3. Thus, the equation for the rate of change of the predator population is

$$\text{(5c)} \qquad \frac{dy_3}{dt} = -fy_3 + gy_2y_3$$

where f and g are positive constants—constants of proportionality for the processes of death and predation. Consequently, a three-component system of differential equations for modeling **a prey-predator system in which the**

young prey are protected is

$$(5)\qquad \begin{aligned} \frac{dy_1}{dt} &= ay_2 - by_1 - cy_1 \\ \frac{dy_2}{dt} &= by_1 - dy_2 - ey_2y_3 \\ \frac{dy_3}{dt} &= -fy_3 + gy_2y_3 \end{aligned}$$

where a, b, c, d, e, f, and g are positive constants.

10.5.2.1 Exercises

1. Find and classify all critical points of the prey-predator model with internal prey competition, system (1). Which critical points are in the first quadrant?

2. Use your computing software to solve the prey-predator model with internal prey competition, system (1), for the initial conditions $x(0) = 3$, $y(0) = 2$ on the interval $[0, 5]$ for $r = 2$, $H = .5$, $s = 2$, $Q = 1$, and (i) $C = 1.5$ and (ii) $C = .5$.

   For cases (i) and (ii):

   a. Display a graph of $x(t)$ and $y(t)$ on $[0, 5]$.

   b. Display a phase-plane graph of y versus x.

   Answer the following questions for cases (i) and (ii):

   Is the solution periodic?

   What happens to the prey population, $x(t)$, as t increases?

   What happens to the predator population, $y(t)$, as t increases?

   Where are the critical points (x^*, y^*) with $x^* \geq 0$ and $y^* \geq 0$?

   What do you think happens for any initial condition (x_0, y_0) where $x_0 > 0$ and $y_0 > 0$?

3. Without using any computer software, decide how the results of exercise 2 will be affected in the following cases:

   a. Only the prey population is harvested with harvesting coefficient $h_1 < r = 2$. That is, for constants as given in exercise 2, what is the effect of adding the term $-h_1x$ to the first equation of system (1)?

   b. Only the predator population is harvested with harvesting coefficient h_2. That is, for constants as given in exercise 2, what is the effect of adding the term $-h_2y$ to the second equation of system (1)?

c. Both prey and predator populations are harvested with harvesting coefficients h_1 and h_2, respectively.

4. a. For $h_1 = h_2 = 0$ find and classify all critical points of the internal prey and internal predator competition with harvesting model, system (2), in the first quadrant.

   b. For $h_1 \neq 0$ and $h_2 \neq 0$ find and classify all critical points of system (2) in the first quadrant.

   c. How does harvesting affect the prey and predator populations? Is the answer the same as for the Lotka–Volterra prey-predator model?

5. Use computer software to solve the internal prey and internal predator competition with harvesting model, system (2), for the initial conditions $x(0) = 3$, $y(0) = 2$ on the interval $[0, 5]$ for $r = 3$, $C = 2$, $H = s = P = Q = 1$, and the following five values for the harvesting coefficients:

   a. $h_1 = h_2 = 0$ b. $h_1 = 1,\ h_2 = 0$ c. $h_1 = 0,\ h_2 = 1$

   d. $h_1 = h_2 = .5$ e. $h_1 = h_2 = .25$

   For each case,

   (i) Display a graph of $x(t)$ and $y(t)$ on the interval $[0, 5]$.

   (ii) Display a phase-plane graph of y versus x.

   (iii) What happens to $x(t)$ and $y(t)$ as t increases?

6. Use your computing software to solve the three species model, system (3), for the initial conditions $y_1(0) = 5$, $y_2(0) = 4$, $y_3(0) = 3$ on the interval $[0, 5]$ for the following values of the constants.

   a. $a = d = g = 1,\ b = c = e = .25$ and $f = h = i = .1$

   b. $a = d = g = 1,\ b = f = h = i = .1$ and $c = e = .25$

   In each case, display a single graph showing $y_1(t)$, $y_2(t)$, and $y_3(t)$ on the interval $[0, 5]$. What happens to $y_1(t)$, $y_2(t)$, and $y_3(t)$ as t increases?

7. a. Find the critical points of the three-species model, system (4).

   b. Use computer software to solve system (4) for the initial conditions $y_1(0) = 1$, $y_2(0) = 1$, $y_3(0) = 1$ on the interval $[0, 6]$ for the following values of the constants.

   (i) $a = 2,\ b = c = d = e = f = g = h = 1$

   (ii) $a = 2,\ b = .5,\ c = d = e = f = g = h = 1$

   In each case, display a single graph showing $y_1(t)$, $y_2(t)$, and $y_3(t)$ on the interval $[0, 6]$. What happens to $y_1(t)$, $y_2(t)$, and $y_3(t)$ as t increases?

8. a. Find the critical points of the prey-predator model in which the young prey are protected, system (5).

   b. Use computer software to solve system (5) for the initial conditions $y_1(0) = 1$, $y_2(0) = 1$, $y_3(0) = 1$ on the interval $[0, 10]$ for the following values of the constants.

      (i) $a = 2, \; b = c = d = .5, \; e = f = g = 1$

      (ii) $a = 2, \; b = c = d = e = f = g = 1$

   In each case, display a single graph showing $y_1(t)$, $y_2(t)$, and $y_3(t)$ on the interval $[0, 10]$. What happens to $y_1(t)$, $y_2(t)$, and $y_3(t)$ as t increases?

10.5.3 Leslie's Prey-Predator Model

In 1948, P. H. Leslie proposed the following prey-predator model

$$\begin{aligned} \frac{dx}{dt} &= ax - bx^2 - cxy \\ \frac{dy}{dt} &= dy - \frac{ey^2}{x}. \end{aligned} \tag{1}$$

The first equation in system (1) for the rate of change of the prey population, x, is the same as the Lotka–Volterra prey-predator model with internal prey competition. That is, in the absence of predation, the prey population is assumed to grow according to the logistic law model and has maximum population size b/a. The second equation in system (1) for the rate of change of the predator population, y, resembles the logistic law equation except the second term has been modified to take into account the density of the prey. If $y/x = d/e$, the predator population is at its equilibrium value. If there are many prey per predator, y/x is small ($y/x < d/e$) and, therefore, the predator population grows nearly exponentially. And if there are few prey per predator, y/x is large ($y/x > d/e$) and the predator population decreases.

10.5.3.1 Exercise

1. a. Find the critical point of Leslie's prey-predator model, system (1) in the first quadrant and determine if it is stable or unstable.

   b. Use computer software to solve system (1) with $a = 2$, $b = e = .5$, $c = d = 1$ on the interval $[0, 5]$ for the initial conditions $x(0) = 1$, $y(0) = 1$. Display a single graph showing $x(t)$ and $y(t)$. Display a phase-plane graph of y versus x. Estimate $\lim_{t\to\infty} x(t)$ and $\lim_{t\to\infty} y(t)$.

10.5.4 Leslie–Gower Prey-Predator Model

In 1960, P. H. Leslie and J. C. Gower studied the following prey-predator model

$$(1) \qquad \begin{aligned} \frac{dx}{dt} &= ax - cxy \\ \frac{dy}{dt} &= dy - \frac{ey^2}{x} \end{aligned}$$

where x is the prey population, y is the predator population, and a, c, d, and e are positive constants. The first equation of this system is the same as in the Lotka–Volterra prey-predator model. Thus, in the absence of predators the prey population is assumed to grow according to the Malthusian law.

10.5.4.1 Exercise

1.Use computer software to solve the Leslie–Gower prey-predator model, system (1), with $a = 1$, $c = .1$, $d = 1$, and $e = 2.5$ on the interval $[0, 6]$ for initial populations of $x(0) = 80$ and $y(0) = 20$. Display a graph of $x(t)$ and $y(t)$ and display a phase-plane graph of y versus x. Estimate $\lim_{t\to\infty} x(t)$ and $\lim_{t\to\infty} y(t)$.

10.5.5 Different Uptake Function

Thus far, we have assumed that the rate of change of the prey population, x, due to predation by the predator population, y, is proportional to the product xy. However, since the predator population's collective appetite and food requirement can be satisfied when there is an abundance of prey, when x is large relative to y, the rate of change in the prey population should approach a function which is proportional to y alone. It has been suggested that the **uptake function** xy of the Lotka–Volterra prey-predator model be replaced by the uptake function $xy/(1 + kx)$. Thus, the new prey-predator model to be considered is

$$(1) \qquad \begin{aligned} \frac{dx}{dt} &= ax - \frac{bxy}{1 + kx} \\ \frac{dy}{dt} &= -cy + \frac{dxy}{1 + kx} \end{aligned}$$

where a, b, c, d, and k are positive constants.

10.5.5.1 Exercises

1. a. Find the critical point of the system

$$(2)\qquad \begin{aligned} \frac{dx}{dt} &= 4x - \frac{8xy}{1+2x} \\ \frac{dy}{dt} &= -2y + \frac{8xy}{1+2x} \end{aligned}$$

in the first quadrant.

b. Determine the stability characteristics of the critical point.

c. Use computer software to solve system (2) on the interval $[0, 5]$ for the initial conditions $x(0) = 1$, $y(0) = 1$. Display a single graph with $x(t)$ and $y(t)$ on the interval $[0, 5]$ and display a phase-plane graph of y versus x. Estimate $\lim_{t\to\infty} x(t)$ and $\lim_{t\to\infty} y(t)$.

2. In the absence of predators, the prey population of system (2) grows according to the Malthusian law. Suppose this growth assumption is changed to logistic law growth, and the prey-predator system becomes

$$(3)\qquad \begin{aligned} \frac{dx}{dt} &= 4x - 4x^2 - \frac{8xy}{1+2x} \\ \frac{dy}{dt} &= -2y + \frac{8xy}{1+2x}. \end{aligned}$$

a. Find the critical point of system (3) in the first quadrant. How does it compare with the critical point of system (2)?

b. Determine the stability characteristics of the critical point of system (3).

c. Use computer software to solve system (3) on the interval $[0, 5]$ for the initial conditions $x(0) = 1$, $y(0) = 1$. Display a graph of $x(t)$ and $y(t)$ and a phase-plane graph of y versus x. Estimate $\lim_{t\to\infty} x(t)$ and $\lim_{t\to\infty} y(t)$.

3. Use your computing software to solve the prey-predator system

$$(4)\qquad \begin{aligned} \frac{dx}{dt} &= 4x - 4x^2 - \frac{16xy}{1+4x} \\ \frac{dy}{dt} &= -2y + \frac{16xy}{1+4x} \end{aligned}$$

on the interval $[0, 10]$ for the following initial conditions:

a. $x(0) = .25, \quad y(0) = .4$ b. $x(0) = .25, \quad y(0) = .675$

c. $x(0) = .25, \quad y(0) = 1$

In each case, produce a phase-plane graph of y versus x. What do you notice about the phase-plane graphs?

10.5.6 May's Prey-Predator Model

The following prey-predator model was proposed by R. M. May

$$(1) \qquad \begin{aligned} \frac{dx}{dt} &= ax - bx^2 - \frac{cxy}{x+k} \\ \frac{dy}{dt} &= dy - \frac{ey^2}{x}. \end{aligned}$$

As before, x is the prey population, y is the predator population, and a, b, c, d, e, and k are positive constants.

10.5.6.1 Exercise

1. Use computer software to solve system (1) with $a = c = 12$, $b = 1.2$, $d = e = 2$, and $k = 1$ on the interval $[0, 5]$ for the following initial conditions:

a. $x(0) = 2, \quad y(0) = 1$ b. $x(0) = 2, \quad y(0) = 1.35$

c. $x(0) = 2, \quad y(0) = 2$

In each case, produce a phase-plane graph of y versus x. What do you notice about the phase-plane graphs?

10.5.7 Competing Species Models

Two similar species sometimes compete with one another for the same limited food supply and living space. Thus, each species removes from the environment resources that promote the growth of the other. This situation usually occurs when two different species of predators are in competition with one another for the same prey. For this reason the models to be discussed are sometimes called **competitive hunters models**. When two predator species compete with one another, one species nearly always becomes extinct while the other, more efficient species, survives. This biological phenomenon is called the **principle of competitive exclusion**.

Let us make the following assumptions regarding two competing species with population sizes x and y.

1. There is sufficient prey to sustain any level of the predator populations.
2. The rates of change of the predator populations depend quadratically on x and y.
3. If one predator species is absent, there is no change in its population size.
4. In the absence of one predator species, the other predator species will grow according to the Malthusian law.

These assumptions lead to the **competitive hunters model**

$$(1) \qquad \begin{aligned} \frac{dx}{dt} &= Ax - Bxy \\ \frac{dy}{dt} &= Cy - Dxy \end{aligned}$$

where A, B, C, and D are positive constants. If the growth assumption is changed from the Malthusian law to the logistic law, then the competitive hunters model becomes

$$(2) \qquad \begin{aligned} \frac{dx}{dt} &= ax - bxy - rx^2 \\ \frac{dy}{dt} &= cy - dxy - sy^2 \end{aligned}$$

where a, b, c, d, r, and s are positive constants.

10.5.8 Exercises

1. a. Show that the competitive hunters model, system (1), has critical points at $(0, 0)$ and $(C/D, A/B)$.

   b. Show that both critical points are unstable.

   c. Show if $x > C/D$ and $y < A/B$, then the x population will increase indefinitely and the y population will become extinct.

   d. Show if $x < C/D$ and $y > A/B$, then the x population will become extinct and the y population will increase indefinitely.

2. Use computer software to solve the competitive hunters model, system (1), with $A = 2$, $B = 1$, $C = 3$, and $D = 1$ on the interval $[0, 5]$ for the following initial conditions: a. $x(0) = 1,\ y(0) = .5$ b. $x(0) = 1.5,\ y(0) = .5$ c. $x(0) = 10,\ y(0) = 9$ d. $x(0) = 10,\ y(0) = 8$ In each case, determine which species becomes extinct.

3. Show that the logistic law, competitive hunters model, system (2), has critical points at $(0, 0)$, $(0, c/s)$, $(a/r, 0)$, and $(p/D, q/D)$ where $p = as - bc$, $q = cr - ad$, and $D = rs - bd$.

4. Consider the competitive hunters model

$$(3) \qquad \begin{aligned} \frac{dx}{dt} &= 2x - 4xy - 4x^2 \\ \frac{dy}{dt} &= 2y - 2xy - 2y^2. \end{aligned}$$

a. Find the critical points of system (3) in the first quadrant.

b. Use computer software to solve system (3) on the interval $[0, 5]$ for the following initial conditions:

(i) $x(0) = .3, \quad y(0) = .1$ (ii) $x(0) = 2, \quad y(0) = .1$

In each case, estimate $\lim_{t\to\infty} x(t)$ and $\lim_{t\to\infty} y(t)$ and determine which species becomes extinct.

5. a. Find the critical points of the competitive hunters model

$$(4) \qquad \begin{aligned} \frac{dx}{dt} &= 4x - .5xy - .2x^2 \\ \frac{dy}{dt} &= 4y - .25xy - \frac{y^2}{3}. \end{aligned}$$

b. Use your computing software to solve system (4) on the interval $[0, 5]$ for the initial conditions:

(i) $x(0) = .1, \ y(0) = 1$ (ii) $x(0) = 1, \ y(0) = .1$

(iii) $x(0) = 20, \ y(0) = 5$ (iv) $x(0) = 20, \ y(0) = 10$

In each case, estimate $\lim_{t\to\infty} x(t)$ and $\lim_{t\to\infty} y(t)$ and determine which species becomes extinct.

6. a. Find the critical points of the competitive hunters model

$$(5) \qquad \begin{aligned} \frac{dx}{dt} &= 2x - 2xy - 2x^2 \\ \frac{dy}{dt} &= 4y - 2xy - 6y^2. \end{aligned}$$

b. Use computer software to solve system (5) on the interval $[0, 5]$ for the initial conditions:

(i) $x(0) = .1, \ y(0) = .1$ (ii) $x(0) = .1, \ y(0) = 1$

(iii) $x(0) = 1, \ y(0) = .1$ (iv) $x(0) = 1, \ y(0) = 1$

In each case, estimate $\lim_{t\to\infty} x(t)$ and $\lim_{t\to\infty} y(t)$. Does either species become extinct?

7. This exercise is designed to illustrate the radical changes which constant-effort harvesting of one of the competing species can bring about. One of the species, for example, might have a fur which humans find very

desirable. Consider the following competitive hunters model in which the x species is harvested with positive harvesting constant H.

$$(6) \qquad \begin{aligned} \frac{dx}{dt} &= 10x - xy - x^2 - Hx \\ \frac{dy}{dt} &= 8y - xy - 2y^2. \end{aligned}$$

For a. $H = 0$, b. $H = 4$, and c. $H = 8$

(i) Find the critical points of system (6) in the first quadrant and determine the stability characteristics of each.

(ii) Use your computing software to solve system (6) on the interval $[0, 5]$ for the initial conditions: 1. $x(0) = 1$, $y(0) = 1$ 2. $x(0) = 1$, $y(0) = 10$ 3. $x(0) = 10$, $y(0) = 1$ 4. $x(0) = 10$, $y(0) = 10$

For each initial condition estimate $\lim_{t\to\infty} x(t)$ and $\lim_{t\to\infty} y(t)$ and determine which, if any, species becomes extinct.

10.6 Epidemics

In Chapter 3, we briefly discussed the history of epidemics and epidemiology, the scientific study of epidemics. We also introduced and studied some of the simpler models for epidemics there. In this section, we will formulate and analyze a few somewhat more complicated models for epidemics. One underlying assumption which we shall make throughout this section is that the population which can contract the disease has a constant size N. That is, we will assume there are no births or immigrations to increase the population size and no emigrations to decrease the population size. Since the time span of many epidemics, such as the flu, is short in comparison to the life span of an individual, the assumption of a constant population size is fairly reasonable. Next, we assume the population is divided into three mutually exclusive sets:

1. The **susceptibles** is the set of individuals who do not have the disease at time t but who may become infected at some later time.

2. The **infectives** is the set of individuals at time t who are infected with and capable of transmitting the disease.

3. The **removeds** is the set of individuals who have been removed at time t either by death, by recovering and obtaining immunity, or by being isolated for treatment.

We will denote the number of susceptibles, infectives, and removeds at time t by $S(t)$, $I(t)$, and $R(t)$, respectively. Under the assumptions we have made

$$S(t) + I(t) + R(t) = N \qquad \text{for all time} \quad t. \tag{1}$$

An assumption which was first made in 1906 by William Hamer, and which has been included in every deterministic epidemic model ever since, is that the rate of change of the number of susceptibles is proportional to the product of the number of susceptibles and the number of infectives. Thus, it is assumed that

$$\frac{dS}{dt} = -\beta S(t)I(t) \qquad \text{for all} \quad t \tag{2}$$

where the positive constant of proportionality β is called the **infection rate**. The product $S(t)I(t)$ represents the rate of contact between susceptibles and infectives while the product $\beta S(t)I(t)$ represents the proportion of contacts which result in the infection of susceptibles.

Next, we assume that the rate of change of the removeds is proportional to the number of infectives. Thus, we assume

$$\frac{dR}{dt} = rI(t) \qquad \text{for all} \quad t \tag{3}$$

where the positive constant r is called the **removal rate**.

Solving equation (1) for I, we get $I = N - S - R$ and differentiating, we find upon substitution from (2) and (3)

$$\frac{dI}{dt} = -\frac{dS}{dt} - \frac{dR}{dt} = \beta S(t)I(t) - rI(t) = (\beta S(t) - r)I(t).$$

Hence, a system of three first-order differential equations for modeling an epidemic is

$$\begin{aligned} \frac{dS}{dt} &= -\beta SI \\ \frac{dI}{dt} &= (\beta S - r)I \\ \frac{dR}{dt} &= rI \end{aligned} \tag{4}$$

where β and r are positive constants. Appropriate initial conditions for this system are the initial number of susceptibles $S(0) = S_0 > 0$, the initial number of infectives $I(0) = I_0 > 0$, and the initial number of removeds $R(0) = R_0 = 0$.

By analyzing the equations of system (4), we can determine some interesting general facts regarding this epidemic model. Since $\beta > 0$, $S(t) \geq 0$, and $I(t) \geq 0$, we have $dS/dt = -\beta SI \leq 0$ for all t. In fact, $dS/dt < 0$ unless $S = 0$ or $I = 0$. Because $dS/dt < 0$, the number of susceptibles, $S(t)$, is a strictly decreasing function of time until S becomes 0 or I becomes 0. Likewise, since $r > 0$ and $I(t) \geq 0$, we have $dR/dt = rI \geq 0$ and, in fact, $dR/dt > 0$ unless $I = 0$. So the number of removeds is a strictly increasing function of time until the number of infectives becomes 0. Since $I(t) \geq 0$ the sign of the rate of change of the number of infectives, $dI/dt = (\beta S - r)I$, depends on the sign of $\beta S - r$. Let us assume $I \neq 0$. When $\beta S - r > 0$ (that is, when $S > r/\beta$), we have $dI/dt > 0$ and the number of infectives increases. When $\beta S - r < 0$ (that is, when $S < r/\beta$), we have $dI/dt < 0$ and the number of infectives decreases. The quantity $p = r/\beta$ is called the **relative removal rate**. Since $S(t)$ is a strictly decreasing function, $0 \leq S(t) \leq S_0$ where S_0 is the initial number of susceptibles. If S_0 is less than $p = r/\beta$, no epidemic occurs since $dI/dt = (\beta S - r)I \leq (\beta S_0 - r)I < 0$ which implies $I(t)$ is a strictly decreasing function. That is, if $S_0 < r/\beta$, the number of infectives decreases monotonically to zero from the initial value of I_0. On the other hand, if $S_0 > r/\beta$, the number of infectives increases from the initial value of I_0 to a maximum value which occurs when the number of susceptibles has decreased to the value r/β at some time $t^* > 0$. For $t > t^*$, it follows that $S(t) < r/\beta$ and the number of infectives decreases. This result is what epidemiologists call the **threshold phenomenon**. That is, there is a critical value which the number of initial susceptibles must exceed before an epidemic can occur. The threshold theorem for system (4), stated below, was proven by W. O. Kermak and A. G. McKendrick in 1927:

Threshold Theorem for Epidemics, System (4)

If $S_0 < r/\beta$, then $I(t)$ decreases monotonically to zero. If $S_0 > r/\beta$, then $I(t)$ increases monotonically to a maximum value and then decreases monotonically to zero. The limit, $\lim_{t\to\infty} S(t)$, exists and is the unique solution, x, of

$$S_0 e^{-\beta(N-x)/r} = x. \tag{5}$$

In order to prevent epidemics, medical personnel try to decrease the number of susceptibles S or to increase the critical value r/β. This is sometimes accomplished by inoculation and by early detection of the disease followed by quarantine procedures.

Example 1 Computer Solution of an Epidemic Model

Solve the epidemic model

$$\begin{aligned} \frac{dS}{dt} &= -.005SI \\ \frac{dI}{dt} &= .005SI - .7I \\ \frac{dR}{dt} &= .7I \end{aligned} \tag{6}$$

on the interval $0 \leq t \leq 10$ days for the initial conditions

(i) $S_0 = 500, \quad I_0 = 5, \quad R_0 = 0$ (ii) $S_0 = 100, \quad I_0 = 100, \quad R_0 = 0$

In each case, display S, I, and R on a single graph and display a phase-plane graph of I versus S.

Solution

(i) Identifying S with y_1, I with y_2, and R with y_3, we ran our computing software by setting $n = 3$, $f_1(t, y_1, y_2, y_3) = -.005y_1y_2$, $f_2(t, y_1, y_2, y_3) = .005y_1y_2 - .7y_2$, and $f_3(t, y_1, y_2, y_3) = .7y_2$. We input the interval of integration as $[0, 10]$ and input the initial conditions: $y_1(0) = 500$, $y_2(0) = 5$, and $y_3(0) = 0$. After integration was completed, the graph of S, I, and R on the rectangle $0 \leq t \leq 10$ and $0 \leq S, I, R \leq 500$ shown in Figure 10.17 was displayed on the monitor. Notice that S decreases monotonically from 500 to approximately 15, and R increases monotonically from 0 to 485. Also observe that I increases monotonically from 5 to a maximum value of approximately 190 when t is approximately 3 days, and then I decreases monotonically to 5 as t approaches 10 days. Since I increases before decreasing, an epidemic occurs.

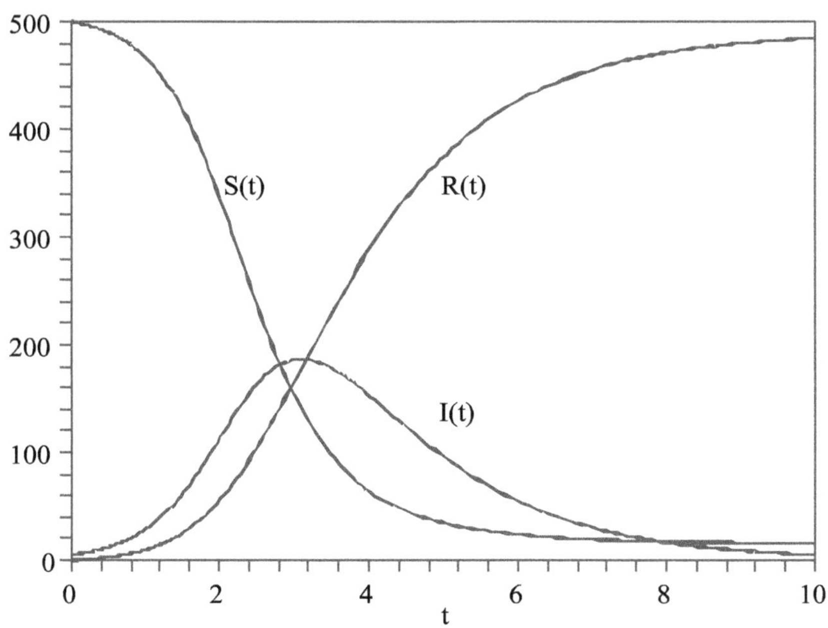

FIGURE 10.17
A Graph of $S(t)$, $I(t)$, and $R(t)$ for System (6) for Initial Conditions (i).

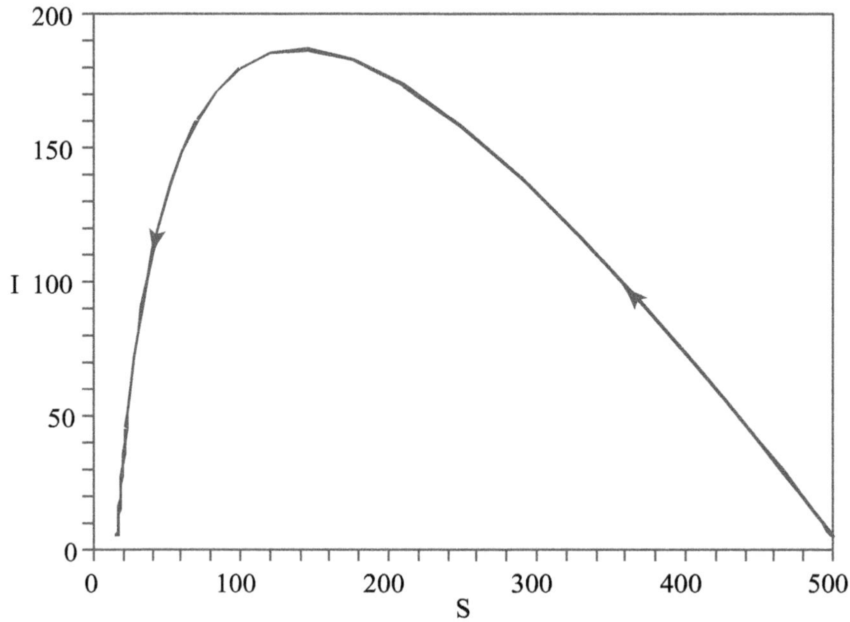

FIGURE 10.18
A Phase-Plane Graph of I versus S for System (6) for Initial Conditions (i).

Since we still wanted to display a phase-plane graph of I versus S, we indicated to our computing system that we wanted $S(t)$ assigned to the horizontal axis and $I(t)$ assigned to the vertical axis and that we wanted the graph displayed on the rectangle $0 \leq S \leq 500$ and $0 \leq I \leq 200$. This phase-plane graph is shown in Figure 10.18. Observe from the graph that S approaches 15 as I approaches 0. Substituting $S_0 = 500$, $\beta = .005$, $N = S_0 + I_0 + R_0 = 505$, $x = 15$, and $r = .7$ into the left-hand side of equation (5), we find

$$S_0 e^{-\beta(N-x)/r} = 500e^{-3.5} = 15.099.$$

Since this value is approximately x, we have verified that the limiting value of the number of susceptibles at the end of the epidemic is approximately 15. That is, $\lim_{t\to\infty} S(t) = 15$. Thus, for this epidemic model, system (6) with initial conditions (i), we have verified equation (5) of the threshold theorem.

(ii) Since we now wanted to solve the same system of differential equations with different initial conditions, we entered the new initial conditions: $S(0) = 100$, $I(0) = 100$, and $R(0) = 0$. A graph of S, I, and R is displayed in Figure 10.19. Notice that S decreases monotonically from the value of 100 to approximately 30, I decreases monotonically from

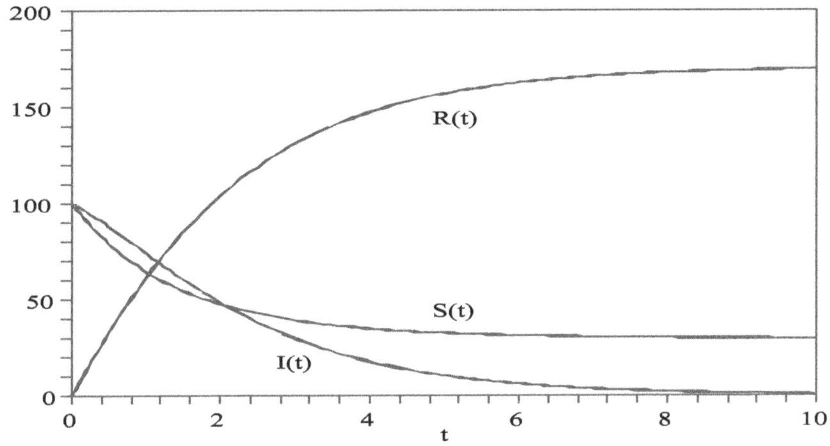

FIGURE 10.19
A Graph of $S(t)$, $I(t)$, and $R(t)$ for System (6) for Initial Conditions (ii).

the value of 100 to nearly 0, and R increases monotonically from 0 to approximately 170. Since I decreases monotonically to 0, no epidemic occurs in this instance. Verify that $x = 30$ is the approximate solution of equation (5) for the given initial values. Hence, in this case, as $I \to 0$, $S \to 30$. Thus, there are 30 susceptibles remaining when the epidemic ends. A phase-plane graph of I versus S is shown in Figure 10.20.

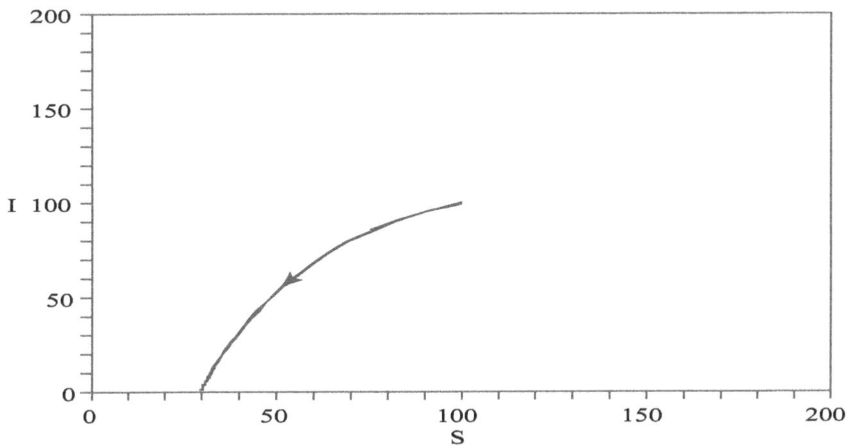

FIGURE 10.20
A Phase-Plane Graph of I versus S for System (6) for Initial Conditions (ii).

■

10.6.1 Exercises

1. Consider the epidemic model

$$
\begin{aligned}
\frac{dS}{dt} &= -.01SI \\
\frac{dI}{dt} &= .01SI - 3I \qquad (7)\\
\frac{dR}{dt} &= 3I
\end{aligned}
$$

a. Calculate the relative removal rate for this model.

b. Use your computing software to solve system (7) on the interval $[0, 5]$ where t is in days for the initial conditions:

$$\text{(i) } S_0 = 300, \quad I_0 = 10, \quad R_0 = 0 \quad \text{(ii) } S_0 = 500, \quad I_0 = 10, \quad R_0 = 0$$

In each case, display S, I, and R on a single graph and display a phase-plane graph of I versus S.

c. From the solution or phase-plane graph of I versus S for system (7) for the initial conditions (i) and (ii), estimate the remaining number of susceptibles at the end of the epidemic. That is, estimate $S_\infty = \lim_{t \to \infty} S(t)$ which is approximately $S(5)$ and verify that S_∞ satisfies equation (5).

2. Suppose when an epidemic begins medical personnel inoculate members of the susceptible group at a rate α which is proportional to the number of susceptibles. Then at each instant in time αS individuals are subtracted from the susceptible group and added to the removed group. So one model for an epidemic with inoculation is

$$
\begin{aligned}
\frac{dS}{dt} &= -\beta SI - \alpha S \\
\frac{dI}{dt} &= \beta SI - rI \\
\frac{dR}{dt} &= rI + \alpha S
\end{aligned}
$$

where α, β, and r are positive constants.

Use computer software to solve the following epidemic with the inoculation model

$$\begin{aligned} \frac{dS}{dt} &= -.01SI - .1S \\ \frac{dI}{dt} &= .01SI - 3I \\ \frac{dR}{dt} &= 3I + .1S \end{aligned}$$

on the interval $[0, 5]$ where t is in days for the initial conditions:

(i) $S_0 = 300, \quad I_0 = 10, \quad R_0 = 0$ (ii) $S_0 = 500, \quad I_0 = 10, \quad R_0 = 0$

In each case, display S, I, and R on a single graph and display a phase-plane graph of I versus S. Compare your results with the results of exercise 1. Does an epidemic occur in case (ii) as it did in exercise 1? What is $\lim_{t\to\infty} S(t)$ in each case?

3. A rapidly increasing number of infectives can frighten members of the susceptible group and cause them to aggressively seek inoculation. So suppose that the inoculation rate is proportional to the product of the number of susceptibles and the number of infectives instead of simply proportional to the number of susceptibles. In this case, a model for an epidemic with inoculation is

$$\begin{aligned} \frac{dS}{dt} &= -\beta SI - \alpha SI \\ \frac{dI}{dt} &= \beta SI - rI \\ \frac{dR}{dt} &= rI + \alpha SI \end{aligned}$$

where α, β, and r are positive constants. Use computer software to solve the following epidemic with inoculation model

$$\begin{aligned} \frac{dS}{dt} &= -.01SI - .05SI \\ \frac{dI}{dt} &= .01SI - 3I \\ \frac{dR}{dt} &= 3I + .05SI \end{aligned}$$

on the interval $[0, 5]$ where t is in days for the initial conditions:

$$\text{(i) } S_0 = 300, \quad I_0 = 10, \quad R_0 = 0 \quad \text{(ii) } S_0 = 500, \quad I_0 = 10, \quad R_0 = 0$$

Display a graph with S, I, and R. Does an epidemic occur in case (ii)? In each case, what is $\lim_{t\to\infty} S(t)$?

4. If the disease is deadly, then the fear of the members of the susceptible group might increase the inoculation rate to the point that it is proportional to the product of the number of susceptibles and the square of the number of infectives. In this instance, a model for an epidemic with inoculation is

$$\begin{aligned} \frac{dS}{dt} &= -\beta SI - \alpha SI^2 \\ \frac{dI}{dt} &= \beta SI - rI \\ \frac{dR}{dt} &= rI + \alpha SI^2 \end{aligned}$$

where α, β, and r are positive constants.

Solve the following epidemic with the inoculation model

$$\begin{aligned} \frac{dS}{dt} &= -.01SI - .05SI^2 \\ \frac{dI}{dt} &= .01SI - 3I \\ \frac{dR}{dt} &= 3I + .05SI^2 \end{aligned}$$

on the interval $[0, 5]$ where t is in days for the initial conditions $S_0 = 500$, $I_0 = 10$, $R_0 = 0$. Display a graph with S, I, and R. Does an epidemic occur? What is $\lim_{t\to\infty} S(t)$?

5. The virus AIDS, or acquired immune deficiency syndrome, had no name when it first appeared in the United States in the late 1970s. The new disease was identified in 1981 but was not isolated and named until 1984. The AIDS virus is spread almost exclusively through sexual contact or blood contact. Today with the careful screening of all blood donations, there is little risk of contracting AIDS through blood transfusion. At present, there is no cure for AIDS. The disease seems to have begun in the United States in the homosexual male population and has since spread to the bisexual male population, the heterosexual male population and the heterosexual female population. Seventy-three percent of the AIDS victims in the United States are sexually active homosexual and bisexual males and

1% of the victims are sexually active heterosexuals. According to the Centers for Disease Control and Prevention, by the end of 2002, there were an estimated 877,275 cases of AIDS in the adult and adolescent population in the United States and an additional 9300 cases of AIDS in children under age 13. Of the adult cases, 718,002 were males and 159,271 were females. It is estimated that by the end of 2002, that a total of 501,669 Americans had died of AIDS, which includes 496,354 adults and adolescents and 5316 children under the age of 15. Stated in terms of percentages, more than 56.5% percent of the adult and adolescent American population who contracted AIDS prior to the end of 2002 had died by the end of 2002.

The following model for the spread of AIDS through sexual contact assumes the rate of increase in infection of a particular group is equal to a sum of terms each of which is proportional to the product of the number of members in the interacting groups which can lead to infection minus the removals. At time t let

S_1 be the number of homosexual males

S_2 be the number of bisexual males

S_3 be the number of heterosexual males

S_4 be the number of heterosexual females

y_1 be the number of homosexual males infected with AIDS

y_2 be the number of bisexual males infected with AIDS

y_3 be the number of heterosexual males infected with AIDS

y_4 be the number of heterosexual females infected with AIDS

A model for the spread of AIDS within these groups through sexual contact is

$$
\begin{aligned}
y_1' &= a_1 y_1(S_1 - y_1) + a_2 y_2(S_1 - y_1) - r_1 y_1 \\
y_2' &= b_1(S_2 - y_2) + b_2 y_2(S_2 - y_2) + b_3 y_4(S_2 - y_2) - r_2 y_2 \\
y_3' &= c y_4(S_3 - y_3) - r_3 y_3 \\
y_4' &= d_1 y_2(S_4 - y_4) + d_2 y_3(S_4 - y_4) - r_4 y_4
\end{aligned}
\tag{8}
$$

where a_1, a_2, b_1, b_2, b_3, c, d_1, d_2, r_1, r_2, r_3, and r_4 are positive constants. The constants r_i are the removal rates for populations y_i. The term $a_1 y_1(S_1 - y_1)$ represents the rate of increase in infection in the homosexual male population due to sexual contact between infected homosexual males y_1 and uninfected homosexual males $(S_1 - y_1)$. The term $a_2 y_2(S_1 - y_1)$ represents the rate of increase in infection in the homosexual male population due to sexual contact between infected bisexual males y_2 and uninfected homosexual males $(S_1 - y_1)$, and so forth.

a. Show if all removal rates are zero (that is, if $r_1 = r_2 = r_3 = r_4 = 0$), the only critical points are $(y_1, y_2, y_3, y_4) = (0, 0, 0, 0)$ and $(y_1, y_2, y_3, y_4) = (S_1, S_2, S_3, S_4)$. That is, if there are no removals due to death, isolation, or recovery, then either no one has AIDS or eventually everyone contracts AIDS.

b. Use computer software to solve system (8) on the interval $[0, 1]$ for $S_1 = 100$, $S_2 = 50$, $S_3 = 1000$, $S_4 = 1300$, $a_1 = .1$, $a_2 = b_1 = .05$, $b_2 = b_3 = c = d_1 = d_2 = .01$, and $r_1 = r_2 = r_3 = r_4 = .02$ for the initial conditions $y_1(0) = 1$, $y_2(0) = y_3(0) = y_4(0) = 0$. Display y_1, y_2, y_3, and y_4 on a single graph.

10.7 Pendulums

In this section, we will determine critical points for several kinds of pendulums and pendulum systems and study their behavior by examining phase-plane graphs. Since electrical and other mechanical systems give rise to similar systems of differential equations, the results which we obtain here for pendulums will apply to those electrical and mechanical systems also.

10.7.1 Simple Pendulum

A simple pendulum consists of a rigid, straight rod of negligible mass and length ℓ with a bob of mass m attached at one end. The other end of the rod is attached to a fixed support, S, so that the pendulum is free to move in a vertical plane. Let y denote the angle (in radians) which the rod makes with the vertical extending downward from S—an equilibrium position of the system. We arbitrarily choose y to be positive if the rod is to the right of the downward vertical and negative if the rod is to the left of the downward vertical as shown in Figure 10.21.

We will assume the only forces acting on the pendulum are the force of gravity, a force due to air resistance which is proportional to the angular velocity of the bob, and an external force, $f(t)$, acting on the pendulum system. Under these assumptions it can be shown by applying Newton's second law of motion that the position of the pendulum satisfies the initial value problem

$$(1) \qquad my'' + cy' + k \sin y = f(t); \qquad y(0) = c_0, \quad y'(0) = c_1$$

where $c \geq 0$ is a constant of proportionality, $k = mg/\ell$, g is the constant of acceleration due to gravity, c_0 is the initial angular displacement of the pendulum from the downward vertical, and c_1 is the initial angular velocity of the pendulum.

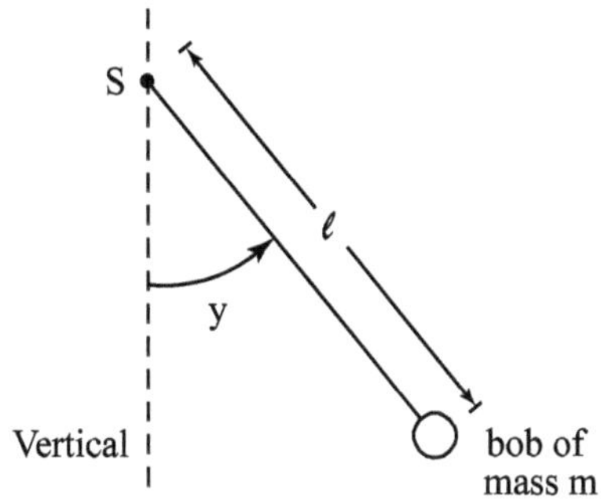

FIGURE 10.21
Simple Pendulum.

10.7.2 Linearized Simple Pendulum with No Damping and No Forcing Function

If we assume there is no damping ($c = 0$) and no forcing function ($f(t) = 0$), then the differential equation of (1) becomes

$$my'' + \frac{mg}{\ell} \sin y = 0.$$

Dividing this equation by m and choosing $\ell = g$ in magnitude, we obtain the nonlinear differential equation

(2) $$y'' + \sin y = 0.$$

This equation is nonlinear because of the factor $\sin y$. The Taylor series expansion of $\sin y$ about $y = 0$ is

$$\sin y = y - \frac{y^3}{3!} + \frac{y^5}{5!} - \frac{y^7}{7!} + \cdots .$$

From this expansion, we see that an approximation which can be made in order to linearize the differential equation (2) is to replace $\sin y$ by y. This approximation is valid only for small angles, say $|y| < .1$ radians $= 5.73°$. So for small y, the solution of the linear initial value problem

(3) $$y'' + y = 0; \qquad y(0) = c_0, \quad y'(0) = c_1$$

approximately describes the motion of a simple pendulum. Letting $y_1 = y$ and $y_2 = y'$, we can rewrite the initial value problem (3) as the following equivalent vector initial value problem

(4) $$\begin{aligned} y_1' &= y_2; & y_1(0) &= c_0 \\ y_2' &= -y_1; & y_2(0) &= c_1. \end{aligned}$$

In system (4), y_1 is the angular displacement of the pendulum and y_2 is its angular velocity. Thus, $y_1(t)$ is the solution of the IVP (3) and $y_2(t)$ is the derivative of the solution. Simultaneously setting $y_2 = 0$ and $-y_1 = 0$, we find that the only critical point of system (4) is the origin. The general solution of (4) is

$$\begin{aligned} y_1(t) &= c_0 \cos t + c_1 \sin t \\ y_2(t) &= -c_0 \sin t + c_1 \cos t. \end{aligned}$$

(Verify this fact.) Since this solution is periodic with period 2π, the origin is a neutrally stable critical point. A phase-plane graph of y_2 versus y_1 (the angular velocity of the pendulum versus its angular displacement) for the initial conditions: (i) $y_1(0) = 0$, $y_2(0) = 1.5$, (ii) $y_1(0) = 0$, $y_2(0) = 2$, and (iii) $y_1(0) = 0$, $y_2(0) = 2.5$ is shown in Figure 10.22. The given initial conditions correspond to the pendulum being at rest in the downward vertical position at time $t = 0$ and being struck sharply to impart an initial angular velocity. The maximum angular displacement, $\max |y_1(t)|$, for the given three initial conditions is 1.5, 2, and 2.5, respectively. Notice that these values are much larger than .1—the maximum value for which we assumed the approximation of $\sin y$ by y to be valid. The reason for choosing initial conditions so large that $\max |y_1(t)|$ is much larger than .1 is so that the linear approximation $\sin y \approx y$ is not valid. Then we can compare the solutions of the linear system (4) with the solutions of the corresponding nonlinear system (6) which we obtain next.

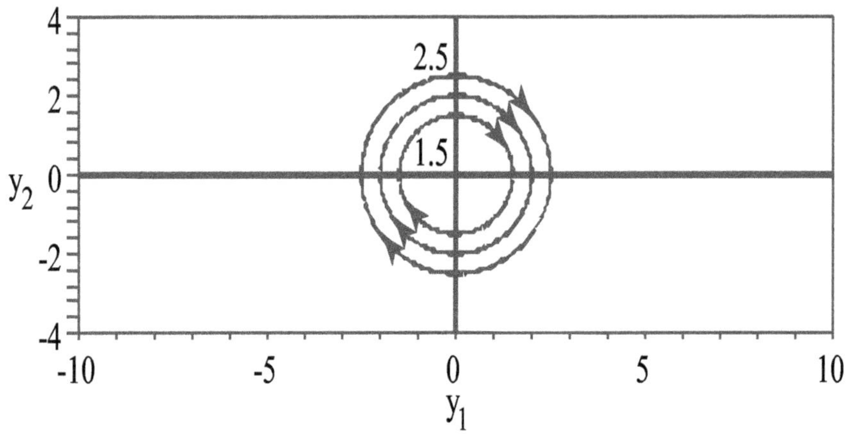

FIGURE 10.22
Phase-Plane Portrait for System (4).

Simple Pendulum with No Damping and No Forcing Function

If we assume there is no damping ($c = 0$) and no forcing function ($f(t) = 0$) and if we choose $\ell = g$ in magnitude, then the equation of motion for a simple pendulum is

$$(5) \qquad y'' + \sin y = 0; \qquad y(0) = c_0, \quad y'(0) = c_1.$$

The general solution of this initial value problem involves elliptic integrals. Letting $y_1 = y$ and $y_2 = y'$ as before, we can rewrite (5) as

$$(6) \qquad \begin{aligned} y_1' &= y_2; & y_1(0) &= c_0 \\ y_2' &= -\sin y_1; & y_2(0) &= c_1. \end{aligned}$$

Simultaneously setting $y_2 = 0$ and $-\sin y_1 = 0$, we find system (6) has an infinite number of critical points located at $(n\pi, 0)$ where n is any integer. Physically the critical points $(2n\pi, 0)$ correspond to the pendulum being at rest ($y_2 = y' = 0$) and hanging vertically downward from the support S, ($y_1 = y = 2n\pi$). Likewise, the physical interpretation of the critical points $((2n+1)\pi, 0)$ is that the pendulum is at rest ($y_2 = 0$) balanced vertically above the support S, ($y_1 = (2n+1)\pi$). From these physical considerations we deduce that the critical points $(2n\pi, 0)$ are stable and the critical points $((2n+1)\pi, 0)$ are unstable. Displayed in Figure 10.23 is a phase-plane graph of y_2 versus y_1 for system (6) for the initial conditions $y_1(0) = 0$ and (i) $y_2(0) = 1.5$, (ii) $y_2(0) = 2$, (iii) $y_2(0) = 2.5$, (iv) $y_2(0) = -2$, and (v) $y_2(0) = -2.5$. From this figure we see that if the pendulum is at rest in a vertically downward position at time $t = 0$ and is struck sharply from the left side, imparting an angular velocity of 1.5 ($y'(0) = 1.5$), then the pendulum executes periodic motion (harmonic motion) about the stable critical point $(0, 0)$. If the initial angular velocity is 2, then the pendulum swings to the right ($y'(0) = 2 > 0$) and balances itself vertically above the support. That is, the solution approaches the unstable critical point $(\pi, 0)$. (Of course, this is not apt to happen in any laboratory experiment!) If the initial angular velocity is 3, then the pendulum rotates indefinitely in a counterclockwise direction about the support. The pendulum has the smallest angular velocity when it is vertically above the support and it has the largest angular velocity when it is vertically below the support. When $y_2(0) = -2$ the pendulum swings to the left and balances itself above the support. And when $y_2(0) = -3$ the pendulum rotates indefinitely in a clockwise direction about the support. Compare these results with those obtained previously when the linear approximation $\sin y \approx y$ was made—that is, compare Figures 10.22 and 10.23.

Simple Pendulum with Damping but No Forcing Function

If we assume there is a damping force due to air resistance or friction at the point of suspension which is proportional to the angular velocity ($c \neq 0$) and

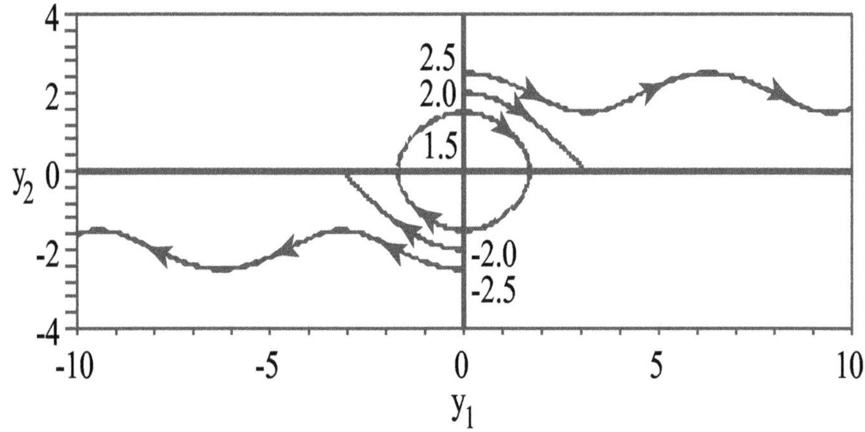

FIGURE 10.23
Phase-Plane Portrait for System (6).

if we assume there is no forcing function ($f(t) = 0$), then the differential equation (1) for the motion of a simple pendulum is

$$my'' + cy' + \frac{mg}{\ell}\sin y = 0.$$

Dividing this equation by m, choosing $\ell = g$ in magnitude, and setting $C = c/m$, this equation becomes

$$y'' + Cy' + \sin y = 0.$$

Letting $y_1 = y$ and $y_2 = y'$, we obtain the following equivalent system of first-order equations

(7)
$$\begin{aligned} y_1' &= y_2 \\ y_2' &= -\sin y_1 - Cy_2. \end{aligned}$$

10.7.3 Simple Pendulum with Constant Forcing Function

If we assume the forcing function $f(t) = a$, a constant, then from (1) the equation of motion for the pendulum is

$$my'' + cy' + \frac{mg}{\ell}\sin y = a.$$

Dividing this equation by m, setting $\ell = g$ in magnitude, and letting $C = c/m$ and $A = a/m$, we can write this equation as

$$y'' + Cy' + \sin y = A$$

or as the equivalent first-order system

$$(8)\qquad \begin{aligned} y_1' &= y_2 \\ y_2' &= -\sin y_1 - Cy_2 + A \end{aligned}$$

where $y_1 = y$ and $y_2 = y'$.

10.7.4 Variable Length Pendulum

A variable-length pendulum consists of a mass m attached to one end of an inextensible string (a string which does not stretch). The string is pulled over some support, S, such as a thin rod. The length of the pendulum, $\ell(t)$, can be varied with time by pulling on the string. A diagram of a variable-length pendulum is shown in Figure 10.24. As before we let y denote the angle the pendulum makes with the downward vertical from the support S. The equation of motion for this pendulum is

$$(9)\qquad m\ell(t)y'' + (2m\ell'(t) + c\ell(t))y' + mg\sin y = f(t)$$

where m is the mass of the bob, $\ell(t)$ is the length of the pendulum, c is the damping constant, g is the gravitational constant, and $f(t)$ is the forcing function. Dividing by $m\ell(t)$ and letting $y_1 = y$ and $y_2 = y'$, we can rewrite (9) as the following equivalent first-order system

$$(10)\qquad \begin{aligned} y_1' &= y_2 \\ y_2' &= -(2\frac{\ell'(t)}{\ell(t)} + \frac{c}{m})y_2 - \frac{g}{\ell(t)}\sin y_1 + \frac{f(t)}{m\ell(t)}. \end{aligned}$$

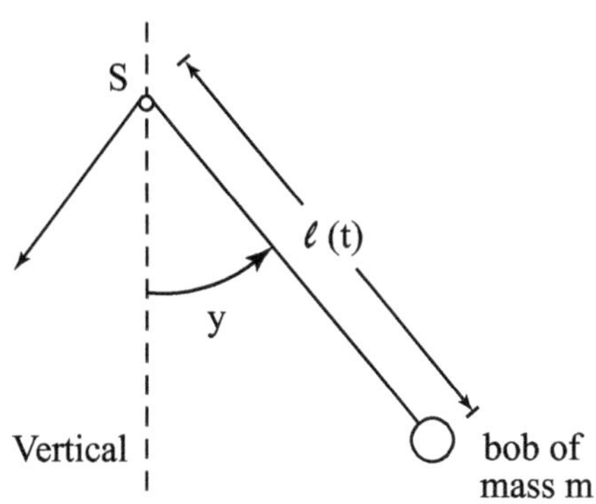

FIGURE 10.24
Variable Length Pendulum.

10.7.5 Foucault Pendulum

The French astronomer and physicist, Jean Bernard Léon Foucault is, perhaps, best remembered as the first person to demonstrate the rotation of the earth without using a point of reference outside the earth—such as stars or the sun. In 1851, Foucault suspended a 62 pound ball on a 220 feet long steel wire from the dome of the Pantheon in Paris. A pin protruded from the bottom of the ball and was adjusted to draw a mark through a circle of wet sand beneath the ball. Foucault pulled the ball to one side of the circle of sand and released it. With each swing the pendulum made a mark in the sand and appeared to rotate in a clockwise direction. (The pendulum only appears to change its plane of oscillation while swinging. It is actually the floor under the pendulum that moves counterclockwise due to the rotation of the earth.) Thus, Foucault demonstrated his prediction that the pendulum would revolve approximately 270° in a 24 hour period and proved that the earth revolved. Foucault presented an intuitive explanation for the motion of his pendulum. When damping is absent or compensated for, the equations of motion of a Foucault pendulum are

$$\begin{aligned} x'' &= 2\omega y' \sin\phi - \frac{gx}{\ell} \\ y'' &= -2\omega x' \sin\phi - \frac{gy}{\ell} \end{aligned} \tag{11}$$

where $\omega = 7.29 \times 10^{-5}$ rads/sec is the angular velocity of the earth's rotation, ϕ is the latitude of the pendulum, $g = 9.8$ m/sec^2 is the gravitational constant, and ℓ is the length of the pendulum. The x-axis and y-axis form a rectangular coordinate system on the floor beneath the suspended pendulum. The origin of the coordinate system lies directly below the equilibrium position of the pendulum.

10.7.6 Spring Pendulum

A spring pendulum consists of a spring of natural length L_0 suspended by one end from a fixed support S. A bob of mass m is attached to the other end of the spring. We assume the spring is stiff enough to remain straight and free to move in a vertical plane. Let $L(t)$ be the length of the spring at time t and let $\theta(t)$ be the angle (in radians) the spring makes with the downward vertical from S. See Figure 10.25.

If the spring obeys Hooke's law with spring constant k, then applying Newton's second law of motion along and perpendicular to the spring it can be shown that L and θ simultaneously satisfy the two second-order differential

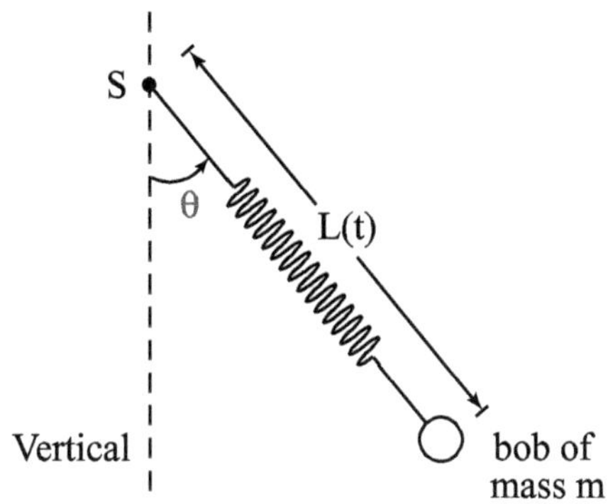

FIGURE 10.25
Spring Pendulum.

equations

$$
(12) \qquad \begin{aligned} L'' - L(\theta')^2 - g\cos\theta + \frac{k(L - L_0)}{m} &= 0 \\ L\theta'' + 2L'\theta' + g\sin\theta &= 0. \end{aligned}
$$

Letting $y_1 = L(t)$, $y_2 = L'(t)$, $y_3 = \theta(t)$, and $y_4 = \theta'(t)$, we can rewrite system (12) as the following system of four first-order differential equations

$$
(13) \qquad \begin{aligned} y_1' &= y_2 \\ y_2' &= y_1 y_4^2 + g\cos y_3 - \frac{k(y_1 - L_0)}{m} \\ y_3' &= y_4 \\ y_4' &= \frac{-2y_2 y_4 - g\sin y_3}{y_1}. \end{aligned}
$$

A spring pendulum has two "natural frequencies." The first frequency $\omega_1 = \sqrt{g/L_0}$ corresponds to the natural frequency of the pendulum with the spring replaced by a rod of fixed length, L_0. The second frequency $\omega_2 = \sqrt{k/m}$ corresponds to the natural frequency of a spring-mass system which oscillates vertically without swinging back and forth.

10.7.7 Exercises

1. All solutions of system (4), the linearized simple pendulum model with no damping and no forcing function, are periodic with period 2π. For initial conditions $y_1(0) = 0$ and $-2 < y_2(0) < 2$ the solutions of system (6) are

periodic. Is the period 2π or does the period vary with $y_2(0)$? Find out by using your computing software to solve system (6) on the interval $[0, 10]$ for $y_1(0) = 0$ and (a) $y_2(0) = .1$ and (b) $y_2(0) = 1.9$. Print solution values for $y_1(t)$ on the monitor and determine the period, or graph $y_1(t)$ and estimate the period.

2. a. Determine the critical points of system (7), the simple pendulum model with damping but no forcing function. Are they the same as for system (6)?

   b. Use your computing software to solve system (7) on the interval $[0, 10]$ for $C = .1$, $y_1(0) = 0$, and (i) $y_2(0) = 2$, (ii) $y_2(0) = 2.5$, and (iii) $y_2(0) = 3$. In each case, display a phase-plane graph of y_2 versus y_1. Does the solution corresponding to initial conditions (i) approach the unstable critical point $(\pi, 0)$ as it did in the undamped case? Give a physical interpretation for the phase-plane graph for the initial conditions (ii) and (iii). What critical point does the solution approach? What does this mean the pendulum has done?

   c. Solve system (7) on the interval $[0, 10]$ for $C = .5$, $y_1(0) = 0$, and (i) $y_2(0) = 3$ and (ii) $y_2(0) = 3.5$. In each case, display a phase-plane graph of y_2 versus y_1 and interpret the results physically.

3. a. Show that for $|A| < 1$ the critical points of system (8), the simple pendulum with constant forcing function model, are $(c_n, 0)$ and $(d_n, 0)$ where $c_n = 2n\pi + \arcsin A$, $d_n = (2n+1)\pi - \arcsin A$, and n is any integer. So, for $|A| < 1$ and $A \neq 0$, the critical points are not equally spaced along the y_1-axis. Show that as $A \to 1^-$, $c_n \to (2n\pi + \pi/2)^-$, and $d_n \to (2n\pi + \pi/2)^+$—that is, as A approaches 1 from below, the critical points $(c_n, 0)$ move toward $(2n\pi + \pi/2,\ 0)$ from the left and the critical points $(d_n, 0)$ move toward $(2n\pi + \pi/2,\ 0)$ from the right.

   b. Show that for $|A| > 1$ system (8) has no critical points.

4. Numerically solve system (8), the simple pendulum with constant forcing function model, on the interval $[0, 10]$ when there is no damping ($C = 0$) for (a) $A = -.5$ and (b) $A = -1$ for the initial conditions $y_1(0) = 0$ and (i) $y_2(0) = 2.5$ and (ii) $y_2(0) = 3.5$. Display a phase-plane graph of y_2 versus y_1 for each solution.

5. Solve system (8), the simple pendulum with constant forcing function model, on the interval $[0, 10]$ when $C = .5$ for (a) $A = -.5$ and (b) $A = -1$ for the initial conditions $y_1(0) = 0$ and (i) $y_2(0) = 2.5$ and (ii) $y_2(0) = 3.5$. Display a phase-plane graph of y_2 versus y_1 for each solution.

6. Use computer software to solve system (10), the variable length pendulum model, on the interval $[0, 10]$ with no damping ($c = 0$) and no forcing function ($f(t) = 0$) for the initial conditions $y_1(0) = 0$, $y_2(0) = 1.5$ for $\ell(t) = g(1 + .1 \sin \omega t)$ where (i) $\omega = .5$, (ii) $\omega = .75$, and (iii) $\omega = .9$. In each case, display a graph of $y_1(t)$. Is the solution periodic? Display a phase-plane graph of y_2 versus y_1.

7. Numerically solve system (10), the variable length pendulum model, on the interval $[0, 10]$ with no forcing function ($f(t) = 0$), with damping coefficient $c = .1m$, and with $\ell(t) = g(1+.1 \sin t)$ for the initial conditions $y_1(0) = 0$ and (i) $y_2(0) = 1.5$ and (ii) $y_2(0) = 2.5$. Display graphs of $y_1(t)$ and phase-plane graphs of y_2 versus y_1.

8. Solve system (10), the variable length pendulum model, on the interval $[0, 10]$ for the initial conditions $y_1(0) = 0$ and $y_2(0) = 2.5$ when $\ell(t) = g(1+.1 \sin t)$, when (i) $c = 0$ and (ii) $c = .1m$, and $f(t) = mg \sin \beta t$ where (a) $\beta = .5$ and (b) $\beta = 1$. Display graphs of $y_1(t)$ and phase-plane graphs of y_2 versus y_1.

9. Let $y_1 = x$, $y_2 = x'$, $y_3 = y$, and $y_4 = y'$. Write system (11), the Foucault pendulum model, as a system of four first-order differential equations. Numerically solve the resulting system on the interval $[0, 10]$ for initial conditions $y_1(0) = 1$, $y_2(0) = 0$, $y_3(0) = 0$, and $y_4(0) = 0$ if $\ell = 9.8$ m and (a) $\phi = 0°$, (b) $\phi = 45°$, (c) $\phi = -45°$, (d) $\phi = 60°$, and (e) $\phi = 90°$. The period of a pendulum with length $\ell = 9.8$ m is 2π, or approximately 6.28 seconds. From the value of $y_3(6.28)$ estimate how long it will take the plane of swing of the pendulum to appear to rotate 360° in cases (a)-(e).

10. When the spring pendulum is at rest in the equilibrium position $y_1 = L_0 + mg/k$ and $y_2 = y_3 = y_4 = 0$. Solve system (13) on the interval $[0, 10]$ for $L_0 = 1.02$ decimeters, $g = .98$ decimeters/second^2, and $m/k = 1$ second^2 for the following initial conditions:

    (i) $y_1(0) = 2, \; y_2(0) = 0, \; y_3(0) = .1$, and $y_4(0) = 0$.
    (The spring pendulum is set into motion by pulling the mass to the right so that $\theta = .1$ radian and then releasing the mass without compressing or elongating the spring.)

    (ii) $y_1(0) = 2.1, \; y_2(0) = 0, \; y_3(0) = .05$, and $y_4(0) = 0$.
    (The spring pendulum is set into motion by pulling the mass down .1 decimeter from its equilibrium position and moving it to the right so that $\theta = .05$ radians and then releasing the mass.)

    For both sets of initial conditions display a graph of the length of the spring as a function of time—$y_1(t) = L(t)$; display a graph of the angle the spring makes with the vertical as a function of time—$y_3(t) = \theta(t)$; display a phase-plane graph of y_2 versus y_1; and display a phase-plane graph of y_4 versus y_3. Is the motion of the spring pendulum periodic?

10.8 Duffing's Equation (Nonlinear Spring-Mass Systems)

In Chapter 6 we saw that the equation of motion for a mass on a spring which satisfied Hooke's law (a **linear spring**) is

$$my'' + ky = 0 \tag{1}$$

where m is the mass and $k > 0$ is the spring constant. The restoring force for linear springs is ky. Springs which do not obey Hooke's law are called **nonlinear springs**. The restoring force for hard nonlinear springs is $ky + py^3$ where $p > 0$ while the restoring force for soft nonlinear springs is $ky - py^3$. Thus, the equation of motion for a nonlinear spring is

$$my'' + ky + py^3 = 0 \tag{2}$$

where m and k are positive constants and $p \neq 0$. Adding an external force of the form $a \sin \omega t$ to drive the spring-mass system, we obtain the equation

$$my'' + ky + py^3 = a \sin \omega t. \tag{3}$$

Dividing by m, letting $K = k/m$, $P = p/m$ and $A = a/m$, we obtain Duffing's equation for the motion of a nonlinear spring with a periodic forcing function

$$y'' + Ky + Py^3 = A \sin \omega t. \tag{4}$$

Including a damping term which is proportional to the velocity of the mass, yields

$$y'' + Cy' + Ky + Py^3 = A \sin \omega t \tag{5}$$

where the damping constant is $C \geq 0$.

10.8.1 Exercise

1. Let $y_1 = y$ and $y_2 = y'$ and write equation (5) as an equivalent system of two first-order differential equations.

   For each of the following cases, display a graph of $y_1(t) = y(t)$ on $[0, 10]$ and display a phase-plane graph of y_2 versus y_1.

   a. Let $P = 0$ (a linear spring) and $K = A = \omega = 1$. Use your computing software to solve the system on the interval $[0, 10]$ for the initial conditions $y_1(0) = 1$ and $y_2(0) = 0$ with damping constants (i) $C = 0$ (no damping) and (ii) $C = .5$.

   b. Let $P = .1$ (a hard nonlinear spring) and $K = A = \omega = 1$. Numerically solve the system on the interval $[0, 10]$ for the initial conditions $y_1(0) = 1$ and $y_2(0) = 0$ with damping constants (i) $C = 0$ and (ii) $C = .5$.

c. Let $P = -.1$ (a soft nonlinear spring) and $K = A = \omega = 1$. Solve the system on the interval $[0, 10]$ for the initial conditions $y_1(0) = 1$ and $y_2(0) = 0$ with damping constants (i) $C = 0$ and (ii) $C = .5$.

10.9 Van Der Pol's Equation

In 1921, E. V. Appleton and B. van der Pol initiated research on the oscillations produced by electrical circuits which contain triode generators. Their research led to the study of the following nonlinear differential equation, now known as **van der Pol's equation**

$$x'' + \epsilon(x^2 - 1)x' + x = 0. \tag{1}$$

During 1926-27, van der Pol developed methods for solving equation (1). The original electrical circuits studied by Appleton and van der Pol in the 1920s contained vacuum tubes. Today circuits which produce similar oscillations occur on semiconductor devices. Van der Pol's equation also arises quite often in nonlinear mechanics.

Letting $y_1 = x$ and $y_2 = x'$, we can rewrite equation (1) as the following equivalent system of two first-order equations

$$\begin{aligned} y_1' &= y_2 \\ y_2' &= -\epsilon(y_1^2 - 1)y_2 - y_1. \end{aligned} \tag{2}$$

10.9.1 Exercises

1.Numerically solve system (2) on the interval $[0, 10]$ for (a) $\epsilon = .5$ and (b) $\epsilon = 2$ for the initial conditions $y_2(0) = 0$ and (i) $y_1(0) = .5$, (ii) $y_1(0) = 2$, and (iii) $y_1(0) = 3$. In each case, display a phase-plane graph of y_2 versus y_1. Notice in cases (a) and (b), there is a periodic solution to van der Pol's equation. This periodic solution is called a **limit cycle**, since solutions which begin inside the periodic solution "spiral outward" toward the periodic solution and solutions which begin outside the periodic solution "spiral inward" toward the periodic solution.

10.10 Mixture Problems

We introduced and studied mixture problems in Chapters 3 and 9. Recall the following underlying assumptions for solving mixture problems:

> The rate of change of the amount of substance in any container at time t is equal to the sum over all inputs to the container of the concentration of each input times the rate of input minus the concentration of the substance in the particular container times the sum of the rates of flow of outputs from the container.

Example 1 A Three Pond Mixture Problem
Three small ponds containing 200 gallons of pure water are formed by a spring rain. Water containing .73 lbs/gallon of salt enters pond A at the rate of 100 gal/hr. Water evaporates from pond A at the rate of 30 gal/hr and flows into pond B at 70 gal/hr. Water evaporates from pond B at 20 gal/hr and flows into pond C at 50 gal/hr. Water evaporates from pond C at 25 gal/hr and flows out of the system at 25 gal/hr. Find the amount of salt in each pond as a function of time. What is the limiting amount of salt in each pond?

Solution

Let $y_1(t)$ be the amount of salt in pond A at time t. The concentration of salt in the inflow to pond A is $c_i = .73$ lbs/gal and the inflow rate is $r_i = 100$ gal/hr. Since the evaporation rate plus the outflow rate is also 100 gal/hr, the volume of pond A remains constant. Hence, the concentration of salt in the outflow from pond A to pond B at any time is $c_o = y_1(t)/200$ (lbs/gal). Applying our underlying principle to pond A, we find

$$y_1'(t) = c_i r_i - c_o r_o$$

where the outflow rate from pond A to pond B is $r_o = 70$ gal/hr. Thus, the amount of salt in pond A satisfies

$$y_1'(t) = (.73)(100) - \left(\frac{y_1}{200}\right)(70) = 73 - .35y_1.$$

A similar analysis shows that the amount of salt, $y_2(t)$, in pond B satisfies

$$y_2'(t) = \left(\frac{y_1}{200}\right)(70) - \left(\frac{y_2}{200}\right)(50) = .35y_1 - .25y_2$$

and the amount of salt, $y_3(t)$, in pond C satisfies

$$y_3'(t) = \left(\frac{y_2}{200}\right)(50) - \left(\frac{y_3}{200}\right)(25) = .25y_2 - .125y_3.$$

Initially, all three ponds contain no salt, so the initial conditions are $y_1(0) = 0$, $y_2(0) = 0$, and $y_3(0) = 0$. We used our computing software to solve the system

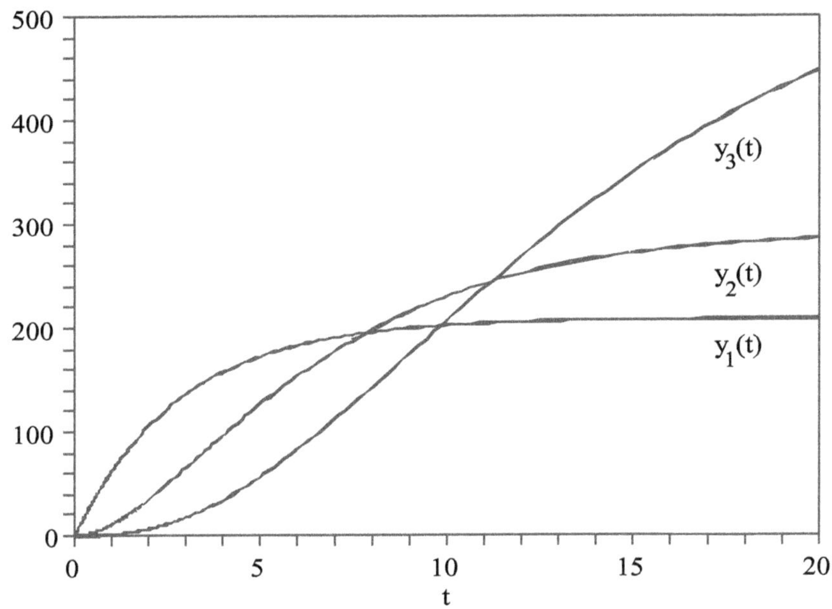

FIGURE 10.26
Solution Graph for System (1).

initial value problem

$$(1)\qquad \begin{aligned} y_1' &= 73 - .35y_1; & y_1(0) &= 0 \\ y_2' &= .35y_1 - .25y_2; & y_2(0) &= 0 \\ y_3' &= .25y_2 - .125y_3; & y_3(0) &= 0 \end{aligned}$$

on the interval $[0, 20]$. A graph showing the amount of salt in each pond on the interval $[0, 20]$ is shown in Figure 10.26. ■

The amount of salt in pond A begins to increase first. The amount of salt in pond B begins to increase next and finally the amount of salt in pond C begins to increase. Equating the right-hand sides of the three differential equations in system (1) to zero, we find the critical point of system (1) to be $(208.57, 292, 584)$. This critical point is asymptotically stable, so as $t \to \infty$, $y_1 \to 208.57$, $y_2 \to 292$ and $y_3 \to 584$.

Pollution in the Great Lakes In Chapter 9 we derived the following equations for the concentration of pollution in the Great Lakes. (See equation 19 and the accompanying table of constants in Chapter 9.)

$$
(2) \qquad
\begin{aligned}
y_1' &= \frac{15C_1 - 15y_1}{2900} \\
y_2' &= \frac{38C_2 - 38y_2}{1180} \\
y_3' &= \frac{15C_3 + 15y_1 + 38y_2 - 68y_3}{850} \\
y_4' &= \frac{17C_4 + 68y_3 - 85y_4}{116} \\
y_5' &= \frac{14C_5 + 85y_4 - 99y_5}{393}
\end{aligned}
$$

Here $y_i(t)$ is the concentration of pollution in lake i at time t and C_i is the concentration of pollutant in the inflow to lake i. Subscript 1 corresponds to Lake Superior, subscript 2 corresponds to Lake Michigan, subscript 3 corresponds to Lake Huron, subscript 4 corresponds to Lake Erie, and subscript 5 corresponds to Lake Ontario. In system (2) the unit of measure of the independent variable, x (time), is years.

10.10.1 Exercise

1. Solve system (2) on the interval $[0, 200]$, assuming all the Great Lakes have the same initial pollution concentration of $y_i(0) = .5\%$ and

   a. the inflow concentration of pollutant for all lakes is reduced to $C_i = .2\%$.

   b. the inflow concentration of pollutant for all lakes is reduced to zero—$C_i = 0$.

(Enter $y_i(0)$ as .5 and enter C_i as .2 or 0, so your results will be expressed in percent. Otherwise, you may experience some numerical difficulties.)

Display a graph for the concentration of pollution in all lakes for both cases and compare the results. In each case, how long does it take for the concentration of pollution in Lake Ontario to be reduced to .3%? Which lake has its level of pollution reduced to .25% first? last?

10.11 The Restricted Three-Body Problem

The general n-body problem of celestial mechanics is to determine the position and velocity of n homogeneous spherical bodies for all time, given their initial positions and velocities. Although no general closed-form solution to this problem exists, much is known regarding various special cases when n is small. The two-body problem for spheres of finite size was solved by Isaac Newton in about 1685. A discussion of this problem as well as the first treatment of the three-body problem appears in Book I of Newton's *Principia*. The restricted three-body problem considers the properties of motion of an infinitesimally small body when it is attracted by two finite bodies which revolve in circles about their center of mass and when the infinitesimally small body remains in the plane of motion of the other two bodies. This problem was first discussed by Joseph Louis Lagrange (1736-1813) in his prize winning memoir *Essai sur le Problme des Trois Corps*, which he submitted to the Paris Academy of Sciences in 1772.

For our purposes we can think of the infinitesimally small mass as a satellite, space-station, or spaceship and the two large bodies as the earth and moon or as the Sun and Jupiter. To be specific, let us consider an earth-moon-spaceship system. Let E denote the mass of the earth and M denote the mass of the moon. The unit of mass is chosen to be the sum of the masses of the earth and moon—hence, $E + M = 1$. It is customary to let μ represent the mass of the smaller of the two large bodies, so $\mu = M$. The distance between the two large masses—E and M, in this instance—is selected as the unit of length. The unit of time is chosen so that the gravitational constant is 1. This choice means the two large bodies complete one circular revolution in 2π units of time. We will present the equations of motion of the infinitesimally small body (the spaceship) in a special two-dimensional rectangular coordinate system—the **barycentric coordinate system**. The origin of this system is at the center of mass of the earth and moon. The x-axis passes through the gravitational centers of the earth and moon. In this coordinate system, the earth is located at $(-\mu, 0)$ and the moon is located at $(1 - \mu, 0)$. The location of the spaceship is $(x(t), y(t))$. See Figure 10.27.

Using Newton's second law of motion and the inverse square law of motion, it can be shown that the equations of motion for the spaceship are

$$\text{(1)} \qquad \begin{aligned} x'' &= 2y' + x - \frac{(1-\mu)(x+\mu)}{r^3} - \frac{\mu(x-1+\mu)}{s^3} \\ y'' &= -2x' + y - \frac{(1-\mu)y}{r^3} - \frac{\mu y}{s^3} \end{aligned}$$

where $r = ((x+\mu)^2 + y^2)^{1/2}$ and $s = ((x-1+\mu)^2 + y^2)^{1/2}$. Thus, r is the distance of the spaceship from the earth and s is its distance from the moon. If we let $y_1 = x$, $y_2 = x'$, $y_3 = y$, and $y_4 = y'$, we obtain the following equivalent system of four first-order equations

$$\text{(2.1)} \quad y_1' = y_2$$

$$\text{(2.2)} \quad y_2' = 2y_4 + y_1 - \frac{(1-\mu)(y_1+\mu)}{((y_1+\mu)^2 + y_3^2)^{3/2}} - \frac{\mu(y_1 - 1 + \mu)}{((y_1 - 1 + \mu)^2 + y_3^2)^{3/2}}$$

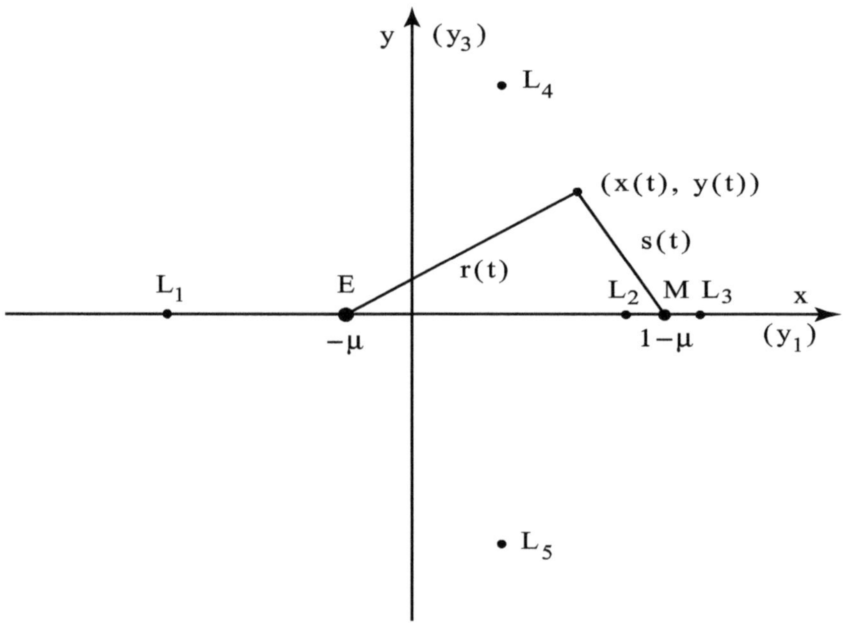

FIGURE 10.27
The Restricted Three-Body Problem and Associated Critical Points.

(2.3) $y_3' = y_4$

(2.4) $$y_4' = -2y_2 + \left[1 - \frac{1-\mu}{((y_1+\mu)^2 + y_3^2)^{3/2}} - \frac{\mu}{((y_1 - 1 + \mu)^2 + y_3^2)^{3/2}}\right] y_3.$$

The critical points of system (2) can be found by setting the right-hand sides of equations (2.1)-(2.4) equal to zero. Doing so, we find from (2.1) that $y_2 = 0$ and from (2.3) that $y_4 = 0$. Substituting $y_2 = 0$ into the right-hand side of equation (2.4), we see either (i) $y_3 = 0$ or (ii) the quantity in square brackets in (2.4) is zero. Assuming (i) $y_3 = 0$ and substituting into the right-hand side of (2.2), we find y_1 must satisfy

(3) $$f(z) = z - \frac{(1-\mu)(z+\mu)}{|z+\mu|^3} - \frac{\mu(z+\mu-1)}{|z+\mu-1|^3} = 0$$

for $(y_1, 0, 0, 0)$ to be a critical point of system (2). The function $f(z)$ is defined and continuous on $(-\infty, \infty)$ for $z \neq -\mu$ and $z \neq 1-\mu$. At $z = -\mu$ and $z = 1-\mu$, the function $f(z)$ has a vertical asymptote. Figure 10.28 is a graph of $f(z)$ on the interval $[-2.5, 2.5]$ for $\mu = .012129$. Notice that $f(z)$ has three real zeros—one in the interval $(-\infty, -\mu)$, since $\lim_{t\to-\infty} f(z) = -\infty$; one in the interval $(-\mu, 1-\mu)$; and one in the interval $(1-\mu, \infty)$, since $\lim_{t\to\infty} f(z) = \infty$.

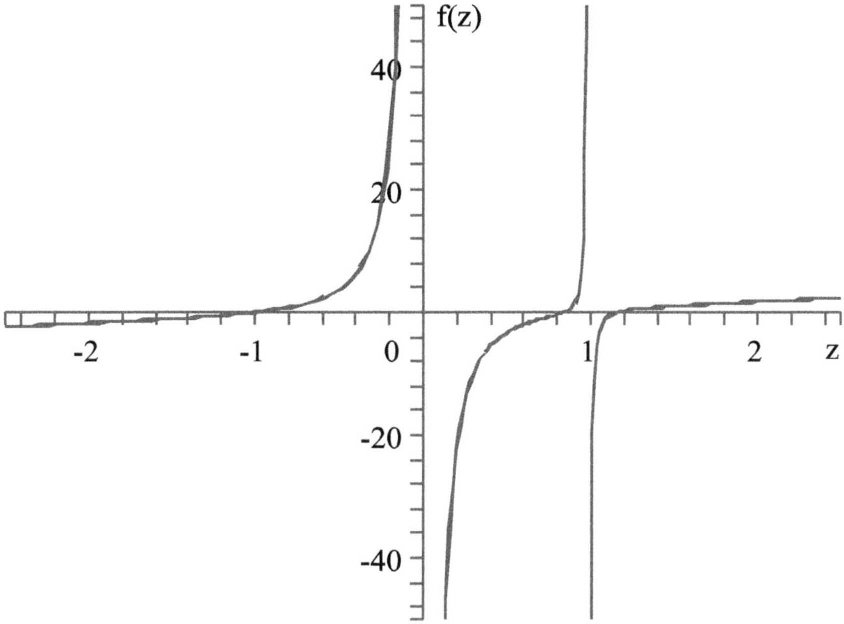

FIGURE 10.28
A Graph of Equation (3) for $\mu = .012129$.

For $z < -\mu$, $|z+\mu| = -(z+\mu)$ and $|z+\mu-1| = -(z+\mu-1)$. Substituting these expressions into equation (3), multiplying by $(z+\mu)^2(z+\mu-1)^2$ and simplifying, we find z must satisfy the quintic equation

$$(4) \qquad z^5 + 2(2\mu-1)z^4 + (6\mu^2 - 6\mu + 1)z^3 + (4\mu^3 - 6\mu^2 + 2\mu + 1)z^2 \\ +(\mu^4 - 2\mu^3 + \mu^2 + 4\mu - 2)z + (3\mu^2 - 3\mu + 1) = 0.$$

For $\mu = .012129$ this equation is

$$(5) \; z^5 - 1.951484z^4 + .9281087z^3 + 1.023382z^2 - 1.951340z + .9640543 = 0.$$

We used a computing program to solve this equation. There is one real root, $a = -1.00505$, and two pairs of complex conjugate roots. Thus, system (2) has a critical point at $(y_1, y_2, y_3, y_4) = (-1.00505, 0, 0, 0)$. This critical point is represented in the xy-plane (the y_1y_3-plane) shown in Figure 10.27 by L_1 : $(a, 0)$. For $-\mu < z < 1-\mu$, $|z+\mu| = z+\mu$ and $|z+\mu-1| = -(z+\mu-1)$. Substituting these expressions into equation (3), multiplying by $(z+\mu)^2(z+\mu-1)^2$ and simplifying, we see z must satisfy

$$(6) \qquad z^5 + 2(2\mu-1)z^4 + (6\mu^2 - 6\mu + 1)z^3 + (4\mu^3 - 6\mu^2 + 4\mu - 1)z^2 \\ +(\mu^4 - 2\mu^3 + 5\mu^2 - 4\mu + 2)z + (2\mu^3 - 3\mu^2 + 3\mu - 1) = 0.$$

For $\mu = .012129$, this equation is

$$(7)\ z^5 - 1.951484z^4 + .9281087z^3 - .9523595z^2 + 1.952216z - .9640508 = 0.$$

We used a computing software to solve equation (7). We found a single real root of $b = .837022$ and two pairs of complex conjugate roots. Hence, there is a critical point of system (2) at $(b, 0, 0, 0)$. This critical point is represented in the y_1y_3-plane shown in Figure 10.27 by $L_2 : (b, 0)$.

For $z > 1 - \mu$, $|z + \mu| = z + \mu$ and $|z + \mu - 1| = z + \mu - 1$. Substituting into equation (3) and simplifying, we find z must satisfy

$$(8) \qquad z^5 + 2(2\mu - 1)z^4 + (6\mu^2 - 6\mu + 1)z^3 + (4\mu^3 - 6\mu^2 + 2\mu - 1)z^2$$
$$+(\mu^4 - 2\mu^3 + \mu^2 - 4\mu + 2)z - (3\mu^2 - 3\mu + 1) = 0.$$

For $\mu = .012129$ this equation is

$$(9)\ z^5 - 1.951484z^4 + .9281087z^3 - .9766175z^2 + 1.951628z - .9640543 = 0.$$

Using the computer algebra system to solve equation (9), we found the single real root $c = 1.15560$ and two pairs of complex conjugate roots. The critical point $(c, 0, 0, 0)$ of system (2) is represented in Figure 10.27 by $L_3 : (c, 0)$.

The three critical points of system (2) corresponding to the points L_1, L_2, and L_3 of Figure 10.27 were first discovered by Leonhard Euler (1707-1783). They are all unstable critical points. In 1772, Lagrange discovered the stable critical points L_4 and L_5 shown in Figure 10.27. The three points E, M, and L_4 form an equilateral triangle as do the three points E, M, and L_5. Hence, L_4 is located at $((1 - 2\mu)/2, \sqrt{3}/2)$ and L_5 is located at $((1 - 2\mu)/2, -\sqrt{3}/2)$ in y_1y_3-space.

10.11.1 Exercises

1. Verify that $((1 - 2\mu)/2,\ 0,\ \pm\sqrt{3}/2,\ 0)$ are critical points of system (2).

2. Find the y_1-coordinate of L_1, L_2, and L_3 for the Sun-Jupiter system given that $\mu = .001$. (The Trojan asteroids drift around the stable critical points L_4 and L_5 of the Sun-Jupiter system.)

3. Generate numerical solutions to system (2) on the interval $[0, 10]$ for the initial conditions:

   a. Near L_1 with zero velocity: $y_2(0) = y_3(0) = y_4(0) = 0$

   (i) $y_1(0) = -1.1$ (ii) $y_1(0) = -.9$

   b. Near L_2 with zero velocity: $y_2(0) = y_3(0) = y_4(0) = 0$

   (i) $y_1(0) = .8$ (ii) $y_1(0) = .9$

c. Near L_3 with zero velocity: $y_2(0) = y_3(0) = y_4(0) = 0$

(i) $y_1(0) = 1.1$ (ii) $y_1(0) = 1.2$

d. Near L_4 with zero velocity: $y_1(0) = .5$, $y_2(0) = 0$, $y_3(0) = .9$, $y_4(0) = 0$

In each case, display a phase-plane graph of y_3 versus y_1 (y versus x). What happens to the spaceship in each instance? (All five points L_1, L_2, L_3, L_4, and L_5 have been considered as sites for locating permanent space stations. Which locations do you think would be better? Why?)

4. In this example a spaceship is initially on the side of the earth opposite the moon. The pilot shuts off the rocket engines to begin a non-powered flight. The purpose of the exercise is to see the effects of the "burnout" position and velocity on the trajectory of the spaceship. Numerically solve system (2) with $\mu = .012129$ on the interval $[0, 50]$ for the following three initial conditions:

a. $y_1(0) = -1.2625$, $y_2(0) = 0$, $y_3(0) = 0$, $y_4(0) = 1.05$

b. $y_1(0) = -1.26$, $y_2(0) = 0$, $y_3(0) = 0$, $y_4(0) = 1.05$

c. $y_1(0) = -1.2625$, $y_2(0) = 0$, $y_3(0) = 0$, $y_4(0) = 1.00$

For the initial conditions a., b., and c. display a phase-plane graph of y_3 versus y_1. Compare the three graphs. What do you conclude? Is the initial position of the spaceship very important? Is the initial velocity of the spaceship very important?

(Note: The specified accuracy of the numerical integration technique is also very important. Use a good double-precision integration routine to solve system (2) with a prescribed accuracy of at least 10^{-12}.)

11

Applications to Combinatorics

11.1 Power Series

The reader has seen power series in various Calculus courses. For this reason, we will not include a general introduction to their theory here, but we will remind the reader of some important facts.

Definition 11.1 Let a be a complex number. A *power series* centered at a is an infinite series of the form

$$\sum_{n\geq 0} f_n(x-a)^n = f_0 + f_1(x-a) + f_2(x-a)^2 + \cdots, \tag{1}$$

where the f_i are complex numbers.

The power series above is certainly convergent if $x = a$, since then all summands are 0 except the first one. In Calculus, the reader has learned various techniques to decide if a power series is convergent or not for a certain value of x, such as the ratio test or the root test. Using these tools, it is not difficult to prove that the set of real numbers x for which (1) is convergent is always a *circle* centered at a. *If we restrict our attention to power series with real coefficients which are defined on the set of real numbers, then the word "circle" becomes the word "interval" in the present discussion.* In this chapter, we will sometimes assume that the coefficients of our power series are real numbers, but a more general treatment is possible.

If the power series (1) is convergent in an interval centered at a, then we write

$$f(x) = \sum_{n\geq 0} f_n(x-a)^n \tag{2}$$

for the function that is defined by the convergent sum on the right-hand side.

Definition 11.2 The function $f(x)$ is *analytic* at a if there is a nontrivial circle C centered at a so that in C, the function f can be obtained as the sum of a convergent power series.

DOI: 10.1201/9781003678014-11

Example 1 The function $f(x) = e^x$ is analytic at $a = 0$, since as we learned in Calculus, the equality

$$e^x = \frac{x^n}{n!}$$

holds for all real numbers x. So in this case, I can be any nontrivial interval centered at 0. ■

Readers familiar with complex analysis will note that $f(z) = \sum_{n\geq 0} \frac{z^n}{n!}$ holds for all complex numbers z as well, but we will not need that fact in our discussion.

If f is not analytic at a, then we say that a is a *singular point* or *singularity* of f.

We also point out that if $f(x)$ is analytic in a neighborhood of a, then there is a *unique* power series centered at a so that (2) holds. Therefore, we are justified when we talk about *the* power series form of f in a neighborhood of a, since there is one and only one such form.

Combinatorialists are often interested in power series that are not convergent (or not known to be convergent) anywhere other than their center. When we want to emphasize this, we call our power series *formal power series.* Derivatives of formal power series can be *defined* as those of convergent power series are computed.

Definition 11.3 Let $g(x) = \sum_{n\geq 0} g_n x^n$ be a formal power series. Then the derivative of $g(x)$ is the formal power series

$$g'(x) = \sum_{n\geq 1} n g_n x^{n-1} = \sum_{n\geq 0} (n+1) g_{n+1} x^n.$$

Example 2 Let $f(x) = \sum_{n\geq 0} n! x^n$. It is then easy to prove that $f(x)$ is not convergent unless $x = 0$. However, as a formal power series, the derivative of f exists, and

$$f'(x) = \sum_{n\geq 1} n! \cdot n x^{n-1} = \sum_{n\geq 0} (n+1)!(n+1) x^n. \blacksquare$$

11.1.1 Exercises

These exercises serve as a reminder of some basic operations with power series that reader encountered in Calculus.

1. Find the closed form of the formal power series $f(x) = \sum_{n\geq 2} \binom{n}{2} x^n$.

2. Find the closed form of the formal power series $f(x) = \sum_{n\geq 3} \binom{n}{3} x^n$.

3. What is the coefficient of x^n in the power series form of $(e^{-3x})/(1-x)$?

4. What is the coefficient of x^n in the power series form of $\sqrt{1-x}$?

5. Define the *antiderivative* of the formal power series $f(x) = \sum_{n\geq 0} a^n x^n$ as the formal power series $F(x) = \sum_{n\geq 0} \frac{a_{n+1}}{n+1} x^{n+1}$. Note that then $F'(x) = f(x)$ as formal power series. Find the coefficients of the formal power series $F(x) = \ln(1/(1-x))$.

6. Compute the coefficient of x^n in

$$\frac{1}{1-x} \cdot \ln\left(\frac{1}{1-x}\right).$$

11.2 Set Partitions

11.2.1 Subsets of a Finite Set: A Combinatorial Introduction

The part of mathematics that investigates the number of ways in which a complicated task can be carried out, or the number of structures of a certain kind and size, is called *enumerative combinatorics.* In this chapter, we will see some applications of differential equations to that fascinating field. To that end, we need to determine the number of k-element subsets of an n-element set. That is, we want to know in how many ways we can choose k objects from a set of n different objects. The order of selection does not matter.

It turns out that it is easiest to answer this question by first pretending that the order of the selected objects *does* matter. Then we have n choices for the first object, $n-1$ choices for the second object, and so on, $n-k+1$ choices for the kth object. So, if the order of the selected objects *did* matter, the number of ways for us to proceed would be $n(n-1)\cdots(n-k+1)$. However, if we count like this, we count each k-element subset S many times, namely one time for each order in which its k elements can be selected. That number is $k(k-1)\cdots 1 = k!$, since we have k choices for the first element of S that is selected, then $k-1$ choices for the second element of S that is selected, and so on. Therefore, counting in this way, we count each k-element subset of the n-element set $k!$ times. Therefore, we proved the following.

Proposition 11.1 The number of k-element subsets of an n-element set is

$$\frac{n(n-1)\cdots(n-k+1)}{k(k-1)\cdots 1} =: \binom{n}{k}. \tag{3}$$

Example 1 The number of 3-element subsets of a 9-element set is

$$\binom{9}{3} = \frac{9\cdot 8\cdot 7}{3\cdot 2\cdot 1} = \frac{504}{6} = 84. \blacksquare$$

Note that the numbers $\binom{n}{k}$ are called *binomial coefficients*. Also note that multiplying both the numerator and the denominator of the fraction in (3) by $(n-k)!$, we get the identity

$$\binom{n}{k} = \frac{n!}{k!(n-k)!}. \tag{4}$$

This shows that $\binom{n}{k} = \binom{n}{n-k}$ for all n and k. This is not surprising, because each k-element subset of an n-element set can be matched to its complement, and that complement has $n-k$ elements.

11.2.2 The Notion of Set Partitions

There are 12 people at a gathering. We ask them to split into some groups. There are almost no rules– the groups can be of any size. The set of groups is not ordered– that is, there will be no such thing as the first group, the second group, and so on. The order of the people inside each group also does not matter. In how many ways is this possible?

If there were just three people instead of 12, we could solve this problem by listing all possibilities. Let us denote the three people by the digits 1, 2, and 3. Then there are the following five possible arrangements.

- $\{1,2\}$, $\{3\}$.
- $\{1,3\}$, $\{2\}$.
- $\{2,3\}$, $\{1\}$.
- $\{1,2,3\}$.
- $\{1\}$, $\{2\}$, $\{3\}$.

The objects we introduced above are *set partitions*. That is, a partition of the set $[n] = \{1, 2, \cdots, n\}$ is a collection of nonempty subsets of $[n]$ so that each element of $[n]$ belongs to exactly one of those subsets. The subsets are called the *blocks* of the set partition. In the above listing, the first three partitions of $[3]$ have two blocks each, the fourth one has just one block, and the last one has three blocks.

So, in the language of enumerative combinatorics, the question we asked is "How many set partitions does a 12-element set have?". In general, how many set partitions does $[n]$ have? This number is so important that it has a name. We call it the nth *Bell number* and we denote it by $B(n)$.

Determining $B(n)$ is not easy. The reader should verify by hand that $B(1) = 1$, $B(2) = 2$, and $B(4) = 15$. We have just seen that $B(3) = 5$. A little bit of more work yields that $B(5) = 52$, but the brute force approach gets too time-consuming fast, even for computers.

At this point, the reader might protest by saying something like this. Fine, set partitions are interesting. But what do they have to do with our subject, differential equations? In this whole book, we dealt with functions that were usually continuous, and even differentiable, while set partitions are discrete objects.

The answer to that question is that enumerative combinatorialists have an extremely powerful tool, *generating functions.* A generating function is a power series associated with a sequence of numbers, hence it is a tool to turn a discrete problem to a problem involving continuous functions. In many situtations, we can set up differential equations that involve these generating functions. Then we can solve them and obtain an explicit form for these generating functions, which reveals a lot of information about the sequence that we are investigating.

11.2.3 Generating Functions

Let $a_0, a_1, a_2, \cdots$ be a sequence of real numbers. The *exponential generating function* of the sequence $\{a_n\}$ is the power series

$$A(x) = \sum_{n\geq 0} a_n \frac{x^n}{n!} = a_0 + a_1 x + a_2 \frac{x^2}{2!} + \cdots . \tag{5}$$

Knowing the exponential generating function of a sequence means knowing the elements of the sequence themselves. Indeed, if we know $A(x)$, then (5) shows that a_n is just $n!$ times the coefficient of x^n in $A(x)$.

Example 2 Let us assume that the sequence $c_0, c_1, c_2, \cdots$ has exponential generating function $C(x) = e^{3x}$. Then

$$C(x) = e^{3x} = \sum_{n\geq 0} \frac{(3x)^n}{n!} = \sum_{n\geq 0} \frac{3^n}{n!} x^n,$$

and so $c_n = 3^n$. ■

It often happens that the formula for a generating function is actually simpler than the formula for its coefficients.

Example 3 Let $f_0, f_1, f_2, \cdots$ be the sequence whose exponential generating function is

$$F(x) = \frac{e^{-x}}{1-x}.$$

Then

$$\begin{aligned} F(x) &= e^{-x} \cdot \frac{1}{1-x} \\ &= \left(\sum_{k \geq 0} \frac{(-1)^k x^k}{k!} \right) \cdot \left(\sum_{m \geq 0} x^m \right) \\ &= \sum_{n \geq 0} \left(\sum_{k=0}^{n} \frac{(-1)^k}{k!} \right) x^n, \end{aligned}$$

so $f_n = n! \sum_{k=0}^{n} x \frac{(-1)^k}{k!}$. ■

In the above example, we see that the formula for f_n is more complex than the formula for $F(x)$.

We will see that this happens for the Bell numbers as well. That is, we will get a closed formula for their exponential generating function, even though a closed formula for the Bell numbers themselves does not exist.

11.2.4 A Recurrence Relation for the Bell Numbers

We start with a recurrence relation for the Bell numbers. Let us assume that we already know the numbers $B(1), B(2), \cdots, B(n)$, and we want to compute the number $B(n+1)$. We also set $B(0) = 1$.

Consider a partition of the set $[n+1]$. In that partition, there is a block that contains the maximal element $n+1$. That block may have some other elements; let us denote the number of those elements by i. Here $0 \leq i \leq n$. Then there are $\binom{n}{i}$ possibilities for those i elements. The rest of the partition is just a partition of the remaining $n-i$ elements. Therefore, the total number of partitions of $[n+1]$ in which the entry $n+1$ shares a block with i other elements is $\binom{n}{i} B(n-i)$. As this argument holds for every possible value of i, we have just proved that

$$B(n+1) = \sum_{i=0}^{n} \binom{n}{i} B(n-i). \tag{6}$$

Setting $j = n - i$, we get the formula

$$B(n+1) = \sum_{j=0}^{n} \binom{n}{n-j} B(j) = \sum_{j=0}^{n} \binom{n}{j} B(j), \tag{7}$$

where we use the fact that $\binom{n}{j} = \binom{n}{n-j}$ that we explained following (4).

Note that recurrence relation (7) that we have just proved holds for $n \geq 0$, while $B(0) = 1$.

11.2.5 The Exponential Generating Function of the Bell Numbers

Now let

$$B(x) = \sum_{n \geq 0} B(n)\frac{x^n}{n!} = 1 + x + 2\frac{x^2}{2} + 5\frac{x^3}{6} + 15\frac{x^4}{24} + \cdots \tag{8}$$

be the exponential generating function of the Bell numbers. We want to use (7) to determine a closed form for $B(x)$. To that end, let us multiply both sides of (7) by $x^n/n!$, and sum the obtained equalites over all $n \geq 0$. We get

$$\sum_{n \geq 0} B(n+1)\frac{x^n}{n!} = \sum_{n \geq 0} \left(\sum_{j=0}^{n} \binom{n}{j} B(j) \right) \frac{x^n}{n!}. \tag{9}$$

It is easy to notice (see Definition 11.3) that the left-hand side is just $B'(x)$. The right-hand side is a little bit harder. We claim that the right-hand side is equal to the product $B(x)e^x$. Indeed,

$$\begin{aligned}
B(x)e^x &= \left(\sum_{j \geq 0} B(j)\frac{x^j}{j!} \right) \cdot \left(\sum_{i \geq 0} \frac{1}{i!}x^i \right) \\
&= \sum_{n \geq 0} \left(\sum_{j=0}^{n} B(j)\frac{1}{j!(n-j)!} \right) x^n \\
&= \sum_{n \geq 0} \left(\sum_{j=0}^{n} B(j)\frac{n!}{j!(n-j)!} \right) \frac{x^n}{n!} \\
&= \sum_{n \geq 0} \left(\sum_{j=0}^{n} B(j)\binom{n}{j} \right) x^n.
\end{aligned}$$

Therefore, (9) simplifies to

$$B'(x) = e^x B(x), \tag{10}$$

which is a linear differential equation for $B(x)$, with initial condition $B(0) = 1$. Solving this initial value problem, we get that

$$B(x) = e^{e^x - 1}. \tag{11}$$

So this is the generating function of the Bell numbers. This single equation contains a wealth of information about the Bell numbers. If we want to know what the 18th Bell number is, we just have to remember (8), ask our favorite software package to compute the coefficient of x^{18} in $B(x)$, and then multiply it by 18!.

Note the surprising nature of formula (11. We started out with an enumeration problem involving positive integers only, and we ended up with a power series that solves that problem, but the power series is a *doubly* exponential function!

11.2.6 Products of Exponential Generating Functions

In Example 3, we have seen how to multiply two power series together. When those two power series happen to be exponential generating functions, then their product has a particularly interesting form.

Proposition 11.2 Let $A(x) = \sum_{n\geq 0} a_n \frac{x^n}{n!}$ and $B(x) = \sum_{n\geq 0} b_n \frac{x^n}{n!}$ be two exponential generating functions. Then their product is

$$A(x)B(x) = \sum_{n\geq 0} \frac{1}{n!} \left(\sum_{k=0}^{n} \binom{n}{k} a_k b_{n-k} \right) x^n.$$

Proof: The coefficient of x^n in the product $A(x)B(x)$ is

$$\sum_{n\geq 0} \frac{a_k}{k!} \cdot \frac{b_{n-k}}{(n-k)!} = \frac{1}{n!} \sum_{k=0}^{n} \frac{n!}{k!(n-k)!} a_k b_{n-k}, \tag{12}$$

since if we select the degree-k summand from $A(x)$, then we must select the degree-$(n-k)$ summand from $B(x)$, in order to get a product of degree n. ■

Note that the expression $\sum_{n\geq 0} \left(\sum_{k=0}^{n} \binom{n}{k} a_k b_{n-k}\right)$, that is, the coefficient of $x^n/n!$ in $A(x)B(x)$, has a clear combinatorial meaning. If a_k is the number of ways to carry out a task on a k-element set, and b_{n-k} is the number of ways to carry out another task on an $(n-k)$-element set, then $\sum_{n\geq 0} \left(\sum_{k=0}^{n} \binom{n}{k} a_k b_{n-k}\right)$ is the number of ways to partition an n-element set into two blocks (empty blocks allowed), carry out the first task on the first block, and the second task on the second block. This proves the following corollary.

Corollary 11.1 (Product formula) If $A(x)$ is the exponential generating function for the number of ways to carry out a task on a set, and $B(x)$ is the exponential generating function for the number of ways to carry out another task on a set, then $A(x)B(x)$ is the exponential generating function for the number of ways to first partition the set into two disjoint blocks L_1 and L_2, then to carry out the first task on L_1 and the second task on L_2.

There is nothing magic about partitioning a set into *two* blocks. We can partition $[n]$ into k blocks, and carry out a task on each block. If the number of ways of carrying out the individual tasks have exponential generating functions $A_1(x), A_2(x), \cdots, A_k(x)$, then the number of ways to carry out of the combined task has exponential generating function

$$A_1(x)A_2(x)\cdots A_k(x).$$

This follows from repeated applications of Corollary 11.1. Indeed, we can view a threefold task as a twofold task where the first task is itself a twofold task. This leads to the following more interesting corollary, which will have applications in the next section.

Corollary 11.2 Let $A(x)$ be the exponential generating function for the number of ways to carry out a task on a nonempty set. Let $H(x)$ be the exponential generating function for the number of ways to take a partition of the set $[n]$ into any number of *nonempty* blocks, then order the blocks linearly, and carry out the first task on each block. Then

$$H(x) = \frac{1}{1 - A(x)}.$$

Proof: If we partition n into k blocks, then Corollary 11.1 implies that the combined task has generating function $A(x)^k$. Summing over all k, we get that the combined generating function $H(x)$ is given by

$$H(x) = 1 + A(x) + A(x)^2 + \cdots = \sum_{k\geq 0} A(x)^k = \frac{1}{1 - A(x)}. \qquad \blacksquare$$

The reader may wonder why the infinite sum $\sum_{k\geq 0} A(x)^k$ is even defined, that is, why the coefficient of x^n is a finite number for all n. The answer to that question is that $A(x)$ is the exponential generating function for the number of ways to carry out a task on a *nonempty* set. Therefore, $a_0 = 0$, so $A(x) = xB(x)$ for some power series $B(x)$. Therefore, $A(x)^k = x^k B(x)^k$, that is, the smallest exponent that appears in $A(x)^k$ is k. Therefore, for each positive integer n, there are only a finite number of values of k for which $A(x)^k$ contains a summand x^n. Indeed, that only happens when $k \leq n$. So yes, the coefficient of x^n in $\sum_{k\geq 0} A(x)^k$ is finite for all n.

Note that in Corollary 11.2, the requirement that the blocks must be nonempty is combinatorially necessary. Indeed, if we allowed empty blocks, the number of ways to proceed would be infinite, since we could add any number of empty blocks, since the number of blocks is not set in advance. This is not the case in Corollary 11.1, and that is why empty blocks are allowed there.

11.2.7 Exercises

1. Solve (10) as a separable equation.
2. Solve (10) as a linear equation.
3. Let $B_2(n)$ be the number of partitions of $[n]$ into blocks that are of size at least two. Let $B_2(x)$ be the exponential generating function of the sequence $B_2(n)$, with $n \geq 2$.

   Prove that
   $$e^x B_2(x) = B(x),$$
   where $B(x)$ is the exponential generating function of the Bell numbers. Then use the identity you just prove to find the closed form of $B_2(x)$.
4. Let $S(n,k)$ be the number of partitions of $[n]$ into k blocks. The numbers $S(n,k)$ are called *Stirling numbers of the second kind.* Prove that for a fixed k,
   $$\sum_{n \geq k} S(n,k)\frac{x^n}{n!} = (e^x - 1)^k . \qquad (13)$$
5. Use formula (13) to prove an explicit formula for $S(n,2)$.
6. Use formula (13) to prove an explicit formula for $S(n,3)$.

11.3 Trees

In combinatorics, a *graph* is a diagram consisting of some points called *vertices* some of which are connected by lines called *edges*. Graph theory is a very important part of both combinatorics and computer science. A graph with no cycles in it is called a *tree*. Finally, a tree with a distinguished vertex, the root, is called a *rooted tree*. The parent of a vertex v is the first vertex w on the unique path from v to the root. In that situation, v is a *child* of w. Note that a each vertex has one parent (except the root, that has none), while a vertex may have several children.

Example 1 The two graphs on the left are not trees because they contain cycles. The two graphs on the right are trees.

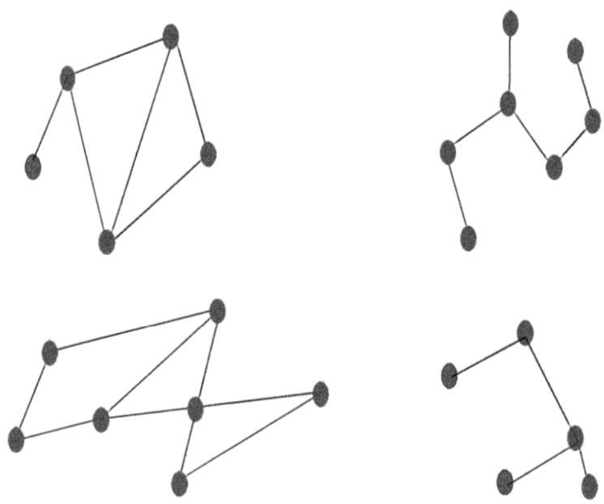

FIGURE 11.1
The two graphs on the right are trees. ■

Example 2 Figure 11.2 shows a rooted tree whose root is vertex 6, which is the parent of vertices 4 and 5. Vertex 4 is the parent of vertices 3 and 2, while vertex 3 is the parent of vertex 1. Vertices 1, 2, and 5 have no children. We call such vertices *leaves*.

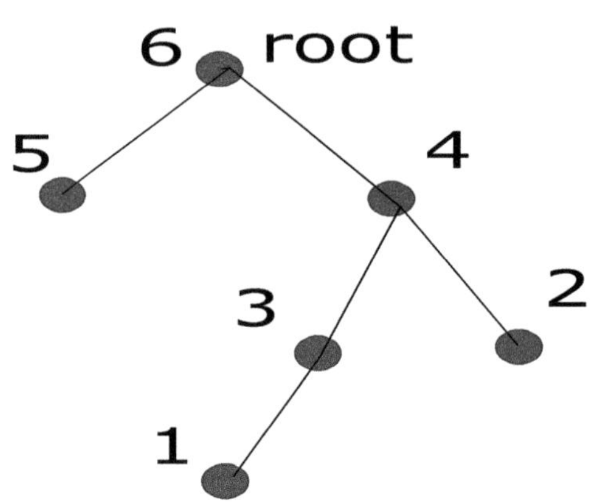

FIGURE 11.2
A rooted tree ■

11.3.1 Decreasing Binary Trees

A decreasing binary tree on vertex set $[n] = \{1, 2, \cdots, n\}$ is a tree whose vertices are labeled with the elements of $[n]$ so that each label is used exactly once, and every vertex has a label that is smaller than the label of its parent. The vertex labeled n is the root. Each vertex has at most two children, and each child is a left child or a right child of its parent, even if it is an only child. Figure 11.3 illustrates this tree variety.

Example 3 The six decreasing binary trees on vertex set [3] are shown in Figure 11.3.

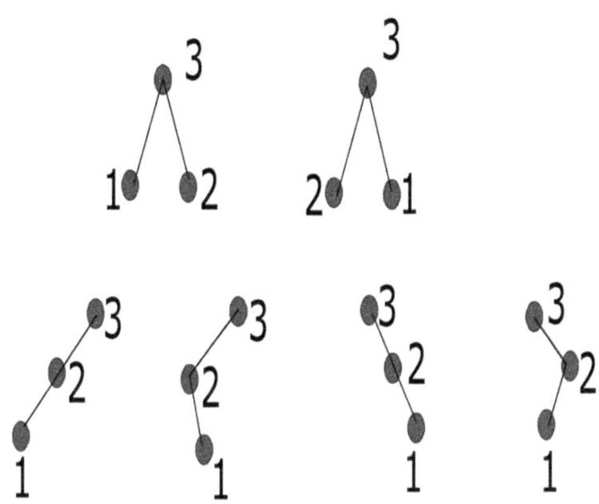

FIGURE 11.3
The six decreasing binary trees on vertex set [3]. ■

Let f_n be the number of all decreasing binary trees on vertex set $[n]$. Let $F(x) = \sum_{n \geq 0} f_n \frac{x^n}{n!}$ be the exponential generating function of the sequence f_n, with $f(0) = 1$. So

$$F(x) = 1 + x + 2\frac{x^2}{2} + 6\frac{x^3}{6} + \cdots .$$

We will use a differential equation to determine the explicit form of $F(x)$, and then compute the coefficients f_n.

The key idea is the following. If we remove the root of a decreasing binary tree on vertex set $[n]$, there are two ways to describe the obtained structure.

1. As the removed root has label n, we get a structure S on vertex set $[n-1]$. The number of these structures *of size* $n-1$ is f_n, since removing the root vertex did not result in the loss of any information. There is exactly one decreasing binary tree so that removing its root results in S, namely the one where n is connected of the roots of the vertices at the top of the one

or two trees in S. Therefore, the exponential generating function of the numbers of these structures S is

$$\sum_{n\geq 1} f_n \frac{x^{n-1}}{(n-1)!} = F'(x).$$

2. Removing the root gives us an *ordered pair* of decreasing binary trees, one of which may be empty. This is the situation discussed in Corollary 11.1. We partition a set, in this case the set $[n-1]$ into two disjoint blocks, then we take a decreasing binary tree on this block. So, using the notation of that corollary, $A(x) = B(x) = F(x)$, and the generating function of the combined task is $F(x)^2$.

As the two ways of considering the structure obtained by removing the root of a decreasing binary tree lead to the same objects, the obtained generating functions should be the same. That is, we have just proved that

$$F'(x) = F(x)^2, \qquad (1)$$

with initial condition $F(0) = 1$. You can solve this equation either as a separable equation or as a Bernoulli equation. In either way, the solution is

$$F(x) = \frac{1}{1-x} = \sum_{n\geq 0} x^n = \sum_{n\geq 0} n!\frac{x^n}{n!}.$$

So $f_n = n!$, and therefore, the number of decreasing binary trees on vertex set $[n]$ is $n!$.

At this point, the reader may be thinking something along the lines of "the number $n!$ is familiar to me; it is the number of ways to arrange n distinct objects in a line. In other words, it is the number of *permutations* of length n. The alert reader is right, of course, and the reader is invited to find a bijection (a one-to-one and onto correspondence) between decreasing binary trees on vertex set $[n]$ and permutations of length n.

11.3.2 Plane Decreasing Trees

A *plane decreasing tree* on n vertices is a rooted tree in which the vertices are labeled with the elements of $[n]$ so that each label is used exactly once, and the label of each vertex is less than that of its parent. Again, the order of children matters. A vertex can have any number of children. However, our trees are not binary, that is, if a vertex has only one child, then that child does *not* have a direction; it is simply a child, not a left child or a right child.

Example 4 The 15 plane decreasing trees on vertex set [4] are shown in Figure 11.4.

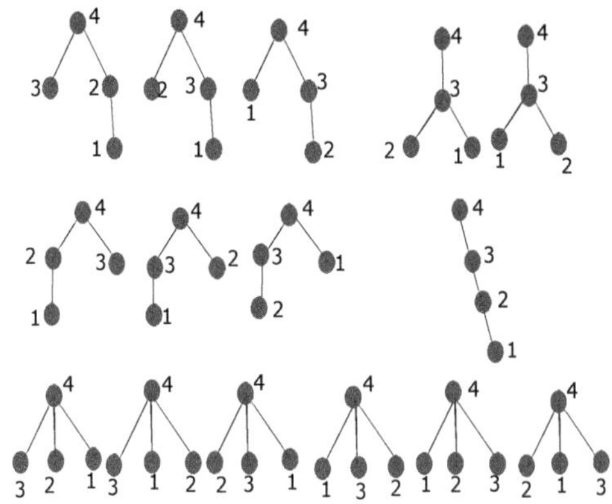

FIGURE 11.4
The 15 plane decreasing trees on vertex set $[4]$. ■

Let $P(x)$ be the exponential generating function of the number of such trees on vertex set $[n]$, and let us set $P(0) = 0$. The reader is invited to verify that $P(x)$ starts with the summands

$$P(x) = x + \frac{x^2}{2!} + 3\frac{x^3}{3!} + 15\frac{x^4}{4!} + \cdots .$$

Let p_n be the number of decreasing plane trees of size n, and let us proceed as in the previous example, that is, let us remove the root of a plane decreasing tree, and find two power series counting the obtained structure.

1. As the removed root has label n, we get a structure P on vertex set $[n-1]$. The number of these structures *of size* $n-1$ is p_n, since removing the root vertex did not result in the loss of any information. There is exactly one plane decreasing tree so that removing its root results in P, namely the one where n is connected to the roots of the vertices at the top of each tree in P. Therefore, the exponential generating function of the numbers of these structures P is

$$\sum_{n \geq 1} p_n \frac{x^{n-1}}{(n-1)!} = P'(x).$$

2. Removing the root gives us a *sequence* of plane decreasing trees. This is the situtation described in Corollary 11.2, implying that the obtained structures are counted by the generating function

$$\frac{1}{1 - P(x)}.$$

As the last two displayed power series are generating functions of the same structures, they have to be equal, that is ,

$$P'(x) = \frac{1}{1 - P(x)} \tag{15}$$

Solving this separable equation, we obtain

$$P(x) = 1 - \sqrt{1 - 2x}.$$

We can now use the Binomial theorem to compute the coefficients of $P(x)$, and therefore, the numbers p_n. Recall that the Binomial theorem states that

$$(1 + z)^\alpha = \sum_{n \geq 0} \binom{\alpha}{n} z^n. \tag{16}$$

Note that $\sqrt{1 - 2x} = (1 - 2x)^{1/2}$, and that

$$\binom{1/2}{n} = \frac{\frac{1}{2}\left(\frac{1}{2} - 1\right) \cdot \left(\frac{1}{2} - 2\right) \cdots \left(\frac{1}{2} - (n-1)\right)}{n!} = \frac{(-1)^{n-1} \cdot (2n-3)!!}{2^n \cdot n!},$$

where we use the notation $(2n - 3)!! = (2n - 3) \cdot (2n - 5) \cdot \cdots 3 \cdot 1$ for odd positive integers $2n - 3$. This expression is called $2n - 3$ *semifactorial.* We set $(-3)!! = -1$ and $(-1)! = 1$ in order to keep the above formula valid for $n = 0$ and $n = 1$.

Setting $\alpha = 1/2$ and $z = -2x$ in (16), we get that

$$\begin{aligned} \sqrt{1 - 2x} &= \sum_{n \geq 0} \binom{1/2}{n} (-2x)^n \\ &= \sum_{n \geq 0} \frac{(-1)^{n-1} \cdot (2n-3)!!}{2^n \cdot n!} (-1)^n 2^n x^n \\ &= \sum_{n \geq 0} -\frac{(2n-3)!!}{n!} x^n. \end{aligned}$$

Therefore,

$$P(x) = 1 - \sqrt{1 - 2x} = \sum_{n \geq 1} \frac{(2n-3)!!}{n!} x^n.$$

So $p_n = (2n - 3)!!$ is the number of plane decreasing trees on vertex set $[n]$.

The reader is invited to explain why we set $P(0) = 0$, and not $P(0) = 1$ in this example.

11.3.3 Non-Plane 1-2 Trees

Non-plane 1-2 trees are trees in which each vertex has a label smaller than its parent, each vertex has at most two children, but left or right does not

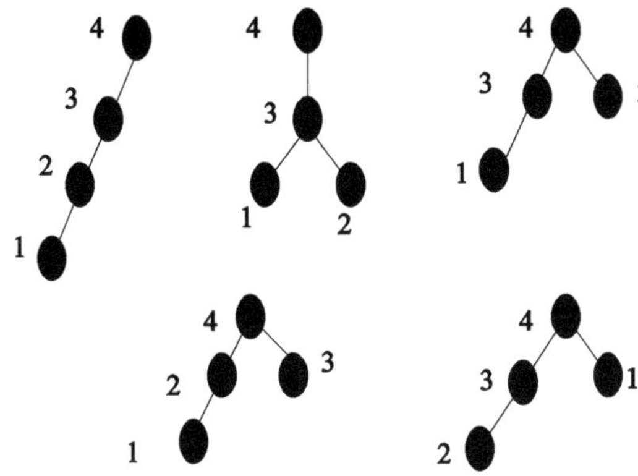

FIGURE 11.5
The five rooted non-plane 1-2 trees on vertex set [4].

matter. In other words, the set of children of a given vertex is not ordered. See Figure 11.5 for an illustration.

Let e_n be the number of such trees on vertex set $[n]$, and let $E(x)$ be the exponential generating function of the sequence of the numbers e_n, with $e_0 = 1$. So

$$E(x) = 1 + x + \frac{x^2}{2!} + 2\frac{x^3}{3!} + 5\frac{x^4}{4!} + \cdots .$$

Let us remove the root of such a tree on vertex set n. As before, we get a structure on vertex set $[n-1]$, but no information was lost, since the removed vertex was n. Therefore, number of possible structures of size $n-1$ is e_n, and their generating function is

$$\sum_{n\geq 1} e_n \frac{x^{n-1}}{(n-1)!} = E'(x).$$

On the other hand, removing the root, we get a partition of $[n-1]$ into two blocks, and a non-plane 1-2 tree on each of those blocks. If the order of these two trees mattered, we could simply apply Corollary 11.1. However, we are dealing with non-plane trees, so the order of the two subtrees does not matter, (there is no left or right here); therefore, the generating function counting such unordered pairs of trees is not $E(x)^2$, but $(E(x)^2+1)/2$. The correction term $+1$ is needed because every pair of subtrees is counted twice, *except* the pair of two empty subtrees, which is only counted once. Indeed, if one of the subtrees is L and the other subtree is R, then they are counted once as the ordered pair (L, R), and once as the ordered tree (R, L). As ordered pairs, $(L, R) = (R, L)$ if and only if $L = R$, and in our case that can only happen if L and R are both empty; otherwise, the set of labels in L is the complement of the set of labels in R in $[n-1]$.

This leads to the differential equation

$$E'(x) = (E(x)^2 + 1)/2, \tag{17}$$

with initial condition $E(0) = 1$.

Solving (17), we get

$$E(x) = \tan\left(\frac{z}{2} + \frac{\pi}{4}\right) = \tan x + \sec x. \tag{18}$$

This is a spectacular result. We obtain power series of two trigonometric functions as the generating function of a combinatorial enumeration problem! Also note that $\tan x$ is an odd function, while $\sec x$ is an even f unction. We will revisit these two power series and their mentioned properties in the exercises. Note that the coefficients of $E(x)$ are called the *Euler numbers.*

11.3.4 Plane 1-2 Trees

Plane 1-2 trees are trees in which each vertex has a label smaller than its parent, each vertex has at most two children, and left or right *does* matter. In other words, the set of children of a given vertex is ordered. They are different from decreasing binary trees in that if a child is an only child of its parent, then it does not have a direction, that is, it is not a left child or a right child; it is simply a child. See Figure 11.6 for an illustration.

Let r_n be the number of such trees on vertex set $[n]$, and let $R(x)$ be the exponential generating function of the sequence of the numbers r_n, with $r_0 = 1$. So

$$R(x) = 1 + x + \frac{x^2}{2!} + 3\frac{x^3}{3!} + 9\frac{x^4}{4!} + \cdots .$$

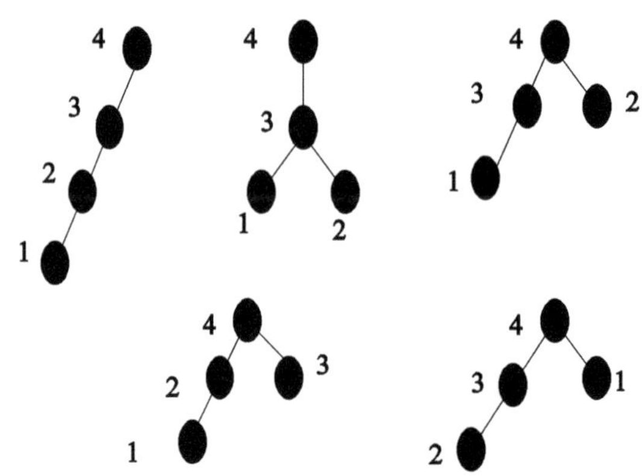

FIGURE 11.6
The nine rooted plane 1-2 trees on vertex set [4].

Let us remove the root of such a tree on vertex set n. As before, we get a structure on vertex set $[n-1]$, but no information was lost, since the removed vertex was n. Therefore, number of possible structures of size $n-1$ is r_n, and their generating function is

$$\sum_{n\geq 1} r_n \frac{x^{n-1}}{(n-1)!} = R'(x).$$

On the other hand, removing the root, we get a partition of $[n-1]$ into two blocks, and a plane 1-2 tree on each of those blocks. As the order of these two trees matters, we can use Corollary 11.1. That yields the generating function $R(x)^2$ for the number of ordered pairs (L, R) of subtrees of the root. However, note that the ordered pair $(L, \emptyset)$ of subtrees yields the same plane 1-2 tree as the pair $(\emptyset, L)$. Indeed, if the root has exactly one child, then that child is not a left child or a right child. So the generating function $R(x)$ counts each tree in which the root has exactly one child twice. Therefore, the correct generating function for the number of structures we get after removing the root is

$$R(x)^2 - (R(x) - 1).$$

Here the -1 correction term is needed since the 1-vertex tree in which the root has no children is not counted twice by $R(x)^2$.

As the last two displayed generating functions count the same structures, they must be equal, that is,

$$R'(x) = R(x)^2 - R(x) + 1, \tag{19}$$

with initial condition $R(0) = 1$.

Solving equation (19), we get

$$R(x) = \frac{1}{2} + \frac{\sqrt{3}}{2}\tan\left(\frac{\sqrt{3}}{2}x + \frac{\pi}{6}.\right) \tag{20}$$

11.3.5 Exercises

1. Let t_n be the number of decreasing binary trees on vertex set $[n]$ in which every vertex has either 0 or 2 children. Let $T(x)$ be the exponential generating function of the numbers t_n. Prove that

$$T'(x) = 1 + T(x)^2, \tag{21}$$

with $T(0) = 0$.

2. Solve equation (21).

3. Let s_n be the number of decreasing binary trees on vertex set $[n]$ in which each vertex has either 0 or 2 children, except for the rightmost vertex, which has one child. Let $S(x)$ be the exponential generating function of the numbers s_n. Prove that

$$S'(x) = T(x)S(x), \tag{22}$$

with $S(0) = 1$. Here $T(x)$ was defined in the previous exercise.

4. Solve equation (22).
5. Solve equation (17)
6. Prove the trigonometric identity appearing in (18).
7. Solve equation (15).
8. Solve equation (19).

Notes

The fundamental theorem of analytic combinatorics states that if $f(x)$ is a power series that is analytic at zero and has nonnegative real coefficients (like all combinatorial generating functions do), then the *exponential order* of the coefficients of f is equal to $1/R$, where R is the convergence radius of f, that is, $R = |s|$, where s is a *dominant singularity* of r. That is, f is not analytic at s, but f is analytic inside the circle $|x| < |s|$. The exponential order of the coefficients of $f(x) = \sum_{n\geq 0} f_n x^n$ is defined as

$$\limsup_n \sqrt[n]{f_n}.$$

This theorem enables us to find the exponential order of the coefficients of a power series even in cases when we do not have a closed formula for those coefficients. For instance, we have seen in (18) that the Euler numbers have exponential generating function $E(x) = \sec x + \tan x$. Both $\sec x$ and $\tan x$ have dominant singularities at $x = \pi/2$ and $x = -\pi/2$, hence the exponential order of the coefficients $E_n/n!$ of $E(x)$ is 2π. Further tools in complex analysis, such as the residue theorem, make an even more precise estimate possible, showing that

$$\frac{E_n}{n!} \sim \frac{4}{\pi} \cdot \left(\frac{2}{\pi}\right)^n. \tag{23}$$

A

Appendix

The oldest and simplest algorithm for generating a numerical approximation to a solution of a differential equation was developed by Leonhard Euler in 1768. Given a specific point (x_0, y_0) on the solution of the differential equation $y' = f(x, y)$, Euler wrote the equation for the tangent line to the solution through (x_0, y_0)—namely, $y = y_0 + f(x_0, y_0)(x - x_0)$. To obtain an approximation to the solution through (x_0, y_0) at x_1, Euler took a small step along the tangent line and arrived at the approximation $y_1 = y_0 + f(x_0, y_0)(x_1 - x_0)$ to the solution at x_1, $y(x_1)$. Continuing to generate points successively in this manner and by connecting the points (x_0, y_0), (x_1, y_1), $(x_2, y_2), \ldots$ in succession, Euler produced a polygonal path which approximated the solution. This first numerical algorithm for solving the initial value problem $y' = f(x, y)$; $y(x_0) = y_0$ is called **Euler's method** or, due to its particular geometric construction, the **tangent line method**.

Euler's method is a single-step method. In single-step methods, only one solution value, (x_0, y_0), is required to produce the next approximate solution value. On the other hand, multistep methods require two or more previous solution values to produce the next approximate solution value. In 1883, more than a century after Euler developed the first single-step method, the English mathematicians Francis Bashforth (1819-1912) and John Couch Adams (1819-1892) published an article on the theory of capillary action which included multistep methods that were both explicit methods and implicit methods. In 1895, the German mathematician Carl David Tolmé Runge (1856-1927) wrote an article in which he developed two single-step methods. The second-order method was based on the midpoint rule while the third-order method was based on the trapezoidal rule. In an article which appeared in 1900, Karl Heun (1859-1929) improved Runge's results by increasing the order of the method to four. And in 1901, Martin Wilhelm Kutta (1867-1944) completed the derivation for the fourth-order methods by finding the complete set of eight equations the coefficients must satisfy. He also specified the values for the coefficients of the classic fourth-order Runge–Kutta method and those of a fifth-order method.

DOI: 10.1201/9781003678014-A

Prior to 1900, most calculations were performed by hand with paper and pencil. Euler's method and Runge–Kutta methods are single-step methods. Euler's method is of order one and requires only one f function evaluation per step. The classic Runge–Kutta method is fourth-order and requires four f function evaluations per step. Adams–Bashforth, Adams–Moulton, and predictor-corrector methods require only two f function evaluations per step; however, since these methods are multistep methods, they require starting values obtained by some other method. By the 1930s significant numerical integration techniques had been developed; however, their effective implementation was severely limited by the need to perform the computations by hand or with the aid of primitive mechanical calculators.

In the late 19th century and early 20th century, several commercially viable mechanical calculators capable of adding, subtracting, multiplying, and dividing were invented and manufactured. Electric motor driven calculators began to appear as early as 1900. These mechanical and electrical computing devices improved the speed and accuracy of generating numerical solutions of simple differential equations. In 1936, the German civil engineer Konrad Zuse (1910-1995) built the first mechanical binary computer, the Z1, in the living room of his parents' home. From 1942 to 1946 the first large scale, general purpose electronic computer was designed and built by John W. Mauchly (1907-1980) and J. Presper Eckert (1919-1995) at the University of Pennsylvania. The computer was named ENIAC, which is an acronym for "Electronic Numerical Integrator and Computer." ENIAC, which used vacuum tube technology, was operated from 1946 to 1955. After many technological inventions such as the transistor and integrated circuitry, the first hand-held, battery-powered, pocket calculator capable of performing addition, subtraction, multiplication, and division was introduced by Texas Instruments in 1967. The first scientific pocket calculator, the HP-35, was produced in 1972 by Hewlett Packard.

In the 1960s and 1970s several sophisticated computer programs were developed to solve differential equations numerically. Since then significant advances in graphical display capabilities have occurred also. Consequently, at the present time there are many computer software packages available to generate numerical solutions of differential equations and to graphically display the results.

In this appendix, we present some of the simpler single-step, multistep, and predictor-corrector methods for computing numerical approximations to the solution of the initial value problem

$$(1) \qquad y' = f(x, y); \quad y(x_0) = y_0$$

and for estimating the error of the computed approximations. We discuss the advantages and disadvantages of each type of method. Then we present and discuss desirable features for computer software to solve the IVP (1). Next, we explain how to use computer software to generate a numerical approximation to the solution of the first-order IVP (1). Finally, we illustrate and interpret the various kinds of results which computer software may produce. Furthermore, we reiterate the importance of performing a thorough mathematical analysis, which includes applying the fundamental theorems to the problem, prior to generating a numerical approximation.

Before we start generating numerical approximations to the solution of the IVP (1), we need to have a basic understanding of how a digital computer represents and processes numbers. First of all, it is impossible to represent all real numbers exactly in a digital computer. In most cases, a digital computer represents and stores real numbers as floating-point quantities using a scheme similar to scientific notation. Since a digital computer is a finite device, **only a finite set of rational numbers can be represented exactly** and they are not equally spaced throughout the range of representable values. When attempting to perform an arithmetic operation whose result would be a number whose magnitude is larger than the largest number representable on the computer **overflow** occurs and most computers terminate execution immediately. Likewise, **underflow** occurs and execution is terminated when attempting to perform an arithmetic operation whose result would be a number whose magnitude is less than the smallest nonzero number representable. In addition, contrary to your experience with real numbers, the floating-point operations of addition and multiplication are not commutative, and the distributive law fails to hold.

When a number which the computer cannot represent exactly is entered into the computer or is calculated within the computer, the computer selects the nearest number in its representable set by rounding-off or chopping-off the number. The error created by rounding or chopping after the final digit is called the *round-off error*. For example, suppose we have a calculator which uses base 10 and has four digits accuracy. If our calculator rounds, it represents 2/3 as .6667; whereas, if our calculator chops, it represents 2/3 as .6666. Now suppose that our calculator actually rounds and we use it to compute

$$\frac{x^2 - \frac{4}{9}}{x - \frac{2}{3}}$$

for $x = .6666$. Calculating, we get

$$\frac{(.6666)^2 - .4444}{.6666 - .6667} = \frac{.4444 - .4444}{-.0001} = 0.$$

Factoring and cancelling before using our calculator, we find

$$\frac{x^2 - \frac{4}{9}}{x - \frac{2}{3}} = \frac{(x - \frac{2}{3})(x + \frac{2}{3})}{x - \frac{2}{3}} = x + \frac{2}{3}.$$

Then using our calculator to evaluate this last expression with $x = .6666$, we obtain $.6666 + .6667 = 1.333$. This example illustrates that round-off error can create some very serious computing problems. Round-off error depends upon the computer and the coding of the algorithm. In general, round-off error is difficult to analyze and, indeed, somewhat unpredictable. In order to minimize round-off error, it is a good idea to use high-precision arithmetic and to reduce the number of computations as much as possible.

Throughout this appendix, we will let $\phi(x)$ denote the unique, explicit solution of the initial value problem (1) $y' = f(x, y)$; $y(x_0) = y_0$ on the interval I which contains x_0. By the fundamental existence and uniqueness theorem and the continuation theorem requiring that f and f_y be continuous in some finite rectangle in the xy-plane will guarantee that the IVP (1) has a unique solution $\phi(x)$ on some interval I containing x_0. Recall that the solution $\phi(x)$ must be defined, continuous, and differentiable at least once on the interval I; that $\phi(x)$ must satisfy the differential equation $\phi'(x) = f(x, \phi(x))$ on I; and that $\phi(x)$ must satisfy the initial condition $\phi(x_0) = y_0$.

A "numerical solution" of the IVP (1) is a discrete approximation of the solution. A numerical solution is, in fact, a finite set of ordered pairs of rational numbers, (x_i, y_i) for $i = 0, 1, 2, \ldots, n$, whose first coordinates, x_i, are distinct points in the interval I and whose second coordinates, y_i, are approximations to the solution ϕ at x_i—that is, $y_i \approx \phi(x_i)$. A numerical method for solving the IVP (1) is an algorithm which chooses points x_i in I such that $x_0 < x_1 < x_2 < \cdots < x_n$ and determines corresponding values $y_0, y_1, y_2, \ldots, y_n$ such that y_i approximates $\phi(x_i)$. Figure A.1 illustrates the relationship between the solution $\phi(x)$ of the IVP (1) and a numerical approximation $F = \{(x_i, y_i) \mid i = 0, 1, \ldots, n\}$. The solution $\phi(x)$ is represented by the solid curve which appears in Figure A.1. The numerical approximation is represented by the set of dots $\{(x_i, y_i) \mid i = 0, 1, \ldots, n\}$ which appear in the graph.

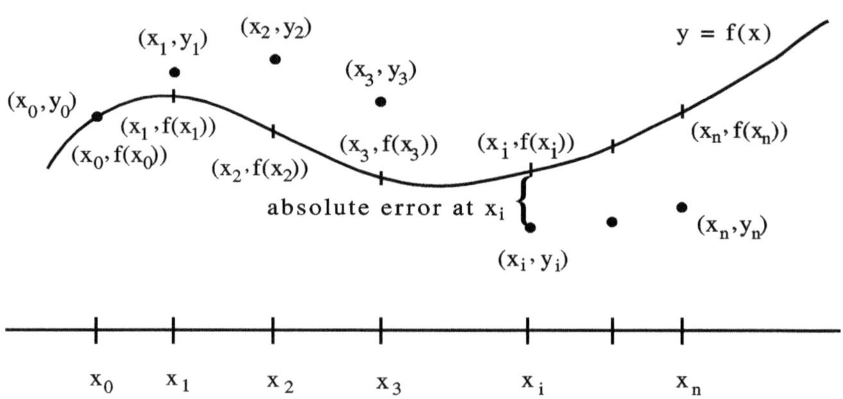

FIGURE A.1
Solution $\phi(x)$ of the IVP(1) and a Numerical Approximation.

Several different measures of error are used when specifying the accuracy of a numerical solution. Three are defined below.

The **absolute error at** x_i is $|\phi(x_i) - y_i|$.

The **relative error at** x_i is $|\phi(x_i) - y_i|/|\phi(x_i)|$ provided $\phi(x_i) \neq 0$.

The **percentage relative error at** x_i is $100 \times |\phi(x_i) - y_i|/|\phi(x_i)|$ provided $\phi(x_i) \neq 0$.

Observe that the absolute error at x_i is the vertical distance between the points $(x_i, \phi(x_i))$ and (x_i, y_i). See Figure A.1.

Most differential equations and initial value problems cannot be solved explicitly or implicitly; and, therefore, we must be satisfied with obtaining a numerical approximation to the solution. Thus, we need to know how to generate a numerical approximation to the solution of the IVP (1) $y' = f(x, y);\ y(x_0) = y_0$. In the differential equation $y' = f(x, y)$ replace y' by dy/dx, multiply by dx, and integrate both sides of the resulting equation from x_0 to x to obtain

$$\int_{x_0}^{x} dy = \int_{x_0}^{x} f(t, y(t))\, dt \quad \text{or} \quad y(x) - y(x_0) = \int_{x_0}^{x} f(t, y(t))\, dt.$$

Adding $y(x_0) = y_0$ to the last equation, we find the symbolic solution to the IVP (1) on $[x_0, x]$ to be

$$(2) \qquad y(x) = y_0 + \int_{x_0}^{x} f(t, y(t))\, dt.$$

When $f(x, y)$ in (1) is a function of the independent variable alone—that is, when the initial value problem is $y' = f(x);\ y(x_0) = y_0$—we can approximate the function $f(x)$ on the interval $[x_0, x_1]$, where x_1 is a specific point, by step functions or some polynomial in x, say $p_1(x)$, and then using this approximation integrate (2) over $[x_0, x_1]$ to obtain an approximation y_1 to the solution $\phi(x_1)$. Next, we approximate $f(x)$ on $[x_1, x_2]$ by some function $p_2(x)$ and integrate over $[x_1, x_2]$ to obtain $y_2 = y_1 + \int_{x_1}^{x_2} p_2(t)\, dt$ which is an approximation to the solution $\phi(x_2)$, and so on.

When $f(x, y)$ in (1) is a function of the dependent variable y, the value of the approximate solution y_1 at x_1 depends on the unknown solution $\phi(x)$ on the interval $[x_0, x_1]$ and the function $f(x, y)$ on the rectangle

$$R_1 = \{(x, y) \mid x_0 \leq x \leq x_1,\ y \in \{\phi(x) \mid x_0 \leq x \leq x_1\}\}.$$

Thus, we must approximate $\phi(x)$ on $[x_0, x_1]$ and $f(x, y)$ on R_1 in order to be able to integrate (2) over the interval $[x_0, x_1]$ and obtain an approximate solution y_1. In this case, additional approximate values $y_2, \ldots, y_n$ are obtained in a like manner.

Now suppose we have generated the numerical approximations (x_0, y_0), $(x_1, y_1), \ldots, (x_n, y_n)$. If the algorithm for generating the numerical approximation at x_{n+1} depends only on (x_n, y_n) and our approximation of $f(x, y)$ at (x_n, y_n), then the algorithm is called a **single-step**, **one-step**, **stepwise**, or **starting method**. On the other hand, if the algorithm for generating the numerical approximation at x_{n+1} depends on (x_n, y_n), (x_{n-1}, y_{n-1}),. . ., (x_{n-m}, y_{n-m}) where $m \geq 1$, then the procedure for generating the numerical approximation is known as a **multistep** or **continuing method**.

A.1 Single-Step Methods

For single-step methods, it is convenient to symbolize the numerical approximation to the exact solution $y = \phi(x)$ of the IVP (1) $y' = f(x, y)$; $y(x_0) = y_0$ by the recursive formula

$$y_{n+1} = y_n + \psi(x_n, y_n, h_n) \tag{3}$$

where $h_n = x_{n+1} - x_n$. The quantity h_n is called the **stepsize at** x_n, and it can vary with each step. However, for computations performed by hand, it is usually best to keep the stepsize constant—that is, set $h_n = h$, a constant, for $n = 0, 1, \ldots$.

A.1.1 Taylor Series Method

If $y(x)$ has $m + 1$ continuous derivatives on an interval I containing x_0, then by **Taylor's formula with remainder**,

$$\begin{aligned} y(x) &= y(x_n) + y^{(1)}(x_n)(x - x_n) + \frac{y^{(2)}(x_n)}{2}(x - x_n)^2 + \cdots \\ &\quad + \frac{y^{(m)}(x_n)}{m!}(x - x_n)^m + \frac{y^{(m+1)}(\xi)}{(m+1)!}(x - x_n)^{m+1} \end{aligned} \tag{4}$$

where ξ is between x and x_n. In particular, if $y(x)$ is a solution to the IVP (1) and $y(x)$ has $m+1$ continuous derivatives, then from the differential equation in (1)

$$y^{(1)}(x_n) = f(x_n, y(x_n))$$

and by repeated implicit differentiation and use of the chain rule, we obtain

$$\begin{aligned} y^{(2)}(x_n) &= f^{(1)}(x_n, y(x_n)) = f_x + f_y y^{(1)} = f_x + f_y f \\ y^{(3)}(x_n) &= f^{(2)}(x_n, y(x_n)) = f_{xx} + f_{xy} y^{(1)} + (f_{yx} + f_{yy} y^{(1)}) y^{(1)} + f_y y^{(2)} \\ &= f_{xx} + 2 f_{xy} f + f_{yy} f^2 + f_x f_y + f_y^2 f \end{aligned}$$

where $f^{(1)}$ denotes df/dx, where $f^{(2)}$ denotes $d^2 f/dx^2$, and where f and its partial derivatives are all evaluated at $(x_n, y(x_n))$. We could continue in this manner and eventually write any derivative of y evaluated at x_n, $y^{(k)}(x_n)$, up to and including order $m+1$, in terms of f and its partial derivatives evaluated at $(x_n, y(x_n))$. However, it is apparent that the evaluation of each successive higher order derivative by this technique usually becomes increasingly difficult unless the function f is very simple. Hence, one chooses m in equation (4) to be reasonably small and approximates $y(x_{n+1})$ by

$$(5) \quad y_{n+1} = y_n + f(x_n, y_n)h_n + \frac{f^{(1)}(x_n, y_n)}{2}h_n^2 + \cdots + \frac{f^{(m-1)}(x_n, y_n)}{m!}h_n^m.$$

This single-step method of numerical approximation to the solution of the IVP (1) is known as the **Taylor series expansion method of order m** and the **discretization error**, **truncation error**, or **formula error** for this method is given by

$$(6) \qquad E_n = \frac{f^{(m)}(\xi, y(\xi))}{(m+1)!}h_n^{m+1}$$

where $\xi \in (x_n, x_{n+1})$.

If all calculations were performed with infinite precision, discretization error would be the only error present. **Local discretization error** is the error that would be made in one step, if the previous values were exact and there were no round-off error. Ignoring round-off error, **global discretization error** is the difference between the solution $\phi(x)$ of the IVP (1) and the numerical approximation at x_n—that is, the global discretization error is $e_n = y_n - \phi(x_n)$.

A derivation of the series that bears his name was published by the English mathematician Brook Taylor (1685-1731) in 1715. However, the Scottish mathematician James Gregory (1638-1675) seems to have discovered the series more than forty years before Taylor published it. And Johann Bernoulli had published a similar result in 1694. The series was published without any discussion of convergence and without giving the truncation error term—equation (6).

Example 1 Third-Order Taylor Series Approximation to the Solution of the IVP: $\mathbf{y' = y + x;\ \ y(0) = 1}$

a. Find an approximate solution to the initial value problem

$$(7) \qquad y' = y + x = f(x, y); \quad y(0) = 1$$

on the interval $[0, 1]$ by using a Taylor series expansion of order 3 and a constant stepsize $h_n = .1$.

b. Estimate the maximum local discretization error on the interval $[0, 1]$.

Solution

a. From the differential equation in (7), we see that

$$y^{(1)} = f(x, y) = y + x, \qquad \text{so} \quad f(x_n, y_n) = y_n + x_n.$$

Differentiating the equation $y^{(1)} = f(x,y) = y + x$ three times, we find

$$\begin{aligned} f^{(1)}(x,y) &= y^{(1)} + 1 = y + x + 1, \quad \text{so} \quad f^{(1)}(x_n,y_n) = y_n + x_n + 1 \\ f^{(2)}(x,y) &= y^{(1)} + 1 = y + x + 1, \quad \text{so} \quad f^{(2)}(x_n,y_n) = y_n + x_n + 1 \\ f^{(3)}(x,y) &= y^{(1)} + 1 = y + x + 1, \quad \text{so} \quad f^{(3)}(x_n,y_n) = y_n + x_n + 1. \end{aligned}$$

Substituting these expressions into equation (5) with $m = 3$ and $h_n = .1$, we obtain the recursive formula

$$\begin{aligned} y_{n+1} &= y_n + (y_n + x_n)(.1) + \frac{y_n + x_n + 1}{2}(.01) + \frac{y_n + x_n + 1}{6}(.001) \\ &= .00516667 + .105167x_n + 1.10517y_n. \end{aligned}$$

In this case, each constant was rounded to six significant digits. The following table is the third-order Taylor series approximation to the IVP (7) on the interval $[0,1]$ obtained using a constant stepsize of $h = .1$. All calculations were performed using six significant digits.

Third-Order Taylor Series Approximation to the IVP: $\mathbf{y' = y + x;\ y(0) = 1}$ on $\mathbf{[0,1]}$ with Stepsize $\mathbf{h = .1}$

x_n	y_n
.0	1.00000
.1	1.11034
.2	1.24279
.3	1.39969
.4	1.58361
.5	1.79739
.6	2.04417
.7	2.32742
.8	2.65098
.9	3.01908
1.0	3.43641

b. The differential equation of the IVP (7) is linear and, in this instance, we can find the exact solution. Therefore, we could use the exact solution when estimating the maximum local discretization error. However, since we will not normally be able to obtain the exact solution, we will do what one must usually do in practice. We will use the information obtained from the numerical approximation to estimate the error. Examining the Taylor series numerical approximation values above, we see that $|y| < 3.45$ for $x \in [0,1]$.

For $x \in [0,1]$, we assume that $|y| < 7$ (which is slightly more than twice the largest y value appearing above) and using the triangle inequality, we see that

(8) $$|f^{(3)}(x,y)| \le |y| + |x| + 1 < 9 \quad \text{for} \quad x \in [0,1].$$

Using this upper bound in equation (6) with $m = 3$ and $h_n = .1$, we obtain the following estimate for the maximum local discretization error on the interval $[0,1]$.

$$|E| \le \frac{1}{4!} \max_{0 \le x \le 1} |f^{(3)}(x,y)|(.1)^4 < \frac{1}{4!} 9(.1)^4 < .0000375. \blacksquare$$

The following example illustrates how to estimate an appropriate stepsize for a Taylor series approximation given a specific accuracy requirement per step.

Example 2 Stepsize Selection for the Third-Order Taylor Series Method When using a Taylor series expansion of order 3 with constant stepsize h to approximate the solution of the IVP (7) $y' = y + x;\ y(0) = 1$ on the interval $[0,1]$, how small must the stepsize be in order to ensure six decimal place accuracy per step?

Solution

Setting $m = 3$ in equation (6), taking the absolute value of the resulting equation, and using the inequality (8), we find on the interval $[0,1]$ that the local discretization error satisfies

$$|E| \le \frac{1}{4!} \max_{0 \le x \le 1} |f^{(3)}(x,y)| h^4 < \frac{1}{4!}(9)h^4.$$

If we require h to satisfy

(9) $$|E| < \frac{1}{4!}(9)h^4 < .5 \times 10^{-6},$$

then the local discretization error will have six decimal place accuracy. Solving the right-hand inequality in (9) for h, we find

$$h < \left(\frac{4}{3} \times 10^{-6}\right)^{\frac{1}{4}} < 0.033.$$

Hence, any constant stepsize less than 0.033 will achieve the desired accuracy per step. $\blacksquare$

The computational disadvantage of using a Taylor series expansion to approximate the solution to an initial value problem is fairly obvious. For any given function $f(x,y)$ one must calculate the derivatives $f^{(1)}, f^{(2)}, \ldots, f^{(m-1)}$

and then evaluate all of theses derivatives at (x_n, y_n). However, the Taylor series expansion is of theoretical value, since most other numerical approximation schemes were derived by attempting to achieve a given order of accuracy without having to calculate higher order derivatives. As a matter of fact, an approximation technique is said to be of order m, if the local discretization error is proportional to the local discretization error of a Taylor series expansion of order m. Hence, when developing a numerical approximation scheme, the object is to produce an error which is proportional to h^{m+1} without having to compute any derivatives of $f(x, y)$.

A.1.2 Runge–Kutta Methods

Among the more popular single-step numerical approximation methods are those developed by the German mathematicians Carl David Tolmé Runge (1856-1927), Karl Heun (1859-1929), and Martin Wilhelm Kutta (1867-1944). Runge was an applied mathematician who studied spectral lines of elements and Diophantine equations. He devised and published his numerical technique in 1895. In 1900, Karl Heun published a paper concerning the improvement of Runge's method and in 1901 Kutta extended Runge's method to systems of equations. Kutta is also well-known for his contributions to airfoil theory.

Improved Euler's Method In deriving Euler's method for approximating the solution of the IVP (1) $y' = f(x, y);\ y(x_0) = y_0$, we noted that the IVP (1) is equivalent to the integral equation

$$y(x) = y_0 + \int_{x_0}^{x} f(t, y(t))\,dt$$

and we replaced the integrand $f(t, y(t))$ over the entire interval $[x_0, x_1]$ by its approximate value at the left endpoint, $f(x_0, y_0)$. Upon integrating from x_0 to x_1, we obtained Euler's formula for approximating the solution to the IVP (1) at x_1. A more accurate approximation may be obtained, if, instead of approximating the integrand by its approximate value at the left endpoint of the interval of integration, we approximate it by the average of its approximate values at the left endpoint and the right endpoint. Thus, when solving the general IVP $y' = f(x, y);\ y(x_n) = y_n$ on the interval $[x_n, x_{n+1}]$ which is equivalent to the integral equation

$$(10) \qquad y(x_{n+1}) = y_n + \int_{x_n}^{x_{n+1}} f(t, y(t))\,dt$$

we replace the integrand $f(t, y(t))$ by the constant

$$\frac{1}{2}(f(x_n, y_n) + f(x_{n+1}, y_{n+1})).$$

Substituting this expression into (10) and integrating, we obtain the following expression for the approximation of the solution to the IVP (1) at x_{n+1}:

$$y_{n+1} = y_n + \frac{1}{2}(f(x_n, y_n) + f(x_{n+1}, y_{n+1}))(x_{n+1} - x_n).$$

This equation involves the unknown y_{n+1} as an argument of f on the right-hand side; and, therefore, will generally be difficult or impossible to solve explicitly for y_{n+1}. Instead of trying to solve this equation for y_{n+1}, we simply replace the y_{n+1} appearing on the right-hand side by the approximation we obtain using Euler's method—namely, $y_{n+1} = y_n + f(x_n, y_n)(x_{n+1} - x_n)$. The following recursive formula which results is known as the **improved Euler's method**, the **second-order Runge–Kutta method**, or the **Heun method** for producing an approximation to the solution of the IVP (1):

$$(11) \qquad y_{n+1} = y_n + \frac{1}{2}[f(x_n, y_n) + f(x_{n+1}, y_n + f(x_n, y_n)h_n)]h_n$$

where $h_n = (x_{n+1} - x_n)$. The local discretization error for this method is

$$(12) \qquad E_n^I = -\frac{1}{12} f^{(2)}(\xi, y(\xi))h_n^3 = -\frac{1}{12} y^{(3)}(\xi)h_n^3$$

where $h_n = (x_{n+1} - x_n)$. Thus, the local discretization error for the improved Euler's method is proportional to the cube of the stepsize; whereas, the local discretization error for Euler's method is proportional to the square of the stepsize. The greater accuracy of the improved Euler's method must be paid for by an increase in the total number of computations which must be performed and the number of f function evaluations per step. Notice that f must be evaluated twice for each step when using the improved Euler's method; whereas, f is only evaluated once per step when using Euler's method.

Example 3 Improved Euler's Approximation of the Solution to the IVP: $\mathbf{y' = y + x;\; y(0) = 1}$

a. Find an approximate solution to the initial value problem

$$(7) \qquad y' = y + x = f(x, y); \quad y(0) = 1$$

on the interval $[0, 1]$ using the improved Euler's method and a constant stepsize $h = .1$.

b. Use equation (12) to estimate the maximum local discretization error on $[0, 1]$. ■

Solution

a. Table A.1 is the improved Euler's approximation to the IVP (7) on the interval $[0, 1]$ obtained using a constant stepsize of $h = .1$. The value $S_1 = f(x_n, y_n) = y_n + x_n$ is an approximation to the slope to the exact solution of (7) at x_n (the left endpoint of the interval of integration $[x_n, x_{n+1}]$). And $S_2 = f(x_{n+1}, y_n + f(x_n, y_n)h) = y_n + f(x_n, y_n)h + x_{n+1}$ is an approximation to the slope of the exact solution at x_{n+1} (the right endpoint of the interval of integration). All calculations were performed using six significant digits.

TABLE A.1
Improved Euler's Approximation to the IVP **(7)** $\mathbf{y' = y + x; \;\; y(0) = 1}$ **on** $\mathbf{[0,1]}$ **with Stepsize** $\mathbf{h = .1}$

n	x_n	y_n	S_1	S_2	$\frac{h}{2}(S_1+S_2)$	$y_{n+1} = y_n + \frac{h}{2}(S_1+S_2)$
0	.0	1.0	1.0	1.2	.11	1.11
1	.1	1.11	1.21	1.431	.13205	1.24205
2	.2	1.24205	1.44205	1.68625	.156415	1.39846
3	.3	1.39846	1.69846	1.96831	.183339	1.58180
4	.4	1.58180	1.98180	2.27998	.213089	1.79489
5	.5	1.79489	2.29489	2.62438	.245963	2.04085
6	.6	2.04085	2.64085	3.00494	.282289	2.32314
7	.7	2.32314	3.02314	3.42546	.322430	2.64557
8	.8	2.64557	3.44557	3.89013	.366785	3.01236
9	.9	3.01236	3.91236	4.40359	.415797	3.42815
10	1.0	3.42815				

b. We shall assume, as we did in the Taylor series expansion example, that $|y(x)| < 7$ on $[0,1]$. Since $y^{(3)} = y + x + 1$ and since $h = .1$, we see from equation (12) that the maximum local discretization error on $[0,1]$ satisfies

$$|E_n^I| = |\phi(x_n) - y_n| \le \frac{1}{12} h^3 \max_{x \in [0,1]} |y^{(3)}| \le \frac{1}{12} (.1)^3 (9) \approx .00075.$$

Analyzing the form of the recursive formula (11) might lead one to try to devise a more general recursion of the form

$$\text{(13)} \qquad y_{n+1} = y_n + [af(x_n, y_n) + bf(x_n + ch_n, y_n + df(x_n, y_n)h_n)]h_n$$

in which the constants a, b, c, and d are to be determined in such a manner that (13) will agree with a Taylor series expansion of as high an order as possible. As we have seen the Taylor series expansion for $y(x_{n+1})$ about x_n is

$$\text{(14)} \qquad y(x) = y(x_n) + fh_n + \frac{1}{2}(f_x + f_y f)h_n^2 + \\ \frac{1}{6}(f_{xx} + 2f_{xy}f + f_{yy}f^2 + f_x f_y + f_y^2 f)h_n^3 + O(h_n^4)$$

where f and its partial derivative are all evaluated at (x_n, y_n) and $O(h_n^4)$ indicates that the error made by omitting the remainder of the terms in the expansion is proportional to the fourth power of the stepsize.

Let $k_1 = f(x_n, y_n)$ and $k_2 = f(x_n + ch_n, y_n + dk_1h_n)$. Using the Taylor series expansion for a function of two variables to expand k_2 about (x_n, y_n), we obtain

$$(15) \qquad k_2 = f(x_n + ch_n, y_n + dk_1h_n) = f(x_n, y_n) + ch_nf_x + dk_1h_nf_y + \frac{c^2h_n^2}{2}f_{xx} + cdh_n^2k_1f_{xy} + \frac{d^2h_n^2k_1^2}{2}f_{yy} + O(h^3).$$

Substituting $k_1 = f$ into (15), substituting the resulting equation into (13), and rearranging in ascending powers of h_n, we find

$$(16) \qquad y_{n+1} = y_n + (a+b)fh_n + b(cf_x + df_yf)h_n^2 + b(\frac{c^2}{2}f_{xx} + cdff_{xy} + \frac{d^2}{2}f^2f_{yy})h_n^3 + O(h_n^4).$$

Comparing (14) with (16), we see that for the corresponding coefficients of h_n and h_n^2 to agree, we must have

$$(17) \qquad a + b = 1, \quad bc = \frac{1}{2}, \quad \text{and} \quad bd = \frac{1}{2}.$$

Thus, we have three equations in four unknowns. Hence, we might hope to be able to choose the constants in such a manner that the coefficients of h_n^3 in (14) and (16) agree. However, for these coefficients to agree, we must have

$$\frac{bc^2}{2} = \frac{1}{6}, \quad bcd = \frac{1}{3}, \quad \frac{bd^2}{2} = \frac{1}{6}, \quad \text{and} \quad f_xf_y + f_y^2f = 0.$$

Obviously, the last equality is not satisfied by all functions f.

There are an infinite number of solutions to the simultaneous equations (17). The choice $a = \frac{1}{2}$, $b = \frac{1}{2}$, $c = 1$, and $d = 1$ yields the improved Euler's method (11). Choosing $a = 0$, $b = 1$, $c = \frac{1}{2}$, and $d = \frac{1}{2}$ results in the following recursion which is known as the **modified Euler's method**:

$$(18) \qquad y_{n+1} = y_n + f(x_n + \frac{1}{2}h_n,\ y_n + \frac{1}{2}f(x_n, y_n)h_n)h_n.$$

Fourth-Order Runge–Kutta Method If one tries to develop a general recursion of the form

$$(19) \quad y_{n+1} = y_n + h_n[a_1f(x_n, y_n) + a_2f(x_n + b_1h_n, y_n + b_1h_nk_1) + a_3f(x_n + b_2h_n, y_n + b_2h_nk_2) + a_4f(x_n + b_3h_n, y_n + b_3h_nk_3)]$$

where $k_1 = f(x_n, y_n)$ and $k_i = f(x_n + b_{i-1}h_n, y_n + b_{i-1}h_nk_{i-1})$ for $i = 2, 3, 4$ by determining the constants $a_1, a_2, a_3, a_4, b_1, b_2$, and b_3 in such a manner that (19) will agree with a Taylor series expansion of as high an order as possible, one obtains a system of algebraic equations in the constants. In this case, as before, there are an infinite number of solutions to the system of equations.

The choice of constants which leads to the classical fourth-order Runge–Kutta recursion is

$$a_1 = a_4 = \frac{1}{6}, \quad a_2 = a_3 = \frac{1}{3}, \quad b_1 = b_2 = \frac{1}{2}, \quad \text{and} \quad b_3 = 1.$$

One usually finds the recursion for y_{n+1} written as

$$\begin{aligned} k_1 &= f(x_n, y_n) \\ k_2 &= f(x_n + \frac{h_n}{2}, \, y_n + \frac{h_n k_1}{2}) \\ k_3 &= f(x_n + \frac{h_n}{2}, \, y_n + \frac{h_n k_2}{2}) \\ k_4 &= f(x_n + h_n, \, y_n + h_n k_3) \end{aligned}$$

(20) $$y_{n+1} = y_n + h_n \frac{(k_1 + 2k_2 + 2k_3 + k_4)}{6}.$$

Hence, the fourth-order Runge–Kutta method may be viewed as a weighted average of four approximate values of the slope of the exact solution $f(t, \phi(t))$ at different points within the interval of integration $[x_n, x_{n+1}]$. The value k_1 is an approximation of the slope of the exact solution at the left endpoint of the interval of integration. The value k_2 is an approximation of the slope of the exact solution at the midpoint of the interval of integration which is obtained by using Euler's method to approximate $\phi(x_n + h_n/2)$. The value k_3 is another approximation of the slope at the midpoint of the interval of integration. And k_4 is an approximation of the slope at the right endpoint x_{n+1}. The local discretization error of the fourth-order Runge–Kutta method is proportional to h_n^5 and if f is a function of x alone, then the fourth-order Runge–Kutta recursion (20) reduces to Simpson's rule. Because of its relative high order of accuracy, the fourth-order Runge–Kutta method is one of the most commonly used single-step methods.

Example 4 Fourth-Order Runge–Kutta Approximation of the Solution to the IVP: $\mathbf{y' = y + x; \;\; y(0) = 1}$
Find an approximate solution to the initial value problem

(7) $$y' = y + x = f(x, y); \quad y(0) = 1$$

on the interval $[0, 1]$ using the fourth-order Runge–Kutta method and a constant stepsize $h = .1$. ■

Solution

Table A.2 contains the fourth-order Runge–Kutta approximation to the IVP (7) on the interval $[0,1]$ obtained using a constant stepsize of $h = .1$. All calculations were performed using six significant digits.

A tabular comparison of the methods we have used in this section to approximate the solution of the IVP (7) $y' = y + x;\ y(0) = 1$ on the interval $[0,1]$ with a constant stepsize of $h = .1$ is displayed in Table A.3. From this table it is obvious the fourth-order Runge–Kutta method is the most accurate method for approximating the solution to this particular initial value problem. However, since the fourth-order Runge–Kutta method requires four f function evaluations per step while the improved Euler's method requires two f function evaluations per step and Euler's method only requires one f function evaluation per step, the fourth-order Runge–Kutta method required approximately twice the computing time of the improved Euler's method and four times the computing time of Euler's method. Consequently, one might anticipate that the approximation to the solution of the IVP (7) generated using Euler's method with a constant stepsize $h = .025$, using the improved Euler's method with $h = .05$, and using the fourth-order Runge–Kutta method with $h = .1$ would require approximately the same amount of computing time and have approximately the same accuracy, since each method would then require 40 evaluations of the function f. Performing the necessary calculations, we obtain the results shown in Table A.4.

TABLE A.2
Fourth-Order Runge–Kutta Approximation to the IVP **(7)** $\mathbf{y' = y + x;\ y(0) = 1}$ **on** $\mathbf{[0,1]}$ **with Stepsize** $\mathbf{h = .1}$

x_n	y_n	k_1	k_2	k_3	k_4	$h(k_1 + 2k_2 + 2k_3 + k_4)/6$
.0	1.0	1.0	1.1	1.105	1.2105	.110341
.1	1.11034	1.21034	1.32086	1.32638	1.44298	.132463
.2	1.24280	1.44280	1.56494	1.57105	1.69991	.156911
.3	1.39971	1.69971	1.83470	1.84145	1.98386	.183931
.4	1.58364	1.98364	2.13283	2.14028	2.29767	.213792
.5	1.79744	2.29744	2.46231	2.47055	2.64449	.246794
.6	2.04423	2.64423	2.82644	2.83555	3.02778	.283266
.7	2.32750	3.02750	3.22887	3.23894	3.45139	.323574
.8	2.65107	3.45107	3.67362	3.68475	3.91954	.368122
.9	3.01919	3.91919	4.16515	4.17745	4.43694	.417355
1.0	3.43655					

TABLE A.3
Approximations to the Solution of the IVP (7) $\mathbf{y' = y + x;\ y(0) = 1}$ on $[\mathbf{0}, \mathbf{1}]$ with Stepsize $\mathbf{h = .1}$

x_n	Taylor Series Order 3	Euler's Method	Improved Euler's Method	Fourth-Order Runge–Kutta	Exact Solution
.0	1.0	1.0	1.0	1.0	1.0
.1	1.11034	1.1	1.11	1.11034	1.11034
.2	1.24279	1.22	1.24205	1.24280	1.24281
.3	1.39969	1.362	1.39846	1.39971	1.39972
.4	1.58361	1.5282	1.58180	1.58364	1.58365
.5	1.79739	1.72102	1.79489	1.79744	1.79744
.6	2.04417	1.94312	2.04085	2.04423	2.04424
.7	2.32742	2.19743	2.32314	2.32750	2.32751
.8	2.65098	2.48718	2.64557	2.65107	2.65108
.9	3.01908	2.81590	3.01236	3.01919	3.01921
1.0	3.43641	3.18748	3.48215	3.43655	3.43656

TABLE A.4
Approximations of the IVP (7) $\mathbf{y' = y + x;\ y(0) = 1}$

x_n	Euler $h = .025$	Improved Euler $h = .05$	Runge–Kutta $h = .1$	Exact Solution
.0	1.0	1.0	1.0	1.0
.1	1.10762	1.11025	1.11034	1.11034
.2	1.23680	1.24261	1.24280	1.24281
.3	1.38977	1.39939	1.39971	1.39972
.4	1.56900	1.58317	1.58364	1.58365
.5	1.77722	1.79678	1.79744	1.79744
.6	2.01743	2.04335	2.04423	2.04424
.7	2.29297	2.32637	2.32750	2.32751
.8	2.60749	2.64964	2.65107	2.65108
.9	2.96504	3.01742	3.01919	3.01921
1.0	3.37009	3.43437	3.43655	3.43656

These results illustrate that it is not only the number of f function evaluations which are used in producing an approximation to the solution but also the manner in which these function evaluations are combined which ultimately determines the accuracy of the approximate solution.

Stepsize Selection When the solution of an initial value problem is increasing or decreasing "slowly" a "large" stepsize may be taken by a numerical approximation technique. However, when the solution is increasing or decreasing "rapidly" a "small" stepsize must be taken by the numerical method in order to maintain accuracy. Stepsize control techniques usually involve comparing current local error estimates with previous local error estimates (absolute error or relative error estimates). Single-step methods have no computational difficulty in selecting an appropriate stepsize except for the initial step, since there is no previous results with which to compare. If the initial stepsize selected is too large, then the lack of accuracy which occurs at the first step will corrupt the accuracy of the entire numerical approximation. Therefore, most initial stepsize selection algorithms are very conservative and tend to select a stepsize which is somewhat smaller than actually required.

Many computer software packages permit the user to select the first stepsize. However, most users, including experienced users, often have no real informed idea as to what the best first stepsize should be. A software program which depends on the user's best guess of the first stepsize to start a numerical approximation method is not reliable. In many of our previous examples, we selected a constant stepsize of $h = .1$. If the computer we are using represents numbers internally as a power of the base 2, we have, in some respect, made a bad choice for the stepsize, since .1 cannot be represented exactly in that computer. Observe that

$$\begin{aligned}(.1)_{10} &= (0.0001100110011\ldots)_2 = (0.0121212\ldots)_4\\ &= (0.063146314\ldots)_8 = (0.1999999\ldots)_{16}\end{aligned}$$

where the subscript denotes the base. Rounding-off or chopping-off any of these infinite expansions after a specified number of digits and then adding ten of the same rounded or chopped numbers does not result in the number 1 exactly!

One technique which has been used to select stepsize, either initially or at each step, is to take the step twice with two different order numerical approximation methods and compare the results. If the results compare favorably, then the step is accepted. If not, the stepsize is decreased, two approximations obtained using the smaller stepsize are made and compared, etc.

A numerical approximation method which selects the stepsize to be used at each step is called an **adaptive method**. One of the more popular adaptive, single-step methods was published in 1968 and 1969 by Erwin Fehlberg in two NASA Technical Reports, R287 and R315. In these reports Fehlberg developed adaptive Runge–Kutta methods ranging in order from 2 to 9. In his procedure, two Runge–Kutta methods of different order are run simultaneously. The user specifies the desired upper bound for the local discretization error. At each step, the algorithm produces an estimate y_{n+1} of the exact solution. A second and more accurate method uses the same function evaluations as the first method plus one or two additional function evaluations to produce a more accurate estimate z_{n+1} of the exact solution. The difference $|z_{n+1} - y_{n+1}|$ yields an estimate of the local discretization error. If the prescribed error bound is satisfied, the step is taken and a new stepsize is estimated for the next step. If the prescribed error bound is not satisfied, the stepsize is reduced and another attempt is made to satisfy the error bound using the smaller stepsize. In 1974, L. F. Shampine and H. A. Watts implemented the Runge–Kutta-Fehlberg 4(5) method, RKF45. This method requires six function evaluations per each successful step. Four function values are combined to produce a fourth-order Runge–Kutta estimate y_{n+1} and all six function values are combined to produce a fifth-order Runge–Kutta estimate z_{n+1}. The RKF45 numerical approximation method is available in several commercial computer software packages.

A.1.3 Exercises

1. Consider the initial value problem $y' = x^2 - y;\ y(0) = 1$.

    a. Derive the Taylor series expansion formula of order 3 for this initial value problem.

    b. Use the formula derived in part a. and a constant stepsize $h = .1$ to calculate an approximate solution on the interval $[0, 1]$.

    c. Estimate the maximum discretization error per step on the interval $[0, 1]$ for the stepsize $h = .1$.

    d. How small must the stepsize be in order to ensure six decimal place accuracy per step?

2. a. Compute an approximate solution to the initial value problem $y' = x^2 - y;\ y(0) = 1$ on the interval $[0, 1]$ using Euler's method and a constant stepsize of $h = .1$.

    b. Find an upper bound for the total discretization error at $x = 1$.

    c. How small must the stepsize be to ensure six decimal place accuracy per step?

    d. How small must the stepsize be to ensure six decimal place accuracy over the interval $[0, 1]$?

3. Use the improved Euler's formula with a stepsize $h = .1$ to generate a numerical approximation to the solution of the IVP $y' = x^2 - y;\ y(0) = 1$ on the interval $[0, 1]$.

4. Use the modified Euler's formula with a stepsize $h = .1$ to generate a numerical approximation to the solution of the IVP $y' = x^2 - y;\ y(0) = 1$ on the interval $[0, 1]$.

5. Use the fourth-order Runge–Kutta formula with a stepsize $h = .1$ to generate a numerical approximation to the solution of the IVP $y' = x^2 - y;\ y(0) = 1$ on the interval $[0, 1]$.

6. a. Find the exact solution of the initial value problem $y' = x^2 - y; y(0) = 1$.

   b. Compare the various approximate solutions generated in Exercises 1- with each other, and the exact solution by producing a table of values.

7. Consider the general recursive formula (13). Suppose that in addition to satisfying equations (17), we require that the coefficients of $f^2 f_{yy}$ in equations (14) and (16) be equal. What is the solution of the resulting system of four equations in the four unknowns a, b, c, and d?

8. Generate numerical solutions to the IVP $y' = y/x + 2;\ y(1) = 1$ on the interval $[1, 2]$ with a stepsize of $h = .05$ using

   a. Euler's method.

   b. improved Euler's method.

   c. modified Euler's method.

   d. the fourth-order Runge–Kutta method.

   e. Find the explicit solution of the IVP $y' = y/x + 2;\ y(1) = 1$. On what interval does the solution exist in?

   f. Produce a table of values comparing the numerical solution values generated in parts a-d with the exact solution values on the interval $[1, 2]$.

A.2 Multistep Methods

Let $z(x)$ be a function which is defined on some interval containing the points $x_0, x_1, \ldots, x_n$. It is well known that there exists only one polynomial, $p(x)$, of degree less than or equal to n for which $p(x_i) = z(x_i)$ for $i = 0, 1, \ldots, n$. The polynomial $p(x)$ is called the **interpolating polynomial**. There are many ways to write an expression for the interpolating polynomial; however, we shall not present any of those expressions here. We are only interested in the fact that an interpolating polynomial exists and that it is unique.

Consider again the IVP (1) $y' = f(x, y);\ y(x_0) = y_0$. Suppose that we have used some single-step method to produce the approximations y_i to the exact solution $\phi(x_i)$ for $i = 1, 2, \ldots, n$, where the x_i's are equally spaced. Each single-step method computes $y_i' = f(x_i, y_i)$ in order to produce y_i. Henceforth, let $y_i' = f(x_i, y_i) = f_i$. In deriving single-step methods, we integrated the differential equation $y' = f(x, y)$ of the IVP (1) over the interval $[x_n, x_{n+1}]$ to obtain the integral equation

$$y(x_{n+1}) = y(x_n) + \int_{x_n}^{x_{n+1}} f(t, y(t))\, dt$$

and then we approximated f on the interval $[x_n, x_{n+1}]$ and integrated. Multistep methods are usually derived by integrating the differential equation $y' = f(x, y)$ from x_{n-p} to x_{n+q} where $p, q \geq 0$ and by approximating the integrand f on the interval $[x_{n-p}, x_{n+q}]$ by the interpolating polynomial $p(x)$ which interpolates f at the $m + 1$ points $x_{r-m}, x_{r-m+1}, \ldots, x_{r-1}, x_r$, where $r = n$ or $r = n + 1$. See Figure A.2. If $r = n$, the resulting formula is said to

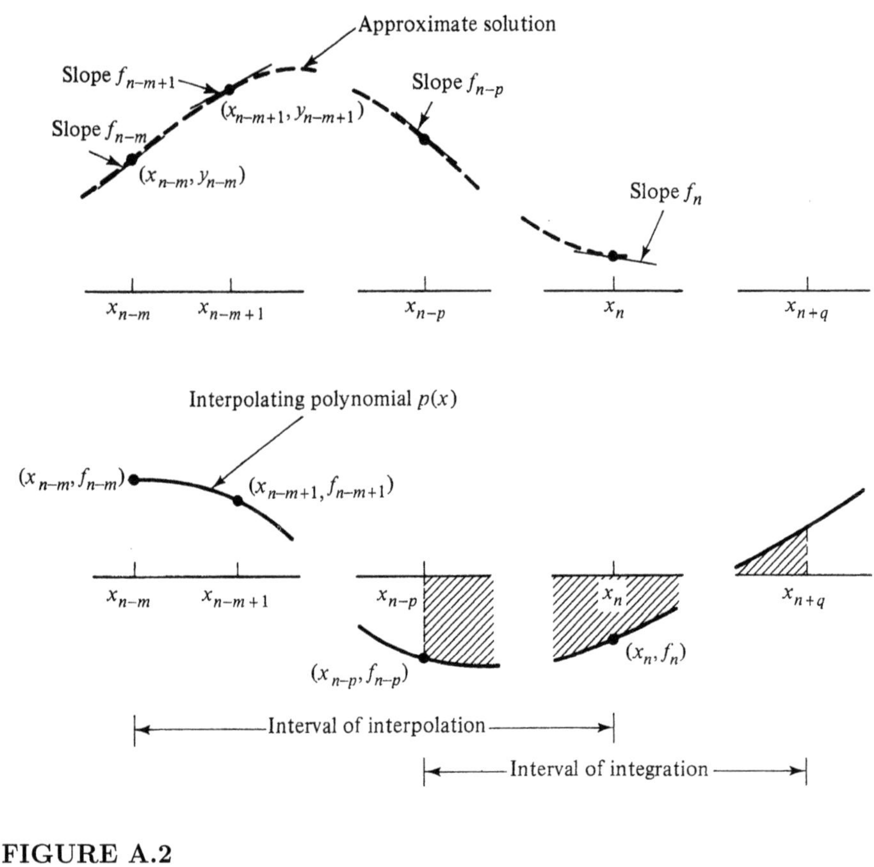

FIGURE A.2
Multistep Integration Diagram for $r = n$.

be **open**; whereas, if $r = n+1$, the resulting formula is called **closed**. Open formulas are explicit formulas for y_{n+1}. Closed formulas, on the other hand, are implicit formulas for y_{n+1}.

A.2.1 Adams–Bashforth Methods

The English mathematician and astronomer John Couch Adams (1819-1892) introduced multistep methods prior to the introduction of Runge–Kutta methods. Adams is perhaps better known as an astronomer than a mathematician. In 1845, he predicted the existence and orbit of a new planet—the planet Neptune. His prediction was based upon an analysis of the perturbations of the orbit of Uranus. In 1882, J. C. Adams and Francis Bashforth published the Adams–Bashforth methods for numerical integration in an article on the theory of capillary action.

Choosing $r = n$, $p = 0$, and $q = 1$ results in a set of formulas which are known as Adams–Bashforth formulas. Determining the interpolating polynomial for the first few values of m and integrating results in the following formulas and their local discretization errors. In each case $\xi \in (x_{m-n}, x_n)$. The formula for $m = 0$ is a single-step method—Euler's method. Notice that each method utilizes f evaluated at $m+1$ points and has a local discretization error of order h^{m+2}. Hence, each formula requires $m+1$ starting values.

m	Adams–Bashforth Formulas	Error
0	$y_{n+1} = y_n + hf_n$	$\dfrac{h^2 y^{(2)}(\xi)}{2}$
1	$y_{n+1} = y_n + \dfrac{h(3f_n - f_{n-1})}{2}$	$\dfrac{5h^3 y^{(3)}(\xi)}{12}$
2	$y_{n+1} = y_n + \dfrac{h(23f_n - 16f_{n-1} + 5f_{n-2})}{12}$	$\dfrac{9h^4 y^{(4)}(\xi)}{24}$
3	$y_{n+1} = y_n + \dfrac{h(55f_n - 59f_{n-1} + 37f_{n-2} - 9f_{n-3})}{24}$	$\dfrac{251h^5 y^{(5)}(\xi)}{720}$

Example 5 Adams–Bashforth Approximations of the Solution to the IVP: $\mathbf{y' = y + x;\ \ y(0) = 1}$

Find an approximate solution to the initial value problem

(7) $$y' = y + x = f(x,y); \quad y(0) = 1$$

on the interval $[0,1]$ using a constant stepsize $h = .1$ and Adams–Bashforth formulas for $m = 1, 2, 3$. Use the starting values obtained from the fourth-order Runge–Kutta method. ■

TABLE A.5
Adams–Bashforth Approximations to the Solution of (7) $\mathbf{y}' = \mathbf{y}+\mathbf{x};\ \mathbf{y}(\mathbf{0}) = \mathbf{1}$ on $[\mathbf{0}, \mathbf{1}]$ with Stepsize $\mathbf{h} = .\mathbf{1}$

x_n	Runge–Kutta Starting Values	Adams–Bashforth Formulas $m = 1$ y_n	$m = 2$ y_n	$m = 3$ y_n	Exact Solution
.0	1.0				1.0
.1	1.11034				1.11034
.2	1.24280	1.24189			1.24281
.3	1.39971	1.39766	1.39963		1.39972
.4		1.58021	1.58345	1.58364	1.58365
.5		1.79236	1.79711	1.79742	1.79744
.6		2.03720	2.04374	2.04420	2.04424
.7		2.31816	2.32682	2.32745	2.32751
.8		2.63903	2.65018	2.65100	2.65108
.9		3.00397	3.01804	3.01911	3.01921
1.0		3.41762	3.43509	3.43644	3.43656

Solution

Table A.5 contains the Adams–Bashforth approximations for $m = 1, 2, 3$ to the solution of the IVP (7) on the interval $[0, 1]$ obtained using a constant stepsize of $h = .1$. For $m = 1$ the initial condition, $y(0) = 1$, and one fourth-order Runge–Kutta value $y_1 = 1.11034$ were used to start the Adams–Bashforth method. For $m = 2$ the initial condition, $y(0) = 1$, and two fourth-order Runge–Kutta values $y_1 = 1.11034$ and $y_2 = 1.24280$ were used to start the Adams–Bashforth method.

A.2.2 Nystrom Methods

A set of formulas called Nystrom formulas results by selecting $r = n$, $p = 1$, and $q = 1$. These formulas were derived by the Finnish mathematician E. J. Nystrom and published in 1925. The formula for $m = 0$—which is the same as the formula for $m = 1$—is known as the **midpoint rule** and has a local discretization error of $h^3 y^{(3)}(\xi)/6$, where $\xi \in (x_{n-1}, x_{n+1})$. The midpoint rule has the simplicity of Euler's method and has a smaller local error; however, it requires one starting value and is generally less stable than Euler's method. (The term "stable" is discussed in Section A.3.) After calculating the interpolating polynomial and integrating one obtains the following formulas.

m	Nystrom Formulas
0	$y_{n+1} = y_{n-1} + 2hf_n$
1	$y_{n+1} = y_{n-1} + 2hf_n$
2	$y_{n+1} = y_{n-1} + \dfrac{h(7f_n - 2f_{n-1} + f_{n-2})}{3}$
3	$y_{n+1} = y_{n-1} + \dfrac{h(8f_n - 5f_{n-1} + 4f_{n-2} - f_{n-3})}{3}$

A.2.3 Adams–Moulton Methods

The Adams–Bashforth methods and the Nystrom methods employ polynomials which interpolate f at x_n and the preceding points $x_{n-1}, \ldots, x_{n-m}$. If we find an interpolating polynomial which interpolates f at $x_{n-1}, x_n, \ldots, x_{n-m}$ and integrate from x_n to x_{n+1}, we obtain a set of formulas known as Adams–Moulton formulas. In our notation $r = n + 1$, $p = 0$, and $q = 1$. The first few Adams–Moulton formulas and their associated local discretization errors follow.

m	Adams–Moulton Formulas	Error
0	$y_{n+1} = y_n + \dfrac{h(f_{n+1} + f_n)}{2}$	$\dfrac{h^3 y^{(3)}(\xi)}{12}$
1	$y_{n+1} = y_n + \dfrac{h(5f_{n+1} + 8f_n - f_{n-1})}{12}$	$\dfrac{h^4 y^{(4)}(\xi)}{24}$
2	$y_{n+1} = y_n + \dfrac{h(9f_{n+1} + 19f_n - 5f_{n-1} + f_{n-2})}{12}$	$\dfrac{19h^5 y^{(5)}(\xi)}{720}$

In each case $\xi \in (x_{m-n}, x_{n+1})$. All of these formulas are implicit formulas for y_{n+1}, since y_{n+1} appears on the right-hand side of each equation in $f_{n+1} = f(x_{n+1}, y_{n+1})$. The formula for $m = 0$ is called the **trapezoidal scheme**. Forest Ray Moulton (1872-1952) was an American astronomer who improved the Adams formula substantially while computing numerical solutions to ballistic problems during World War I.

A.2.4 Exercises

1. a. Compute a numerical approximation to the initial value problem $y' = x^2 - y;\ y(0) = 1$ on the interval $[0, 1]$ using the Adams-Bashforth formula for $m = 1$. Use a constant stepsize $h = .1$ and use the exact solution values for starting values.

   b. Estimate the maximum local discretization error on the interval $[0, 1]$.

   c. How small must the stepsize be to ensure six decimal place accuracy per step?

2. a. Use the midpoint rule and a stepsize $h = .1$ to compute a numerical approximation to the initial value problem $y' = x^2 - y;\ y(0) = 1$ on the interval $[0, 1]$. Use the exact solution values for starting values.

   b. Estimate the maximum local discretization error on the interval $[0, 1]$.

   c. How small must the stepsize be to ensure six decimal place accuracy per step?

3. Consider the initial value problem $y' = x^2 - y;\ y(0) = 1$.

   a. For the given initial value problem solve the Adams–Moulton formula for $m = 0$ explicitly for y_{n+1}.

   b. Use the formula derived in part a. to produce a numerical solution to the given initial value problem on the interval $[0, 1]$ using a stepsize of $h = .1$.

4. Make a table of values and compare the numerical approximations produced in Exercises 1-3 with the exact solution of the initial value problem $y' = x^2 - y;\ y(0) = 1$ on the interval $[0, 1]$.

A.3 Predictor-Corrector Methods

The Adams–Moulton formulas of the previous section are implicit formulas. In general, $f(x, y)$ will be nonlinear and it will be impossible to solve the implicit formula explicitly for y_{n+1}. However, we can try to determine y_{n+1} by iteration. That is, we obtain, in some manner, a first approximation to y_{n+1}, call it $y_{n+1}^{<0>}$, and then we successively calculate $f_{n+1}^{<k>} = f(x_{n+1}, y_{n+1}^{<k>})$ and use this approximation of f_{n+1} in the implicit formula to successively calculate $y_{n+1}^{<k+1>}$ for $k = 0, 1, \ldots$. Under fairly general conditions on the function f, it can be shown that for sufficiently small values of the stepsize h, the sequence $< y_{n+1}^{<k>} >_{k=1}^{\infty}$ converges to a solution y_{n+1} of the implicit formula and that the solution y_{n+1} is unique. [Note that y_{n+1} will be the solution of the implicit formula; but, in general, it will not be the solution of the differential equation at x_{n+1}, $\phi(x_{n+1})$.] An explicit formula that is used to obtain the first approximation, $y_{n+1}^{<0>}$, is called a **predictor formula** and an implicit formula used in the iteration procedure to calculate $y_{n+1}^{<k>}$ for $k = 1, 2, \ldots$ is called a **corrector formula**. One usually chooses the predictor and corrector formulas of the iteration procedure so that the order of the local discretization error of each formula is nearly the same. Generally, the corrector formula is chosen so that the local error is smaller than the local error of the predictor formula. Both single-step and multistep predictor-corrector methods may be devised. A simple single-step, predictor-corrector method, for instance, might employ Euler's formula for the predictor and the trapezoidal scheme (Adams–Moulton formula with $m = 0$) for the corrector. Hence, one would have the following iteration procedure:

(21p) $$y_{n+1}^{<0>} = y_n + h f_n(x_n, y_n)$$

(21c) $$y_{n+1}^{<k>} = y_n + \frac{h(f(x_{n+1}, y_{n+1}^{<k-1>}) + f(x_n, y_n))}{2}, \quad k = 1, 2, \ldots$$

The local discretization error for the predictor formula (21p) is $h^2 y^{(2)}(\xi)/2$ and the local error of the correction formula (21c) is $h^3 y^{(3)}(\xi)/12$.

The following multistep, predictor-corrector, iteration procedure was derived and published in 1926 by the American mathematician William E. Milne (1890-1971). The explicit predictor formula was derived by choosing $r = n$, $p = 3$, $q = 1$, and $m = 3$ and the implicit corrector was derived by choosing $r = n + 1$, $p = 2$, $q = 1$, and $m = 2$ or $m = 3$ (which yield the same formula). Hence, the Milne predictor-corrector iteration is

(22p) $$y_{n+1}^{<0>} = y_{n-3} + \frac{4h(2f_n - f_{n-1} + 2f_{n-2})}{3}$$

(22c) $$y_{n+1}^{<k>} = y_{n-1} + \frac{h(f(x_{n+1}, y_{n+1}^{<k-1>}) + 4f_n + f_{n-1})}{3}, \quad k = 1, 2, \ldots$$

The local discretization error of each of these formulas is of order h^5.

Another commonly used predictor-corrector method for which each formula has local error of order h^5 but a slightly smaller error coefficient than Milne's method employs the Adams–Bashforth formula with $m = 3$ as the predictor and the Adams–Moulton formula with $m = 2$ as the corrector. This iteration procedure is

(23p)

$$y_{n+1}^{<0>} = y_n + \frac{h(55f_n - 59f_{n-1} + 37f_{n-2} - 9f_{n-3})}{24}$$

(23c)

$$y_{n+1}^{<k>} = y_n + \frac{h(9f(x_{n+1}, y_{n+1}^{<k-1>}) + 19f_n - 5f_{n-1} + f_{n-2})}{24}, \quad k = 1, 2, \ldots$$

In order to approximate the solution of an initial value problem using a predictor-corrector algorithm, one needs to specify (1) the stepsize, h, to be taken; (2) the maximum absolute iteration error, $E = |y_{n+1}^{<k>} - y_{n+1}^{<k-1>}|$, or the maximum relative iteration error, $\epsilon = |y_{n+1}^{<k>} - y_{n+1}^{<k-1>}|/|y_{n+1}^{<k>}|$, to be allowed per step; (3) the maximum number of iterations, K, to be taken per step; and (4) what to do if K is reached before the error requirement E or ϵ is satisfied. And, of course, if the predictor-corrector algorithm being utilized involves a multistep formula, one must obtain starting values using some single-step method. In selecting the stepsize, the maximum iteration error per step, and the maximum number of iterations per step, one must keep in mind that these are not independent but are related through the algorithm local error formula which in turn depends upon the differential equation. Usually, the maximum iteration error desired per step is chosen, the maximum number of iterations per step is set at a small number—often two or three, and the stepsize is then determined so that it is consistent with the maximum iteration error, the maximum number of iterations per step, the algorithm, and the differential equation.

Using the predictor-corrector formulas (21), (22), and (23), we generated numerical approximations to the solution of the IVP (7) $y' = y + x$; $y(0) = 1$ on the interval $[0, 1]$ using a constant stepsize $h = .1$. Where necessary (for equations (22) and (23)), we used the exact solution values as the starting values. We set the maximum absolute iteration error, E, equal to 5×10^{-6} and recorded the number of iterations per step, k, required to achieve this accuracy. The results of our calculations are shown in Table A.6. Observe that the single-step predictor-corrector formula (21) is not able to maintain accuracy with four iterations per step while the multistep predictor-corrector formulas (22) and (23) are able to do so with only two iterations per step.

Convergence and Instability A numerical method for approximating a solution to an initial value problem is said to be **convergent** if, assuming there is no round-off error, the numerical approximation approaches the exact solution as the stepsize approaches zero. All the numerical methods presented in this text are convergent. However, this does *not* mean that as the stepsize approaches zero, the numerical approximation will *always* approach the exact solution, since round-off error will always be present. Sometimes the error of a numerical approximation turns out to be larger than predicted by the local discretization error estimate. And, furthermore as the stepsize is decreased, the error for a particular fixed value of the independent variable may become larger instead of smaller. This phenomenon is known as **numerical instability**. Numerical instability is a property of both the numerical method and the initial value problem. That is, a numerical method may be unstable for some initial value problems and stable for others. Numerical instability usually arises because a first-order differential equation is approximated by a second or higher order difference equation. The approximating difference equation will have two or more solutions—the fundamental solution which approximates the exact solution of the initial value problem, and one or more

TABLE A.6
Predictor-Corrector Approximations to the Solution of (7) $\mathbf{y' = y + x;\ y(0) = 1}$ on $[\mathbf{0}, \mathbf{1}]$ with Stepsize $\mathbf{h} = .\mathbf{1}$

	Equations (21)		Equations (22)		Equations (23)		Exact
x_n	y_n	k	y_n	k	y_n	k	Solution
.0							1.00000
.1	1.11053	4					1.11034
.2	1.24321	4					1.24281
.3	1.40039	4					1.39972
.4	1.58464	4	1.58365	2	1.58365	2	1.58365
.5	1.79882	4	1.79744	2	1.79744	2	1.79744
.6	2.04606	4	2.04424	2	2.04424	2	2.04424
.7	2.32985	4	2.32751	2	2.32751	2	2.32751
.8	2.65405	4	2.65108	2	2.65108	2	2.65108
.9	3.02290	4	3.01921	2	3.01921	2	3.01921
1.0	3.44109	4	3.43656	2	3.43657	2	3.43656

parasitic solutions. The parasitic solutions are so named because they "feed" upon the errors (both round-off and local discretization errors) of the numerical approximation method. If the parasitic solutions remain "small" relative to the fundamental solution, then the numerical method is stable; whereas, if a parasitic solution becomes "large" relative to the fundamental solution, then the numerical method is unstable. For h sufficiently small, single-step methods do not exhibit any numerical instability for any initial value problems. On the other hand, multistep methods may be unstable for some initial value problems for a particular range of values of the stepsize or for all stepsize. In practice, one chooses a particular numerical method and produces a numerical approximation for two or more reasonably "small" stepsizes. If the approximations produced are essentially the same, then the numerical method is probably stable for the problem under consideration and the results are probably reasonably good also. If the results are not similar, then one should reduce the stepsize further. If dissimilar results persist, then the numerical method is probably unstable for the problem under consideration and a different numerical method should be employed.

Example 6 Numerical Instability Use Euler's method ($y_{n+1} = y_n + hf_n$) and the midpoint rule ($y_{n+1} = y_{n-1} + 2hf_n$) with a constant stepsize $h = .1$ to produce numerical approximations on the interval $[0, 2.4]$ to the initial value problem $y' = -3y + 1;\ y(0) = 1$. Compare the numerical results with the exact solution.

TABLE A.7
Euler and Midpoint Rule Approximations to the IVP $\mathbf{y' = -3y + 1;\ y(0) = 1}$ on the Interval $\mathbf{[0, 2.4]}$ with $\mathbf{h = .1}$

	Exact Solution	Euler's Method		Midpoint Rule	
x_n	$\phi(x_n)$	y_n	$y_n - \phi(x_n)$	y_n	$y_n - \phi(x_n)$
.0	1.000000	1.000000	.000000	1.000000	.000000
.2	.699208	.870000	.170792	.703673	.004465
.4	.534129	.596300	.062171	.540668	.006539
.6	.443533	.462187	.018654	.452303	.008770
.8	.393812	.396472	.002660	.406768	.013868
1.0	.366525	.364271	−.002254	.387669	.021144
1.2	.351549	.348493	−.003056	.388130	.036581
1.4	.343330	.340762	−.002568	.408319	.064989
1.6	.338820	.336973	−.001847	.455502	.116682
1.8	.336344	.355117	−.001227	.546665	.210321
2.0	.334986	.334207	−.000779	.714628	.379642
2.2	.334241	.333762	−.000479	1.019856	.685615
2.4	.333831	.333543	−.000288	1.572232	1.238401

Solution

We easily find the exact solution of the given linear initial value problem to be $y(x) = (2e^{-3x} + 1)/3$. Table A.7 contains the values of the exact solution, Euler's approximation, Euler's approximation minus the exact solution, the midpoint rule approximation, and the midpoint rule approximation minus the exact solution. Notice that near the initial value, $x_0 = 0$, the midpoint rule approximation is more accurate than the Euler's method approximation. This is due to the fact that the midpoint rule has a smaller local discretization error. But notice that as x increases, the error of the midpoint rule approximation increases rapidly. This occurs because the parasitic solution associated with the midpoint rule is beginning to overwhelm the fundamental solution. For $x > .8$ the Euler's approximation is more accurate than the midpoint rule approximation. ■

In summary, single-step methods, such as the fourth-order Runge–Kutta method, have the advantages of being self-starting, numerically stable, and requiring a small amount of computer storage. They have the disadvantages of requiring multiple function evaluations per step and providing no error estimates except for Runge–Kutta-Fehlberg methods. Multistep methods have the advantage of requiring only one function evaluation per step but have the disadvantages of requiring starting values, occasionally being numerically unstable, providing no error estimate, and requiring more computer storage than single-step methods. Predictor-corrector methods provide error estimates at each step and require only a few function evaluations per step. The amount of computer storage and the numerical stability of predictor-corrector algorithms depend upon whether the formulas employed are single-step or multistep.

Many people prefer to use a fourth or higher order single-step method such as a Runge–Kutta method to numerically solve simple initial value problems on a one-time basis over a "small" interval. Such a method is usually selected because it is self-starting and numerically stable but—most usually—because the user has a better understanding of and confidence in such a method.

However, when a numerical method is to be used to solve the same or similar complex initial value problems many times or the solution is to be produced over a "large" interval, then some form of a multistep, predictor-corrector method should be chosen. Such a method of solution should be selected because it normally requires less computing time to produce a given accuracy and because an estimate of the error at each step is built into the method.

In this appendix, we presented several single-step, multistep, and predictor-corrector methods for approximating the solution of an initial value problem. In our examples, we always used a constant stepsize; however, commercially available software permits the user to specify the initial value problem, the interval on which it is to be solved, and an error bound—either absolute or relative error. If the solution values vary by orders of magnitude, then it is more appropriate to specify a relative error bound. It is essential that the software you choose to use have an automatic stepsize selection and error control routine. In addition, the software should have the following features as well. If the algorithm is a multistep method, the software should determine the necessary starting values itself. If the algorithm is a variable-order method, which Adams and predictor-corrector methods sometimes are, then the algorithm should select and change the order of the method automatically. Also the software should calculate approximate values of the solution at any set of points the user chooses rather than just at points selected by the algorithm during its integration procedure.

Answers to Selected Exercises

Chapter 1 Introduction

Exercises 1.2 Definitions and Terminology

	Order	Linear or Nonlinear
1.	1	Linear
3.	1	Nonlinear
5.	1	Nonlinear
7.	1	Linear
9.	2	Nonlinear
11.	2	Linear
13.	3	Nonlinear

15. $(-\infty, -1),\ (1, \infty)$ 17. $y = 0$

19. a. No, $y = 1/x$ is not differentiable at 0.
    b. Yes, $y = 1/x$ is differentiable on $(0, \infty)$ and satisfies the differential equation on $(0, \infty)$.
    c. Yes, $y = 1/x$ is differentiable on $(-\infty, 0)$ and satisfies the differential equation on $(-\infty, 0)$.

21. a. No, $y = \sqrt{x}$ is not defined and not differentiable on $(-1, 0)$.
    b. Yes, $y = \sqrt{x}$ is differentiable on $(0, \infty)$ and satisfies the differential equation on $(0, \infty)$.
    c. No, $y = \sqrt{x}$ is not defined and not differentiable on $(-\infty, 0)$.

Intervals on which the solution exists.

23. $(-\infty, 0),\ (0, \infty)$ 25. $(0, \infty)$ 27. $(-\infty, 0),\ (0, \infty)$

29. y_1 is a solution on $(-\infty, \infty)$. y_2 is a solution on $(0, \infty)$.

31. $r = -3$ 33. $r = -1$ 35. $r = 1/2$ 37. $r = -4,\ r = -1$

39. a. $r = 0$ b. $r = 2/29$ c. $r = 1/29$

Exercises 1.3

7. a. $y = 2x \ln x / \ln 2$

   b. No unique solution. There are an infinite number of solutions of the form $y = c_1 x \ln x$ where c_1 is arbitrary.

   c. No solution.

9. $y = 2x - x^2 + x^3$

Exercises 1.4 A Nobel Prize Winning Application

1. 59.5%; 35.8%; 675.7 years   3. $133\frac{1}{3}$ grams; 2.4 years

5. Approximately 3941 years before 1950 or 1991 B.C.

Chapter 2 The Initial Value Problem $\mathbf{y' = f(x, y);\ y(c) = d}$

Exercises 2.1 Direction Fields

1. $A \leftrightarrow b$, $B \leftrightarrow a$ 3. $E \leftrightarrow e$, $F \leftrightarrow f$ 5. $I \leftrightarrow j$, $J \leftrightarrow i$

7. The direction field is defined in the entire xy-plane. The function $y = 0$ is a solution on $(-\infty, \infty)$. In the first and third quadrants, the solutions are strictly increasing, and in the second and fourth quadrants, the solutions are strictly decreasing. Thus, relative minima occur on the positive y-axis and relative maxima occur on the negative y-axis. If a solution is positive for $x > 0$, then $y(x) \to +\infty$ as $x \to +\infty$. If a solution is negative for $x > 0$, then $y(x) \to -\infty$ as $x \to +\infty$. If a solution is positive for $x < 0$, then $y(x) \to +\infty$ as $x \to -\infty$. If a solution is negative for $x < 0$, then $y(x) \to -\infty$ as $x \to -\infty$.

9. The direction field is undefined on the y-axis where $x = 0$. The function $y = 0$ is a solution for $x \neq 0$. In the first and third quadrants, the solutions are strictly increasing, and in the second and fourth quadrants, the solutions are strictly decreasing. If a solution is positive for $x > 0$, then $y(x) \to +\infty$ as $x \to +\infty$. If a solution is negative for $x > 0$, then $y(x) \to -\infty$ as $x \to +\infty$. If a solution is positive for $x < 0$, then $y(x) \to +\infty$ as $x \to -\infty$. If a solution is negative for $x < 0$, then $y(x) \to -\infty$ as $x \to -\infty$.

11. The functions $y = 0$ and $y = 3$ are solutions. For $y > 3$ solutions are strictly increasing and asymptotic to $y = 3$ as $x \to -\infty$. For $0 < y < 3$ solutions are strictly decreasing, asymptotic to $y = 3$ as $x \to -\infty$ and asymptotic to $y = 0$ as $x \to +\infty$. For $y < 0$ solutions are strictly decreasing and asymptotic to $y = 0$ as $x \to +\infty$.

13. The function $y = 0$ is a solution. All other solutions are increasing. There are no relative minima or relative maxima. Solutions below $y = 0$ are asymptotic to $y = 0$ as $x \to +\infty$. Solutions above $y = 0$ are asymptotic to $y = 0$ as $x \to -\infty$.

15. The direction field is undefined for $y \leq -x$. Solutions increase for $y > -x + 1$ and decrease for $-x < y < -x + 1$. Relative minima occur where $y = -x + 1$.

17. The direction field is undefined on and outside the circle $x^2 + y^2 = 15$. All solutions inside the circle are strictly increasing.

Exercises 2.2 Fundamental Theorems

1. $(-3, d)$ and $(5, d)$ for all d 3. $(0, d)$ for all d and $(c, 0)$ for all c

5. (c, d) for all c and $-2 < d < 1$ 7. $(c, 0)$ for all c

9. $(c, 2)$ and $(c, -2)$ for all c. 11. $(c, 1)$ for all c

13. (c, d) where $c \geq 0$ and $d < 4$ and (c, d) where $c \leq 0$ and $d > 4$

15. $(-\infty, \infty)$ 17. $(-\infty, \infty)$ 19. $(-\infty, \infty)$ 21. $(\pi/2, 3\pi/2)$

23. $(0, 2)$ 27. No 31. No, because f_y is undefined at $(6, -9)$.

Exercises 2.3 Solution of Simple First-Order Differential Equations

Exercises 2.3.1 Solution of $\mathbf{y' = g(x)}$

1. $y = \frac{2}{3}x^{3/2} + C \quad (0, \infty)$ 3. $y = 2\ln|x - 3| + C \quad (-\infty, 3),\ (3, \infty)$

5. $y = \ln|x + 1| + C \quad (-\infty, -1),\ (1, \infty)$ 7. $x \ln x - x + C \quad (0, \infty)$

9. $y = -\cot(x/2) + C \quad (2n\pi, 2(n+1)\pi)$ where n is an integer.

11. $y = \text{Arctan}\, x + C \quad (-\infty, \infty)$

13. $y = \ln|x + \sqrt{x^2 - 1}| + C \quad (-\infty, -1),\ (1, \infty)$

15. $y = \frac{1}{2}\ln|x^2 - 4x + 5| + 7\text{Arctan}\,(x - 2) + C \quad (-\infty, \infty)$

17. $y = \frac{3}{2}x^2 + x - \frac{1}{2}\ ; \quad (-\infty, \infty)$ 19. $y = -2\cos x - 1\ ; \quad (-\infty, \infty)$

21. $y = 1 + \ln|x - 1|\ ; \quad (1, \infty)$

23. $y = 1 + \frac{1}{2}\ln 3 + \frac{1}{2}\ln\left|\dfrac{x-1}{x+1}\right|\ ; \quad (1, \infty)$

25. $y = -\ln|\cos x|\ ; \quad (-\pi/2, \pi/2)$

Exercises 2.3.2 Solution of the Separable Equation $\mathbf{y' = g(x)/h(y)}$

1. $y = 2x^3/3 + C$ 3. $1 + s^2t^2 = Cs^2$ 5. $e^{-y} = -e^x + C$

7. $y = Ce^{x^3/3+x}$ 9. $Cy\sqrt{1+x^2} = 1$ 11. $x\ln x + \ln|\ln y| = x + C$

13. $y = -e^{3x}$; $(-\infty, \infty)$ 15. $y = 1 + e^{-x}$; $(-\infty, \infty)$

17. $y = -2x$; $(-\infty, 0)$ 19. $y = 2/(1+e^{2x})$; $(-\infty, \infty)$

21. $y = -\ln(\frac{x^2}{2} + 1)$; $(-\infty, \infty)$ 23. $r = r_0e^{-2t^2}$; $(-\infty, \infty)$

Exercises 2.3.3 Solution of the Linear Equation $\mathbf{y' = a(x)y + b(x)}$

1. $y = e^{3x}$; $(-\infty, \infty)$ 3. $y = x^4 + Cx^3$; $(-\infty, 0)$, $(0, \infty)$

5. $i = 2 + Ce^{-t}$ $(-\infty, \infty)$

7. $y = C\sin x - \cos x$ for $x \neq n\pi$ where n is an integer.

9. $r = \frac{\theta^2}{6} + \frac{C}{\theta^4}$ $(-\infty, 0)$, $(0, \infty)$ 11. $y = (5e^{4x} - 1)/4$; $(-\infty, \infty)$

13. $y = -2x$; $(-\infty, 0)$ 15. $y = x(1 + \int_{-1}^{x} \frac{\sin t^2}{t}\, dt)$; $(-\infty, 0)$

17. $y = (x - \pi/2)\sin x$; $(0, \pi)$ 21. $40°F$ 23. A

25. $x^2 - y^2 = C$ 27. $y = x(C - x)$ 29. $y^3 = 1 - Ce^{-x^2/2}$

Exercises 2.4 Numerical Solution

Exercises 2.4.1 Euler's Method

1. a.

x_n	Euler's Method
.0	1.000000
.1	.900000
.2	.811000
.3	.733900
.4	.669510
.5	.618559
.6	.581703
.7	.559533
.8	.552579
.9	.561321
1.0	.586189

b. $|y_{10} - y(1)| \leq .081606(e-1) < .140222$

c. $h \leq 10^{-3}/\sqrt{1.632121} \approx .000783$

d. $h \leq 10^{-6}/(1.632121(e-1)) \approx .000000357$

Exercises 2.4.2 Pitfalls of Numerical Methods

1. The given differential equation is linear with $a(x) = 0$ and $b(x) = 1/(x-1)$. These functions are both continuous on the intervals $(-\infty, 1)$ and $(1, \infty)$. Since $0 \in (-\infty, 1)$ and the differential equation is linear, the solution exists and is unique on the interval $(-\infty, 1)$. [Note: The unique, explicit solution on $(-\infty, 1)$ is $y = \ln|x-1| + 1$.]

3. The differential equation is linear with $a(x) = 1/x$ and $b(x) = 0$. Since $a(x)$ and $b(x)$ are both continuous on $(-\infty, 0)$ and $(0, \infty)$ and since $-1 \in (-\infty, 0)$, there exists a unique solution to both initial value problems on the interval $(-\infty, 0)$.

   a. The unique, explicit solution on $(-\infty, 0)$ is $y = -x$.

   b. The unique, explicit solution on $(-\infty, 0)$ is $y = x$.

5. The given differential equation is nonlinear. The functions $f(x, y) = y^2$ and $f_y(x, y) = 2y$ are continuous on the entire plane, so there exists a unique solution until $x \to -\infty$, $x \to +\infty$, $y \to -\infty$, or $y \to +\infty$.

   a. The unique, explicit solution on $(-\infty, 0)$ is $y = -1/x$.

   b. The unique, explicit solution on $(-\infty, \infty)$ is $y = 0$.

   c. The unique, explicit solution on $(-\infty, 3)$ is $y = -1/(x-3)$.

7. The differential equation is nonlinear and $f(x, y) = -3x^2/(2y)$ and $f_y(x, y) = 3x^2/(2y^2)$ are both continuous for $y \neq 0$.

   a. and b. Thus, there exists a unique solution until $x \to -\infty$, $x \to +\infty$, $y \to +\infty$, or $y \to 0^+$.

   b. Hence, there exists a unique solution until $x \to -\infty$, $x \to +\infty$, $y \to 0^-$, or $y \to -\infty$.

   c. The unique, explicit solution on $(-\infty, 0)$ is $y = \sqrt{-x^3}$.

   d. The unique, explicit solution on $(-\infty, \sqrt[3]{-3/4})$ is $y = \sqrt{-x^3 - 3/4}$.

   e. There is no solution, because the differential equation is undefined for $y = 0$.

   f. The unique, explicit solution on $(-\infty, 0)$ is $y = -\sqrt{-x^3}$.

9. The differential equation is nonlinear and $f(x, y) = 3xy^{1/3}$ is continuous on the entire plane, so a solution exists until $x \to -\infty$, $x \to +\infty$, $y \to -\infty$, or $y \to +\infty$. The function $f_y(x, y) = xy^{-2/3}$ is continuous for $y \neq 0$.

a. a., b., c., and e. The solution exists and is unique at least until $y = 0$ where multiple solutions may exist.

b. The solution exists, but it may not be unique.

c. The unique, explicit solution on $(-\infty, \infty)$ is $y = (x^2 + (9/4)^{1/3} - 1)^{3/2}$.

d. The unique, explicit solution on $(-\infty, 0)$ is $y = -x^3$.

e. The unique, explicit solution on $(-\infty, -\sqrt{1-(1/4)^{1/3}})$ is $y = (x^2 + (1/4)^{1/3} - 1)^{3/2}$.

f. The solution is not unique. The function $y = 0$ is a solution on $(-\infty, \infty)$ and $y = \pm(x^2 - 1)^{3/2}$ are solutions on $(-\infty, -1)$.

g. The unique, explicit solution on $(-\infty, 0)$ is $y = x^3$.

11. The differential equation is nonlinear and $f(x, y) = y/(y - x)$ and $f_y(x, y) = -x/(y - x)^2$ are both continuous for $x \neq y$.

a. a., c., and d. Thus, the solution exists and is unique until the solution reaches the line $y = x$.

b. There is no solution, because the differential equation is undefined at $(1, 1)$.

c. The unique solution on $(0, \infty)$ is $y = 2x$.

d. The unique solution on $(0, \infty)$ is $y = 0$.

e. The unique solution on $(-\infty, \infty)$ is $y = x - \sqrt{x^2 + 3}$.

13. The differential equation is nonlinear. The function $f(x, y) = x\sqrt{1 - y^2}$ is real and continuous for $-1 \leq y \leq 1$, so a solution exists so long as it remains in $\{(x, y) | -1 \leq y \leq 1\}$. Since $f_y(x, y) = -xy/\sqrt{1 - y^2}$ is real and continuous for $-1 < y < 1$, the solution exists and is unique so long as it remains in $\{(x, y) | -1 \leq y \leq 1\}$. If the solution reaches $y = 1$ or $y = -1$, it may no longer be unique.

a. The solution is not unique. Both $y = 1$ and $y = \sin(\frac{x^2+\pi}{2})$ are solutions on $(-\infty, \infty)$.

b. The unique solution on $(-\infty, \infty)$ is $y = \sin(\frac{x^2}{2} + \arcsin .9)$.

c. The unique solution on $(-\infty, \infty)$ is $y = \sin(\frac{x^2}{2} + \frac{\pi}{6})$.

d. The unique solution on $(-\infty, \infty)$ is $y = \sin(\frac{x^2}{2})$.

Chapter 3 APPLICATIONS OF THE INITIAL VALUE PROBLEM: $\mathbf{y' = f(x, y); \quad y(c) = d}$

Exercises 3.1 Calculus Revisited

1. .8413004 3. .1393828

5. Does not exist. (Note that the integral is an improper integral.)

7. 1.118413 9. .1857833 11. .9096077 13. 1.352778

15. .8818989 17. 1.54787 19. 8.933486 21. 1.586823

23. a. $4.235612\pi = 13.30657$ b. $2.878877\pi = 9.0442589$

25. a. $4.591169\pi = 14.42358$ b. $12.00143\pi = 37.70360$

27. a. $S_x = 86.96795\pi = 273.2179$ $V_x = 400\pi/3 \approx 418.879$
b. $S_y = 75.08322\pi = 235.8809$ $V_y = 320\pi/3 \approx 335.1032$

29. a. $S_x = 24.83054\pi = 78.00744$ $V_x = 12.58826\pi = 39.54720$
b. $S_y = 11.909792\pi = 37.415715$ $V_y = 6.294044\pi = 19.77332$

31. a. $V_x = .4561457\pi = 1.433024$ b. $V_y = .178585\pi = .5610416$

33. a. $V_x = 2.355243\pi = 7.399214$ b. $V_y = .9472964\pi = 2.976019$

35. $A = 12\pi^3 \approx 372.07532$ $s = 24.439576$

37. $A = 18.84954$ $s = 16$ 39. $A = 12.56623$ $s = 99.95802$

41. $A = 8.337780$ $s_i = 2.682437$ $s_0 = 10.68242$ 43. 78.53941

45. $A = 20\pi \approx 62.83185$ $s = 28.36161$

47. $A = 2\pi \approx 6.283186$ $s = 16$ 49. $s = 39.47825$

Exercises 3.2 Learning Theory Models

1. $y(t) = A + Ce^{-kt}$

3. 15.8 min., 41.6 min. Both values are larger than the values in Example 1.

Exercises 3.3 Population Models

1. $P(t) = (2.5 \times 10^8)10^{\frac{t-1650}{300}}$ where t is the calendar year. 2250.

3. 2.1×10^8 people $= 210$ million people

year	1800	1810	1820	1830	1840	1850	1860	1870	1880	1890	1900
prediction	5.31	7.19	9.71	13.05	17.43	23.09	30.28	39.21	49.99	62.55	76.63

5. a. $\lim_{t\to+\infty} P(t)$ appears to approach approximately 450.
   b. $\lim_{t\to+\infty} P(t)$ appears to approach approximately 300.
   c. $\lim_{t\to+\infty} P(t) = 0$.

7. a. $P(t)$ decreases. b. $P(t)$ increases.

Exercises 3.4 Simple Epidemic Models

1. a. Duration $t^* \approx 9.84$; $I(t^*) \approx 205$
   b. Duration $t^* \approx 8.48$; $I(t^*) \approx 76$

Exercises 3.5 Falling Bodies.

1. $dv/dt = g - cv^2$. Terminal velocity ≈ 11.31 ft/sec.

3. $dv/dt = g - c\sqrt{v}$. No terminal velocity.

Exercises 3.6 Mixture Problems

1. About 3.7 sec 3. 10.4 years, 17.3 years, 26.3 years

Exercises 3.7 Curves of Pursuit

1. a. The boat lands .25 miles below the man.
   b. and c. The boat lands at the man.

Exercises 3.8 Chemical Reactions

1. b. 2 moles/liter c. 0 moles/liter

3. 3 moles/liter c. .376 sec

5. b. 4.5 moles/liter
   c. $\lim_{t\to\infty} C_A = 4.5$ moles/liter $\lim_{t\to\infty} C_B = .5$ moles/liter

Miscellaneous Exercises

5. 2.07 days 500

Chapter 4 N-TH ORDER LINEAR DIFFERENTIAL EQUATIONS

Exercises 4.1 Basic Theory

1. $(-\infty, \infty)$ 3. $(0, 3)$ 5. $(-\infty, 1)$

7. Because they are linear combinations of the solutions e^x and e^{-x}.

11. a. $c_1 = -3, \quad c_2 = 2$ b. $c_1 = 4, \quad c_2 = -3, \quad c_3 = 1$

 c. $c_1 = 1, \quad c_2 = c_3 = -1$ d. $c_1 = 0, \quad c_2 = -2, \quad c_3 = -1, \quad c_4 = 1$

13. b. $y = c_1 e^x + c_2 e^{-x}$ c. $y = \frac{1}{2}e^x - \frac{1}{2}e^{-x}$

15. b. $y = c_1 x + c_2 x \ln x$ c. $y = 2x - 3x \ln x$

17. b. $y'' + 9y = 0$ d. $y_c = c_1 \sin 3x + c_2 \cos 3x$

 e. $y = c_1 \sin 3x + c_2 \cos 3x + 3x + 2$ f. $y = 6 \sin 3x + 21 \cos 3x + 3x + 2$

Exercises 4.2 Roots of Polynomials

1. $x_1 = 1.36881, \quad x_2 = -1.68440 + 3.43133i, \quad x_3 = -1.68440 - 3.43133i$

3. $x_1 = .644399, \quad x_2 = -1.87214 + 3.81014i, \quad x_3 = -1.87214 - 3.81014i,$

 $x_4 = 3.09987$

7. a. $7.42445 \pm 22.7486i, \quad -19.4245 \pm 14.0689i, \quad 24$

 b. $1, \quad 1.50473 \pm .516223i, \quad 1.01797 \pm 2.60909i, \quad -6.04539$

9. $1 + i, \quad 1 + i, \quad 4 - 3i, \quad 4 + 3i, \quad 3.999 + 3i$

Exercises 4.3 Homogeneous Linear Equations with Constant Coefficients

1. $y = c_1 e^{-2x} + c_2 e^{2x}$

3. $y = c_1 + c_2 e^{4x}$

5. $y = c_1 e^{-x} \cos x + c_2 e^{-x} \sin x$

7. $y = c_1 e^{.5x} + c_2 e^{-1.5x}$

9. $y = c_1 e^{-x} + (c_2 + c_3 x) e^{3x}$

11. $y = c_1 e^{-2x} + c_2 e^{3x} \cos 2x + c_3 e^{3x} \sin 2x$

13. $y = c_1 \cos 2x + c_2 \sin 2x + c_3 e^{-2x} + c_4 e^{2x}$

15. $y = (c_1 + c_2 x) e^{x} \cos x + (c_3 + c_4 x) e^{x} \sin x$

17. $y = (c_1 + c_2 x) e^{x/2} + (c_3 + c_4 x) e^{-x/3}$

19. $y = c_1 e^{x} + (c_2 + c_3 x) e^{-2x} + c_4 e^{2x} \cos 3x + c_5 e^{2x} \sin 3x$

21. a. $y = c_1 \cos(\sqrt{\alpha} x) + c_2 \sin(\sqrt{\alpha} x)$ b. $y = c_1 + c_2 x$

    c. $y = c_1 e^{\sqrt{-\alpha} x} + c_2 e^{-\sqrt{-\alpha} x}$

23. $y = (c_1 + c_2 x) e^{\alpha x}$

25. $y = c_1 e^{-\alpha x} \cos \beta x + c_2 e^{-\alpha x} \sin \beta x$

27. $r^3 - (3 + 4i) r^2 - (4 - 12i) r + 12 = 0;\quad 2i,\quad 2i,\quad 3;$

    $y = c_1 e^{2ix} + c_2 x e^{2ix} + c_3 e^{3x} = (c_1 + c_2 x) \cos 2x + i(c_1 + c_2 x) \sin 2x + c_3 e^{3x}$

29. We conclude $e^{ix} = \cos x + i \sin x$ for all $x \in (-\infty, \infty)$.

Exercises 4.4 Nonhomogeneous Linear Equations with Constant Coefficients

1. $y = c_1 \cos 2x + c_2 \sin 2x$

3. $y = c_1 \cos 2x + c_2 \sin 2x + \frac{x}{4} \sin 2x$

5. $y = c_1 e^{-x} + c_2 x e^{-x} + 3 - x + 3e^{x}/4$

7. $y = c_1 e^{x} + c_2 \cos x + c_3 \sin x + x \cos x - x \sin x$

9. $y = (c_1 + c_2 x) e^{x} + (c_3 + c_4 x) e^{2x} + x^2 e^{x} - 2x^2 e^{2x}$

11. $y = (c_1 + c_2 x) \cos x + (c_3 + c_4 x) \sin x + 3 + \frac{1}{9} \cos 2x$

13. $y = c_1 e^{2x} + c_2 e^{2x} \cos 3x + c_3 e^{2x} \sin 3x - 2x e^{2x} \sin 3x$

15. $y = (c_1 + c_2 x + c_3 x^2) e^{2x} \cos x + (c_4 + c_5 x + c_6 x^2) e^{2x} \sin x - \frac{1806}{2105} e^{x} \cos x +$

    $\frac{768}{2105} e^{x} \sin x$

Exercises 4.5 Initial Value Problems

1. $y = e^{-2x}$
3. $y = -e^{-2x} + 3e^{2x} - e^{3x}$
5. $y = 2(x + e^{2x} - e^{-2x})$
7. $y = -2\cos x - 4\sin x + 2e^{2x}$
9. $y = -2e^x + 3\cos x + 4\sin x + xe^x$

Chapter 5 THE LAPLACE TRANSFORM METHOD

Exercises 5.1 The Laplace Transform and Its Properties

1. a. $2bs/(s^2 + b^2)^2, \quad s > 0$ b. $2b(s-a)/((s-a)^2 + b^2)^2, \quad s > a$

3. e^{-3s}/s 5. $(e^{-4s} - 2e^{-2s} + 1)/s^2$

7. a. $5/s$ b. e/s
   c. $3/s^2 - 2/s$ d. $2/(s+1)^2 + 1/(s+1)$
   e. $2/(s^2 - 4s + 13) - 2s/(s^2 + 1)$ f. $e^2/(s-3)$

9. a. $3x^2/2$ b. $4xe^{-2x}$
   c. $-2x\cos\sqrt{3}x$ d. $e^{-x} + x - 1$
   e. $e^x\cos 2x$ f. $e^x\sin 2x$
   g. $4(\cos x - 1)$ h. $2e^{-x}\cos x + 3e^{-x}\sin x$
   i. $x + 2\sinh x$ j. $3e^{2x}\cosh x$
   k. $(e^{2x}\sin\sqrt{5}x)/\sqrt{5}$ l. $(e^{2x}\cos\sqrt{5}x) + 2(e^{2x}\sin\sqrt{5}x)/\sqrt{5}$

Exercises 5.2 Using the Laplace Transform and Its Inverse to Solve Initial Value Problems

1. $y = Ae^x$
3. $y = Ae^{-2x} + 2$
5. $y = A\cos 3x + B\sin 3x - \dfrac{1}{3}x\cos 3x$
7. $y = Ae^x + Bxe^x + C + Dx - x^2e^x + x^3e^x/6 - 9x^2 - 2x^3 - x^4/4$

9. $y = e^x + 2xe^x$

11. $y = (-33\cos 3x + 8\sin 3x + 3x + 6)/27$

13. $y = e^x \cos x - x^2/2 - x$

Exercises 5.3 Convolution and the Laplace Transform

1. $h(x) = (1 - \cos 3x)/9$

2. $h(x) = [(3x - 1)e^{2x} + e^{-x}]/9$

3. $h(x) = (2\sin 2x - \cos 2x + e^x)/5$

4. $y = -3 + 5e^{2x}$

5. $y = (1 - \cos 3x)/9$

6. $y = e^{2x} - e^{-x}$

7. $y = 3xe^x - 3e^x + \cos x$

Exercises 5.4 The Unit Function and Time-Delay Function

1. a. $f_1(x) = 2u(x) - u(x-1)$
   b. $f_2(x) = u(x-2) - u(x-4)$
   c. $f_3(x) = u(x-1)(x-1)^2$
   d. $f_4(x) = u(x-1)(x^2 - 2x + 3)$
   e. $f_5(x) = u(x-\pi)\sin 3(x-\pi)$
   f. $f_6(x) = u(x)x + u(x-1)(x-1)$
   g. $f_7(x) = [u(x) - u(x-1)]x$

3. a. $u(x-1)e^{-2(x-1)}$
   b. $x - u(x-2)(x-2)$
   c. $u(x-\pi)\cos 3(x-\pi)$
   d. $u(x-\pi)\cosh 3(x-\pi)$
   e. $u(x-2)e^{-(x-2)}\sin(x-2)$
   f. $u(x-3)e^{-(x-3)}[\cos(x-3) - \sin(x-3)]$
   g. $u(x-3)(e^{x-3} - e^{-3(x-3)})/4$
   h. $u(x-1)(x-1)e^{x-1}$

Exercises 5.5 Impulse Function

1. $y = u(x-2)e^{-3(x-2)}$

3. $y = [u(x-\pi)\sin 3(x-\pi) - u(x-3\pi)\sin 3(x-3\pi)]/3$

5. $y = [10u(x-\pi)e^{x-\pi}\sin 2(x-\pi) + 4\cos x - 2\sin x + 16e^x\cos 2x$
   $-7e^x \sin 2x]/20$

7. $y = [f(\pi)u(x-\pi)\sin ax]/a$

Chapter 6 APPLICATIONS OF LINEAR DIFFERENTIAL EQUATIONS WITH CONSTANT COEFFICIENTS

Exercises 6.1 Second-Order Differential Equations

Exercises 6.1.1 Free Motion

1. $P = 4.49$ sec $\quad F = .22$ cycles/sec

3. The period of oscillation of both undamped pendula is $\omega = \sqrt{g/\ell}$, which is independent of the mass of the bob. That is, the undamped pendula oscillate at the same rate. If both pendula are subject to the same damping, then the one with the larger bob oscillates faster.

5. a. a. $y = \sin(\sqrt{19.6}t + \pi/2)$ $\quad$ b. $A = 1$ $\quad P = 1.42$ sec
   b. $\phi = \pi/2 = 90°$
   c. $y' = -4.43$ rad/sec, $\quad y'' = 0$; $\quad y' = 4.43$ rad/sec, $\quad y'' = 0$

7. a. $y = .499e^{-.5t}\sin(4.01t)$ $\quad$ b. .041 rad

9. $c = .0651$ kg/sec $\quad$ 11. $m = .8$ kg $\quad$ 13. $m = 4$ kg

15. $c < 9.1447616$ kg/sec, $\quad c = 9.1447616$ kg/sec, $\quad c > 9.1447616$ kg/sec

17. a. simple harmonic motion $\quad$ b. overdamped motion
    c. critically damped motion $\quad$ d. damped oscillatory motion
    e. overdamped oscillatory motion

Exercises 6.1.2 Forced Motion

1. a. $\omega^* = 5.715476$ cycles/sec
   b. $y = .043307\sin(5.715476t) + .2\cos(5.715476t) - .247524\sin 10t$
   c. $F_R = .9065468$ cycles/sec

3. a. $q_p(t) = EC$ b. EC c. 0

5. a. $i_p(t) = E(1 - e^{Rt/L})/R, \qquad \lim_{t\to+\infty} i_p(t) = E/R$
   b. $i_p(t) = E(L\omega^* \sin \omega^* t + R \cos \omega^* t)/(R^2 + (L\omega^*)^2)$

Exercises 6.2 Higher Order Differential Equations

1. a. $y_1 = c_1 \cos(2.07302t) + c_2 \sin(2.07302t) + c_3 \cos(9.56122t) + c_4 \sin(9.56122t)$
   b. $y_2 = 1.3764107c_1 \cos(2.07302t) + 1.3764107c_2 \sin(2.07302t) - .2075804c_3 \cos(9.56122t) - .2075804c_4 \sin(9.56122t)$
   c. $c_1 = -.0815925, \qquad c_2 = .140808, \qquad c_3 = .1815925, \qquad c_4 = .0008474028$

3. $y_1 = .0567167 \cos(3.94038t) + .0184145 \sin(3.94038t) - .0067167 \cos(9.08148t) + .0085272 \sin(9.08148t)$
   $y_2 = .0743975 \cos(3.94038t) + .0241549 \sin(3.94038t) + .0256022 \cos(9.08148t) - .0325033 \sin(9.08148t)$

5. $y_1 = c_1 e^{-.366290t} + c_2 e^{-4.84005t} + c_3 e^{-15.7937t} + u$
   $y_2 = 1.211237c_1 e^{-.366290t} - .2666667c_2 e^{-4.84005t} - 3.666667c_3 e^{-15.7937t} + u$
   $y_3 = 1.278117c_1 e^{-.366290t} - .9074786c_2 e^{-4.84005t} + 2.388889c_3 e^{-15.7937t} + u$

7. $y = c_1 + c_2 x + c_3 e^{x/120} + c_4 x e^{x/120} + 1.2 \times 10^{-6} w_0 x^2 e^{x/120}$

9. $y = c_1 e^{.387298x} \cos(.387298x) + c_2 e^{.387298x} \sin(.387298x) + c_3 e^{-.387298x} \cos(.387298x) + c_4 e^{-.387298x} \sin(.387298x) + \dfrac{w_0 L^4 \cos(\pi x/L)}{EI\pi^2(\pi^2 - .09L^2)}$

Chapter 7 SYSTEMS OF FIRST-ORDER DIFFERENTIAL EQUATIONS

Exercises 7.1 Properties of Systems of Differential Equations

5. a. linear b. $(0, \infty)$
   b. $(-\infty, \infty)$ Because y_1, y_2, y_1', and y_2' are all defined and continuous at $x = 0$, but the system of differential equations (25) is not defined at $x = 0$. No, because the initial value problem (25) is undefined at $x = 0$.

7. a. nonlinear
   b. There is a solution on some subinterval of $(-\pi/2, \pi/2)$ containing the point $x = 0$.

9. a. nonlinear
   b. There is a solution on some subinterval of $(-3, 4)$ containing the point $x = 0$.
   c. $x \to -3^+, \quad x \to 4^-, \quad y_1 \to -2^+, \quad y_1 \to +\infty, \quad y_2 \to -\infty, \quad y_2 \to 2^-.$

Exercises 7.2 Writing Systems As Equivalent First-Order Systems

1. Let $u_1 = y$, $u_2 = y^{(1)}$, $u_3 = y^{(2)}$, and $u_4 = y^{(3)}$. Then

$$\begin{aligned} u_1' &= u_2 \\ u_2' &= u_3 \\ u_3' &= u_4 \\ u_4' &= -3xu_1^2 + u_2^3 - e^x u_3 u_4 + x^2 - 1 \end{aligned}$$

$$u_1(1) = -1, \quad u_2(1) = 2, \quad u_3(1) = -3, \quad u_4(1) = 0$$

3. Let $u_1 = y$ and $u_2 = y'$. Then

$$\begin{aligned} u_1' &= u_2 \\ u_2' &= -\frac{c}{m}u_2 - \frac{k}{m}\sin u_1 \end{aligned}$$

$$u_1(0) = 1, \quad u_2(0) = -2$$

5. Let $u_1 = y$, $u_2 = y'$, $u_3 = z$, and $u_4 = z'$. Then

$$\begin{aligned} u_1' &= u_2 \\ u_2' &= 2u_1 - 3u_4 \\ u_3' &= u_4 \\ u_4' &= 3u_2 - 2u_3 \end{aligned}$$

$$u_1(0) = 1, \quad u_2(0) = -3, \quad u_3(0) = -1, \quad u_4(0) = 2$$

7. Let $u_1 = y$, $u_2 = z$, and $u_3 = z'$. Then

$$\begin{aligned} u_1' &= xu_1 + u_2 \\ u_2' &= u_3 \\ u_3' &= -x^2 u_1 + u_3 - 3e^x \end{aligned}$$

$$u_1(1) = -2, \quad u_2(1) = 3, \quad u_3(1) = 0$$

9. $(-\infty, \infty)$ 10. (i) $(0, \pi/2)$ (ii) $(\pi/2, 3\pi/2)$

11. (i) $(-1, 2)$ (ii) $(2, \infty)$ 12. $(0, \infty)$ 13. $(-\infty, \infty)$

14. $(-\infty, \infty)$ 15. $(-\infty, \infty)$

Chapter 8 LINEAR SYSTEMS OF FIRST-ORDER DIFFERENTIAL EQUATIONS

Exercises 8.1 Matrices and Vectors

1. $\begin{pmatrix} -1 & 7 \\ -4 & 9 \end{pmatrix}$ 3. $\begin{pmatrix} -3 \\ -11 \end{pmatrix}$ 5. Cannot compute.

7. $\begin{pmatrix} 5 \\ -3 \\ 2 \end{pmatrix}$ 9. Cannot compute. 11. $\begin{pmatrix} 2 & -4 \\ -1 & 2 \end{pmatrix}$

13. Cannot compute. 15. $\begin{pmatrix} -1 \\ -12 \end{pmatrix}$ 17. 4 19. −8 21. 1

23. −1 25. −16 27. $2x^3$ 29. Yes 33. Linearly dependent

35. Linearly dependent

Exercises 8.2 Eigenvalues and Eigenvectors

1. $\lambda_1 = -1, \quad \lambda_2 = 2, \quad \mathbf{v}_1 = \begin{pmatrix} -1.78885 \\ .447214 \end{pmatrix}, \quad \mathbf{v}_2 = \begin{pmatrix} 1.49071 \\ -1.49071 \end{pmatrix}$

3. $\lambda_1 = 1, \quad \lambda_2 = i, \quad \lambda_3 = -i, \quad \mathbf{v}_1 = \begin{pmatrix} 0 \\ 1 \\ 0 \end{pmatrix}, \quad \mathbf{v}_2 = \begin{pmatrix} .5 - .5i \\ -.5 + .5i \\ i \end{pmatrix},$

$\mathbf{v}_3 = \begin{pmatrix} .5 + .5i \\ -.5 - .5i \\ -i \end{pmatrix}$

5. $\lambda_1 = 5, \quad \lambda_2 = 3, \quad \lambda_3 = 2, \quad \mathbf{v}_1 = \begin{pmatrix} -.870022 \\ 1.74004 \\ .580015 \end{pmatrix},$

$\mathbf{v}_2 = \begin{pmatrix} 3.08415 \\ -6.16831 \\ -3.08415 \end{pmatrix}, \mathbf{v}_3 = \begin{pmatrix} -4.47214 \\ 4.47214 \\ 4.47214 \end{pmatrix}$

7. $\lambda_1 = 0, \quad \lambda_2 = 64, \quad \lambda_3 = 0, \quad \lambda_4 = 0, \quad \mathbf{v}_1 = \begin{pmatrix} -.995495 \\ .031603 \\ .047405 \\ .094809 \end{pmatrix},$

$$\mathbf{v}_2 = \begin{pmatrix} -.164805 \\ -.329610 \\ -.494415 \\ -.988829 \end{pmatrix}, \quad \mathbf{v}_3 = \begin{pmatrix} -.044944 \\ -.157303 \\ .764045 \\ -.471910 \end{pmatrix}, \quad \mathbf{v}_4 = \begin{pmatrix} -4.00895 \\ 1.03203 \\ .048052 \\ .096093 \end{pmatrix}$$

Exercises 8.3 Linear Systems with Constant Coefficients

1. $\begin{pmatrix} y_1' \\ y_2' \end{pmatrix} = \begin{pmatrix} 2 & -3 \\ 1 & 4 \end{pmatrix} \begin{pmatrix} y_1 \\ y_2 \end{pmatrix} + \begin{pmatrix} 5e^x \\ -2e^{-x} \end{pmatrix}$

3. 3. $\begin{pmatrix} y_1' \\ y_2' \\ y_3' \end{pmatrix} = \begin{pmatrix} 0 & 2 & 0 \\ 3 & 0 & 0 \\ -1 & 0 & 2 \end{pmatrix} \begin{pmatrix} y_1 \\ y_2 \\ y_3 \end{pmatrix}$

5. b. $\mathbf{y}(x) = c_1\mathbf{y}_1(x) + c_2\mathbf{y}_2$ d. $\mathbf{y}(x) = c_1\mathbf{y}_1(x) + c_2\mathbf{y}_2 + \mathbf{y}_p(x)$

7. $\mathbf{y} = c_1 \begin{pmatrix} -.408248 \\ -.408248 \\ -.816497 \end{pmatrix} e^{-x} + c_2 \begin{pmatrix} 1.73205 \\ 1.73205 \\ 1.73205 \end{pmatrix} e^{x} + c_3 \begin{pmatrix} 1.41421 \\ 2.82843 \\ 1.41421 \end{pmatrix} e^{2x}$

9. $\mathbf{y} = c_1 \begin{pmatrix} 1.66410 \\ .55470 \\ .55470 \end{pmatrix} e^{4x} + c_2 \begin{pmatrix} 0 \\ 1.20185 \\ -1.20185 \end{pmatrix} e^{x} + c_3 \begin{pmatrix} -1.2 \\ -.4 \\ 1.4 \end{pmatrix} e^{-x}$

11. $\mathbf{y} = c_1 \begin{pmatrix} -.894427 \\ -.447214 \\ -.447214 \end{pmatrix} e^{-3x} + c_2 \begin{pmatrix} .745356 \\ -.745356 \\ .745356 \end{pmatrix} + c_3 \begin{pmatrix} -.511827 \\ .244087 \\ .755913 \end{pmatrix} e^{-3x}$

13.

$$\mathbf{y} = c_1\{e^{2x}\cos x\begin{pmatrix}1\\0\\0\\0\end{pmatrix} - e^{2x}\sin x\begin{pmatrix}0\\1\\0\\0\end{pmatrix}\}$$

$$+ c_2\{e^{2x}\sin x\begin{pmatrix}1\\0\\0\\0\end{pmatrix} + e^{2x}\cos x\begin{pmatrix}0\\1\\0\\0\end{pmatrix}\}$$

$$+ c_3\{e^{3x}\cos 4x\begin{pmatrix}0\\0\\-1\\0\end{pmatrix} - e^{3x}\sin 4x\begin{pmatrix}0\\0\\0\\1\end{pmatrix}\}$$

$$+ c_4\{e^{3x}\sin 4x\begin{pmatrix}0\\0\\-1\\0\end{pmatrix} + e^{3x}\cos 4x\begin{pmatrix}0\\0\\0\\1\end{pmatrix}\}$$

$$= \begin{pmatrix} c_1e^{2x}\cos x + c_2e^{2x}\sin x \\ -c_1e^{2x}\sin x + c_2e^{2x}\cos x \\ -c_3e^{3x}\cos 4x - c_4e^{3x}\sin 4x \\ -c_3e^{3x}\sin 4x + c_4e^{3x}\cos 4x \end{pmatrix}$$

15.

$$\mathbf{y} = c_1\{e^{3x}\cos 2x\begin{pmatrix}1\\0\\0\\0\end{pmatrix} - e^{3x}\sin 2x\begin{pmatrix}0\\1\\0\\0\end{pmatrix}\}$$

$$+ c_2\{e^{3x}\sin 2x\begin{pmatrix}1\\0\\0\\0\end{pmatrix} + e^{3x}\cos 2x\begin{pmatrix}0\\1\\0\\0\end{pmatrix}\}$$

$$+ c_3\begin{pmatrix}0\\0\\1\\0\end{pmatrix}e^{x} + c_4\begin{pmatrix}0\\0\\0\\1\end{pmatrix}e^{2x} = \begin{pmatrix} c_1e^{3x}\cos 2x + c_2e^{3x}\sin 2x \\ -c_1e^{3x}\sin 2x + c_2e^{3x}\cos 2x \\ c_3e^{x} \\ c_4e^{2x} \end{pmatrix}$$

Chapter 9 APPLICATIONS OF LINEAR SYSTEMS WITH CONSTANT COEFFICIENTS

Exercises 9.1 Coupled Spring-Mass Systems

1. a.

$$\mathbf{u} = c_1\{\cos 2.13578x \begin{pmatrix} -.467992 \\ .000000 \\ .131401 \\ .000000 \end{pmatrix} - \sin 2.13578x \begin{pmatrix} .000000 \\ -.999527 \\ .000000 \\ .280644 \end{pmatrix}\}$$

$$+ c_2\{\sin 2.13578x \begin{pmatrix} -.467992 \\ .000000 \\ .131401 \\ .000000 \end{pmatrix} + \cos 2.13578x \begin{pmatrix} .000000 \\ -.999527 \\ .000000 \\ .280644 \end{pmatrix}\}$$

$$+ c_3\{\cos .662153x \begin{pmatrix} .732913 \\ .000000 \\ 1.30515 \\ .000000 \end{pmatrix} - \sin .662153x \begin{pmatrix} .000000 \\ .485301 \\ .000000 \\ .864212 \end{pmatrix}\}$$

$$+ c_4\{\sin .662153x \begin{pmatrix} .732913 \\ .000000 \\ 1.30515 \\ .000000 \end{pmatrix} + \cos .662153x \begin{pmatrix} .000000 \\ .485301 \\ .000000 \\ .864212 \end{pmatrix}\}$$

b.

$$\mathbf{u} = c_1[(\cos \beta x)\mathbf{v}_1 - (\sin \beta x)\mathbf{v}_2] + c_2[(\sin \beta x)\mathbf{v}_1 + (\cos \beta x)\mathbf{v}_2]$$
$$+ c_3[(\cos \gamma x)\mathbf{v}_3 - (\sin \gamma x)\mathbf{v}_4] + c_4[(\sin \gamma x)\mathbf{v}_3 + (\cos \gamma x)\mathbf{v}_4]$$

where $\beta = 1.93185$, $\gamma = .517638$ and

$$\mathbf{v}_1 = \begin{pmatrix} -.517362 \\ .000000 \\ .189368 \\ .000000 \end{pmatrix}, \quad \mathbf{v}_2 = \begin{pmatrix} .000000 \\ -.999466 \\ .000000 \\ .365830 \end{pmatrix},$$

$$\mathbf{v}_3 = \begin{pmatrix} 1.11595 \\ 0.00000 \\ 1.52442 \\ 0.00000 \end{pmatrix}, \text{ and } \quad \mathbf{v}_4 = \begin{pmatrix} .000000 \\ .577659 \\ .000000 \\ .789096 \end{pmatrix}$$

3. a.

$$
\begin{aligned}
u_1' &= u_2 \\
u_2' &= -\frac{(k_1+k_2)u_1}{m_1} - \frac{d_1 u_2}{m_1} + \frac{k_2 u_3}{m_1} \\
u_3' &= u_4 \\
u_4' &= \frac{k_2 u_1}{m_2} - \frac{k_2 u_3}{m_2} - \frac{d_2 u_4}{m_2}
\end{aligned}
$$

b.

$$
\mathbf{u}' = \begin{pmatrix} 0 & 1 & 0 & 0 \\ -\frac{k_1+k_2}{m_1} & -\frac{d_1}{m_1} & \frac{k_2}{m_1} & 0 \\ 0 & 0 & 0 & 1 \\ \frac{k_2}{m_2} & 0 & -\frac{k_2}{m_2} & -\frac{d_2}{m_2} \end{pmatrix} \mathbf{u}
$$

c.

$$
\begin{aligned}
\mathbf{u} = {} & c_1[(e^{\alpha x}\cos\beta x)\mathbf{v}_1 - (e^{\alpha x}\sin\beta x)\mathbf{v}_2] \\
& + c_2[(e^{\alpha x}\sin\beta x)\mathbf{v}_1 + (e^{\alpha x}\cos\beta x)\mathbf{v}_2] + c_3 e^{ax}\mathbf{v}_3 + c_4 e^{bx}\mathbf{v}_4
\end{aligned}
$$

where $\alpha = -.75,\ \beta = 1.31083,\ a = -.164102,\ b = -1.33590$ and

$$
\mathbf{v}_1 = \begin{pmatrix} -.571024 \\ .000000 \\ .445842 \\ .000000 \end{pmatrix}, \quad \mathbf{v}_2 = \begin{pmatrix} .326715 \\ -.993552 \\ -.255092 \\ .775742 \end{pmatrix}, \quad \mathbf{v}_3 = \begin{pmatrix} -.512927 \\ .0841723 \\ -.656944 \\ .167806 \end{pmatrix},
$$

$$
\text{and } \mathbf{v}_4 = \begin{pmatrix} -.422975 \\ .565052 \\ -.541737 \\ .723705 \end{pmatrix}
$$

5. a.

$$\mathbf{u}' = \begin{pmatrix} 0 & 1 & 0 & 0 \\ -\dfrac{k_1+k_2}{m_1} & -\dfrac{d_1}{m_1} & \dfrac{k_2}{m_1} & 0 \\ 0 & 0 & 0 & 1 \\ \dfrac{k_2}{m_2} & 0 & -\dfrac{k_2+k_3}{m_2} & -\dfrac{d_2}{m_2} \end{pmatrix} \mathbf{u}$$

b. (i)

$$\mathbf{u} = c_1 e^{-2x} \begin{pmatrix} .161444 \\ -.322888 \\ -.161444 \\ .322888 \end{pmatrix} + c_2 e^{-4x} \begin{pmatrix} -.433951 \\ 1.73580 \\ .433951 \\ -1.73580 \end{pmatrix}$$

$$+ c_3 e^{-x} \begin{pmatrix} .170219 \\ -.170219 \\ .170219 \\ -.170219 \end{pmatrix} + c_4 e^{-5x} \begin{pmatrix} -.163779 \\ .818897 \\ -.163779 \\ .818897 \end{pmatrix}$$

b. (ii)

$$\mathbf{u} = c_1 [e^{-2x} \cos x \begin{pmatrix} .16 \\ .00 \\ -.16 \\ .00 \end{pmatrix} - e^{-2x} \sin x \begin{pmatrix} -.32 \\ .80 \\ .32 \\ -.80 \end{pmatrix}]$$

$$+ c_2 [e^{-2x} \sin x \begin{pmatrix} .16 \\ .00 \\ -.16 \\ .00 \end{pmatrix} + e^{-2x} \cos x \begin{pmatrix} -.32 \\ .80 \\ .32 \\ -.80 \end{pmatrix}]$$

$$+ c_3 e^{-x} \begin{pmatrix} .307729 \\ -.307729 \\ .307729 \\ -.307729 \end{pmatrix} + c_4 e^{-3x} \begin{pmatrix} -.359035 \\ 1.07711 \\ -.359035 \\ 1.07711 \end{pmatrix}$$

b. (iii)

$$\mathbf{u} = c_1[(e^{\alpha x}\cos\beta x)\mathbf{v}_1 - (e^{\alpha x}\sin\beta x)\mathbf{v}_2] + c_2[(e^{\alpha x}\sin\beta x)\mathbf{v}_1$$

$$+ (e^{\alpha x}\cos\beta x)\mathbf{v}_2] + c_3[(e^{\gamma x}\cos\delta x)\mathbf{v}_3 - (e^{\gamma x}\sin\delta x)\mathbf{v}_4]$$

$$+ c_4[(e^{\gamma x}\sin\delta x)\mathbf{v}_3 + (e^{\gamma x}\cos\delta x)\mathbf{v}_4]$$

where $\alpha = -1,\ \beta = 1.73205,\ \gamma = -1,\ \delta = 1$, and

$$\mathbf{v}_1 = \begin{pmatrix} .360288 \\ .000000 \\ -.360288 \\ .000000 \end{pmatrix}, \qquad \mathbf{v}_2 = \begin{pmatrix} -.208013 \\ .832050 \\ .208013 \\ -.832050 \end{pmatrix},$$

$$\mathbf{v}_3 = \begin{pmatrix} .300463 \\ .000000 \\ .300463 \\ .000000 \end{pmatrix}, \text{ and } \quad \mathbf{v}_4 = \begin{pmatrix} -.300463 \\ .600925 \\ -.300463 \\ .600925 \end{pmatrix}$$

Exercises 9.2 Pendulum Systems

7. a. Let $u_1 = y_1$, $u_2 = y_1'$, $u_3 = y_2$, and $u_4 = y_2'$. Then

$$\begin{pmatrix} u_1' \\ u_2' \\ u_3' \\ u_4' \end{pmatrix} = \begin{pmatrix} 0 & 1 & 0 & 0 \\ -\left(\frac{g}{\ell_1} + \frac{m_2}{m_1\ell_1}\right) & 0 & \frac{m_2}{m_1\ell_1} & 0 \\ 0 & 0 & 0 & 1 \\ \frac{g}{\ell_2} + \frac{m_2}{m_1\ell_2} & 0 & -\left(\frac{g}{\ell_2} + \frac{m_2}{m_1\ell_2}\right) & 0 \end{pmatrix} \begin{pmatrix} u_1 \\ u_2 \\ u_3 \\ u_4 \end{pmatrix}.$$

b.

$$\begin{aligned} \mathbf{u} = c_1[(\cos\beta x)\mathbf{v}_1 - (\sin\beta x)\mathbf{v}_2] + c_2[(\sin\beta x)\mathbf{v}_1 + (\cos\beta x)\mathbf{v}_2] \\ + c_3[(\cos\gamma x)\mathbf{v}_3 - (\sin\gamma x)\mathbf{v}_4] + c_4[(\sin\gamma x)\mathbf{v}_3 + (\cos\gamma x)\mathbf{v}_4] \end{aligned}$$

where $\beta = 4.42568,\ \ \gamma = 6.26525$ and

$$\mathbf{v}_1 = \begin{pmatrix} -.0399569 \\ .0000000 \\ -.0798053 \\ .0000000 \end{pmatrix}, \qquad \mathbf{v}_2 = \begin{pmatrix} .000000 \\ -.176837 \\ .000000 \\ -.353193 \end{pmatrix},$$

$$\mathbf{v}_3 = \begin{pmatrix} -.000152981 \\ 0.000000 \\ 0.225342 \\ 0.000000 \end{pmatrix}, \text{ and } \quad \mathbf{v}_4 = \begin{pmatrix} 0.00000 \\ -0.000958463 \\ 0.00000 \\ 1.41182 \end{pmatrix}$$

Exercises 9.4 Mixture Problems

9. a. $\mathbf{q}_c = c_1 e^{-.0633975t} \begin{pmatrix} .806898 \\ .590690 \end{pmatrix} + c_2 e^{-.236603t} \begin{pmatrix} -.357759 \\ .977416 \end{pmatrix}$

b. $\mathbf{q} = \mathbf{q}_c + \begin{pmatrix} 200 \\ 100 \end{pmatrix}$

c. $\mathbf{q} = -221.49499 e^{-.0633975t} \begin{pmatrix} .806898 \\ .590690 \end{pmatrix}$

$+31.542932 e^{-.236603t} \begin{pmatrix} -.357759 \\ .977416 \end{pmatrix} + \begin{pmatrix} 200 \\ 100 \end{pmatrix}$

d. 200 lbs. 100 lbs. They are equal.

11.

$$\mathbf{q}(t) = \begin{pmatrix} q_1(t) \\ q_2(t) \end{pmatrix} = 88.721 e^{-.0863104t} \begin{pmatrix} -.921753 \\ .387778 \end{pmatrix}$$

$$+420.797 e^{-.00868956t} \begin{pmatrix} -.209652 \\ -.996690 \end{pmatrix} + \begin{pmatrix} 200 \\ 400 \end{pmatrix}$$

Limiting amounts: Tank A, 200 lbs.; Tank B, 400 lbs.

13.

$$\mathbf{q}(t) = \begin{pmatrix} q_A(t) \\ q_B(t) \\ q_C(t) \end{pmatrix} = -37333.35 e^{-.00025t} \begin{pmatrix} 0 \\ 0 \\ 1 \end{pmatrix} - 21000 e^{-.0005t} \begin{pmatrix} 0 \\ 1 \\ -2 \end{pmatrix}$$

$$-4285.71 e^{-.0007t} \begin{pmatrix} 1 \\ -3.5 \\ 3.88889 \end{pmatrix} + \begin{pmatrix} 4285.71 \\ 6000.00 \\ 12000.00 \end{pmatrix}$$

Limiting amounts:

Lake A, 4285.71 lbs.; Lake B, 6000 lbs.; Lake C, 12,000 lbs.

15. a.

$$\begin{pmatrix} c_s' \\ c_m' \\ c_h' \\ c_e' \\ c_o' \end{pmatrix} = \begin{pmatrix} -\frac{r_s}{V_s} & 0 & 0 & 0 & 0 \\ 0 & -\frac{r_m}{V_m} & 0 & 0 & 0 \\ \frac{r_s}{V_h} & \frac{r_m}{V_h} & -\frac{r_h}{V_h} & 0 & 0 \\ 0 & 0 & \frac{r_h}{V_e} & -\frac{r_e}{V_e} & 0 \\ 0 & 0 & 0 & \frac{r_e}{V_o} & -\frac{r_o}{V_o} \end{pmatrix} \begin{pmatrix} c_s \\ c_m \\ c_h \\ c_e \\ c_o \end{pmatrix} + \begin{pmatrix} \frac{R_s C_s}{V_s} \\ \frac{R_m C_m}{V_m} \\ \frac{R_h C_h}{V_h} \\ \frac{R_e C_e}{V_e} \\ \frac{R_o C_o}{V_o} \end{pmatrix}$$

b.

$$\mathbf{c}_c(t) = k_1 e^{-.251908t} \begin{pmatrix} 0 \\ 0 \\ 0 \\ 0 \\ 1 \end{pmatrix} + k_2 e^{-.732759t} \begin{pmatrix} 0 \\ 0 \\ 0 \\ 1 \\ -.449796 \end{pmatrix}$$

$$+ k_3 e^{-.08t} \begin{pmatrix} 0 \\ 0 \\ 1 \\ .898045 \\ 1.12987 \end{pmatrix} + k_4 e^{-.00517241t} \begin{pmatrix} 1 \\ 0 \\ .235837 \\ .190011 \\ .166561 \end{pmatrix}$$

$$+ k_5 e^{-.0322034t} \begin{pmatrix} 0 \\ 1 \\ .935336 \\ .782666 \\ .770484 \end{pmatrix}$$

c.

	Time for Pollution to Reach	
	.4%	.3%
Superior	43 years	99 years
Michigan	7 years	16 years
Huron	12 years	25 years
Erie	3 years	16 years
Ontario	5 years	14 years

d. (i)

$$\mathbf{c}(t) = \mathbf{c}_c(t) + \begin{pmatrix} .002 \\ .002 \\ .002 \\ .002 \\ .002 \end{pmatrix}$$

d.

	Time for Pollution to Reach	
	.4%	.3%
Superior	190 years	403 years
Michigan	31 years	65 years
Huron	49 years	140 years
Erie	39 years	72 years
Ontario	36 years	104 years

Chapter 10 APPLICATIONS OF SYSTEMS OF EQUATIONS

Exercises 10.1 Richardson's Arms Race Model

1. Unstable arms race

2. a. (i) equilibrium point in first quadrant (ii) stable arms race
   b. (i) equilibrium point in third quadrant (ii) unstable arms race

5.

r	**s**	**E**	**Possible Arms Race(s)**
0	0	+	md
0	0	−	rar
−	−	+	md
−	−	−	md, sar, rar
+	+	+	sar
+	+	−	rar
−	+	+	md, sar
−	+	−	md, sar, rar

Exercises 10.2 Phase-Plane Portraits

1. $\lambda_1 = 1, \quad \lambda_2 = -1,$ unstable saddle point

3. $\lambda_1 = -1, \quad \lambda_2 = -1,$ asymptotically stable node

5. $\lambda = -1 \pm 2i,$ asymptotically stable spiral point

7. $\lambda_1 = -2, \quad \lambda_2 = -4,$ asymptotically stable node

9. Asymptotically stable node at $(0, 0)$; unstable saddle point at $(1, 1)$

11. Unstable saddle point at $(0, 0)$; neutrally stable center at $(\frac{1}{2}, 1)$

13. Asymptotically stable spiral point at $(1, 1)$; unstable saddle point at $(1, -1)$

15. Unstable saddle point at $(0, 0)$; asymptotically stable node at $(0, -1)$; unstable spiral point at $(2, 1)$

Exercises 10.3 Modified Richardson's Arms Race Models

1. a. $(5, 4)$

   b. The associated linear system

$$\begin{aligned} x' &= -10(x-5) + 5(y-4) \\ y' &= 4(x-5) - 8(y-4) \end{aligned}$$

   has an asymptotically stable node at $(5, 4)$. Therefore, $(5, 4)$ is an asymptotically stable critical point of the nonlinear system (9).

3. a. $(3, 2), \quad (4.22457, 2.98300)$

   b. At $(3, 2)$, the associated linear system

$$\begin{aligned} x' &= -6(x-3) + 9(y-2) \\ y' &= 4(x-3) - 4(y-2) \end{aligned}$$

   has an unstable saddle point. So $(3, 2)$ is an unstable critical point of the nonlinear system (11).
   At $(4.22457, 2.98300)$, the associated linear system

$$\begin{aligned} x' &= -8.44914(x-4.22457) + 9(y-2.98300) \\ y' &= 4(x-4.22457) - 5.966(y-2.98300) \end{aligned}$$

   has an asymptotically stable node. So $(4.22457, 2.98300)$ is an asymptotically stable node of the nonlinear system (11).

5. a. A parabola with vertex at $(r/C, 0)$, axis of symmetry the x-axis, and opens to the right. A parabola with vertex at $(0, s/D)$, axis of symmetry the y-axis, and opens upward. Four. Two.

b. $AB^2x^4 + 2ABsx^2 - CD^2x + As^2 + rD^2 = 0$

c. (i) $(5, 4)$

(ii) The associated linear system

$$\begin{aligned} x' &= -4(x-5) + 8(y-4) \\ y' &= 10(x-5) - 5(y-4) \end{aligned}$$

has an unstable saddle point at $(5, 4)$. So $(5, 4)$ is an unstable critical point of the nonlinear system (15).

d. (i) $(2, 2)$

(ii) The associated linear system is

$$\begin{aligned} x' &= -4(x-2) + 4(y-2) \\ y' &= 4(x-2) - 4(y-2). \end{aligned}$$

The eigenvalues of the associated linear system are 0 and -8. Nothing. Nothing.

(iii) The critical point is unstable.

7. a. (i) unstable b. (i) stable

Exercises 10.4 Lanchester's Combat Models

1. b.

	Winner	Time Over (Days)	Number of Remaining Winning Combatants
(i)	y	1.975	2.35
(ii)	x	2.76	3.78
(iii)	x	.995	2.66

3.

	Winner	Time Over (Days)	Number of Remaining Winning Combatants
(i)	y	1.885	1.47
(ii)	y	1.85	1.70

5. No

Exercises 10.5 Models for Interacting Species

Lotka–Volterra Prey-Predator Model

1. a. $T = 3.148$ years

	minimum	maximum	average
x	1.7500	2.2727	2
y	3.5000	4.5455	4

b. $T = 3.18$ years—slightly longer than T for part a.

	minimum	maximum	average
x	1.9952	2.6343	2.3
y	2.8800	3.9791	3.4

Minimum, maximum, and average prey population increase; while minimum, maximum, and average predator population decrease.

Modified Prey-Predator Models

1. The critical point $(0,0)$ is a saddle point.

   The critical point $(\frac{r}{C},0)$ is in the first quadrant. If $Qr - Cs < 0$, then $(\frac{r}{C},0)$ is an asymptotically stable node. If $Qr - Cs > 0$, then $(\frac{r}{C},0)$ is a saddle point.

   If $Qr - Cs < 0$, then the critical point $(\frac{s}{Q}, \frac{Qr - Cs}{QH})$ is in the fourth quadrant and is a saddle point. If $Qr - Cs \geq 0$, then the critical point $(\frac{s}{Q}, \frac{Qr - Cs}{QH})$ is in the first quadrant and is an asymptotically stable node provided $C^2 s - 4Q(Qr - Cs) \geq 0$; otherwise, the critical point is an asymptotically stable spiral point.

3. a. The average predator population decreases.

   b. The average prey population increases.

   c. The average prey population increases and the average predator population decreases.

5. (iii)

As t increases	a	b	c	d	e
$x(t) \rightarrow$	1.34	1	1.34	1.15	1.36
$y(t) \rightarrow$	0.34	0	0.00	0.00	0.00

7. a. $(0,\ 0,\ 0)$, $\quad (\frac{a}{b},\ 0,\ 0)$, $\quad (0,\ \frac{g}{h},\ \frac{-d}{f})$, $\quad (\frac{d}{e},\ \frac{ae - bd}{ce},\ 0)$

   $(\frac{ah - cg}{bh},\ \frac{g}{h},\ \frac{aeh - bdh - ceg}{bfh})$

   b.

As t increases	(i)	(ii)
$y_1(t) \rightarrow$	1.15	1.9
$y_2(t) \rightarrow$	.93	1.1
$y_3(t) \rightarrow$	.11	.9

Leslie's Prey-Predator Model

1. a. $(\frac{ae}{be+cd}, \frac{ad}{be+cd})$; stable node
   b. $\lim_{t\to\infty} x(t) = .8$ $\lim_{t\to\infty} y(t) = 1.6$

Leslie–Gower Prey-Predator Model

1. a. $\lim_{t\to\infty} x(t) \approx 26$ $\lim_{t\to\infty} y(t) \approx 10$

Different Uptake Function

1. a. $(\frac{1}{2}, 1)$ b. unstable spiral point
   c. $\lim_{t\to+\infty} x(t) = 0$ $\lim_{t\to+\infty} y(t) = 0$

3. The solution of case a. spirals outward toward the solution of case b. The solution of case c. spirals inward toward the solution of case b.

May's Prey-Predator Model

1. The solution of case a. spirals inward toward the solution of case b. The solution of case c. spirals outward toward the solution of case b.

Competing Species Models

5. a. (80/7, 24/7)
   b.

	(i)	(ii)	(iii)	(iv)
$\lim_{t\to\infty} x(t) \approx$	0	20	20	0
$\lim_{t\to\infty} y(t) \approx$	12	0	0	12
extinct species	x	y	y	x

7. a. (i) None
   (ii) $1-4$ $\lim_{t\to\infty} x(t) = 10$ $\lim_{t\to\infty} y(t) = 0$, y becomes extinct.
   b. (i) (4, 2) (ii) $1-4$ $\lim_{t\to\infty} x(t) = 4$ $\lim_{t\to\infty} y(t) = 2$
   c. (i) None
   (ii) $1-4$ $\lim_{t\to\infty} x(t) = 0$ $\lim_{t\to\infty} y(t) = 4$, x becomes extinct.

Exercises 10.6 Epidemics

1. a. 300 c. (i) $S(5) = 232$ (ii) $S(5) = 150$

3. Yes (i) $S(5) = 148$ (ii) $S(5) = 114$

Exercises 10.7 Pendulums

1. The period varies with $y_2(0)$.

9.

$$\begin{aligned} y_1' &= y_2 \\ y_2' &= 2\omega y_4 \sin\phi - g y_1/\ell \\ y_3' &= y_4 \\ y_4' &= -2\omega y_2 \sin\phi - g y_3/\ell \end{aligned}$$

where $y_1 = x,\ \ y_2 = x',\ \ y_3 = y,$ and $y_4 = y'$.

a. $y_3(6.28) = 0,$ The plane of oscillation does not appear to rotate.
b. $y_3(6.28) = -3.238823E - 04;$ 1.41 days
c. $y_3(6.28) = 3.238823E - 04;$ 1.41 days
d. $y_3(6.28) = -3.966742E - 04;$ 1.15 days
e. $y_3(6.28) = -4.580397E - 04;$ .998 days

Exercises 10.8 Duffing's Equation

1. $y_1' = y_2$
$y_2' = -Ky_1 - Py_1^3 - Cy_2 + A\sin\omega t$

Exercises 10.10 Mixture Problems

Pollution in the Great Lakes

1. Pollution in Lake Ontario less than .3%: a. 41 years b. 13.6 years

Pollution Level Reduced to .25%

	first	last
a.	Lake Michigan	Lake Superior
b.	Lake Michigan	Lake Superior

Exercises 10.11 The Restricted Three-Body Problem

3. L_4 or L_5 would be better sites for space stations, since they are stable critical points. L_1, L_2, and L_3 are unstable critical points.

Chapter 11 Applications to Combinatorics

Exercises 11.1

1. It is $x^2/(1-x)^3$, as can be seen by differentiating the power series of $1/(1-x)$ twice.

3. It is $\sum_{k=0}^{n} \frac{(-3)^k}{k!}$.

5. Note that $\ln(1/(1-x)) = \int 1/(1-x)\, dx$, so

$$\ln(1/(1-x) = \int \sum_{n \geq 0} x^n \, dx = \sum_{n \geq 0} \frac{x^{n+1}}{n+1} = \sum_{n \geq 1} \frac{x^n}{n}.$$

Exercises 11.2

1. After rearrangement, we have

$$\begin{aligned} \frac{B'(x)}{B(x)} &= e^x \\ \frac{dB}{B \cdot dx} &= e^x \\ \frac{1}{B}\, dB &= e^x \, dx \\ \ln B &= e^x + C, \end{aligned}$$

   and the initial condition $B(0) = 1$ forces $C = -1$. So $\ln B(x) = e^x - 1$, and therefore, $B(x) = e^{e^x - 1}$.

3. We have $B_2(x) = e^{e^x - x - 1}$.

5. The exponential generating function of the sequence of numbers $S(n, 2)$ is

$$\frac{1}{2}\left(e^x - 1\right)^2,$$

   so $S(n, 2) = 2^{n-1} - 1$.

Exercises 11.3

1. Removing the root, we get two trees of the same kind as the original tree. The only exception is when the original tree had one vertex only. Then we get the empty set.

3. Note that trees like this always have an even number of vertices. Removing the root, we get two trees. The one on the right will be the same kind as the original tree, whereas the one on the left will be the kind of Exercise 1.

5. This is a separable equation. After rearrangement, we have

$$\begin{aligned} \frac{1}{E^2+1}\, dE &= \frac{1}{2}\, dx \\ \tan^{-1} E &= \frac{x}{2} + C \\ E &= \tan\left(\tfrac{x}{2} + C\right), \end{aligned}$$

   and the initial condition $E(0) = 1$ forces $C = \pi/4$.

7. This is a separable equation that leads to $P^2 - 2P + 2x = 0$. This is a quadratic equation in P, having the solutions $P(x) = 1 \pm \sqrt{1-2x}$. We must choose the negative sign to satisfy the initial condition $P(0) = 0$.

Appendix

Exercises A.1

1. a. $y_{n+1} = h_n^3/3 + (1 - h_n + h_n^2/2 - h_n^3/6)y_n + (h_n^2 - h_n^3/3)x_n + (h_n - h_n^2/2 + h_n^3/6)x_n^2$

   c. $E_n \leq .000004167$ d. $h \leq .059$

x_n	1. b. Taylor Series Order 3	3. Improved Euler	5. Fourth-Order Runge-Kutta	Exact Solution
.0	1.000000	1.000000	1.000000	1.000000
.1	.905167	.905500	.905163	.905163
.2	.821277	.821928	.821270	.821269
.3	.749192	.750145	.749182	.749182
.4	.689692	.690931	.689681	.689680
.5	.643483	.644992	.643470	.643469
.6	.611203	.612968	.611189	.611188
.7	.593430	.595436	.593416	.593415
.8	.590687	.592920	.590672	.590671
.9	.603447	.605892	.603431	.603430
1.0	.632137	.634782	.632121	.632121

7. $a = \frac{1}{4}$, $b = \frac{3}{4}$, $c = d = \frac{2}{3}$

Exercises A.2

1. b. .000417 c. .010626

3. a. $y_{n+1} = [y_n(2-h) + h(x_n^2 + x_{n+1}^2)]/(2+h)$

x_n	1. a. Adams-Bashforth $m=0$	3. b. Adams-Moulton $m=0$	Exact solution
.0			1.000000
.1		.905238	.905163
.2	.820888	.821406	.821269
.3	.748513	.749367	.749182
.4	.688781	.689904	.689680
.5	.642389	.643722	.643469
.6	.609970	.611463	.611188
.7	.592094	.593705	.593415
.8	.589278	.590971	.590671
.9	.601991	.603736	.603430
1.0	.630656	.632427	.632121

References

Bailey, N. T. J. *The Mathematical Theory of Infectious Disease and Its Applications*, Hafner Press, New York, 1975.

Barrow, D. et al. *Solving Differential Equations with Maple V Release 4*, Brooks/Cole Publishing Company, Pacific Grove, CA, 1998.

Bell, E. T. *The Development of Mathematics*, 2nd ed., McGraw-Hill, New York, 1945.

Bender, E. A. *An Introduction to Mathematical Modeling*, John Wiley & Sons, Inc., New York, 1978.

Birkhoff, G., and Rota, G.-C. *Ordinary Differential Equations*, 2nd ed., Springer-Verlag, New York, 1983.

Bóna, M. *A Walk Through Combinatorics, 5th edition*, World Scientific, River Edge, NJ, 2023.

Bóna, M. *Combinatorics of Permutations, 3rd edition*, Chapman & Hall/CRC, Boca Raton, FL, 2022.

Bóna, M. *Introduction to Enumerative and Analytic Combinatorics*, Chapman & Hall/CRC, Boca Raton, FL, 2025.

Borrelli, R. L., and Coleman, C. S. *Differential Equations: A Modeling Approach*, Prentice-Hall, Inc., Englewood Cliffs, NJ, 1987.

Boyce, W. E., and DiPrima, R. C. *Elementary Differential Equations and Boundary Value Problems*, 7th ed., John Wiley & Sons, Inc., New York, 2001.

Boyer, C. B. *A History of Mathematics*, Princeton University Press, Princeton, NJ, 1985.

Braun, M. *Differential Equations and Their Applications*, Springer-Verlag, New York, 1983.

Coddington, E. A., and Levinson, N. *Theory of Ordinary Differential Equations*, McGraw-Hill, Inc., New York, 1955.

Danby, J. M. A. *Computing Applications to Differential Equations*, Reston Publishing Co., Reston, VA, 1985.

Edwards, C. H. *The Historical Development of the Calculus*, Springer-Verlag, New York, 1979.

Eves, H. *Great Moments in Mathematics after 1650*, The Mathematical Association of America, 1983.

Eves, H. *Great Moments in Mathematics before 1650*, The Mathematical Association of America, 1983.

Flajolet, F, Sedgewick, R. *Analytic Combinatorics*, Cambridge University Press, Cambridge, UK, 2008.

Garvin, F. *The Maple Book*, Chapman & Hall/CRC, Boca Raton, FL, 2002.

James, I. *Remarkable Mathematicians*, Cambridge University Press, Cambridge, UK, 2002.

Kermack, W. D., and McKendrick, A. G. "A contribution to the mathematical theory of epidemics," *Journal of the Royal Statistical Society*, 115(1927), 700-721.

Lanchester, F. W. *Aircraft in Warfare, the Dawn of the Fourth Arm*, Tiptree, Constable and Co., Ltd., UK, 1916.

Leslie, P. H. "Some further notes on the use of matrices in population mathematics," *Biometrika*, 35(1948), 213-245.

Leslie, P. H., and Gower, J. C. "The properties of a stochastic model for the predator-prey type of interaction between two species," *Biometrika*, 46(1960), 219-234.

May, R. M. *Stability and Complexity in Model Ecosystems*, Princeton University Press, Princeton, NJ, 1973.

May, R. M. "Biological populations with non-overlapping generations: stable points, stable cycles and chaos," *Science*, 186(1974), 645-647.

McCarty, G. *Calculator Calculus*, Page-Ficklin Publications, Palo Alto, CA, 1975.

Nagle, R. K., and Saff, E. B. *Fundamentals of Differential Equations*, The Benjamin/Cummings Publishing Co., Inc., Redwood City, CA, 1989.

Olinick, M. *An Introduction to Mathematical Models in the Social and Life Sciences*, Addison-Wesley Publishing Co., Reading, MA, 1978.

Priestley, W. M. *Calculus: An Historical Approach*, Springer-Verlag, New York, 1979.

Rainville, E. D., and Bedient, P. E. *Elementary Differential Equations*, 7th ed., Macmillan Publishing Co., Inc., New York, 1989.

Richardson, L. F. "Generalized foreign policy," *British Journal of Psychology Monographs Supplements*, 23(1939).

Roberts, C. E. *Ordinary Differential Equations: A Computational Approach*, Prentice-Hall, Inc., Englewood Cliffs, NJ, 1979.

Ross, S. L. *Introduction to Ordinary Differential Equations*, 4th ed., John Wiley & Sons, Inc., New York, 1989.

Shampine, L. F., and Gordon, M. K. *Computer Solution of Ordinary Differential Equations*, W. F. Freeman and Company, San Francisco, 1975.

Simmons, G. F. *Differential Equations with Applications and Historical Notes*, McGraw-Hill, Inc., New York, 1972.

Smith, D. A. *Interface: Calculus and the Computer*, Houghton Mifflin Co., Boston, MA, 1976.

Smith, D. E. *History of Mathematics*, Vols. 1 and 2, Dover Publications, Inc., New York, 1958.

Suzuki, J. *A History of Mathematics*, Prentice-Hall, Inc., Upper Saddle River, NJ, 2002.

Index

For Product Safety Concerns and Information please contact our EU representative GPSR@taylorandfrancis.com
Taylor & Francis Verlag GmbH, Kaufingerstraße 24, 80331 München, Germany

www.ingramcontent.com/pod-product-compliance
Ingram Content Group UK Ltd.
Pitfield, Milton Keynes, MK11 3LW, UK
UKHW022212190926
475702UK00003B/317
9781041151463